Small Gas Engines

fundamentals ◆ service ◆ troubleshooting ◆ repair ◆ applications

by

Alfred C. Roth
Professor Emeritus, College of Technology
Eastern Michigan University
Ypsilanti, Michigan

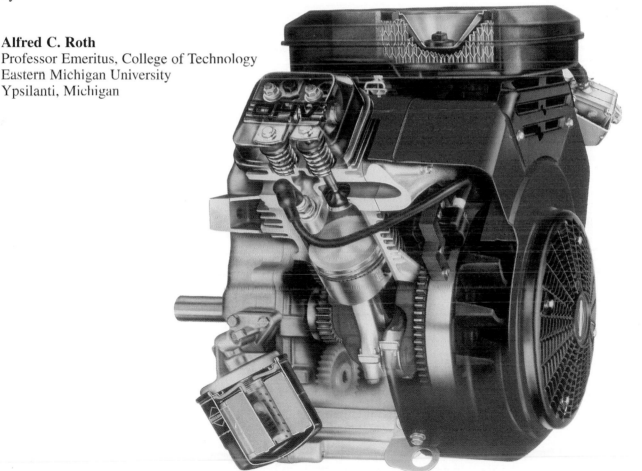

Publisher
THE GOODHEART-WILLCOX COMPANY, INC.
Tinley Park, Illinois

"This book contains the most complete and accurate information available at the time of its publication. The Goodheart-Willcox Company, Inc. cannot assume responsibility for any changes, errors, or omissions in this text. Always refer to the appropriate manufacturer's technical service manual for specific service, repair, or maintenance recommendations.

The Goodheart-Willcox Company, Inc. and the author make no warranty or representation whatsoever, either expressed or implied, with the respect of the equipment, procedures, and applications described or referred to herein, their quality, performance, merchantability, or fitness for a particular purpose.

Further, the Goodheart-Willcox Company, Inc. and the author specifically disclaim any liability whatsoever for direct, indirect, special, incidental, or consequential damages arising out of the use or inability to use any of the equipment, procedures, and applications described or referred herein."

Library of Congress Cataloging-in-Publication Data
Roth, Alfred C.
    Small gas engines : fundamentals, service, trouble-shooting, repair, applications / by Alfred C. Roth.
        p. cm.
    Includes index.
    ISBN 1-56637-574-6
1. Small gas engines. I. Title
TJ790.R719 1999
621.43'3--dc21   98-53017
                                                CIP

Engine illustrations used on the front and back covers courtesy of Briggs & Stratton Corporation; Generac Corporation; Kawasaki Motor Corporation, U.S.A.; and Mercury Marine.

Introduction

Since the first edition of *Small Gas Engines,* many revolutionary developments in other fields have found their way into the design of small gas engines. Some of these include higher horsepower-to-weight ratios, fuel injection, electronic ignition, pressure lubrication, and space-age materials.

Today's engines last longer, use less fuel, run quieter, are more reliable, and are much safer to use. Because of various types of legislation, engine manufacturers have made engines environmentally safer. Legislation on safer levels of emissions has spawned significant changes to the small gas engine. The high level of technology associated with the design and manufacture of modern engines calls for closer tolerances and precise airflow, lubrication, and emissions requirements. As engines have become more sophisticated, they have also become more complex.

The small gas engine now serves as a labor-saving power source for more products than ever before.

- General applications for small gas engines include lawn mowers, edgers, lawn tractors, garden tillers, snow throwers, string trimmers, brush cutters, chain saws, portable pumps, post hole drills, leaf blowers, and much more.
- Industrial and more specialized applications for small gas engines include portable electric generators, conveyor systems, cement mixers, welding generators, lifting mechanisms, portable winches, tampers, horizontal boring units, trenchers, farm equipment, golf course maintainence equipment, and more.
- Recreational applications for the small gas engines include personal watercraft, motorcycles, minibikes, go-carts, all-terrain vehicles (ATVs), ultra-light aircraft, outboard engines, snowmobiles, and many more. Outboard engines are available in a wide range of horsepower ratings for a variety of water sport applications.

Small gas engines play a very important part in almost everyone's life. When the need for power extends out of reach of the electrical cord or beyond the application of an electric motor, the small gas engine is used to ease chores and power recreational vehicles.

Because the applications for small gas engines are numerous, the related career opportunities are increasing. There are jobs as engineers, designers, lawn and garden shop owners, sales distributors, product representatives, service technicians, etc. For the technicians, there is an Outdoor Power Equipment (OPE) certification that can be obtained. This certification shows competence in diagnosing and repairing outdoor power equipment.

Design Flexibility

Small Gas Engines is designed to provide students, do-it-yourselfers, and aspiring technicians with practical information about small gas engine theory, construction, operation, lubrication, maintenance, troubleshooting, service, rebuilding, and repair. This text is written in clear, nontechnical language. Small engine users at all levels are finding it beneficial to have a thorough understanding of engine fundamentals and service procedures. This text provides detailed information about one-, two-, and three-cylinder; two-cycle and four-cycle gasoline engines. Diesel and LPG engines are also covered.

Fonts in this text

Throughout the chapters of this text, different fonts are used to call out and emphasize the meaning or use of a word(s), to identify important term(s), or to identify figure references.

Words or terms that are called out and emphasized always appear in *italic* type. For example, the letter *L* may be stamped on the fastener with left-handed threads.

Important terms always appear in ***bold-italic*** type. These important terms can be found in the terms to know list at the end of the chapter, and most are listed in the glossary at the back of this text. For example, a ***combination wrench*** has a box-end wrench on one end and an open-end wrench on the other.

Figure references in the body of the text and those found with the illustrations always appear in **bold** type. This makes it easy to identify the reference in body and the corresponding illustration.

Chapter components

At the beginning of each chapter a list of *learning objectives* is presented. These objectives are topics covered and goals to be achieved in the chapter. These objectives are in the order that the topics are presented in the chapter. Review the objectives before reading the chapter to get an understanding of the material to come. After completing the chapter, review the objectives once more as a review of chapter material.

A chapter *summary* is found at the end of each chapter. The summary highlights the material just covered in the chapter. Review the summary after completing a chapter. If something in a summary is not understood, go back and review the related section in the chapter.

At the end of each chapter there is *list of terms* to know. These terms are in the order that they appear in the chapter. After completing a chapter, review and define each term. If a term(s) cannot be defined at this time, go back and review the related section in the chapter.

Chapter *review questions* are presented at the end of each chapter. After completing a chapter answer all of the questions on a separate sheet of paper. This is a great way of reviewing the material just presented in the chapter.

A number of *suggested activities* are presented at the end of each chapter. These activities are related experiences that enhance the understanding of the material just covered in the chapter.

Special notices

There are a variety of notices used throughout this text. These notices consist of technical information, hints, cautions, and warnings. The notices are identified by color, icon, and rules around them. The notices are as follows:

A note includes technical information and/or hints that are aimed at increasing knowledge about the systems, procedures, or applications dealing with small gas engines.

A caution identifies the possibility of potential problems, temporary damage, or permanent damage of equipment or tools if the proper procedures or safety measures are not followed. If a caution is not understood, always consult a supervisor or instructor.

A warning identifies the possibility of potential problems that may result in personal injury if the proper procedures or safety measures are not followed. If a warning is not understood, always consult a supervisor or instructor.

Color Code

Colors are used throughout this text to illustrate the different components and materials related to small gas engines. Colors are also used to call attention to significant areas of a figure. Unless otherwise indicated on an individual figure, the following color key is used throughout this text:

Engine specific
- Piston/Crankshaft Components
- Valve-Related Components
- Air
- Fuel
- Air/Fuel Mixture
- Oil
- Combustion
- Exhaust
- Water
- Electrical Part (i.e. copper wire)
- Positive Electrical
- Negative Electrical

General illustration
- Component of Interest
- Secondary Component
- General Parts
- General Parts
- General Parts

Enhancing the Text

To aid in the learning process, a comprehensive workbook and a series of instructional videos have been created. Both of these items are designed to be used with this text.

The workbook contains learning objectives and a variety of related questions that are organized to correlate with the chapters in this text. The workbook includes a number of jobs that are related to a specific preventative maintenance, troubleshooting, service, or repair procedures. Also included in the workbook are sample Outdoor Power Equipment certification tests for two-stroke and four-stroke engines. These sample tests contain questions similar to those found on the OPE certification tests.

The *Small Gas Engine Videos* are correlated to this text. These videos complement the material presented in this text. There are five videos, which can be purchase individually or as a package. The five videos cover the following areas:
- Construction and Operation
- Disassembly
- Inspection, Measurement, Cleaning
- Reassembly
- Troubleshooting and Tune-up

Both the workbook and the videos can be purchased directly from Goodheart-Willcox Company, Inc.

About the Author

Mr. Alfred Roth has extensive experience with the repair and service of small, as well as larger, internal combustion engines. He maintains continuous contact with industrial associations related to the manufacture of small engines, as well as the various engine and implement manufacturers. He is an inventor that holds six patents, and he has been a member of the Engine & Equipment Council/Certification Test Committee, Epsilon Pi Tau, Council on Technology Teacher Education, National Association of Industrial and Technology Teacher Educators, Society of Manufacturing Engineers, American Welding Society, American Foundrymen's Society, American Society for Metals, and Society of Automotive Engineers.

Mr. Alfred Roth received his undergraduate degree from Kent State University, and his masters degree from Ohio University. He has four years of teaching experience at Cuyahoga Falls High School, four years of technical teaching experience at Ohio University, and thirty years of technical teaching experience at Eastern Michigan University. At Eastern Michigan University he taught courses that include metals and machine tool processes, metallurgy, power and energy, engine technology, welding, and industrial drafting.

Mr. Alfred Roth began the aviation ground school and coordinated a flight school at Eastern Michigan University. In addition, he has taught courses in aircraft construction, during which he supervised students in the design, construction, and completion of full size, flyable aircraft. He is a pilot and an active member of the Experimental Aircraft Association.

Mr. Alfred Roth keeps abreast of the constantly changing dynamics of the small engine industry and the expanding use of small engine applications. His enterprise in this field is evident in the quality and scope of this textbook. It is with great interest and dedication that he has authored this textbook for the educational benefit of those who devote their time and energy to study and practice small engine repair and service.

Acknowledgments

For their assistance and contributions to this edition of **Small Gas Engines,** the author and publisher would like to thank the following:

AC Spark Plug Division, GMC
Baldor
Beaird-Poulan, Inc.
Brave Industries, Inc.
Champion Spark Plug Company
Chicago Rawhide Manufacturing Company
Clevite Corporation
Clinton Engine Corporation
Deere & Company
Detecto; Cardinal Scale Manufacturing Company
Dresser Industries Inc.
Duro Metal Products Inc., Stanley-Proto
Dutton-Lainson Company
Eastern Technology Corporation
Easy-Up Industries, Inc.
Engine & Equipment Training Council
Evinrude Motors
Go-Power Corporation
Graymills Corporation
Holo-Krome
Jacobsen Manufacturing Company
Justrite Manufacturing Company
Kohler Company
Kubota Tractor Corporation
Lawn-Boy Power Equipment
Lufkin Rule Company
McCulloch
Merc-O-Tronic Instrument Corporation
Mercury Hi-Performance
Minati Fun Karts
Monarch Instruments
MTD Products, Inc.
Neway Manufacturing Company
Onan Corporation
Outboard Marine Corporation
Park Tool Company
P.A. Sturtevant
Perfect Circle Products
Power-Grip Company
Polaris
Recreatives Industries, Inc.
Rupp Industries, Inc.
Sea-Doo
Simplicity Manufacturing Company
Sioux Tools, Inc.
Snap-on Tools Corporation
Stihl
Sunnen Products Company
Tecumseh Products Company
The Toro Company
Vibra-Tach
Vemco, Inc.
Wisconsin Motors Corporation
Wix Filters
Zenith Division, Bendix Corporation

Special Thanks To...

Briggs & Stratton Corporation; *Milwaukee, Wisconsin* for the engine illustrations used for the front cover, the chapter opener art, and numerous other illustrations throughout the text.

Engine Service Association, Incorporated; *Exton, Pennsylvania* for their contribution of materials and illustrations in Chapter 22.

Generac Corporation; *Waukesha, Wisconsin* for the engine illustration used for the back cover and a number of other illustrations used throughout the text.

J & R Lawn and Garden; *Mokena, Illinois* for the use of much of the lawn equipment illustrated throughout this text and specifically in Chapter 18.

Kawasaki Motors Corporation U.S.A; *Irvine, California* for the engine illustration used for the back cover and a number of other illustrations used throughout the text.

The L. S. Starrett Company; *Athol, Massachusetts* for the use of the precision measuring instruments illustrated throughout this text and specifically in Chapter 2.

Mercury Marine; *Fon du Lac, Wisconsin* for the engine illustration used for the back cover and a number of other illustrations used throughout the text.

Sears, Roebuck and Company–Orland Park Retail Store; *Orland Park, Illinois* for the use of many of the tools illustrated throughout this text and specifically in Chapter 2.

Contents

Chapter 1   Safety in the Small Gas Engine Shop 13

Shop Safety • Keep Work Area Clean • Handle Hazardous Materials Properly • Wear Proper Clothing • Maintain Adequate Ventilation • Use Hand Tools Properly • Use Power Tools Properly • Use Compressed Air Carefully • Lift Properly • Use Proper Electrical Wiring/Grounding Procedures • Operate Engines Safely • Be Prepared for Emergencies • Follow OSHA Requirements

Chapter 2   Tools and Measuring Instruments 21

Wrenches • Pliers • Retaining Ring Pliers • Screwdrivers • Hammers • Punches • Cold Chisels • Gear Pullers • Probe and Pickup Tools • Vise • Cleaning Tank • Measuring Instruments • Micrometers • Cleaning and Calibrating a Micrometer • Using a Micrometer • Reading The Standard Micrometer • Reading a Vernier Micrometer • Digital Micrometers • Micrometer Depth Gauges • Vernier Calipers • Reading a Vernier Caliper • Telescoping Gauges • Small Hole Gauge • Thickness Gauges • Valve Spring Tension Tester • Combination Square • Dial Indicator • Screw Pitch Gauge

Chapter 3   Fasteners, Sealants, and Gaskets 43

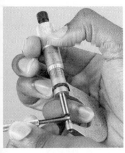

Threaded Fasteners • Set Screws • Self-Tapping Screws • Bolts • Nuts • Lock Nuts • Bolt and Nut Terminology • Bolt Grades • Thread Types • Thread Fit • Thread Designations • Tightening and Loosening Threaded Fasteners • Tightening to Specific Torque Settings • Chasing Threads • Tapping New Threads • Rules for Hand Tapping • Threading With a Die • Washers • Pins • Cotter Pins • Clevis Pins • Dowel Pins • Straight Pins • Grooved Pins • Taper Pins • Retaining Rings • Keys • Thread Adhesives • Sealants • Antiseize Compounds • Gaskets

Chapter 4 **Engine Construction and Principles of Operation** 63

Gasoline Engines • Simple Engine in Operation • Gasoline • Gasoline Must Burn Quickly • Fuel is Atomized • Explosion Must be Contained • Further Improvement • Basis for an Engine • Cylinder Block • Crankshaft and Crankcase • Pistons • Connecting Piston to Crankshaft • Intake and Exhaust Ports • Poppet Valves • Valve Spring Assembly • Valve Lifter or Tappet • Camshaft • Flywheel

Chapter 5 **Two-Cycle and Four-Cycle Engines** 77

Small Engine Identification • Four-Stroke Cycle Engine • Intake Stroke • Compression Stroke • Power Stroke • Exhaust Stroke • Valve Timing • Lubrication • Two-Stroke Cycle Engine • Variations in Design • Principles of Operation • Intake into Crankcase • Fuel Transfer • Ignition-Power • Exhaust • Scavenging and Tuning • Rotary Disc Valve Engine • Reed Valve Engine • Four-Cycle Engine vs. Two-Cycle Engine

Chapter 6 **Measuring Engine Performance** 91

Basic Terminology • Engine Bore and Stroke • Engine Displacement • Compression Ratio • Force • Work • Power • Energy • Horsepower • Horsepower Formula • Kinds of Horsepower • Brake Horsepower • Indicated Horsepower • Frictional Horsepower • Rated Horsepower • Corrected Horsepower • Torque • Torque is not Constant • Torque and Horsepower • Volumetric Efficiency • Practical Efficiency • Mechanical Efficiency • Thermal Efficiency

Chapter 7 **Fuel and Emission Control Systems** 109

Engine Fuels • Gasoline • Liquefied Petroleum Gas and Natural Gas • Combustion of LPG • Advantages of LPG • Disadvantages of LPG • Kerosene and Diesel Fuels • Two-Cycle Fuel Mixtures • Tanks, Lines, and Fittings • Fuel Filters • Fuel Pumps • Mechanical Fuel Pumps • Impulse Diaphragm Fuel Pumps • Pressurized Fuel System • Vapor Return Fuel Systems • Emission Controls • Exhaust Emissions • Environmental and Health Impact of Small Engine Emissions • Proposed Phase 1: Small Spark-Ignition Engine Rule • Role of the Consumer

Chapter 8 **Carburetion** 123

Principles of Carburetion • Air-Fuel Mixture • Carburetor Pressure Differences • Vacuum • Atmospheric Pressure • Venturi Principle • Types of Carburetors • Float-Type Carburetors • Float Bowl Ventilation • Choke System • Throttle System • Load Adjustment • Acceleration System • Acceleration Well • Economizer Circuit • Idling Circuit • Part-Throttle, Full-Throttle Sequence • Primer • Diaphragm-Type Carburetors • Diaphragm Carburetor Operation • Manual Throttle Controls • Governor Throttle Controls • Types of Governors • Governor Features • Air Cleaners and Air Filters • Oil-Wetted Air Cleaner • Dry-Type Air Cleaner • Dual Element Air Cleaners

Chapter 9 Ignition Systems 147

Basic Ignitions System Operation • Fundamental Electrical Principles •
The Electron Theory • Electrical Units of Measurement • Ohm's Law •
Magnetism • Magnets and Electricity • Ignition Coil • Spark Plugs • Spark Plug
Heat Transfer • Measuring Spark Plug Temperature • Types of Electrodes •
Switching Devices • Breaker Points • Electronic Switching Devices • The MBI
Magneto System • The MBI Magneto Cycle • The Stop Switch • Ignition Advance
Systems • Dwell and Cam Angle • Electronic Ignition • Operation of Capacitive
Discharge Ignition (CDI) System • Operation of Transistor Controlled Ignition
(TCI) System • Magneto Ignition Systems Compared • Battery Ignition Systems •
High Voltage Secondary Current • Auto-Transformer Type Ignition Coil •
The Lead-Acid Battery

Chapter 10 Lubrication Systems 175

Principles of Lubrication • Friction • Preventing Wear Due to Friction •
Lubricating Oil • Permits Easy Starting • Lubricates and Prevents Wear • Protect
Against Rust and Corrosion • Keeps Engine Parts Clean • Cools Engine Parts •
Seals Combustion Pressures • Prevents Foaming • Aids Fuel Economy •
Oil Selection • SAE Viscosity Grade • API Engine Oil Service Classification •
Engine Lubrication • Two-Cycle Engine Lubrication Systems • Four-Cycle Engine
Lubrication Systems • Splash Lubrication System • Constant Level Splash System •
Ejection and Barrel Pumps • Positive Displacement Oil Pumps • Full-Pressure
Lubrication Systems • Oil Filter Systems • Bypass Systems • Shunt Filter Systems •
Full-Flow Filter Systems

Chapter 11 Cooling Systems 191

Principles of Engine Cooling • How Air Cooling Works • How Exhaust Cooling
Works • How Water Cooling Works • Water Pumps • Outboard Water Circulation
Systems • Observing Thermostat Operation • Radiators • How Oil Cooling Works

Chapter 12 Preventive Maintenance and Troubleshooting 201

General Preventive Maintenance • Keeping Engines Clean • Checking Oil Level
and Condition • When to Change Oil • Changing Oil • Mixing Oil and Fuel • Air
Cleaner Service • Crankcase Breather Service • Muffler Service • Maintaining
Water Cooling Systems • Systems Cooled with Salt Water • Storing Water-Cooled
Engines • Systematic Troubleshooting • Check the Easiest Things First • Verify
the Fundamental Operating Requirements • Check RPM • Test Compression •
Test Differential Pressure • Service Information • Troubleshooting Charts •
Tolerances and Clearances • Torque Specifications

Chapter 13 Fuel System Service 217

Troubleshooting the Fuel System • Fuel Pump • Carburetor Adjustments • High Speed and Idle Mixture Adjustment • Testing Two-Cycle Engine Reeds • Carburetor Overhaul • Carburetor Removal • Carburetor Disassembly • Inspection of Parts • Float-Type Carburetor Repair • Assembling Float-Type Carburetors • Engine Priming • Troubleshooting Float-Type Carburetors • Diaphragm Carburetor Repair • Assembling Diaphragm Carburetors • Troubleshooting Diaphragm Carburetors • Engine Governor Adjustments

Chapter 14 Ignition System Service 239

Ignition System Service • Spark Plugs • Spark Plug Removal • Analysis of Used Spark Plugs • Cleaning Spark Plugs • Gapping Spark Plugs • Spark Plug Installation • Ignition Testing Procedure • Magneto Service • Removing the Flywheel • Inspecting the Flywheel • Testing Ignition System Components • Adjusting Breaker Point Ignition Systems • Servicing Battery Ignition Systems • Batteries • Wiring • Ammeters • Switches • Checking the DC Starter-Generator Circuit • Commutators and Brushes • Maintaining an Alternator • Alternator Output Tests • Voltage Regulator Service • Distributors

Chapter 15 Engine Inspection, Disassembly, and Cylinder Reconditioning . . . 271

Engine Inspection • Engine Disassembly • Organizing the Job • Cylinder Reconditioning • Cylinder Inspection • Cylinder Measurement • Reboring the Cylinder • Honing the Cylinder

Chapter 16 Piston and Piston Ring Service 281

Piston and Piston Ring Service • Piston Construction • Piston Fit • Cam-Ground Pistons • Piston Thrust Surfaces • Piston Head Size • Piston Head Shape • Piston Ring Design • Piston Ring Construction • Piston Ring Types • Compression Rings • Oil Control Rings • Ring Tension • Ring Movement • Ring Side Clearance • Ring End Gap • Rings and Cylinder Walls • Piston Damage • Piston Ring Wear-In • Piston Pins • Removing Piston Pins • Measuring Piston Pins and Bosses

Chapter 17 **Rod, Bearing, Crankshaft, Valve, and Camshaft Service** **297**

Connecting Rods and Bearings • Friction-Type Rod Bearings • Antifriction Bearings • Rod Bolt Locking Devices • Crankshaft • Crankshaft Balance • Crankshaft Main Bearings • Crankshaft Clearances • Measuring Bearing Clearance • Crankcase Seals • Valve Service • Removing Valve Assembly • Inspecting Valves and Seats • Inspecting Valve Springs • Valve Guides • Inspecting Valve Guides • Valve Guide Reaming • Alternate Valve Guide Replacement • Valve Seat Angle and Width • Lapping Valves • Valve Seat Inserts • Removing a Valve Seat Insert • Installing New Valve Insert (Cast Iron Block) • Installing New Valve Insert (Aluminum Alloy Block) • Valve Lifter-to-Stem Clearance • Refacing Valves • Adjusting Valve Clearance • Overhead Valve Systems • Overhead Valve System Disassembly • Removing Valves • Servicing Overhead Valves, Seats, and Guides • Installing Overhead Valves • Installing Cylinder Head • Installing Rocker Arms • Adjusting Valve Clearance • Ports, Reeds, and Rotary Valves • Camshafts and Gears • Automatic Compression Release

Chapter 18 **Lawn Equipment** **331**

Working Safely • Lawn Mowers • Knowing Mowing Program • Purchasing Considerations • Push-Type and Self-Propelled Mowers • Two-Cycle or Four-Cycle Engines • Optional Features • Conveniences • Engine Starting • Blade Brakes • Procedure for Starting an Engine • Minor Checks • General Maintenance • Blade Sharpening • Spark Plugs • Air Cleaners • Changing Oil • Muffler Replacement • V-Belts • Storing a Power Lawn Mower • Removing a Mower From Storage • Chain Saws • Purchasing Considerations • Safety Features • Rules for Safe Operation • Avoiding Kickback • Chain Saw Maintenance • Fuel and Carburetor • Cylinder Fins • Air Cleaner (Filter) • Fuel Filter • Lubrication • Muffler • Spark Plug • Guide Bar • Storage • String Trimmers and Brushcutters • Purchasing Considerations • Operator Safety • Starting and Safe Operation • String Trimmer and Brushcutter Maintenance • Edger/Trimmers • Purchasing Considerations • Safety Features and Adjustments • Rules for Safe Operation • Edger/Trimmer Maintenance • Edger/Trimmer Storage

Chapter 19 **Lawn and Garden Tractors** **365**

Operating Safely • Protect Children • Protective Clothing • Considerations While Mowing • Avoid Tipping the Mower • Parking a Tractor Safely • Transporting a Tractor • Purchasing Considerations • Mowing and Mulching • Proper Mowing • Mower Deck and Blade Adjustment • Proper Mulching • Leaf Collecting • Plowing, Throwing, and Blowing Snow • Transmissions • Variable Speed Transmission • Hydrostatic Transmissions • Hydrostatic Transmission Fluid Flow • Electrical Safety Systems • Safety Interlocks • Electric Start Systems • Testing the Interlock Electrical System • Testing the Safety Reverse Switch • Testing the Solenoid on Electric Start Tractors • Recoil Start System • Testing the Interlock System on the Recoil Start System • Testing the Safety Reverse Switch on the Recoil Start System • Testing Procedure for Operator Present System (Safety Seat) • Disassembly Procedure • Assembly Procedure • Circuits for Study • General Maintenance • Spark Plugs • Engine Lubrication • Air Filter • Mufflers • Engine Cleaning • Radiator Cleaning • Battery Maintenance • Sharpening and Balancing Mower Blades • Fueling and Refueling • Storing a Tractor • Removing a Tractor from Storage

Chapter 20 **Snow Throwers** **403**

Operating Safely • Protect Children • Preparation for using a Snow Thrower •
Using a Snow Thrower • Maintenance and Storage • Purchasing Considerations •
Machine Size and Type • Small Snow Throwers • Mid-Size Snow Throwers •
Heavy-Duty Snow Throwers • Engine Controls • Electric Starting • Drive Systems •
Track Drive Snow Thrower • Skid Shoes • Skid Shoe Adjustment • Operator
Presence • Snow Thrower Operation • Engine Starting • Snow Throwing •
General Maintenance • Spark Plugs • Snow Thrower Engine Fuel • Four-Cycle
Engine Lubrication • Air Filter • Muffler • Preventing Freeze-Up • Auger and
Auger Gear Box Lubrication • Lubrication • Off-Season Storage • Starting After
Storage

Chapter 21 **Personal Watercraft** **425**

General Boating Terms • Operation Precautions • Preride Checklist • Launch
Ramp Etiquette • Identification Numbers and Placards • Signs, Symbols, Buoys,
and Markers • Courtesy and Common Sense • Navigational Rules • Purchasing
Considerations • PWC Engines • Basic Engine Parts • Two-Stroke Cycle Engine •
Engine Cooling System • Jet Pump Propulsion System • New Engine Break-In •
General Maintenance • Transporting PWC • PWC Storage

Chapter 22 **Career Opportunities and Certification** **463**

Engine Mechanic • Service Manager • Sales Manager • General Manager •
Technician • Sales Representative • Engineer • Executive • Entrepreneur • Teacher •
Certification • Why Get Certified? • What is on a Certification Test? • Who may
take the Test, and Where?

Appendix **468**

Glossary **494**

Index **504**

Safety in the Small Gas Engine Shop

After studying this chapter, you will be able to:
- ▼ Explain why a clean, well-organized shop is extremely important.
- ▼ List several dangers associated with working in a small engine shop.
- ▼ Explain the importance of maintaining and using tools properly.
- ▼ Describe methods for minimizing the risks involved in working with small engines.
- ▼ Explain the function of OSHA.

Shop Safety

Small gas engine work can be rewarding and exciting. However, you may encounter dangerous situations whenever you work in a small engine shop. Special precautions should be taken when working with small engines. It is very important to recognize potential *hazards* and make sure that your work area is safe.

Safety is the responsibility of everyone in the small engine shop. If you notice dangerous shop conditions or unsafe work practices, notify your instructor immediately. Never take unnecessary risks to complete a job. Safe shop practices can prevent serious injury or save a life.

The symbols illustrated in **Figure 1-1** appear throughout the text to signal potentially dangerous situations. The warnings that accompany the warning symbols should be read carefully and followed closely. Failure to follow these warnings can result in serious injury or death. The cautions that accompany the caution symbols should also be read carefully and followed closely. Failure to follow these cautions can result in serious damage to tools or equipment.

Keep work area clean

A clean, well-organized work area is very important to everyone in the shop. Floors should be free from oil and dirt. An oily floor is slippery and can cause serious falls. Always use spill control devices to capture leaks from any type of container. See **Figure 1-2**. Always clean up after working on a project. Pick up all tools and store them properly in a toolbox or workbench. Return all unused supplies to the proper storage area and discard all waste in appropriate containers. Aisles and doorways should be free from obstructions.

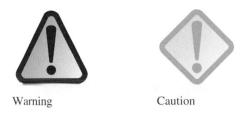

Warning   Caution

Figure 1-1. *There are many safety hazards in the small engine shop. The warning and caution symbol will be used throughout the text to signal potential dangers.*

Figure 1-2. *Spill control pallets catch leaks or spills from containers in an easy-to-clean sump below. (Justrite Manufacturing Co.)*

Keeping the shop area clean can also eliminate fire hazards. When combustible materials are allowed to accumulate in the shop, the possibility of fire increases. Never store used rags in a closet or corner. Rags saturated with gasoline or *solvent* are highly *flammable* and can be easily ignited. An approved container for storing flammable waste is shown in **Figure 1-3**. A clean work area will increase safety and productivity.

Figure 1-3. *Oil-saturated or solvent-saturated waste is extremely flammable and should be stored in a proper container. (Justrite Manufacturing Co.)*

Handle hazardous materials properly

There are many dangerous chemicals used in the small engine shop. Always store chemicals in a safe place. Flammable liquids should be kept in closed safety containers when not in use. See **Figure 1-4**. These containers should be stored in safety cabinets to further minimize risks. Gasoline is an extremely flammable liquid and its vapors can explode if exposed to sparks or flames. Never fill the fuel tank while the engine is running or hot. Heat from the engine could ignite the gasoline.

Some small gas engines are equipped with battery-operated ignition systems. The batteries

A

B

Figure 1-4. *A—Flammable liquids should be stored in closed containers. These safety cans are equipped with flame arrestors, which prevent flames or sparks from entering containers. B—Containers should be stored in safety cabinets. (Justrite Manufacturing Co.)*

used in these systems are similar to those used in automobiles. Handle batteries carefully to avoid splashing acid on clothes, skin, or in eyes. *Hydrogen gas* is produced when the battery is being charged or discharged. If the hydrogen gas is ignited, the battery can explode, throwing acid and fragments from its case in every direction. Always keep sparks and flames away from the battery.

Only use chemicals for their intended purpose. Gasoline should never be used as a cleaning solvent. Gasoline has a low *flashpoint* and can be ignited easily. Many of the chemicals encountered in a small engine shop can cause serious burns. Avoid contact with skin. Wear rubber gloves and safety goggles when working with cleaning solvents.

Wear proper clothing

Proper clothing should be worn when working with small gas engines. Avoid loose-fitting clothing, which can get caught in moving machine parts. Neckties and jewelry should never be worn when working near rotating machinery. Long hair should be worn up or secured under a cap. To avoid serious injury, keep hands, feet, hair, and clothing away from rotating engine parts. Never operate machinery with safety shrouds removed.

Goggles should always be worn when handling and working with chemicals. See **Figure 1-5.** *Safety glasses* should be worn to protect eyes when using drills, grinders, hammers, chisels, or compressed air. A pair of safety shoes is recommended to prevent foot and toe injury.

Ear protection should be worn to protect ears when working with air wrenches, engines under load, or engines running in an enclosed area. Two effective forms of ear protection are *earphone-type protectors* and *earplugs*. See **Figure 1-6.**

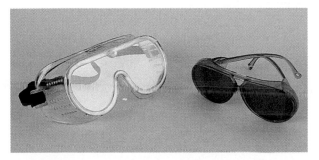

Figure 1-5. *Safety goggles or glasses should be worn when working in the small engine shop.*

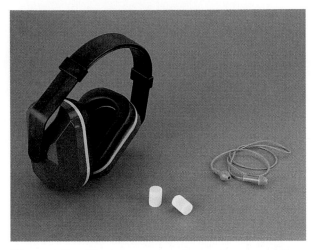

Figure 1-6. *The two common types of ear protection are earphone-type protectors and earplugs. Two variations of earplugs are shown.*

Maintain adequate ventilation

The exhaust gases produced by gasoline engines contain carbon monoxide. *Carbon monoxide* is colorless and odorless. Breathing small amounts of carbon monoxide can cause drowsiness and headaches. Large amounts of carbon monoxide can cause death.

If an engine must be operated in the shop, make sure that a properly maintained ventilation system is running and the doors and windows are open. A lethal amount of carbon monoxide can accumulate in a closed, one-car garage in three minutes. Adequate ventilation is extremely important in the small gas engine shop. In addition to proper ventilation, personal respirators are used to remove contaminants from the air. See **Figure 1-7.**

Solvents used to clean engine parts can release *toxic fumes*. Check warnings on solvent labels and follow instructions carefully. When working with any solvent for an extended period of time, make sure that there is plenty of fresh air.

Use hand tools properly

The safe use of hand tools is often taken for granted in the small engine shop. Many accidents, however, are caused by the improper use of common hand tools.

Keep tools clean. Greasy or oily tools are likely to slip from your hand and may fall into

Figure 1-7. *A respirator should be available to be worn when working in the small engine shop, when working with toxic solvents, or when working with equipment that may produce dust or dirt particles.*

rotating engine parts. The rotating parts can throw the tool, causing serious injury.

Tools should only be used for the job they were designed for. Never, for example, use screwdrivers or files to pry items loose. These tools are not designed for this activity. Most screwdrivers and files are made from hardened steel and may crack or shatter if improperly used.

Keep tools in top shape. Sharpen tools periodically. Dull tools require greater effort to use. Make sure all tools are equipped with appropriate handles. When using a wrench, always pull the handle toward your body. This will help to prevent injury if the tool slips. See **Figure 1-8**. Hammer

Figure 1-8. *To prevent injury, always pull wrench toward your body.*

heads must be securely attached to the handle. If the head is loose, it could fly off during use.

Use power tools properly

Before using a power tool, make sure all guards and shields are in place. Wear safety goggles when operating power tools. If you are not familiar with a tool, read the operating instructions carefully or ask for help before attempting to operate the unit.

Never make adjustments on a power tool when it is running. Shut the tool off and wait for it to stop completely before attempting to service the device in any way.

All power tools should be equipped with a ***dead man switch***. This type of switch automatically shuts the tool off when the operator releases the control button.

Use compressed air carefully

Compressed air is used in the small engine shop to accomplish various tasks. Wear safety goggles when using compressed air. Never use compressed air to clean your clothing or your hair. Flying particles can be blown into your eyes, causing serious injury or blindness. These particles can also penetrate your skin.

Check all connections before turning on a compressed air system. Always hold the hose nozzle tightly when using compressed air. Never set the hose down without shutting off the air nozzle. Pressure in the hose can cause it to whip violently.

Lift properly

Always lift heavy objects carefully. If necessary, ask for help when moving heavy items. Many shops are equipped with small overhead cranes to help move large objects.

To avoid unnecessary back strain, always lift with your legs, not with your back. Keep your back as straight as possible when lifting heavy objects. See **Figure 1-9**.

Avoid carrying items that will obstruct your view. Make several trips if necessary. When carrying long items, use two people so that the item is held level and both ends are attended. Never reach for heavy overhead items. The item may accidentally fall, causing severe head injury. Always use a good quality ladder.

Improper
A

Proper
B

Figure 1-9. *When lifting heavy objects, keep your back straight and use your legs to lift weight. A—Improper lifting procedure. B—Proper way to lift heavy objects.*

Use proper electrical wiring/grounding procedures

Electrical hazards can be found in every small engine shop. Electricity is the most common cause of shop fires. Before using electrical equipment, check wires for fraying or cracking. Make sure all electrical equipment is properly grounded or double insulated. If equipment is not grounded, electrical shock can occur.

All outlets, switches, and junction boxes should be covered. Label circuit breaker (fuse box) clearly so that it can be located in case of an emergency. Breaker switches should also be labeled.

Extension cords should not be used as permanent substitutes for fixed wiring. Extension cords should never run through holes in walls or floors.

Do not overload outlets. Too many components on one circuit can cause excess current to flow in the circuit. Overloaded circuits are a frequent cause of electrical fires.

Operate engines safely

Never operate a small engine at speeds greater than those recommended by the manufacturer. Excessive engine speed can cause parts to break loose from the engine. Severe personal injury can result from flying parts. Never tamper with the governor setting to increase maximum engine speed.

Keep hands, feet, and hair away from rotating engine parts. Small engines develop considerable speed and torque and can cause serious injuries.

Never operate an engine with guards or shrouds removed.

Some small gas engine components get extremely hot. Avoid touching the engine when it is running. Let engine cool before attempting repairs. In addition to causing burns, a hot engine may cause a fire if gasoline is accidentally spilled on hot surfaces.

Avoid touching electrical wires while the engine is running. The high voltage produced by ignition systems can cause electrical shock. Some systems produce more than 30,000 volts (V).

Do not operate an engine without a muffler. Wear ear protection when working on a running engine for a long period of time.

Be prepared for emergencies

In the event of an emergency, it is very important to be prepared. Emergency equipment should be stored in a highly visible place. List emergency numbers next to each telephone in the shop.

All shop areas should be equipped with *fire extinguishers*. These extinguishers should be mounted in highly visible, unobstructed areas. All extinguishers should be inspected monthly. Always keep the area around the extinguisher free from obstructions.

Fire extinguishers are categorized according to the type of fire that each is designed to suppress or extinguish. See **Figure 1-10.** Class A fires involve ordinary combustibles such as wood, cloth, and paper. Class B fires involve flammable liquids such as gasoline and solvents. Class C fires are electrical fires. Be sure to use the proper type of extinguisher. Using the wrong extinguisher can be dangerous. Some fire extinguishers can be used for all types of fires.

First aid kits should be properly stocked and placed in prominent locations. If someone gets hurt, notify your instructor or supervisor immediately. Always seek professional help for serious injuries.

Follow OSHA requirements

The *Occupational Safety and Health Administration (OSHA)* is a governmental organization that establishes rules for safe work practices. All businesses and industries are required to follow OSHA regulations. It is very important to be familiar with OSHA rules and recommendations.

Fire Extinguishers and Fire Classifications

Fires	Type	Use	Operation
Class A Fires Ordinary Combustibles (Materials such as wood, paper, textiles.) *Requires...* *cooling-quenching* **A** (triangle)	**Soda-acid** Bicarbonate of soda solution and sulfuric acid	Okay for use on **A** (triangle) Not for use on **B** (square) **C** (circle) **D** (star)	Direct stream at base of flame.
	Pressurized Water Water under pressure	Okay for use on **A** (triangle) Not for use on **B** (square) **C** (circle) **D** (star)	Direct stream at base of flame.
Class B Fires Flammable Liquids (Liquids such as grease, gasoline, oils, and paints.) *Requires...blanketing or smothering* **B** (square)	**Carbon Dioxide (CO$_2$)** Carbon dioxide (CO$_2$) gas under pressure	Okay for use on **B** (square) **C** (circle) Not for use on **A** (triangle) **D** (star)	Direct discharge as close to fire as possible, first at edge of flames and gradually forward and upward.
Class C Fires Electrical Equipment (Motors, switches, and so forth.) *Requires...* *a nonconducting agent.* **C** (circle)	**Foam** Solution of aluminum sulfate and bicarbonate of soda	Okay for use on **A** (triangle) **B** (square) Not for use on **C** (circle) **D** (star)	Direct stream into the burning material or liquid. Allow foam to fall lightly on tire.
Class D Fires Combustible Metals (Flammable metals such as magnesium and lithium.) *Requires...blanketing or smothering.* **D** (star)	**Dry Chemical**	Multi-purpose type / Ordinary BC type Okay for **A** **B** **C** (multi-purpose) / **B** **C** (ordinary BC) Not okay for **D** (multi-purpose) / **A** **D** (ordinary BC)	Direct stream at base of flames. Use rapid left-to-right motion toward flames.
	Dry Chemical *Granular type material*	Okay for use on **D** (star) Not for use on **A** (triangle) **B** (square) **C** (circle)	Smother flames by scooping granular material from bucket onto burning metal.

Figure 1-10. *Chart illustrates various fire extinguisher types and fire classifications. In the small engine shop, always use an extinguisher designed for use on electrical and chemical fires.*

Summary

Certain precautions must be taken when working on small gas engines. Keeping the work area clean will increase safety and productivity in the shop. Hazardous materials must be handled with care to avoid fires or chemical burns. Dispose of rags that are saturated with solvents in a proper container.

Proper clothing should be worn when working on small engines. Avoid loose-fitting clothing, which can get caught in rotating engine parts.

Adequate ventilation is imperative when working in an enclosed area.

Use all tools properly. Read all instructions before using power tools. Never use compressed air to clean clothing.

Electrical malfunctions are the most common cause of shop fires. Do not overload electrical circuits.

Do not operate an engine at speeds greater than those recommended by the manufacturer. Keep hands and feet away from rotating engine parts. Avoid touching hot engine parts.

Be prepared for emergencies. List emergency phone numbers above each telephone. Keep fire extinguishers in highly visible areas. Make sure a first aid kit is properly stocked and easily accessible.

Know These Terms

hazards
solvent
flammable
hydrogen gas
flashpoint
goggles
safety glasses
earplugs
earphone-type
   protectors

carbon monoxide
toxic fumes
dead man switch
fire extinguisher
first aid kit
Occupational Safety
   and Health
   Administration
   (OSHA)

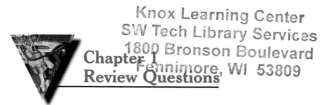

Chapter 1 Review Questions

Answer the following questions on a separate sheet of paper.

1. There are many potential hazards in the small gas engine shop. True or False?
2. Oily or dirty floors can cause people to _____.
3. A flammable liquid frequently used in the small engine shop is _____.
4. Gasoline should always be stored in a closed container. True or False?
5. To prevent injury when working around small engines, avoid wearing _____.
   a. loose clothing
   b. jewelry
   c. neckties
   d. All of the above.
6. Carbon monoxide is _____ and _____.
7. Power tools should not be operated without proper safety shrouds. True or False?
8. Never use compressed air to clean _____ or _____.
9. Always lift with your _____.
   a. arms
   b. legs
   c. back
   d. All of the above.
10. Electrical malfunction is the most common cause of shop fires. True or False?
11. Small engine ignition systems can produce more than 30,000 volts. True or False?
12. Batteries produce _____ _____ when charging or discharging.
13. Fire extinguishers are categorized by the type of fire they are designed to suppress. True or False?
14. OSHA establishes regulations for _____.

Suggested Activities

1. Make several safety posters warning of the potential dangers in a small engine shop and place them throughout your work area. Emphasize good housekeeping and proper storage of hazardous materials.
2. Check guards on all power tools and equipment and discuss the purpose for each guard. Make sure that all guards are correctly mounted and in proper condition.
3. Walk through the shop area and identify potential hazards. Discuss ways to minimize these hazards with your instructor.
4. Locate emergency equipment throughout your work area. Check fire extinguishers for sufficient charge. Make sure that they are designed for use on flammable liquids and electrical equipment. Make sure the first aid kit is properly stocked. Familiarize yourself with all the items in the first aid kit.

This riding lawnmower is a perfect example of how small gas engines are used today. (Deere & Co.)

C H A P T E R

2

Tools and Measuring Instruments

After studying this chapter, you will be able to:
▼ Explain why quality tools and measuring instruments should be used when servicing small gas engines.
▼ Use common hand tools properly.
▼ Summarize the reasons that small engine components must be measured carefully.
▼ Demonstrate several of the common measuring techniques.

Shop Tools

High-quality tools should always be used when servicing a small gas engine. Quality tools allow you to service small engines easily and effectively. To avoid damage to engine parts, always use the tools recommended by the manufacturer. Keep tools clean and in proper working condition.

Some tools are common to most engine work, while others may have only one or two specific applications. Special-purpose tools may be designed by a manufacturer for limited use on only one engine make or model. The following sections will describe how to use common tools. Some special-purpose tools will also be examined, along with examples of their applications.

The tools and measuring instruments described in this chapter are available in US customary and metric sizes. A shop or technician should have sets of both.

Wrenches

There are many types of wrenches available to suit practically every situation encountered when servicing small engines. These include box-end wrenches, open-end wrenches, adjustable wrenches, Allen wrenches, socket wrenches, and torque wrenches. The type of wrench used may depend on the kind of fastener to be installed or removed. Box-end, open-end, adjustable, and socket wrenches are used on hexagon head bolts and nuts. An **open-end wrench** should only be used when it is not possible to encompass the nut or bolt head with a box-end wrench or a socket. See **Figure 2-1A.** A **box-end wrench** can be used where partial or full-turn clearance is available. It may be six or twelve sided and is less likely to slip around the bolt head corners than open-end wrenches or adjustable wrenches. See **Figure 2-1B.**

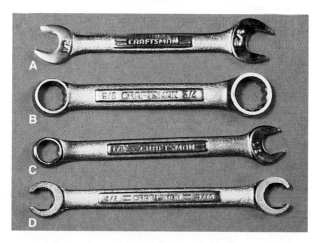

Figure 2-1. *Always select the correct type of wrench for the job at hand. A—Open-end wrench. B—Box-end wrench. C—Combination box/open-end wrench. D—Tubing wrench.*

A *combination wrench* has a box-end wrench on one end and an open-end wrench on the other. See **Figure 2-1C.** A *tubing wrench* is used on metal tubing connection fittings. See **Figure 2-1D.** When applying force to a wrench, always pull the tool instead of pushing on it. See **Figure 2-2.** This will prevent hand and knuckle injury if the wrench should accidentally slip from the bolt head.

An *adjustable wrench* should only be used as a last resort when other wrenches are not available. Adjustable wrenches are used similarly to open-end wrenches and should also be pulled in the correct direction. See **Figure 2-3.** Due to the movable jaw in adjustable wrenches, they are prone to loosening and slipping around the corners of bolts and nuts.

Many varieties of *socket sets* are available. For small engine work, sets with standard and deep-length sockets, one ratchet wrench, one spark plug socket, and several extensions will meet most needs.

See **Figure 2-4.** Sockets are extremely useful when bolts or nuts are recessed in counterbored holes. There are many occasions when, due to obstructions, other wrenches cannot be applied or turned very far. In these cases, a socket and ratchet wrench is the only way a bolt or nut can be removed or tightened. At other times, sockets can simply save time compared with other wrenches.

An *Allen wrench* is used to remove or install hex socket-head screws. They may be the conventional right angle style or straight with tee handles. See **Figure 2-5.** The correct size Allen wrench should be used to avoid slipping in the hex socket recess and damaging the socket. When this happens, the removal of a screw may be difficult, if not impossible.

Some hex socket-head screws have metric size sockets. Never attempt to use metric Allen wrenches and US customary Allen wrenches interchangably. This can deform the socket and render the screw impossible to remove.

A *torque wrench* is used to tighten threaded fasteners to a specific torque setting. *Torque* is the turning force applied to the fastener. **Figure 2-6** shows how to use a *preset* torque wrench. After the desired amount of torque is set, the wrench socket is placed on the bolt head and the handle is drawn until a click can be felt and heard. The socket should be held down firmly with one hand, while pulling on the handle with the other.

The torque wrench shown in **Figure 2-7** uses a pointer that moves up the scale as torque is applied. With this wrench, the mechanic pulls the wrench handle until the pointer reaches the correct torque reading. Torque data charts, like the one in **Figure 2-8,** supply the necessary data to correctly tighten critical parts.

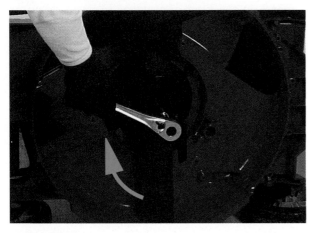

Figure 2-2.   *A wrench should always be pulled in the proper direction to prevent it from slipping off a bolt or nut and rounding the corners.*

Figure 2-3.   *Adjustable wrenches should always be pulled in the correct direction when tightening or loosening bolts and nuts. The movable jaw should always face the turning direction. The jaws should be adjusted to fit the bolt or nut as tightly as possible.*

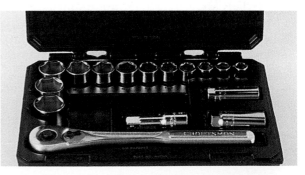

Figure 2-4.   *A socket set is almost a necessity for small engine work. It permits installation and removal of bolts in hard-to-reach places.*

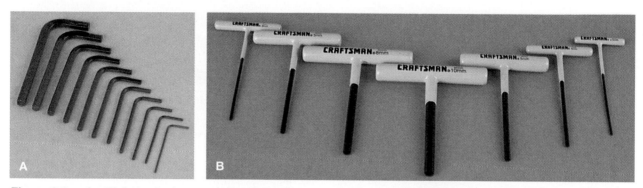

Figure 2-5. A—Right angle, hexagon Allen wrenches are used on hex socket-head screws. B—Tee-handle Allen wrenches.

Support socket with one hand

Figure 2-6. Proper procedure for using a torque wrench is to support socket with one hand and apply turning effort at right angles to the handle.

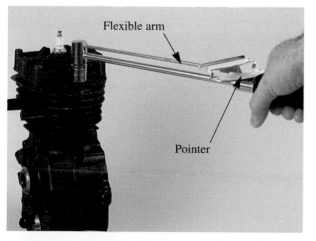

Flexible arm

Pointer

Figure 2-7. A variety of torque wrenches are available for small engine use. This particular wrench uses a pointer and a flexible arm.

Four Cycle Torque Specifications			
		in-lb	*ft-lb*
Cylinder Head Bolts		140-200	12-16
Connecting Rod Lock Nuts	1.5-3.5 H.P.	65-75	5.5-6
	4-6 H.P.	86-100	7-9
Cylinder Cover or Flange to Cylinder		65-110	5.5-9
Flywheel Nut		360-400	30-33

Figure 2-8. Torque specifications are provided in engine manuals for all critical bolts and nuts.

Most small engine torque charts specify both inch-pounds (in-lb) and foot-pounds (ft-lb). Torque wrenches may be calibrated in either of these increments (units). The scale on the wrench is clearly marked whether the reading is given in in-lb or ft-lb.

The torque reading is the product of the length of the wrench handle (in feet or inches) and the applied force. For example, applying one pound of force through a handle one foot long would produce 1 ft-lb or 12 in-lb of torque. In order to convert ft-lb to in-lb, multiply the ft-lb reading by 12.

Pliers

Pliers are extremely useful tools for gripping, bending, pulling, and in some cases, cutting wires. They should not be used to replace wrenches. Because most pliers are not designed for tightening purposes, they usually damage surfaces when used in this way. **Figure 2-9** shows a variety of pliers that are helpful in small engine repair work. *Vise-grip pliers* are designed to apply great clamping

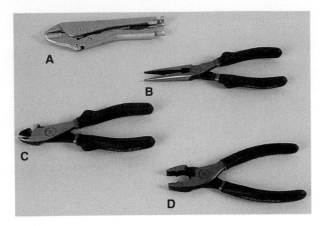

Figure 2-9. *Pliers are used for gripping, bending, pulling, and cutting wires. They should not be used to replace wrenches on nuts and bolts. A—Vise-grip pliers use mechanical advantage to increase grip. B—Needle nose pliers. C— Diagonal, side-cutting pliers. D— Combination slip-joint pliers.*

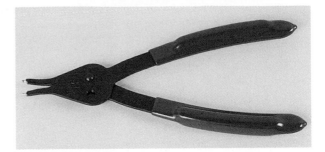

Figure 2-10. *Retaining ring pliers are used to remove and install retaining rings on shafts or in cylindrical holes. Retaining rings are made of spring steel and can fly off if they slip from the nibs of the pliers. Always wear safety glasses when using these pliers.*

pressure and have large gripping teeth. They use mechanical advantage to increase grip. They are sometimes used as a last resort to loosen something that is rusted or frozen in place. Once the part is removed, however, it is usually damaged and needs to be replaced. A good penetrating oil should be used to assist in removing rusted parts. *Needle nose pliers* are very useful for bending small wires and for gripping items that have fallen into small recesses. *Diagonal side-cutting* pliers are used for cutting various types of electrical wire. *Combination slip-joint pliers* are general purpose pliers and are less expensive than other types of pliers. They can also be used for cutting soft, solid wire. Pliers should not be allowed to get hot since they will become distorted and lose the hardness of their cutting jaws.

Retaining ring pliers

Inside and outside retaining rings are used to keep a part from moving axially on a shaft (in a direction parallel to the centerline of a shaft). For example, a sliding gear may be limited from moving too far by a retaining ring set in a groove machined into a shaft. Retaining rings are made from spring steel and must be installed and removed with special *retaining ring pliers,* such as those shown in **Figure 2-10.** To install an outside retaining ring, the nibs of the retaining ring pliers are inserted into the small holes in the ring

and the ring is expanded by squeezing the handles of the pliers. The ring is then slid over the shaft to the machined groove, where it is released. For internal retaining rings, the ring is compressed to fit inside a cylindrical hole when the handles are squeezed. The ring is then inserted into the hole and released into a machined groove. Some pliers are only for inside or outside rings, while others are designed to accommodate both types. Some types have replacement nibs.

 Always wear safety glasses when using these retaining ring pliers because retaining rings can slip off and fly with considerable velocity.

Screwdrivers

Several types of screwdrivers are frequently used when servicing small engines. Standard screwdrivers are available in a variety of shapes and sizes. See **Figure 2-11.** The proper size blade should be used to match the length and thickness

Figure 2-11. *Standard screwdrivers are used for engine work.*

of the screw head slot. See **Figure 2-12.** In addition to installing and removing screws, there are times when screwdrivers can be used to apply leverage to move or separate parts. Quality screwdrivers are very strong and will not be harmed by these applications.

Phillips screwdrivers of various sizes are useful for Phillips head screws and bolts, which have a cross-shaped recess. See **Figure 2-13A.** These screwdrivers are available in a variety of sizes to accommodate various sizes of screws and bolts.

In tight situations, the length of a regular screwdriver may prohibit its use. When this occurs, an *offset screwdriver* can be very useful. See **Figure 2-13B.** Offset screwdrivers are available with standard and Phillips-type heads.

Hammers

Hammers are extremely useful for small engine work. *Ball peen hammers* are used for tapping things into place. They are often used in conjunction with other tools, such as punches and chisels. The ball peen hammer is considered to be a hard-faced hammer because its head is made of steel. See **Figure 2-14.** When using a ball peen hammer for small engine work, care must be taken so that parts are not dented or deformed by the hard faces. The ball peen hammer can be used to tap on a wrench to loosen a stubborn bolt or nut. The ball peen hammer is also used with pin punches to install locking pins in holes or with cold chisels to shear bolts, pins, or sheet metal. Ball peen hammers are available in a variety of sizes and are rated according to weight. A hammer's weight is usually stamped on the side of its head.

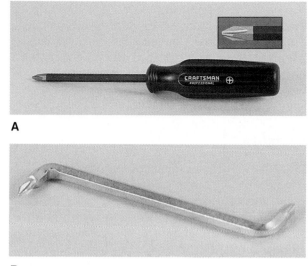
A

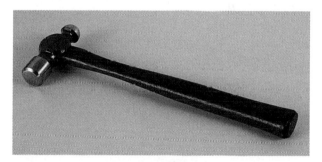
B

Figure 2-13. *Other types of screwdrivers are also used for engine work. A—Phillips screwdriver. B—Offset screwdriver.*

Figure 2-14. *The ball peen hammer has a cylindrical flat face on one end of the head and a ball shaped end on the other. Hammers are classified by head weight (in ounces), which is usually stamped on the head.*

Correct Tip Size
Fills Screw Slot

Too Small of a
Tip Will Damage
Screwdriver

Too Large of a
Tip Will Strip
Screwhead

Figure 2-12. *To avoid damaging screw head, screwdriver blade must fit the slot or recess in the screw properly.*

Soft-faced hammers are used to tap on parts that are easily damaged by hard-faced hammers. Soft-faced hammers are made from a variety of materials that are softer than steel. Lead, copper, brass, leather (rawhide), wood, rubber, and plastic are commonly used to make soft-faced hammers. **Figure 2-15** shows several soft-faced hammers.

Punches

Many types of punches are used for small engine work. A *center punch* has a hardened steel point and is used to make depressions in metal surfaces before drilling. The depression helps to start the drill and keep it in the desired location. **Figure 2-16A** illustrates a center punch with a 90° point angle. A *prick punch* is similar to a center punch. However, the point angle of a prick punch is 60°. The prick punch is used to make a very small depression prior to using the center punch to enlarge the depression for drilling. Center punches and prick punches are driven with a ball peen hammer.

Pin punches of various diameters and lengths are available for driving straight pins, tapered pins,

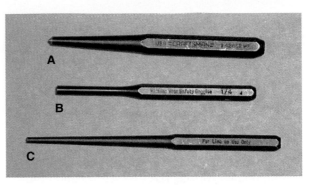

Figure 2-16. *There are several types of punches available for small engine use. A—Center punch. B—Pin punch. C—Drift punch.*

and roll pins in and out of holes. See **Figure 2-16B.** A ball peen hammer is used to apply the driving force to the punch. Sometimes, a rusty bolt can be driven out of its hole with a pin punch.

A *drift punch* is tapered and is used to align holes in mating parts. This allows a bolt or pin to be passed through the mating parts. See **Figure 2-16C.**

Cold chisels

The *cold chisel* is a cutting tool that can shear bolts, pins, rivets, sheet metal, rods, and other materials. See **Figure 2-17.**

The cold chisel is made of a special tool steel, which is hardened and tempered by heat-treatment. See **Figure 2-18.** The cutting edge is very hard and sharpened to an angle between 60° and 90°. A 60° angle is used for shearing sheet metal. A 90° angle is for shearing bolts and rivets. The cutting edge can be sharpened when it becomes dull and may need heat treatment after several grindings. It should have a slightly curved edge for shearing and a straight edge for cutting flat surfaces. See **Figure 2-19.** Although the portion immediately behind the cutting edge is tempered for toughness, the chisel's shank portion is annealed (made soft) to prevent it from shattering. When the shank end of the chisel becomes flared, it should be reground to remove the flared portion. See **Figure 2-20.** The flared part becomes work hardened and highly stressed from being hammered.

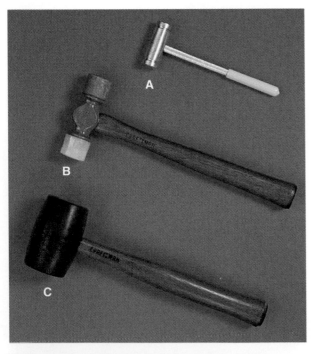

Figure 2-15. *Soft-faced hammers have faces that are softer than the objects they are used on. A—Brass or lead hammer. B—Plastic-faced hammer. C—Rubber mallet.*

 Safety glasses should always be worn when using punches or chisels. Pieces from the flared end of a chisel can fracture and fly at high velocity.

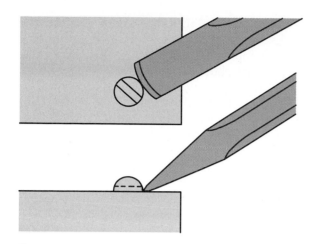

Figure 2-17. *Cold chisels are used for shearing bolts, screws, rivets, sheet metal, rods, or other materials.*

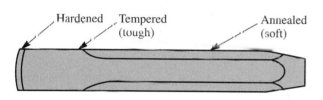

Figure 2-18. *The cold chisel is a heat treated tool with a hardened and tempered cutting edge and an annealed shank.*

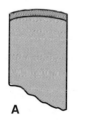 

Figure 2-19. *A—The cutting edge of a cold chisel should be slightly curved for shearing. B—The cold chisel's edge should be straight for cutting flat materials.*

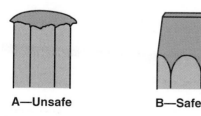

Figure 2-20. *A—The hammered end of a cold chisel becomes work hardened and flared after extended use. B—The flared end is very dangerous and should be ground to a slight taper.*

Gear pullers

A *gear puller* is used for pulling gears and bearings from shafts. See **Figure 2-21.** Gears and bearings are often press fit onto shafts and have to be pulled with considerable force to be removed. The gear puller is designed to provide this pulling force. Gear pullers can be adapted for many applications and are often used to remove flywheels from small gas engines. Refer to *Removing the flywheel* section in Chapter 14 of this text. Also, timing gears can be removed from crankshafts with a gear puller. When roller bearings are pulled, a tool called a *bearing splitter* must be used to avoid damage to the bearings and the outer race.

Probe and pickup tools

The *probe and pickup tools* shown in **Figure 2-22** assist when bolts, screws, washers, or other small items are dropped into crevices where they cannot be reached with the hands or fingers. The mirror probe can help locate items that cannot be seen. Ferrous items (iron bearing) can be removed with the magnetic probe. Nonferrous (nonmagnetic) items can be gripped and removed with the finger pickup tool.

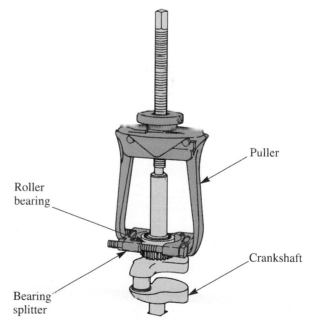

Figure 2-21. *A gear puller is used to remove gears and bearings from shafts. When pulling bearings, a tool called a bearing splitter should be used to avoid damaging the bearing while force is being applied. (Tecumseh Products Co.)*

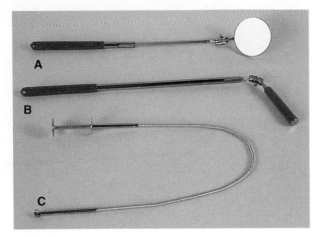

Figure 2-22. *Probe and pickup tools are helpful in locating small parts that may drop into crevices or hard-to-reach places. A—Mirror probe. B—Magnetic pickup tool. C—Finger pickup tool.*

Vise

The *machinist's vise* is extremely useful for holding parts while they are being worked on. See **Figure 2-23.** Some vises can be swiveled for convenient positioning of the work piece. The jaws are hardened steel and have a rough gripping surface. When parts that must not be scratched are clamped in a vise, soft jaw covers should be placed over the steel jaws. Sheet copper, soft aluminum, or wood can be used to pad the jaws. The anvil portion of the vise can be used to flatten sheet metal or to straighten bent parts. Always

be careful not to distort or damage parts by overtightening the vise. Never clamp critical engine components, such as pistons, in a vise. Many engine parts are extremely delicate and the slightest distortion will render them useless.

The machinist's vise can also be used for holding parts or materials that need to be drilled, filed, formed, or sawed. When filing or sawing an object in a vise, adjust the piece so that the work is done close to, but not in contact with, the vise jaws. This tends to minimize vibration.

Cleaning tank

A *cleaning tank* is used to safely clean small engine parts. See **Figure 2-24.** A nontoxic, nonflammable solvent should be used to scrub engine parts after old gaskets and excess grease have been removed. The tank has a pump to recirculate and filter the solvent and direct it through a flexible tube. The tube can be directed at the parts for flushing. After the parts are clean, a low-pressure safety blow gun can be used to remove excess solvent.

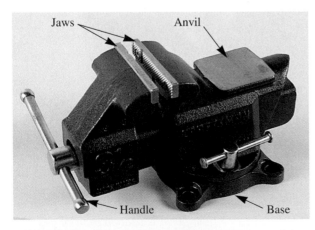

Figure 2-23. *The machinist's vise can be a helpful third hand for holding parts while they are being worked on. Soft copper or aluminum jaw covers can be placed over the steel jaws to protect delicate parts.*

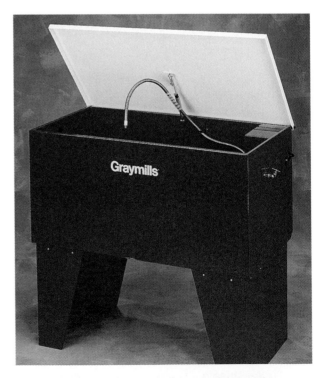

Figure 2-24. *A cleaning solvent tank is used to clean engine parts. This tank is fitted with a fusible link, which will automatically close the lid in case of fire. It also has a built-in pump and filter system for cleaning the solvent. (Graymills Corp.)*

Measuring Instruments

High-quality measuring instruments should always be used when servicing a small gas engine. Many dimensions are critical to proper engine operation. Therefore, it is extremely important to be able to measure various engine parts and clearances accurately. When making engine repairs, measurements must be made to determine if parts are within specified limits or if replacements and adjustments must be made. Quality measuring instruments allow you to service small engines easily and effectively. Keep instruments clean and in proper working condition.

Some measuring instruments are common to most engine work, while others have only one or two specific applications. Special-purpose tools may be manufacturer designed for limited use on only one engine make or model. The following sections describe how to use various measuring instruments.

Micrometers

The **micrometer** is a precision instrument designed to accurately measure pistons, crankshafts, valve stems, and other small engine components. You must be able to read a micrometer correctly to make judgments about the condition of various engine parts.

There are several varieties of micrometers available. Each type is designed for a specific purpose. An outside micrometer is used to measure thicknesses and outside diameters. See **Figure 2-25.** The inside micrometer is designed for taking measurements of internal dimensions. See **Figure 2-26.** A blade micrometer, such as the one shown in **Figure 2-27,** is used to take measurements in narrow slots. **Figure 2-28** illustrates the use of an inside micrometer and an outside micrometer.

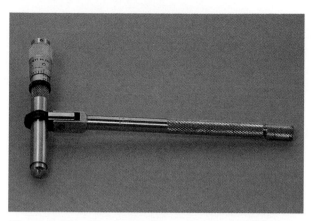

Figure 2-26. *The inside micrometer is designed to measure internal dimensions.*

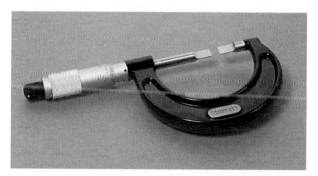

Figure 2-27. *The blade micrometer is designed to enable the spindle and anvil to measure in narrow grooves.*

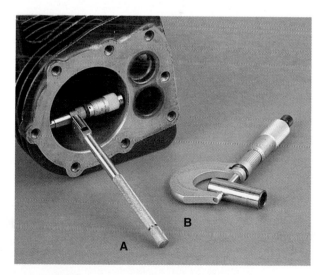

Figure 2-28. *A—An inside micrometer being used to measure the internal diameter of a cylinder. B—An outside micrometer being used to measure a piston pin.*

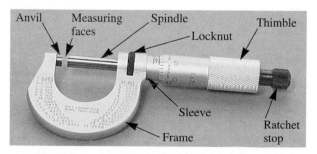

Figure 2-25. *Note the parts of an outside micrometer with ratchet stop. This particular micrometer is graduated in thousandths of an inch (.001).*

Because the micrometer is made of metal, it will expand when heated and contract when cooled. The micrometer *should not* be held in the hand for long periods of time. Body heat can cause inaccuracy in the instrument. Always hold the micrometer properly. **Figure 2-29** shows the proper one-hand method for measuring a valve stem diameter with a micrometer. **Figure 2-30** shows a micrometer being held with two hands. Note the minimal contact between the hands and the micrometer. Measuring instruments are quite expensive and should always be handled with care.

Micrometers are available in several sizes. Most micrometers have the capability to measure in 1 inch increments. For example, a 0-1″ micrometer can be used to measure objects smaller than 1″. A 2-3″ micrometer, on the other hand, is designed to measure objects that are between 2″ and 3″.

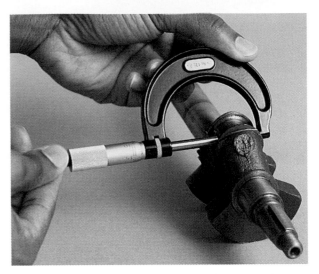

Figure 2-30. *The proper technique for measuring with a micrometer when two hands are necessary. Use very light pressure when turning the thimble of the micrometer.*

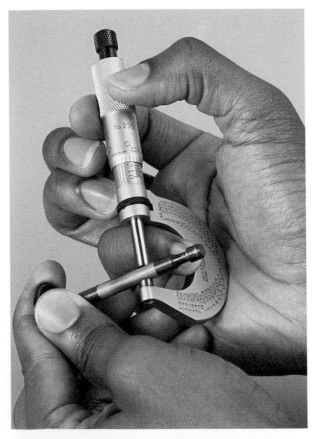

Figure 2-29. *The proper way to hold a micrometer in one hand while holding the piece to be measured in the opposite hand. Note small finger in frame of micrometer.*

Cleaning and calibrating a micrometer

Before attempting to measure any object, make sure that the micrometer's anvil and spindle faces are clean. If necessary, clean the faces by gently closing the micrometer on a piece of clean, white paper and drawing the paper from between the faces. To assure accuracy, the micrometer should always be checked for proper calibration before use. A 0-1″ micrometer can be checked by simply closing it and observing the reading on the sleeve and thimble. If the reading is not zero, clean the anvil and spindle again and retest the micrometer. If it still does not read zero, check the manufacturer's instructions for adjusting the instrument. Micrometers larger than 0-1″ require the use of a gauge block, cylindrical gauge, or gauge pin to verify calibration. These blocks are precision ground to exact dimensions and are used to check the micrometer for accuracy. For example, a 1″ gauge block can be used to check the calibration of a 1-2″ micrometer by simply measuring the block. The micrometer should read exactly one inch. If it does not, it must be recalibrated.

Using a micrometer

To use a micrometer, simply place the object to be measured against the anvil and turn the thimble until the spindle touches the object. If the micrometer has a ratchet-stop, **Figure 2-25,** click it just once after making contact with the work surface. If the micrometer does not have a ratchet-stop, the thimble must be turned very gently so the anvil and spindle faces lightly contact the outer surfaces of the part being measured. Overtightening can permanently damage the micrometer. If it is necessary to remove the micrometer from the part to read it, the lock nut can be used to secure the spindle so that it does not turn during removal.

Standard and Vernier micrometers

Some micrometers provide more accurate readings than others. The *standard micrometer* is graduated in thousandths of an inch (.001″). See **Figure 2-25.** The *Vernier micrometer,* on the other hand, is graduated in ten-thousandths of an inch (.0001″). The Vernier micrometer has an additional scale on its sleeve. See **Figure 2-31.** Part tolerances are specified in three or four decimal place numbers. A Vernier micrometer can be used for either measurement, but the standard micrometer is only accurate to three decimal places.

Reading the standard micrometer

The first step in reading any micrometer is to familiarize yourself with the divisions (graduations) on the sleeve and the thimble.

The micrometer's sleeve graduations reflect the fact that the spindle moves 1/40″ (0.025″) for each revolution of the thimble. Therefore, the micrometer's sleeve is divided into 40 equal spaces. Each line on the sleeve represents 1/40″ (0.025″). Every fourth line is numbered. These numbers represent 1/10″ (.100″). See **Figure 2-32.** Looking at the example in **Figure 2-34A,** the thimble has been rotated four turns, or one tenth of an inch (4 × .025 = .100″).

The micrometer's thimble is divided into 25 equal parts. Each line on the thimble represents 0.001″. **Figure 2-33** illustrates a typical micrometer thimble. **Figure 2-34B** shows a thimble that has been rotated ten full turns (.250″), plus .008″ more, totaling .258″. Now, read the micrometer scale in **Figure 2-34C.**

Reading a Vernier micrometer

To obtain readings to four decimal places, a Vernier micrometer must be used. The Vernier micrometer has an additional scale, called the Vernier scale, located on the top of its sleeve. See **Figure 2-35.** The first three decimal places on a Vernier micrometer are read in the same way as they are on the standard micrometer. The fourth

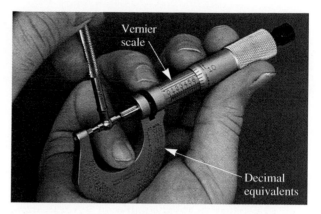

Figure 2-31. *A Vernier micrometer can measure to four decimal places with the special scale that is located on its sleeve. The fourth decimal place is determined by the two lines that coincide with each other.*

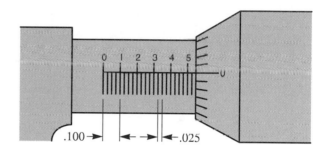

Figure 2-32. *Each of the small spaces on the micrometer's sleeve is equal to one fortieth of an inch (.025). Since there are four small spaces between each of the numbers printed on the sleeve, the distance between numbers is equal to one tenth of an inch (.100). The reading on this micrometer is five-hundred and fifty thousandths of an inch.*

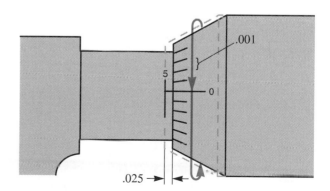

Figure 2-33. *Each small space on the thimble of the micrometer is equal to one one-thousandth of an inch (.001). One complete rotation of the thimble moves the spindle twenty-five thousandths of an inch (.025), or one small space on the sleeve.*

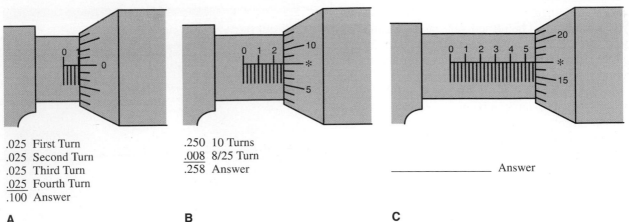

A

.025 First Turn
.025 Second Turn
.025 Third Turn
.025 Fourth Turn
.100 Answer

B

.250 10 Turns
.008 8/25 Turn
.258 Answer

C

_____ Answer

Figure 2-34. *Study the micrometer readings on the 0-1″ micrometers in A and B. Can you read the micrometer in C? (Answer = .567)*

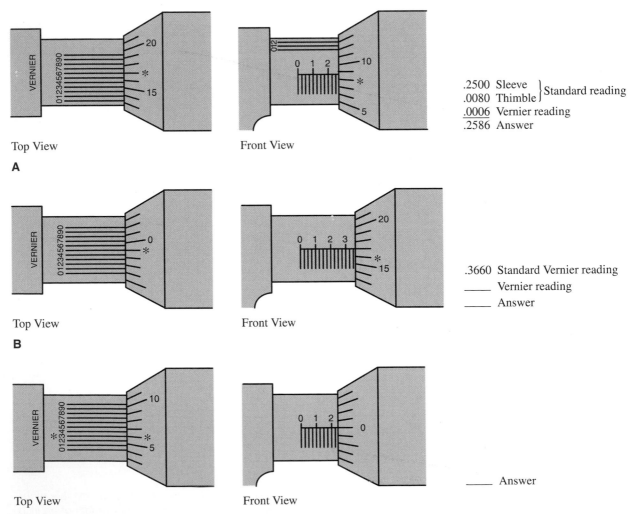

A

Top View   Front View

.2500 Sleeve ⎫ Standard reading
.0080 Thimble ⎬
.0006 Vernier reading
.2586 Answer

B

Top View   Front View

.3660 Standard Vernier reading
_____ Vernier reading
_____ Answer

C

Top View   Front View

_____ Answer

Figure 2-35. *Study the Vernier micrometer reading in A. Complete the reading for B (.3665). Read C (.2253). Assume that these are 0-1″ micrometers.*

decimal number, however, is obtained from the Vernier scale. Unless the zero line on the thimble is aligned with the sleeve's horizontal reference line, only one line of the Vernier scale will be perfectly aligned with one of the lines on the thimble. If the sixth line of the Vernier scale is aligned with one of the lines on the thimble, the fourth decimal place number would be six. **Figure 2-35A** illustrates a 0-1″ Vernier micrometer displaying a reading of .2586″. Can you read the measurements in **Figures 10-35B** and **10-35C?**

Remember, always handle measuring instruments carefully and keep them clean to maximize accuracy and reliability. Once you have practiced a few readings, you will see how easily and how quickly micrometers can be used.

Digital micrometers

Digital micrometers eliminate reading anything but the digital readout. See **Figure 2-36.** This type of micrometer has simplified the measuring procedure. Although the measurement appears in the window, it can also be read from the barrel and thimble.

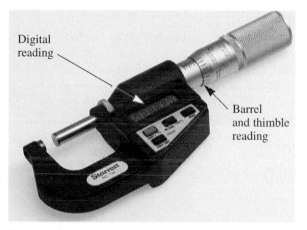

Figure 2-36. *Today, digital micrometers are becoming more and more common.*

Reading a metric micrometer

The metric micrometer closely resembles the standard and Vernier micrometers. The only difference is the markings on the thimble and sleeve. The metric micrometer is marked to measure hundredths of a millimeter. Each line on the thimble equals 0.01mm, and each line on the sleeve equals 0.5mm. Two full revolutions of the thimble, equals 1.00mm on the sleeve.

To read a metric micrometer, note the number of millimeters between the zero line and the thimble. Then, locate the line on the thimble that coincides with the horizontal line on the sleeve. Add these numbers, and this is the reading. See **Figure 2-37.**

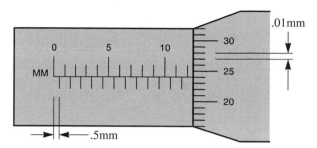

Figure 2-37. The metric micrometer measures to the hundredths of a millimeter. (Answer = 12.24mm)

Micrometer depth gauges

The *micrometer depth gauge* is a measuring device used to measure depths. See **Figure 2-38.** A depth micrometer is read using the same steps used for reading standard and Vernier micrometers.

A micrometer depth gauge is read using almost the same steps as were used for reading the standard and Vernier micrometers. The main difference is the numbers on the sleeve that are under the thimble are used for the reading. It is necessary to see which numbers and spaces are visible on the sleeve and deduce those that are under the sleeve.

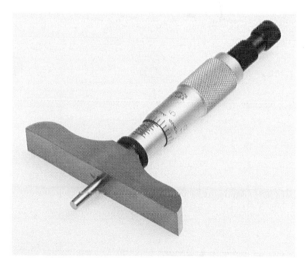

Figure 2-38. *Micrometer depth gauges are used to measure the depths.*

For example, if a number 4, a space, and a partial space are visible, then a 3, two full spaces, and a partial space is under the thimble. Therefore, the reading to this point is .325″. Now a reading from the thimble is taken. So, if the twelfth line on the thimble is aligned with the reference line, then .012 is added to .325 to get a reading of .337″.

Vernier calipers

The **Vernier caliper** is a measuring device used to measure both internal and external measurements. See **Figure 2-39.** To take a measurement, slide the assembly until the jaws almost contact the part being measured. Lock the clamping screw, and adjust the fine adjusting nut. The contact between the jaws and the part must be firm but not tight. Now, lock the slide on the beam, carefully remove the caliper, and make your reading.

Reading a Vernier caliper

Vernier calipers are available in either 25 or 50 division Vernier scale. Both types of scales are graduated in thousandths of an inch (.001″).

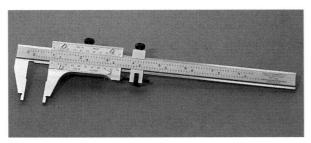

Figure 2-39. *Vernier calipers can be used to make internal and external measurements.*

25 division scale

On a 25 division Vernier caliper, each inch on the beam is graduated into 40 equal parts. Each graduation is .025″ (1/40″). So, every fourth one is numbered and is .100″ (1/10). On the Vernier scale there are 25 divisions numbered 5, 10, 15, 20, and 25.

To read a 25 division Vernier caliper, note how many inches (1, 2, 3, etc.), tenths (.100, .200, .300, etc.), and fortieths (.025, .050, or .075) there are between the zero line on the Vernier scale and the zero line on the beam. Add these numbers together. See **Figure 2-40A.**

Now count the number of graduations (each being .001″ or 1/1000″) that are between the zero line on the Vernier scale and the line that lines up exactly with a line on the beam. Only one will exactly line up. Add this number to the total previously found. This is the reading.

50 division scale

On a 50 division Vernier caliper beam, every other graduation between the inch lines are numbered and are equal to .100″ (1/10″). The unnumbered ones are equal to .050″ (1/20″). The Vernier scale is graduated into 50 parts, each representing .001″ (1/1000″). Every fifth or tenth line is numbered (5, 10, 15, etc. or 10, 20, 30, etc.).

To read 50 division Vernier caliper, note how many inches (1, 2, 3, etc.), tenths (.100, .200, .300, etc.), and twentieths (.050″) there are between the zero line on the Vernier scale and the zero line on the beam. Add these three numbers together. See **Figure 2-40B.**

Now count the number of graduations (each being .001″ or 1/1000″) that are between the zero line on the Vernier scale and the line that lines up

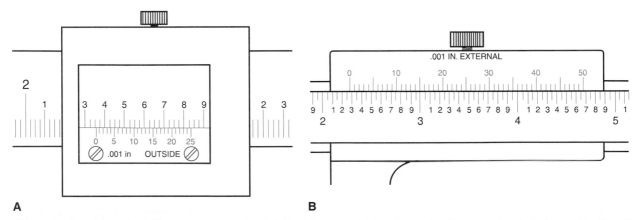

A **B**

Figure 2-40. *A Vernier caliper may have a 25 division (A) or 50 division (B) scale. (Answers A = 2.359, and B = 2.265)*

exactly with a line on the beam. Only one will exactly line up. Add this number to the total previously found. This is the reading.

Digital Vernier caliper

Digital Vernier calipers eliminate reading anything but the digital readout. See **Figure 2-41.** This type of micrometer has simplified the measuring procedure.

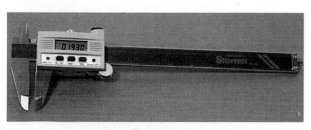

Figure 2-41. *Today, digital Vernier calipers are becoming more and more common.*

Reading a metric Vernier caliper

The metric Vernier caliper closely resembles the standard type. Like the standard types, the metric Vernier calipers are available in 25 or 50 division scales.

Metric 25 division Vernier caliper

On a metric 25 division Vernier caliper, there are 25 equal divisions on the Vernier plate. Each division is equal to 0.02 mm. Every fifth line is numbered (0.10mm, 0.20mm, etc.).The beam is graduated in 0.5mm divisions. Every twentieth division is numbered (10mm, 20mm, etc.).

To read the metric 25 division Vernier caliper, note the number millimeters between the zero line on the beam and the zero on the Vernier plate. Then, locate the line on the Vernier plate that coincides with one of the lines on the beam. Note the value of this line as indicated on the Vernier plate. Add these numbers, and this is the reading. See **Figure 2-42A.**

Metric 50 division Vernier caliper

On a 50 division Vernier caliper, there are 50 equal divisions on the Vernier plate. Each division is equal to 0.02 mm. Every fifth line is numbered (.10mm, .20mm, etc.).The beam is graduated in 0.1 mm divisions. Every tenth division is numbered (10mm, 20mm, etc.).

To read a 50 division Vernier caliper, note the number millimeters between the zero line on the beam and the zero line on the Vernier plate. Locate the line on the Vernier plate that coincides with one of the lines on the beam. Note the value of this line as indicated on the Vernier plate. Add these numbers, and this is the reading. See **Figure 2-42B.**

Telescoping gauges

Telescoping gauges are *transfer-type* measuring instruments. They do not provide a direct dimensional reading. A telescoping gauge, like the one in **Figure 2-43,** can be used to transfer the distance from A to B to a micrometer. See **Figure 2-44.**

Telescoping gauges can be purchased singly, **Figure 2-45,** or in sets, providing a wide range of sizes to accommodate a variety of measurements.

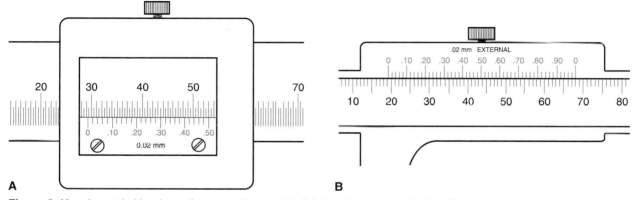

A

B

Figure 2-42. A metric Vernier caliper may have a 25 division (A) or a 50 division (B) scale. (Answers A = 29.28mm, and B = 19.28mm)

See **Figure 2-46.** The spindle faces are curved so that each end has only one point of contact when curved surfaces are being measured. See **Figure 2-47.** To use the telescoping gauge to determine the diameter of a cylinder, loosen the lock screw on the end of the handle so that the telescoping parts can be retracted, locked, and placed in the cylinder. Once the gauge is located in the cylinder,

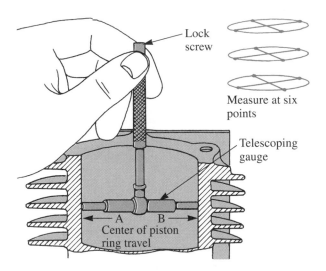

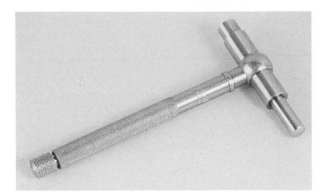

Figure 2-45. *A single telescoping gauge.*

Figure 2-43. *A telescoping gauge can be used to measure the inside diameter of a cylinder. 1—Depress the spindle and lock with lock screw. 2—Place gauge in cylinder and release lock screw. Spindle will spring out to touch walls of cylinder. 3—Tighten lock screw. 4—Tilt gauge handle and remove the gauge. 5—Measure across spindle faces with an outside micrometer.*

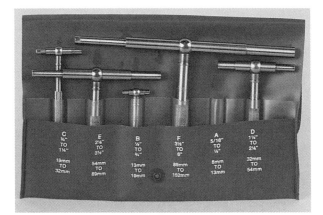

Figure 2-46. *Telescoping gauge set.*

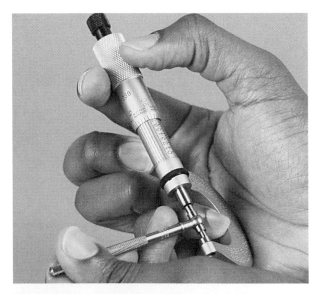

Figure 2-44. *After removing the telescoping gauge from the cylinder, measure the gauge with a micrometer.*

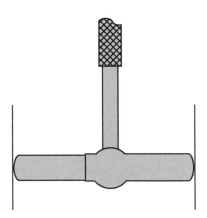

Figure 2-47. *The spindle faces of telescoping gauges are curved so spindle touches cylinder at only one point.*

release the lock screw to allow the telescoping ends to extend. The handle must be held perfectly in line with the centerline of the cylinder being measured. Also, it is essential that the telescoping ends be aligned exactly across the true diameter of the cylinder. Refer to **Figure 15-11.** The general practice to obtain this position is to hold the handle gently with the thumb and forefinger while sliding the gauge up and down the cylinder walls. Spring tension will cause the gauge to seek the true diameter of the cylinder. When the true diameter has been located, use the opposite hand to gently tighten the lock screw, securing the telescoping ends. When removing the gauge from the cylinder, tilt the handle so that it can be removed without changing the setting. See **Figure 2-48.** After the gauge is removed, measure the distance between the telescoping ends with a micrometer. See **Figure 2-44.**

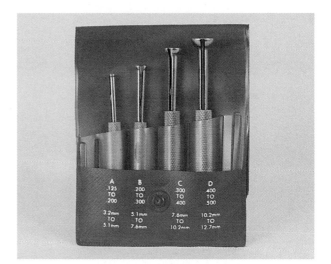

Figure 2-49. *Small hole gauges are used to measure the diameter of holes that are too small for telescoping gauges.*

hole. See **Figure 2-50.** Test for correct fit by moving the gauge back and forth in the hole until you feel a slight drag. Remove the gauge and measure across the ball with a micrometer. See **Figure 2-51.**

Thickness gauges

Thickness gauges are sometimes called *feeler gauges* because they rely on the user's sense of feel for accuracy. They are used to measure small

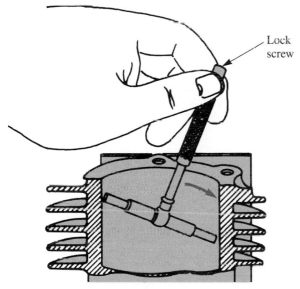

Figure 2-48. *Tilting the telescoping gauge allows easy removal and will not change the position of the spindles.*

Small hole gauge

The *small hole gauge* is similar to the telescoping gauge. It is intended for measuring holes that are too small for the smallest telescoping gauge. Small hole gauges are usually provided in sets to accommodate a variety of hole sizes. See **Figure 2-49.** The small hole gauge is expanded by turning a knurled screw on the end of its handle until the split ball touches the inner walls of the

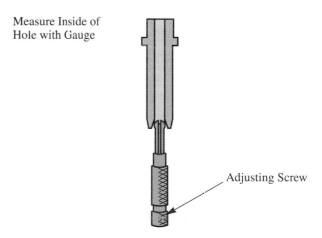

Figure 2-50. *A small hole gauge used to measure the inside diameter of a valve guide. Insert the gauge and expand the split ball end by turning the adjusting screw until contact is felt. Move gauge up and down in the hole as adjustment is made. (Deere & Co.)*

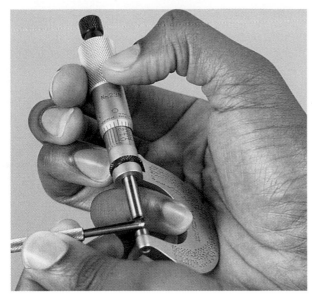

Figure 2-51. *Measuring the small hole gauge with a micrometer.*

Figure 2-53. *Measuring valve stem clearance with a thickness gauge. This gauge is often referred to as a feeler gauge because the sensitivity of feel is required to use the gauge accurately.*

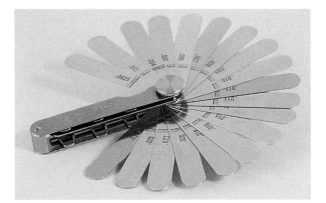

Figure 2-54. *The leaves of a thickness gauge are marked with their exact thickness. When several leaves are used to obtain a desired thickness, each must be flat and clean. Dirty or bent leaves will cause inaccurate readings.*

spaces and gaps between surfaces. Thickness gauges are used to measure crankshaft end play, spark plug gap, breaker point gap, and piston ring gap. See **Figure 2-52. Figure 2-53** illustrates a thickness gauge being used to check valve stem clearance.

Thickness gauges consist of a set of metal leaves that vary in thickness. Each leaf has its decimal and metric thickness etched onto its surface. See **Figure 2-54.** Some leaves are as thin as .0005″ and may be damaged if not handled carefully. It is sometimes necessary to select several leaves to equal a desired thickness. When using a thickness gauge, the leaves must be perfectly clean. Avoid bends and distortions. These condi-

tions will increase the total thickness of the leaves and cause measurement errors.

Figure 2-52 shows a thickness gauge being used to check for proper ring gap. To use the gauge, select a leaf that is thinner than the gap to be measured and place it in the gap. Progressively use thicker leaves until you can feel a slight contact or drag from the gap's edge surfaces. At this point, compare the leaf thickness with the amount of gap specified. If the gap is too small, it may be corrected by carefully filing the ring ends. Refer to **Figure 16-19.** If the gap is too wide, a new ring is required. Do not use leaf-type thickness gauges for measuring gaps on used spark plugs or pitted

Figure 2-52. *A thickness gauge is being used to measure piston ring end gap. Note that the ring is placed in the cylinder.*

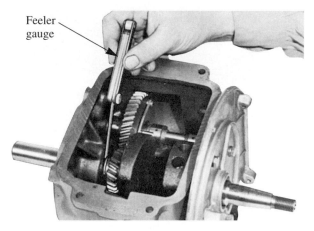

Feeler gauge

Figure 2-55. *Thickness gauge being used to measure crankshaft endplay. Begin with a thin leaf and progress to thicker leaves until one will not enter the space. Then, return to the thickest leaf that will fit space.*

breaker points. **Figure 2-55** shows a thickness gauge being used to measure crankshaft end play.

Valve spring tension tester

The function of the valve springs is to close the valves quickly and securely. If spring tension is inadequate, valve leakage and poor valve timing can occur. The *valve spring tension tester* is designed to test valve spring tension with the aid of a torque wrench. See **Figure 2-56.** The infor-

Valve spring

Figure 2-56. *Valve spring tension tester. A conversion plate located on the tester converts pounds-inch and pounds-foot of torque to pounds of applied force. If specifications are in metric Newton-meters, one Newton-meter is equal to 1.36 pounds-foot.*

mation plate on the base of the tester converts pounds-inch and pounds-foot of torque to pounds of applied force. Some torque specifications are provided in metric Newton-meters (N•m).

Combination square

The *combination square* has many uses. It is often used to measure the length of valve springs and check them for straightness. See **Figure 2-57.** For accuracy, always place spring and square on a flat, machined surface.

Figure 2-57. *Straightness and length of valve springs can be measured on a flat surface with a combination square.*

Dial indicator

A *dial indicator* is a precision instrument that is very useful for measuring the movement of various parts. See **Figure 2-58.** It is also used to check for surface irregularities and run-out. The dial indicator is equipped with a spring-loaded spindle, which is placed against the part to be measured. A needle on the instrument's dial is used to indicate the amount of movement made by the part being tested. The needle rotates over the dial indicator's face, which is calibrated in thousandths of an inch, ten-thousandths of an inch, or hundredths of a millimeter. The indicator's bezel can be revolved to locate the zero line on the dial face under the needle. The zero line coincides with the initial position of the needle and serves as a reference point when checking for movement. The

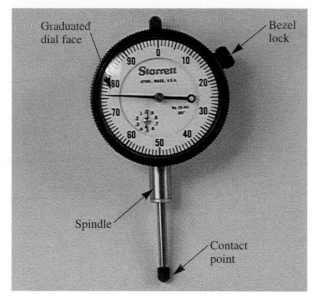

Figure 2-58. *The dial indicator can be used to measure linear movement. Each space on this dial indicator represents one one-thousandth of an inch (.001).*

Figure 2-60. *A dial indicator being used to measure side movement of a valve stem in its valve guide. (Kubota Tractor Co.)*

Screw pitch gauge

The ***screw pitch gauge*** is used to determine the number of threads per inch on bolts, screws, nuts, and in threaded holes. See **Figure 2-61.** Each leaf of the gauge is marked with the number of threads per inch it will match when the leaf is placed onto the screw, bolt, etc. See **Figure 2-62.** Keep trying leaves until one fits exactly into the threads. Read the number on the leaf to determine the number of threads per inch on the item being checked. Screw pitch gauges are available for both standard and metric threads.

dial indicator can be mounted on a variety of devices that are designed to hold it in the desired location. One convenient holding device has a magnetic base that can be turned on and off. When the mounting surface is nonmagnetic, however, a clamp-type holding device must be used. **Figure 2-59** shows a dial indicator being used to measure crankshaft endplay. The side movement of a valve in its guide is being checked in **Figure 2-60.**

Figure 2-61. *The screw pitch gauge has many blades. Each blade is marked with the number of threads per inch or threads per millimeter that it will match. The use of this gauge is a trial and error attempt to match the teeth of a gauge blade to the threads of a screw.*

Figure 2-59. *A dial indicator is often used to measure crankshaft end-play. (Kubota Tractor Co.)*

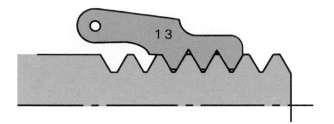

Figure 2-62. *A screw pitch blade matched to the threads of a screw. Threaded bolts must match the threads of a threaded hole to prevent damage. A bolt should never be forced to enter a threaded hole if it will not turn freely.*

Summary

Quality tools and measuring instruments should be used when servicing small engines.

Many types of wrenches are necessary for small engine work. These include the box-end wrench, open-end wrench, adjustable wrench, Allen wrench, socket wrench, and torque wrench. Pliers are used for gripping, bending, pulling, or cutting. They should never be used for tightening purposes. Retaining ring pliers are used to remove inside and outside retaining rings. Always wear safety glasses when working with retaining ring pliers.

Several types and sizes of screwdrivers are used for small engine work, including the straight blade screwdriver, the Phillips screwdriver, and the offset screwdriver.

Hammers are used for tapping things into place. Hard-faced hammers have a head that is made of steel. The head on soft-faced hammers is made of lead, copper, brass, leather, wood, or plastic. Soft-faced hammers are used on parts that would be easily damaged by a hard-faced hammer.

Punches are used to make depressions in metal surfaces before drilling, drive pins from holes, and align holes in mating parts. The cold chisel is used to shear bolts, pins, rivets, sheet metal, and other materials.

Gear pullers are used to remove gears and bearings from shafts.

Probe and pickup tools assist when small items are dropped into areas that cannot be reached with hands and fingers.

A machinist's vise is useful for holding parts while they are being worked on. Sheet copper, aluminum, or wood can be used to pad the vise jaws when working on parts that must not be scratched. A cleaning tank is often used to clean small engine parts.

A micrometer is a precision instrument used to measure crankshafts, pistons, and other components. A standard micrometer is used to measure to three decimal places. A metric micrometer measures to the hundredths of a millimeter. Vernier micrometer can be used to obtain readings to four decimal places. Digital micrometers are becoming more common because they are easier to read.

A Vernier caliper is a precision measuring instrument, which is either standard or metric. It can be used to take internal and external measurements. Digital Vernier calipers are becoming more common because they are easier to read.

Telescoping gauges transfer dimensions to an outside micrometer. A small hole gauge is used to measure holes that are too small for the telescoping gauge. The small hole gauge also transfers dimensions to the outside micrometer. Thickness gauges, often called *feeler gauges,* are used to measure small spaces and gaps between surfaces.

A valve spring tester is used to check valve spring tension. It is usually used in combination with a torque wrench. A combination square can be used to measure the length of valve springs and to check them for straightness.

The dial indicator is used to measure the movement of various parts. Many holding devices are available to hold the indicator in the desired location.

A screw pitch gauge is used to determine the number of threads on a screw, bolt, or nut. They are available for both standard and metric threads.

Know These Terms

open-end wrench	vise grip pliers
box-end wrench	needle nose pliers
combination wrench	diagonal side cutting
tubing wrench	pliers
adjustable wrench	combination slip-joint
socket sets	pliers
Allen wrench	retaining ring pliers
torque wrench	Phillips screwdriver
torque	offset screwdriver
pliers	ball peen hammer

center punch
prick punch
pin punch
drift punch
cold chisel
gear puller
probe and pickup tools
machinist's vise
cleaning tank
micrometer
standard micrometer

Vernier micrometer
micrometer depth gauge
Vernier caliper
telescoping gauge
small hole gauge
thickness gauge
valve spring tension
   tester
combination square
dial indicator
screw pitch gauge

Chapter 2
Review Questions

Answer the following questions on a separate sheet of paper.

1. What kind of wrench is open on one end and boxed on the other?
2. Which kind of wrench should be used only as a last resort if others are not available?
3. To remove a spark plug, what kind of wrench would be best to use to avoid damaging the plug during removal?
4. A torque wrench is used to measure the turning force applied to a threaded fastener. True or False?
5. Why must safety glasses be worn when using retaining ring pliers?
6. To locate a position to be drilled, a hammer and a(n) _____ _____ should be used.
7. When using punches and cold chisels, it is very important to make sure the hammered end does not have a(n) _____ end.
8. Name several types of measuring instruments used to determine if engine parts are within manufacturer's tolerance.
9. Each space on the thimble of a standard micrometer represents _____.
   a. .0001″   c. .01″
   b. .001″   d. .1″
10. Each space on the sleeve of a standard micrometer represents what part of an inch?
   a. .0025″   c. .050″
   b. .100″   d. .025″
11. The discrimination of a Vernier micrometer is _____.
   a. .001″   c. .0001″
   b. .01″   d. .000001″

12. The telescoping gauge is used for measuring _____.
   a. inside diameters
   b. outside diameters
   c. depths of cylinders
   d. valve angles
13. Which measuring instrument is similar to the telescoping gauge?
14. An instrument called a(n) _____ is used to determine the number of threads per inch on bolts or nuts.
15. Telescoping and small hole gauges transfer inside dimensions to an outside _____.
16. Dial indicators are used to check for _____, _____, or _____.

Suggested
Activities

1. Identify the various types of wrenches, pliers, punches, chisels, screwdrivers, and other tools in your shop.
2. Demonstrate the proper way to use wrenches, chisels, screwdrivers, pliers, punches, and other tools in your shop.
3. Clean the anvil and spindle of a micrometer and check its calibration accuracy.
4. Practice measuring engine parts with a standard outside micrometer.
5. Practice measuring engine parts with a Vernier micrometer.
6. Measure a cylinder bore with a telescoping gauge and outside micrometer.
7. Measure a valve guide bore with a small hole gauge and outside micrometer.
8. Check the straightness and length of a small engine valve spring. Does it meet the manufacturer's specifications?
9. Check ring groove and piston ring clearance with a thickness gauge.
10. Check ring end clearance in a cylinder with a thickness gauge.
11. Check the thread pitch of several different size bolts or several different size screws with a screw pitch gauge.
12. Measure the end play of a crankshaft with a dial indicator.

C H A P T E R

Fasteners, Sealants, and Gaskets

After studying this chapter, you will be able to:
- ▼ Identify fasteners used on small gas engines and implements.
- ▼ Remove and install various fasteners correctly.
- ▼ Repair or produce internal and external threads.
- ▼ Properly select and install fasteners.
- ▼ Remove, select, and install gaskets correctly.

Threaded Fasteners

Small gas engines and the implements for which they provide power are held together by fasteners. There are many kinds of fasteners. See

Figure 3-1. Some are common and others are designed to perform special functions. During service fasteners may be exposed to conditions such as heating and cooling, cyclic loading, tensile and shearing loads, corrosion, and vibration.

The helical portion of a screw or bolt, or the *helix* in a hole that it fastens into, is called a *thread*. A thread is an inclined plane that circles around the cylindrical bolt or hole. See **Figure 3-2.** The incline of the bolt or screw must be the same as the incline of the nut or threaded hole into which it is placed. The tightness (tension) of threaded fasteners is very important and will be discussed in the *Torque* section of this chapter.

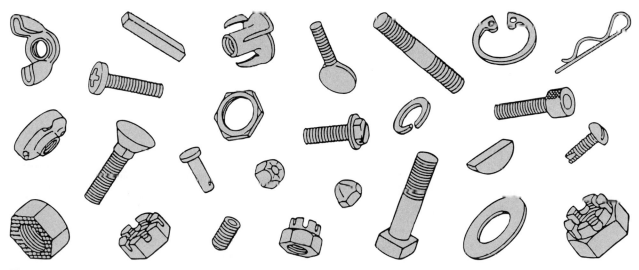

Figure 3-1. *There are many kinds of fasteners to hold parts together.*

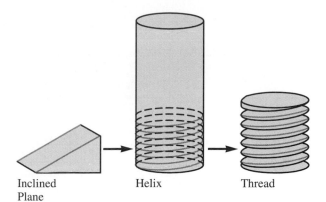

Figure 3-2. *A thread is an inclined plane wrapped around a cylinder and is called a helix.*

Inclined Plane · Helix · Thread

When disassembling an engine, note the location of all parts, fasteners, and washers. Care should be taken during this process, so that the correct parts, fasteners, and washers are replaced during reassembly. When fasteners are badly damaged or worn, replace them with new ones. Lightly rusted fasteners should be cleaned with a wire brush or wire wheel and examined for damage. Lightly apply oil to the thread and shank before installing bolts, nuts, or screws.

Screws

Screws are threaded fasteners that hold parts together by passing through one part and threading into another. See **Figure 3-3.** Flat head screws must be set into a *countersunk* hole so that the head will be flush with the surface. The most common angle for countersunk holes is 82 1/2°. See **Figure 3-4.** However, 60° and 100° countersinks are used for some special applications. Modern practice uses many hexagonal screw heads because of the prevalent use of hexagonal sockets and box wrenches for tightening and loosening. Round head, flat head, and other screw heads require the appropriate screwdriver or Allen wrench to turn them. The most common screw and bolt heads arc shown in **Figure 3-5.** Notice that each head has a particular name that identifies the head type. Screws may be threaded all the way to the head.

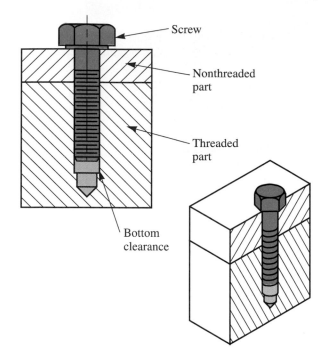

Figure 3-3. *Screw passes through one part and threads into mating part. Note clearance at bottom of hole.*

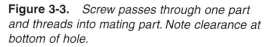

Figure 3-4. *Flat head screw recessed in countersunk hole to be flush with surface.*

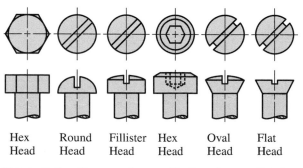

| Hex Head | Round Head | Fillister Head | Hex Head | Oval Head | Flat Head |

Figure 3-5. *Common types of bolt and screw heads.*

Set screws

Set screws are heat-treated, hardened-alloy steel fasteners, which are used to secure such things as pulleys, gears, and shafts. See **Figure 3-6.** When using set screws on a shaft it is necessary to provide a relief groove so that the part can be removed from the shaft when necessary. The common head types are square, slotted *hexagon socket,* and fluted socket. The set screw points may be flat, cup point, cone, half dog, or full *dog.* See **Figure 3-7.** Flat points are used when just clamping friction is enough to hold the part without deforming its surface. Cup and cone point set screws cut into the surface of the shaft or part to prevent its motion or rotation. Dog point set screws are designed to positively lock into a predrilled hole in a shaft matching the diameter of the dog. See **Figure 3-8.**

Self-tapping screws

A variety of *self-tapping screws* are shown in **Figure 3-9.** Self-tapping screws are fasteners that will cut their own threads in a predrilled hole if the hole diameter is of appropriate size. *Tapping* is the

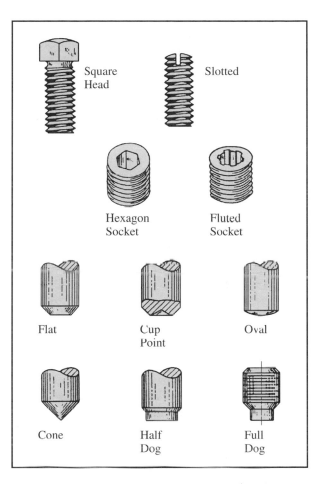

Figure 3-7. *Common types of set screw heads and points. Set screws are hardened steel.*

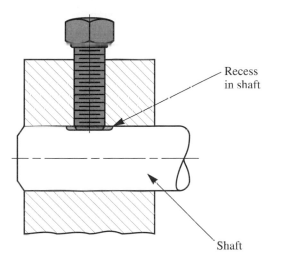

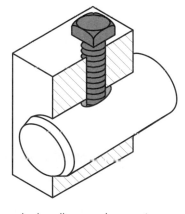

Figure 3-6. *Set screws lock pulleys and gears to shafts to prevent rotation of shaft in hole. Note recessed area on shaft.*

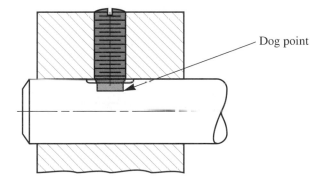

Figure 3-8. *Dog point of screw in hole in shaft.*

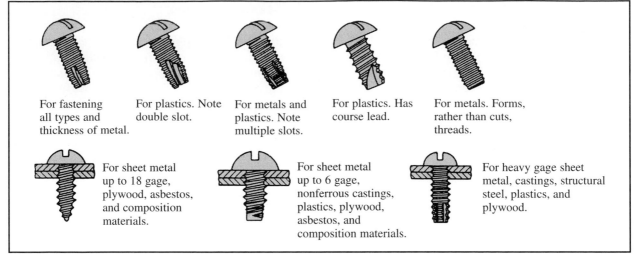

Figure 3-9. *Self-tapping screws cut, or form their own threads.*

process of cutting threads in a hole. ***Threading*** is the process of making external threads on an external cylindrical surface. Self-tapping screws have a grooved or tapered point that forms the threads in the hole. Self-tapping screws are hardened steel because they perform as a cutting tool as well as a fastener.

Bolts

Bolts are threaded fasteners that hold parts together by squeezing them between the head on one end and a nut on the other end. See **Figure 3-10.** Bolts are used with a nut to apply a squeezing force to one or more components. The hole the bolt passes through is not threaded. It should have a small amount of clearance so that the bolt does not have to be driven through the hole. The head of a bolt may be the same shape as the head used on a screw since their use is interchangeable. See **Figure 3-5.** As previously mentioned, screws may be threaded all the way to the head. However, if screws are to be used as bolts, they should not be threaded all the way to the head. The unthreaded portion, or shank, of a bolt should pass through all of the top part and partially through the second component. See **Figure 3-11.** The threaded portion is called the *grip-length* of the bolt.

Often, a lock washer is placed between the face of the nut and the part surface to prevent loos-

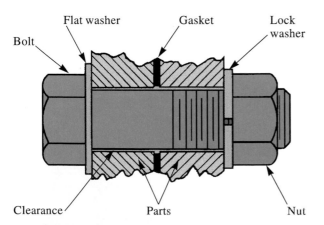

Figure 3-10. *A bolt and nut apply great clamping force. Washers are used with nut and bolt. Note gasket between parts.*

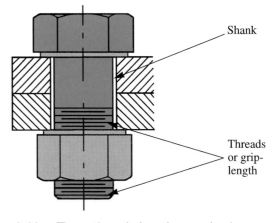

Figure 3-11. *The unthreaded portion, or shank, of a bolt should pass through all of the top part and partially through the second component.*

ening caused by vibration. A flat washer is often used to provide a smooth and larger clamping surface. This is often necessary when fastening soft materials such as aluminum, plastic, or wood. Washers are covered in detail later in this chapter.

Nuts

Nuts vary in shape and size depending upon their intended function. Plain hexagon nuts are most common. Other types used include wing nuts, castle nuts, and various self-locking nuts. See **Figure 3-12.**

Square nuts, like square headed bolts, are not commonly used, but can be found on old implements. A *jam nut* is used in conjunction with a plain *hexagon nut.* The jam nut is a thinner nut used with a plain nut to produce a locking condition of one nut tightening against the other. See **Figure 3-13.** The *castle nut* is used on bolts that have a drilled hole through the threaded end. A cotter pin is used to prevent the castle nut from turning. Cotter pins should always be installed properly as shown in **Figure 3-14.** Castle nuts are

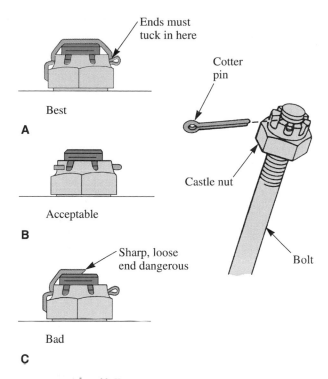

Figure 3-14. *Cotter pin is placed through castle nut and bolt hole to prevent nut from coming off bolt. Installation method A or B is acceptable. Method A is more difficult to produce, but eliminates sharp ends.*

used on bolts or shafts when a component turns or pivots on it. They are also used when *axial clearance* is required, such as with axle shafts having tapered roller or ball bearings. *Acorn nuts* are used to tighten and also cover the sharp thread end of a bolt for safety. It is important when using acorn nuts to be certain that the bolt end does not bottom in the nut before it tightens. Acorn nuts get their name from their likeness in shape to oak tree acorns and are often used to provide a smooth, neat appearance. *Wing nuts* are used when something needs to be frequently adjusted and can be tightened or loosened by hand. There are many special nuts, some of which are shown in **Figure 3-15.** After being tightened, a bolt should be long enough to pass through the parts, any washers, the nut, and protrude from one and one-half to two threads beyond the nut.

Lock nuts

Lock nuts are designed to create friction to reduce the tendency for vibration, or motion to rotate and loosen the nut. **Figures 3-12** and **3-15** illustrate some locking-type nuts.

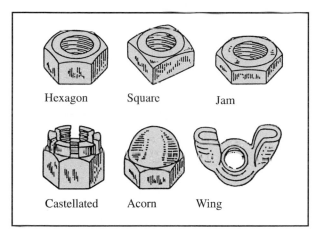

Figure 3-12. *Various kinds of nuts apply clamping pressure on bolts.*

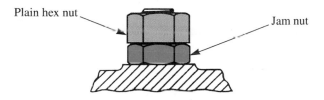

Figure 3-13. *Two nuts used to prevent loosening of bolt.*

Palnut

Single-thread lock nut applied and tightened after regular nut is in place

Self-Retaining

Threads in section above slot are deformed to provide friction grip

Anco

Pin impinges against bolt to hold nut in place

Esna

Fiber collar grips bolt threads. Also available with metallic collar

Tinnerman Speed Nut

Lamson

Raised crown is distorted and heat-treated to give spring grip on bolt thread

Spin-lock

Ratchet-shaped serrated teeth imbedded in work provide friction grip

Flexloc

Segments press against bolt threads because inner diameter of upper part is slightly less than outside diameter of bolt

Figure 3-15. *Some nuts are self locking.*

Bolt and nut terminology

Bolts and nuts come in various sizes (lengths, diameters, and head size), grades (strengths), and thread types. Being familiar with these differences is important when the need arises to replace nuts and bolts. Important bolt dimensions are:

- **Bolt size.** The major (largest) diameter of the bolt threads.

- **Bolt head size.** The dimension across the flats of the hexagon. It is the same as the wrench size.
- **Bolt length.** The distance from the base of the bolt head to the threaded end of the bolt.
- **Thread pitch.** The number of threads per inch on U.S. customary fasteners. On metric fasteners it is the distance between each thread measured in millimeters.

Bolt grades

Bolt grades are related to the minimum tensile (pulling) strength requirements of the bolt. **Figure 3-16** shows the amounts in P.S.I. (pounds per square inch of cross section of the bolt). The bolt heads are often marked with a symbol indicating the grade of the bolt. For example, an SAE Grade 5 bolt has three marks on the head. A Grade 6 bolt has four marks on the head. In every case the number of marks on the head is 2 less than the Grade Number. A Grade 1 and 2 has no marks on the head. Metric bolt heads are marked with 5D, 8G, 10K, 12K. See **Figure 3-16** for corresponding *tensile strengths.*

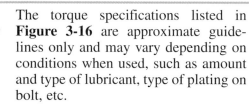

The torque specifications listed in **Figure 3-16** are approximate guidelines only and may vary depending on conditions when used, such as amount and type of lubricant, type of plating on bolt, etc.

It should be noted that the tensile strengths would only be accurate for a bolt having exactly a 1″ cross-sectional area. For example, a 1/4″ diameter, Grade 2 bolt showing a 64,000 P.S.I. strength would support a maximum 3155 lb of weight. The reason is a 1/4″ diameter bolt area is only .0493 of a 1 sq in. Therefore: .0493 × 64,000 P.S.I = 3,155.20 lb. It should not be expected that one 1/4″ bolt will withstand a 64,000 lb load. It should also be understood that the load applied to the bolt due to overtightening (primary load), plus the external (secondary) load exerted upon it could exceed the tensile limits of the bolt. An external load could be applied as a result of pulling, lifting a load, and/or heat expansion.

When a bolt reaches its load bearing limit it becomes weaker, exceeds its *elastic limit,* and begins to stretch plastically. When a bolt stretches plastically it does not return to its original length or shape when the load is released. The bolt may appear to be loose or show signs of leakage at a

SAE Standard/Foot-Pounds							Metric Standard					
Grade of Bolt	**SAE 1 & 2**	**SAE 5**	**SAE 6**	**SAE 8**			**Grade of Bolt**		**5D**	**.8G**	**10K**	**12K**	
Min. Ten. Strength	64,000 P.S.I.	105,000 P.S.I.	133,000 P.S.I.	150,000 P.S.I.			Min. Ten. Strength		71,160 P.S.I.	113,800 P.S.I.	142,200 P.S.I.	170,679 P.S.I.	
Markings on Head	⬢	⬢	⬢	⬢	Size of Socket or Wrench Opening		Markings on Head		5D	.8G	10K	12K	Size of Socket or Wrench Opening
U.S. Standard	Foot Pounds				U.S. Regular		Metric		Foot Pounds				Metric
Bolt Dia.					Bolt Head	Nut	Bolt Dia.	U.S. Dec. Equiv.					Bolt Head
1/4	5	7	10	10.5	3/8	7/16	6mm	.2362	5	6	8	10	10mm
5/16	9	14	19	22	1/2	9/16	8mm	.3150	10	16	22	27	14mm
3/8	15	25	34	37	9/16	5/8	10mm	.3937	19	31	40	49	17mm
7/16	24	40	55	60	5/8	3/4	12mm	.4720	34	54	70	86	19mm
1/2	37	60	85	92	3/4	13/16	14mm	.5512	55	89	117	137	22mm
9/16	53	88	120	132	7/8	7/8	16mm	.6299	83	132	175	208	24mm
5/8	74	120	167	180	15/16	1	18mm	.7090	111	182	236	283	27mm
3/4	120	200	280	296	1-1/8	1-1/8	22mm	.8661	182	284	394	464	32mm

Figure 3-16. *General bolt torque chart. Torque values increase as bolt size and grade increases.*

gasket. Unknowingly, one might retighten the bolt. This stretches it, weakens it more, and will result in the bolt failing in service or breaking during the tightening.

Thread types

Figure 3-17 illustrates various parts of a thread. There are several types, or series, of threads of commercial importance. Only three are of significance for the purpose of this text. The first type is the coarse-thread series designated *Unified National Coarse (UNC)* and National Coarse (NC). These are for general use where they are not subjected to vibration.

The second type is the fine-thread series designated *Unified National Fine (UNF)* and National Fine (NF). These are for work where vibration is a considerable factor, such as automotive and aircraft applications.

Unified coarse and unified fine refer to the number of threads per inch of length on threaded fasteners. Every bolt or nut diameter will have a specific number of threads per inch of length. For example, a 1/2″ diameter Unified National Coarse bolt or nut will always have 13 threads per inch of length. A Unified National Fine thread of the same diameter will always have 20 threads per inch. The 1/2″ is always the *major (largest) diameter* of the thread of the bolt or nut.

The third type is the metric series designated *Metric (M)*. The metric thread is formed with a 60° angle, which is similar to the unified threads. For the metric threads, the root may be rounded and the depth somewhat greater. The International Standards Organization (ISO) has attempted to standardize metric threads. The ISO metric thread series has 25 thread diameters ranging from 1.6 millimeters (mm) to 100mm.

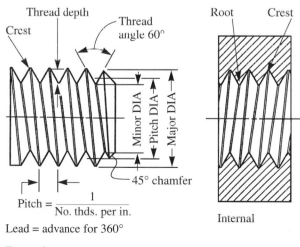

$$Pitch = \frac{1}{No. \; thds. \; per \; in.}$$

Lead = advance for 360°

External

Internal

Figure 3-17. *Thread terminology for external and internal threads.*

Threads are either right-handed or left-handed. A fastener with right-handed threads must be turned clockwise to tighten it. A fastener with left-handed threads must be turned counterclockwise to tighten it. The letter *L* may be stamped on the fastener with left-handed threads.

 The thread of a nut or threaded hole must always be the same series, size, and type as that of the bolt or screw entering it. If they are not the same, thread stripping or damage will occur.

Thread fit

Some thread applications can tolerate loose fitting threads. Other applications may require closer fitting, or tight threads. For example, the head on a gasoline engine may be held to the engine block with bolts that are threaded on both ends called ***stud bolts***. See **Figure 3-18.** One end is threaded into the engine block. The other end receives a nut that tightens against the cylinder head. It is desirable to have the stud bolt remain in the engine block when the nut is removed. The block end requires a tighter fitting thread than the

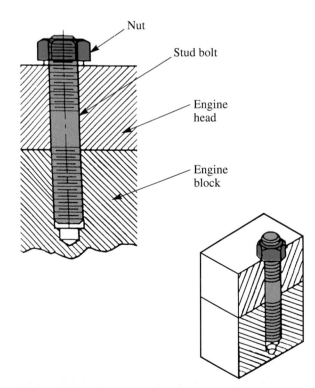

Figure 3-18. *Stud bolts are threaded at both ends.*

end with the nut. If the fit of the nut is too tight, the entire stud may be removed when the nut is turned with a wrench. In some cases, a UNF thread is used in the block and a UNC thread on the nut end.

Unified threads are classified as external or internal, and according to classification of fit as follows:
- ***Class 1 Fit.*** Has the largest manufacturing tolerance. Used where ease of assembly is desired and a loose thread is not objectionable.
- ***Class 2 Fit.*** Used on the largest percentage of threaded fasteners.
- ***Class 3 Fit.*** Will be tight when assembled.

Thread designations

The thread designation is a series of numbers and letters used to describe a bolt and thread. For example, the designation of 1/2-13 UNC-2A × 1 is defined as follows:
- ***1/2.*** Indicates the thread diameter. In this case, the diameter is 1/2″.
- ***13.*** Indicates the threads per inch. In this case, there are 13.
- ***UNC.*** Indicates the series of thread. In this case, it is a Unified National Coarse thread. The letters ***UNF*** is a Unified National Fine thread.
- ***2.*** Indicates the class of fit. In this case, a ***2*** indicates a class 2 fit. The number ***1*** is a class 1 fit and ***3*** is a class 3 fit.
- ***A.*** Indicates that it is an external or internal thread. In this case, an ***A*** indicates an external thread. The letter ***B*** is an internal thread, such as a nut or threaded hole.
- ***1.*** This number indicates the length of the fastener in inches. In this case, the ***1*** is 1″.

A letter *L* at the far right of thread designation indicates left-handed threads. For example, 1/4-20 UNC-2A L is a left-handed thread. If the *L* is not present, the thread is understood to be right-handed.

Metric thread designations are slightly different. For example, the metric designation of M-10 × 1.5 × 25 is defined as follows:
- ***M.*** Indicates that the thread is metric.
- ***10.*** Indicates the diameter of the thread in millimeters. In this case, the diameter is 10mm.
- ***1.5.*** Indicates the distance between threads (pitch). In this case, the pitch is 1.5mm.
- ***25.*** Indicates the length of the fastener in millimeters. In this case, the length is 25mm.

Tightening and loosening threaded fasteners

As previously mentioned, if a bolt, nut, or screw has right-handed threads, the direction for tightening is always clockwise and the direction for loosening is always counterclockwise. For beginners, this may be a problem until they gain this understanding. It can be particularly difficult to understand if the bolt, nut, or screw is in the inverted (upside-down) position on an engine or implement. In fact if not careful, the head of the fastener may be twisted off by attempting to turn it the wrong direction. This may introduce the problem of getting the broken fastener out of the hole. If removing the fastener is successful, then the threads may have to be recut (chased) in the hole. A threading tap is a tool used for chasing the threads. For more details on the chasing procedure, see the *Chasing threads* section in this chapter.

Tightening to specific torque settings

Tightening bolts, nuts, and screws on engines should be done with a torque wrench. They should be tightened to specific torque (turning effort) settings. See torque charts for fractional and metric size bolts in **Figure 3-16**. More information is provided on this subject in *Torque wrenches* section of Chapter 2 and in the *Torque specifications* section of Chapter 12.

Chasing threads

When threads become damaged in a nut or threaded hole, it may be necessary to recut the thread with a tool called a **threading tap**. See **Figure 3-19**. This procedure is called **chasing** the thread. If the hole goes all the way through, then it is called a **through hole**. A nut is an example of a through hole. In the case of a nut, select a tap designated to fit the existing thread. If the existing thread is 1/2-13 UNC, then select a tap designated 1/2-13 UNC. This designation will be stamped on the tap shank of the tap.

For a through hole a taper tap should be used. A **taper tap** has a slender taper at the beginning of the tap that makes it start easier in the threads. Select a tap wrench to turn the tap. Install the square end of the tap in the wrench and tighten it. You will need to decide whether the part can be secured in a vise or whether it can be done in place. A small item can be clamped in a vise to

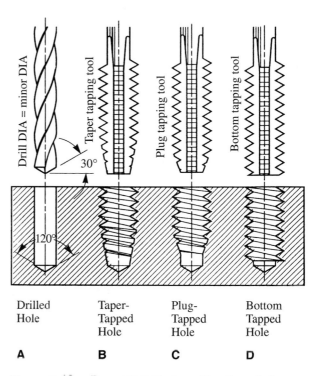

Figure 3-19. *Taps are tools for cutting threads in holes. For blind holes the taper tap should be used first, followed by the plug tap, and then the bottom tap.*

hold it. See **Figure 3-20.** Add a few drops of cutting oil on the tap before beginning. This will improve the cutting action and produce better threads. Align the tap with the hole and turn it clockwise (right-handed threads) until the tap turns freely. Then, reverse the rotation until the tap can be removed.

If bolt or screw threads are damaged, they can be chased with a threading die held in a die holder. Place the bolt head in a vise. Select the correct die according to the thread of the bolt. If the thread type is not known, determine it by measuring the major diameter of the thread, and counting the number of threads per lineal inch. If the bolt diameter is 3/8″ and there are 16 threads per inch, it is a 3/8-16 UNC thread. Select a corresponding die. Now, place the die in a die handle and secure it with the set screws. See **Figure 3-21.** There is a right side and a wrong side for starting the die. The correct side has beginning teeth tapered. This makes starting easier and the cutting edges of the teeth cut in the proper direction. Place some cutting oil on the die teeth and on the bolt threads. Place the die on the end of the bolt and begin turning it clockwise (right-handed threads) to cut and

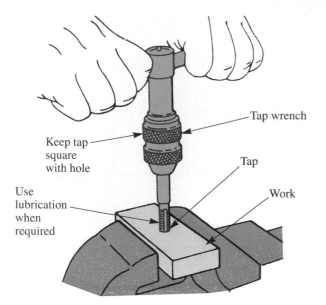

Figure 3-20. *Taps fit into a special tap wrench. Tap must be held straight with the hold as it is turned into the work. Turn wrench ahead 2/3 of one turn, back up 1/3 of one turn. Use a tapping fluid on the tap.*

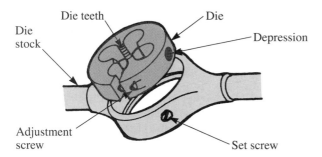

Figure 3-21. *Threading dies are used to cut external threads. The die is inserted in a wrench called a die-stock. The conical teeth of the die should be started on the rod or bolt.*

correct the threads. Now, reverse the die rotation until it can be removed from the bolt.

Tapping new threads

The procedure of tapping new threads is similar to that of chasing threads. Prior to tapping, however, a new hole must be drilled to the proper diameter. The proper diameter is obtained for UNC and UNF threads from a tap drill chart. See **Figure 3-22.** If a tap drill chart is not available, the following formula can be used to obtain the decimal drill size to use.

For Unified thread form (UNC and UNF):

$$\frac{\text{Drill}}{\text{size}} = \frac{\text{Basic major}}{\text{diameter}} - \frac{(1.08253 \times \%)\text{size}}{\text{No. threads per inch}}$$

For practical purposes when hand tapping, between 65% and 75% thread depth is suitable. For example, to find the drill size for a 1/4-20 UNC threaded hole, multiply 1.08253 by .75 to get .8119. Now, divide .8119 by 20 to get .0416. Finally, subtract .0416 from .25(1/4) to get .2084. Choose the drill size that is closest to .2084, which would be a number 4 drill. Refer to the drill size chart in **Figure 3-22** to check this drill size. For tap drill sizes for metric threads refer to **Figure 3-23.**

Rules for hand tapping

1. Use a good tapping fluid, except when tapping gray cast iron, which should be tapped dry.
2. For through holes, start and end with a taper tap.
3. For **blind holes,** start with a taper tap, followed with a **plug tap,** followed with a **bottom tap.**
4. Be careful to start the tap straight in the hole.
5. The tap wrench should be turned clockwise two-thirds of a turn then reversed one-third of a turn to break and clear the chips. Continue this through, or to the bottom of the hole.
6. Never excessively force a tap to turn.

> ⚠ Taps are very hard and brittle, and can break when forced. Removing a broken tap can be difficult, or impossible. The smaller the tap, the easier it breaks.

Threading with a die

Threading is cutting external threads on a rod, bolt, shaft, or pin. A cutting tool called a **die** is used in a die handle called a **die-stock**. See **Figure 3-21.** The threading procedure is the same as for tapping. The diameter of the rod must be the same as the major diameter of the thread. Select the die. If the die is of the split type, it can be opened with the adjustment screw for the first cut and then adjusted down later to the desired fit in the threaded hole. Place the die in the die-stock so that the set screws align with the depressions along the edge of the die. Tighten the set screws. Start the die on the correct side and keep it as straight as possible with the rod. A good cutting fluid should be used on the die teeth.

Decimal Equivalents and Tap Drill Sizes

Fraction or Drill Size			Decimal Equivalent	Tap Size	Fraction or Drill Size		Decimal Equivalent	Tap Size
	Number Size Drills	80	.0135			39	.0995	
		79	.0145			38	.1015	5 — 40
1/64			.0156			37	.1040	5 — 44
		78	.0160			36	.1065	6 — 32
		77	.0180		7/64		.1094	
		76	.0200			35	.1100	
		75	.0210			34	.1110	
		74	.0225			33	.1130	6 — 40
		73	.0240			32	.1160	
		72	.0250			31	.1200	
		71	.0260		1/8		.1250	
		70	.0280			30	.1285	
		69	.0292			29	.1360	8 — 32, 36
		68	.0310			28	.1405	
1/32			.0312		9/64		.1406	
		67	.0320			27	.1440	
		66	.0330			26	.1470	
		65	.0350			25	.1495	10 — 24
		64	.0360			24	.1520	
		63	.0370			23	.1540	
		62	.0380		5/32		.1562	
		61	.0390			22	.1570	
		60	.0400			21	.1590	10 — 32
		59	.0410			20	.1610	
		58	.0420			19	.1660	
		57	.0430			18	.1695	
		56	.0465		11/64		.1719	
3/64			.0469	0 — 80		17	.1730	
		55	.0520			16	.1770	12 — 24
		54	.0550			15	.1800	
		53	.0595	1 — 64, 72		14	.1820	12 — 28
1/16			.0625			13	.1850	
		52	.0635		3/16		.1875	
		51	.0670			12	.1890	
		50	.0700	2 — 56, 64		11	.1910	
		49	.0730			10	.1935	
		48	.0760			9	.1960	
5/64			.0781			8	.1990	
		47	.0785	3 — 48		7	.2010	1/4 — 20
		46	.0810		13/64		.2031	
		45	.0820	3 — 56		6	.2040	
		44	.0860			5	.2055	
		43	.0890	4 — 40		4	.2090	
		42	.0935	4 — 48		3	.2130	1/4 — 28
3/32			.0938		7/32		.2188	
		41	.0960			2	.2210	
		40	.0980			1	.2280	

Figure 3-22. *The correct hole size to drill for a given thread series and diameter can be obtained from a tap drill chart like this.*

(Continued)

Decimal Equivalents and Tap Drill Sizes

Fraction or Drill Size	Decimal Equivalent	Tap Size
Letter Size Drills A	.2340	
$\frac{15}{64}$	.2344	
B	.2380	
C	.2430	
D	.2460	
$\frac{1}{4}$ — E	.2500	
F	.2570	$\frac{5}{16}$ — 18
G	.2610	
$\frac{17}{64}$	.2656	
H	.2660	
I	.2720	$\frac{5}{16}$ — 24
J	.2770	
K	.2810	
$\frac{9}{32}$	.2812	
L	.2900	
M	.2950	
$\frac{19}{64}$	.2969	
N	.3020	
$\frac{5}{16}$	.3125	$\frac{3}{8}$ — 16
O	.3160	
P	.3230	
$\frac{21}{64}$	.3281	
Q	.3320	$\frac{3}{8}$ — 24
R	.3390	
$\frac{11}{32}$	.3438	
S	.3480	
T	.3580	
$\frac{23}{64}$	.3594	
U	.3680	$\frac{7}{16}$ — 14
$\frac{3}{8}$	.3750	
V	.3770	
W	.3860	
$\frac{25}{64}$	.3906	$\frac{7}{16}$ — 20
X	.3970	
Y	.4040	
$\frac{13}{32}$	.4062	
Z	.4130	
$\frac{27}{64}$	.4219	$\frac{1}{2}$ — 13
$\frac{7}{16}$	.4375	
$\frac{29}{64}$	.4531	$\frac{1}{2}$ — 20
$\frac{15}{32}$	.4688	
$\frac{31}{64}$	.4844	$\frac{9}{16}$ — 12
$\frac{1}{2}$	.5000	
$\frac{33}{64}$	.5156	$\frac{9}{16}$ — 18
$\frac{17}{32}$	.5312	$\frac{5}{8}$ — 11
$\frac{35}{64}$	.5469	
$\frac{9}{16}$	.5625	
$\frac{37}{64}$	.5781	$\frac{5}{8}$ — 18
$\frac{19}{32}$	.5938	

Fraction or Drill Size	Decimal Equivalent	Tap Size
$\frac{39}{64}$	.6094	
$\frac{5}{8}$	.6250	
$\frac{41}{64}$	.6406	
$\frac{21}{32}$	.6562	$\frac{3}{4}$ — 10
$\frac{43}{64}$	.6719	
$\frac{11}{16}$	.6875	$\frac{3}{4}$ — 16
$\frac{45}{64}$	.7031	
$\frac{23}{32}$	.7188	
$\frac{47}{64}$	.7344	
$\frac{3}{4}$	.7500	
$\frac{49}{64}$	.7656	$\frac{7}{8}$ — 9
$\frac{25}{32}$	.7812	
$\frac{51}{64}$	.7969	
$\frac{13}{16}$	.8125	$\frac{7}{8}$ — 14
$\frac{53}{64}$	.8281	
$\frac{27}{32}$	.8438	
$\frac{55}{64}$	.8594	
$\frac{7}{8}$	.8750	1 — 8
$\frac{57}{64}$	.8906	
$\frac{29}{32}$	.9062	
$\frac{59}{64}$	.9129	1 — 12
$\frac{15}{16}$	.9375	
$\frac{61}{64}$	.9531	
$\frac{31}{32}$	.9688	
$\frac{63}{64}$	.9844	$1\frac{1}{8}$ — 7
1	1.0000	
$1\frac{3}{64}$	1.0469	$1\frac{1}{8}$ — 12
$1\frac{7}{64}$	1.1094	$1\frac{1}{4}$ — 7
$1\frac{1}{8}$	1.1250	
$1\frac{11}{64}$	1.1719	$1\frac{1}{4}$ — 12
$1\frac{7}{32}$	1.2188	$1\frac{3}{8}$ — 6
$1\frac{1}{4}$	1.2500	
$1\frac{19}{64}$	1.2969	$1\frac{3}{8}$ — 12
$1\frac{11}{32}$	1.3438	$1\frac{1}{2}$ — 6
$1\frac{3}{8}$	1.3750	
$1\frac{27}{64}$	1.4219	$1\frac{1}{2}$ — 12
$1\frac{1}{2}$	1.5000	

Pipe Thread Sizes (NPSC)

Thread	Drill	Thread	Drill
$\frac{1}{8}$ — 27	$\frac{11}{32}$	$1\frac{1}{2}$ — $11\frac{1}{2}$	$1\frac{3}{4}$
$\frac{1}{4}$ — 18	$\frac{7}{16}$	2 — $11\frac{1}{2}$	$2\frac{7}{32}$
$\frac{3}{8}$ — 18	$\frac{37}{64}$	$2\frac{1}{2}$ — 8	$2\frac{21}{32}$
$\frac{1}{2}$ — 14	$\frac{23}{32}$	3 — 8	$3\frac{1}{4}$
$\frac{3}{4}$ — 14	$\frac{59}{64}$	$3\frac{1}{2}$ — 8	$3\frac{3}{4}$
1 — $11\frac{1}{2}$	$1\frac{5}{32}$	4 — 8	$4\frac{1}{4}$
$1\frac{1}{4}$ — $11\frac{1}{2}$	$1\frac{1}{2}$		

Figure 3-22. *Continued*

Metric Tap Drill Size								
	Recommended Metric Drill				Closest Recommended Inch Drill			
Metric Tap Size	Drill Size (mm)	Inch Equivalent	Probable Hole Size (in.)	Probable Percent of Thread	Drill Size	Inch Equivalent	Probable Hole Size (in.)	Probable Percent of Thread
M1.6 × .35	1.25	.0492	.0507	69	—	—	—	—
M1.8 × .35	1.45	.0571	.0586	69	—	—	—	—
M2 × .4	1.60	.0630	.0647	69	#52	.0635	.0652	66
M2.2 × .45	1.75	.0689	.0706	70	—	—	—	—
M2.5 × .45	2.05	.0807	.0826	69	#46	.0810	.0829	67
*M3 × .5	2.50	.0984	.1007	68	#40	.0980	.1003	70
M3.5 × .6	2.90	.1142	.1168	68	#33	.1130	.1156	72
*M4 × .7	3.30	.1299	.1328	69	#30	.1285	.1314	73
M4.5 × .75	3.70	.1457	.1489	74	#26	.1470	.1502	70
*M5 × .8	4.20	.1654	.1686	69	#19	.1660	.1692	68
*M6 × 1	5.00	.1968	.2006	70	#9	.1960	.1998	71
M7 × 1	6.00	.2362	.2400	70	$\frac{15}{64}$	.2344	.2382	73
*M8 × 1.25	6.70	.2638	.2679	74	$\frac{17}{64}$	.2656	.2697	71
M8 × 1	7.00	.2756	.2797	69	J	.2770	.2811	66
*M10 × 1.5	8.50	.3346	.3390	71	Q	.3320	.3364	75
M10 × 1.25	8.70	.3425	.3471	73	$\frac{11}{32}$	.3438	.3483	71
*M12 × 1.75	10.20	.4016	.4063	74	Y	.4040	.4087	71
M12 × 1.25	10.80	.4252	.4299	67	$\frac{27}{64}$	.4219	.4266	72
M14 × 2	12.00	.4724	.4772	72	$\frac{15}{32}$	.4688	.4736	76
M14 × 1.5	12.50	.4921	.4969	71	—	—	—	—
*M16 × 2	14.00	.5512	.5561	72	$\frac{35}{64}$	.5469	.5510	76
M16 × 1.5	14.50	.5709	.5758	71	—	—	—	—
M18 × 2.5	15.50	.6102	.6152	73	$\frac{39}{64}$	.6094	.6144	74
M18 × 1.5	16.50	.6496	.6546	70	—	—	—	—
*M20 × 2.5	17.50	.6890	.6942	73	$\frac{11}{16}$	.6875	.6925	74
M20 × 1.5	18.50	.7283	.7335	70	—	—	—	—
M22 × 2.5	19.50	.7677	.7729	73	$\frac{49}{64}$	.7656	.7708	75
M22 × 1.5	20.50	.8071	.8123	70	—	—	—	—
*M24 × 3	21.00	.8268	.8327	73	$\frac{53}{64}$	.8281	.8340	72
M24 × 2	22.00	.8661	.8720	71	—	—	—	—
M27 × 3	24.00	.9449	.9511	73	$\frac{15}{16}$	.9375	.9435	78
M27 × 2	25.00	.9843	.9913	70	$\frac{63}{64}$	.9844	.9914	70
*M30 × 3.5	26.50	1.0433						
M30 × 2	28.00	1.1024						
M33 × 3.5	29.50	1.1614						
M33 × 2	31.00	1.2205	Reaming Recommended to the Drill Size Shown					
M36 × 4	32.00	1.2598						
M36 × 3	33.00	1.2992						
M39 × 4	35.00	1.3780						

Figure 3-23. *Metric Tap Drill Chart.*

Washers

Flat washers are used to provide a wider bearing surface for a bolt or screw head and/or nut. When tightening a bolt or screw against a softer material such as wood, plastic, and soft metals (like aluminum, copper, or brass), the head may gradually become imbedded in the surface. This may cause the fastener to become loose during use. A *flat washer* tends to prevent imbedding and provides a harder surface for the bolt or screw head to pull against. See **Figure 3-24.**

Lock washers prevent loosening of bolts and screws. There are many kinds to choose from. The most common for locking nuts is the *kantlink* washer. It is made of spring steel and has beveled ends. Due to the slight helix of the washer, it tends to cut into the mating surfaces of the nut and

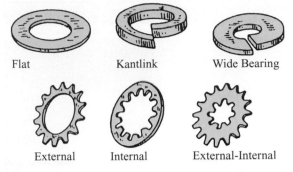

Figure 3-24. *Common washers common of several sizes and shapes.*

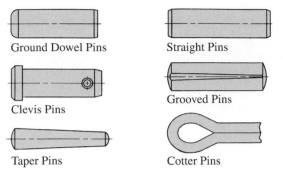

Figure 3-25. *Pins are used to hold parts together in assembly.*

component. When the nut is tightened the washer is compressed flat. The tendency of the nut to reverse rotation causes the toothed ends of the washer to dig into the surfaces. This prevents further loosening of the fastener.

The **wide bearing** lock washer combines the characteristics of a flat washer and a kantlink lock washer. It provides the hard surface for the bolt or screw head to pull against that is characteristic of a flat washer. It also provides toothed ends that are characteristic of the kantlink washer.

Multiple **toothed washers** are stamped from sheet metal and are either *internal*, *external*, or *external-internal* toothed. This type of washer is used under the heads of screws to prevent their backing out. The teeth are twisted to resist rotation in the direction that would cause loosening of the screw.

Pins

Pins are used to either retain parts in a fixed position or to preserve alignment of parts. **Figure 3-25** illustrates several types of pins that may be found on some gasoline engines or their parts.

Cotter pins

Cotter pins are sized by a nominal dimension, such as 3/32″. The hole size for a cotter pin should be slightly larger than the nominal size of the pin. Cotter pins are used to lock castle nuts and secure clevis pins. Cotter pins should be installed properly as shown in **Figure 3-14.** Cotter pins may be made of copper, brass, aluminum, or stainless steel. They can be cut to length with side cutting pliers. They can be bent with combination slip joint, or needle nose pliers. They can be tapped lightly with a soft hammer to form them.

As with old fasteners, tools, part, or equipment, old cotter pins should be replaced with new ones.

Clevis pins

Clevis pins function as an axle so a part can swivel on it. It requires a flat washer and cotter pin to prevent the part from sliding off the pin.

Dowel pins

Dowel pins are used for alignment and usually fit very snugly. They are heat treated and hardened. The dowel pin is pressed into a hole with an interference fit. This means the mating part has a matching hole that fits closely to the pin, but allows the part to be assembled or disassembled easily.

Straight pins

Straight pins are also used for alignment. They fit closely, but are not usually an interference fit.

Grooved pins

Grooved pins are driven into an interference hole. The groove cuts into the wall of the hole and secures the pin. There are seven types of grooved pins. Each type is shaped differently, and each has a different size and shape of groove. See **Figure 3-26.**

Taper pins

Taper pins have a uniform taper of .250 per foot over the length of the pin. Each end is rounded slightly. Taper pins are generally used to fasten pulleys and gears to shafts to prevent rotation on the shaft. Taper pins fit into tapered holes that match the two mating parts. The taper pin is held in the hole by tapping it into the tapered holes, thus wedging it tightly in place. A taper pin should be of such size and length, that when tapped tightly into the mating parts it does not extend beyond the holes. The tapered pin can be removed from the small end by driving it out with a pin pinch and hammer. Pins are designated by pin size number and standard lengths.

Retaining Rings

Retaining rings are circular spring steel that fit externally or internally into a groove of a part. For example, they may be placed in a groove that is machined into the surface of a shaft, or internally in a groove in a cylindrical hole. There are several types of retaining rings. See **Figure 3-27.**

The purpose of an external retaining ring is to prevent movement of a shaft beyond a point through a hole such as in a bearing. At the same time the retaining ring does not prevent the shaft from rotating. See **Figure 3-28.** Internal retaining rings prevent a shaft from traveling beyond the retaining ring located in its groove in the cylindrical part. Most retaining rings must be installed and removed with a retaining ring tool. The tool has nibs that fit into the small holes at the ends of the rings. When the handles of the plier-like tool are squeezed, internal rings are compressed inward and smaller so they can be installed in the cylinder and groove. External rings are forced open so they

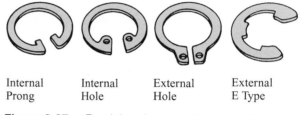

| Internal Prong | Internal Hole | External Hole | External E Type |

Figure 3-27. *Retaining rings may be internal or external.*

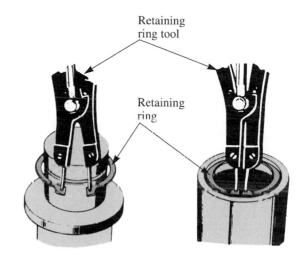

Figure 3-28. *A special plier-like tool is required to install some retaining rings. Nibs are inserted in holes in ring to expand or close ring.*

can be slid over the shaft and into the groove. The reverse action is used for removal. The springs are very strong and require extreme care when installing and removing.

⚠️ Safety glasses with side shields must always be worn when installing or removing retaining rings. They can easily slip off the nibs of the tool, and because they are made of strong spring steel they can fly with considerable velocity.

Keys

Keys are used almost exclusively on shafts that have a component which fits and rotates with the shaft. The recess in which the key rests in the shaft is called the *keyseat*. The groove in the pulley, gear, or collar is the *keyway*. See **Figure 13-29.**

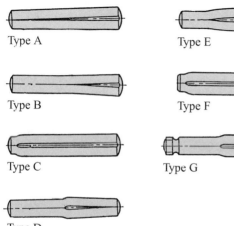

Type A

Type B

Type C

Type D

Type E

Type F

Type G

Figure 3-26. *Groove pins come in several types.*

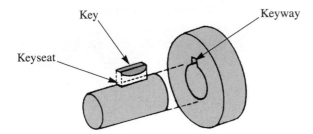

Figure 3-29. *Key rests in keyseat of shaft. Keyway is located in surrounding part.*

An example is the flywheel on the crankshaft of a small gas engine. The key must fit the keyseat and keyway closely to prevent motion between them. Engine flywheel keys keep the engine timing correct. Several key types are used as shown in **Figure 3-30.**

Adhesives and Sealants

Many types of adhesives and sealants can be encountered when working with engines. Both adhesives and sealants are either a liquid or semi-liquid material. They can be sprayed, brushed, or spread on. There are various types that have varying properties. Some of these adhesives and sealants setup hard and others remain pliable. The sections to follow details some of these adhesives and sealants.

Thread adhesives

Thread adhesives can be applied to the threads of nuts, bolts, or screws to prevent them from loosening during service. Adhesive strengths vary from light (removable) to high strength (that might require applied heat to remove). Only a drop of adhesive on the thread prior to fastening is needed. See **Figure 3-31.**

The adhesive cures once the threads are mated. This locks the mating threads together. Complete cure time may vary from 30 minutes to 24 hours. There are other uses for these adhesives such as fastening bearings, bushings, gears, and sleeves on shafts.

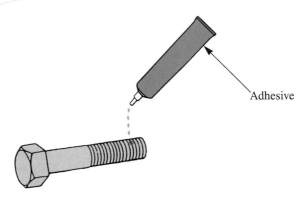

Figure 3-31. *A drop of thread adhesive can be placed on the thread to lock a screw or nut.*

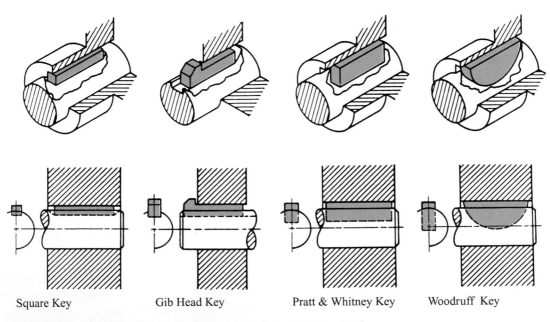

Square Key Gib Head Key Pratt & Whitney Key Woodruff Key

Figure 3-30. *Keys used to connect shaft to pulley, gear, or wheel.*

Always read health warning labels on adhesive containers. Adhesives may contain chemicals that can cause injury to eyes, lungs, and other parts of the body. Always follow the manufacturer's recommended procedures.

Sealants

Most sealants are resistant to oil, water, gas, grease, and salt solutions. Resistance to hot and cold conditions vary. Most sealants can be used for all applications, except for use on the exhaust system. Special high-temperature sealants are used for exhaust systems.

Form-in-place sealants can be used in place of conventional gaskets (gaskets are covered later in this chapter). This type of sealant can be used when the exact replacement gasket is not available. *Room temperature vulcanizing sealant (RTV)* is a form-in-place sealant that is also referred to as silicon sealant.

Anaerobic sealants are similar to RTV, but they can cure in the absence of air. This type of sealant can be used as a thread locking material or between two machined surfaces.

Become familiar with the sealants you use. Know the properties of those sealants and their recommended uses.

Antiseize compounds

Antiseize compounds are applied to threaded fasteners and metal components that are exposed to constant heat. The compound is a lubricant that prevents the metal material from being *cold welded* together. If a threaded fastener and its connecting metal component are cold welded, then removal of the fastener will be impossible. The antiseize compounds should be applied to the threads or the connecting metal. These compounds can be used for the connection of the exhaust system.

Remember, antiseize compounds are lubricants and not sealants.

Gaskets

Gaskets are used between engine parts to seal and prevent leakage of engine oil, coolant, compression, and vacuum. Gaskets are soft, pliable materials such as fiber, rubber, neoprene (synthetic rubber), cork, treated paper, thin steel, or laminated materials. Gaskets are manufactured to the shape of the surfaces between the mating parts with appropriate shapes and locations of holes. See **Figure 3-32.** The particular gasket material used depends upon the functions and conditions of the parts to be joined. This is specified by the engine manufacturer. Some replacement gaskets can be made by hand.

If gaskets are made by hand, make sure they are made of the same material and same thickness as the originals. Incorrect substitution will result in failure of the gasket. This may cause damage to the part and/or engine.

When a gasket is placed between parts and the bolts or screws are tightened, the gasket material is compressed and deformed. Any small dents, gaps, scratches, or other imperfections in the part surfaces are filled by the gasket. This produces a leakproof seal between the parts.

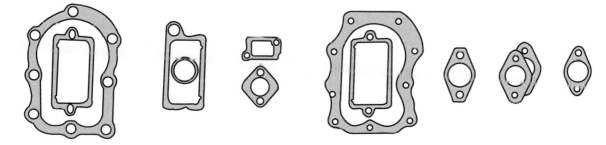

Figure 3-32. *Engine gaskets vary in size, shape, and material. Some gaskets are delicate and should be handled carefully.*

Gasket rules

The following are specific gaskets rules. These rules should always be followed when working with gaskets.

1. **Inspect for leaks before disassembling engine.** Determine if *only* the gasket is leaking or whether the part is cracked or seriously deformed.

2. **Avoid damaging the parts during disassembly.** Care must be taken while removing parts. Do not score, dent, or deform the mating surfaces.

3. **Remove old gasket carefully.** Remove all of the old gasket material from the part surfaces by scraping or wire brushing. Soft metals such as aluminum and brass are damaged easily and require extra care. Use a dull scraper and wire brush lightly.

4. **Wash and dry the parts thoroughly.** Wash the parts in solvent after the gasket has been removed. Blow-dry with compressed air and wipe dry with a clean cloth.

5. **Check new gasket fit.** Compare the new gasket shape to the part surface shape. Lay the new gasket in place and inspect its fit. All holes and sealing surfaces must match precisely. Read any manufacturer's notices about the gasket. It may look symmetrical in shape, but really is not. It is common to install the gasket upside down or in reverse direction. This may cover or partly obscure a small hole in the part.

6. **Use gasket sealant as directed only.** Some gaskets may need a **gasket sealant,** and others may not. Check the service manual for details about the recommended type(s) to use. Use sealant sparingly. Using large amounts of sealant may cause clogging of passages in the parts.

7. **Start all fasteners by hand before tightening.** After the gasket and parts are in place, start the bolts by hand. This assures proper alignment and threading of the bolts. Check for proper bolt lengths at this time.

8. **Tighten the bolts in small steps.** Tighten each bolt a little at a time. Tighten the first one to about half its specified torque. Then tighten the others the same amount. Repeat to about three-fourths torque specification.

9. **Use a crisscross tightening pattern.** Final torquing should follow either a basic crisscross pattern or the factory recommended pattern. This procedure produces even gasket compression and sealing, and prevents possible warping of parts. See **Figure 3-33.**

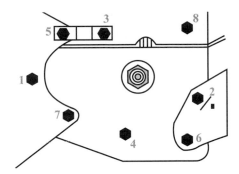

Figure 3-33. *Bolts on an engine head should be tightened in a crisscross manner.*

10. **Do not overtighten fasteners.** Apply only the specified torque. It is easy to distort sheet metal or thin parts by overtightening. Instead of sealing, overtightening can create leakage between the fasteners.

Summary

Many varied fasteners are used in the assembly of small gasoline engines and the implements they drive. Most, but not all, are threaded fasteners. Fasteners are exposed to conditions such as heating and cooling, cyclic loading, tensile and shear loads, corrosion, and vibration. Helical portions of screws and bolts are called threads. When disassembling an engine all parts, fasteners, and washers should be noted so they can be replaced in proper locations. Damaged fasteners should be replaced with new ones. Lightly rusted fasteners may be cleaned and examined for reuse. All threaded fasteners should be lubricated before installing.

Screws hold parts together by passing through one part and threading into another. Set screws are heat treated, hardened-alloy steel used to secure pulleys, gears, and shafts. They have a variety of heads and points. Self-tapping screws are hardened steel and cut their own threads in a predrilled hole of proper size for the screw. The process of cutting threads is called tapping.

Bolts are threaded fasteners that hold parts together by squeezing them between the bolt head on one end and a nut on the other. A lock washer is often placed between the face of the nut and the part surface to prevent loosening caused by vibration. Nuts vary in shape and size depending upon their intended function. Most common are plain hexagon nuts. Other types are wing nuts, castle nuts, acorn nuts, jam nuts, and various self-locking nuts.

Bolt terminology includes major diameter, head size, length, and thread pitch. Bolt grades are related to the minimum tensile strength required of the bolt. Bolt heads are marked with symbols to identify grade. Thread types of significance for small gas engines are Unified National Coarse (UNC), Unified National Fine (UNF), and metric (M). Threads of a bolt or screw must be the same as the mating thread in the hole or nut. Threads are identified by a thread designation such as 1/2–13 UNC–2A for American threads and M10 × 1.5 × 25 for metric threads. Tightening bolts, nuts, and screws on engines should be done with a torque wrench to a specified tightness obtained from a torque chart. Threads can be repaired by chasing them with a threading tap for holes, or a threading die for bolts and screws. When tapping new threads, the proper size hole must first be drilled, as determined from a tap drill chart or by formula. Proper tapping and threading procedures must be applied to avoid breaking a tap in the hole or tearing threads.

Flat washers are used to provide a bearing surface for bolt and screw heads and nuts. Various types of lock washers prevent loosening of bolts and screws.

Cotter pins are used to secure castle nuts and clevis pins. Dowel pins and straight pins are used for alignment of parts and should fit quite snugly. Grooved pins are driven into interference fit holes. The grooves are cut into the walls of the hole to secure the pin. Taper pins have a taper of .250 inch per lineal foot. They are used to fasten pulleys and gears to shafts to prevent slipping around the shaft. Retaining rings are circular spring steel that fit into a groove around a shaft or in a hole. They require a special plier-like tool to install and remove.

Keys are used on shafts that have a gear, pulley, or sleeve that fits and rotates with the shaft. The key rests in a keyseat and keyway.

Many kinds of adhesives and sealants are used on engines and implements. Thread adhesives are used to prevent threaded fasteners from loosening caused by vibration. Some liquid sealants are used to prevent leakage between the parts. It is important to use the correct sealant as specified by the manufacturer. Antiseize compounds are lubricants that prevent parts from locking together.

Gaskets seal between engine parts to prevent leakage of engine oil, coolant, compression, and vacuum. Gaskets are made of soft, pliable materials die cut to fit the shapes of surfaces they seal. The correct material, shape, and thickness is important when replacing a gasket. Following the rules for good procedures when installing gaskets will avoid gasket failure.

Know These Terms

helix	taper tap
thread	blind holes
screws	plug tap
countersunk	bottom tap
set screws	die
hexagon socket	die stock
self-tapping screws	lock washers
tapping	kantlink
threading	toothed washer
bolts	pins
square nuts	cotter pins
jam nuts	clevis pins
hexagon nut	dowel pins
castle nut	straight pins
axial clearance	grooved pins
acorn nuts	interference hole
wing nuts	taper pins
bolt grades	retaining rings
tensile strength	keys
elastic limit	keyseat
Unified National Coarse (UNC)	keyway
	thread adhesives
Unified National Fine (UNF)	form-in-place sealants
	room temperature vulcanizing sealant (RTV)
major diameter	
Metric (M)	
thread fit	anaerobic sealant
stud bolts	antiseize compounds
threading tap	gaskets
chasing	neoprene
through hole	gasket sealant

Chapter 3 Review Questions

Answer the following questions on a separate sheet of paper.

1. The helical portion of a screw or bolt, or the helix in a hole that it fastens into, is called a(n) _____.

2. A type of hardened screw that makes its own threads, usually found in sheet metal, is a(n) _____ screw.
3. The heads found on screws are very different than those found on bolts. True or False?
4. Flat head screws must be set into _____ holes.
5. The most common angle for flat head screws is _____°.
6. Name the four common head types for set screws.
7. Name the six set screw points.
8. The unthreaded part of a bolt or screw is called the _____.
9. The threaded area of a bolt or screw is called the _____.
10. The term bolt implies the use of a(n) _____.
11. What does a palnut do?
12. Jam nuts are usually used with _____ _____ nuts.
13. Answer the following about the thread notation given below:
    3/8-16 UNC 1A
    a. The major diameter of the thread is _____″.
    b. How many threads per lineal inch are there?
    c. Is the thread coarse or fine?
    d. Is the thread internal or external?
    e. Is the thread a loose, average, or close fit?
14. The proper thread notation for a 14 millimeter thread with a 1.5 millimeter pitch on a screw that is 40 millimeters long is _____.
15. A(n) _____ drill is the proper tap drill for a 1/4-28 UNF screw.
16. For cutting threads in a *through-hole* the proper tap to use would be a(n) _____ tap.
17. The cutting tool used to cut external threads is called a threading _____.
18. To provide a wider bearing surface for a bolt head or nut, a(n) _____ or _____ washer should be used.
19. Toothed washers are either internal or external. True or False?
20. What fasteners are used to lock castle nuts?

21. Retaining rings are made of _____ steel.
22. The four styles of keys shown in this chapter are the _____, _____, _____, and _____.
23. _____ _____ are liquids or semi-liquids that can be applied to the threads of nuts, bolts, or screws to prevent them from loosening during service.
24. _____ are used between engine parts to seal and prevent leakage of engine oil, coolant, compression, and vacuum.
25. When tightening bolts on an engine head use a(n) _____ pattern unless a different pattern is specified by the manufacturer.

Suggested Activities

1. Make a collection of fasteners for a display board. Categorize and label each.
2. Explore the shop and identify as many different kinds of screws, bolts, nuts, and washers as you can. List them and the function of each kind.
3. Identify UNC and UNF taps and dies.
4. Identify the correct side to start a threading die.
5. Identify taper, plug, and bottom taps.
6. Chase threads on a damaged bolt. Chase threads on a nut.
7. Select the proper tap drill for a screw and drill a blind hole about 3/4″ deep in a piece of mild steel. Tap threads to the bottom of the hole using the proper procedure and sequence of taps. If you have never tapped threads before, to avoid breakage, select a screw size 3/8″ or more in diameter.
8. Make a display of keys and pins.
9. Display proper and improper installations of cotter pins.
10. Demonstrate proper installation of a gasket.
11. Demonstrate the proper technique for tightening engine head bolts. Demonstrate the proper way to torque engine head bolts.

Engine Construction and Principles of Operation

After studying this chapter, you will be able to:
▼ Explain simple engine operation.
▼ List the qualities of gasoline that make it an efficient fuel for small engines.
▼ Explain why gasoline is atomized in the small engine.
▼ Identify the basic components of a small engine and describe the function of each part.

Gasoline Engines

A gasoline-fueled engine is a mechanism designed to transform the chemical energy of burning fuel into mechanical energy. In operation, it controls and applies this energy to mow lawns, cut trees, propel tractors, and perform many other laborsaving jobs.

A gasoline engine is an internal combustion engine. Gasoline is combined with air and burned inside the engine. In its simplest form, an engine consists of a ported cylinder, piston, connecting rod, and crankshaft. See **Figure 4-1.**

The piston is a *close fit* inside the cylinder, yet it is free to slide on the lubricated walls. One end of the connecting rod is attached to the piston; the other end is fastened to an offset crankpin, or journal, of the crankshaft. As the piston moves up and down, the connecting rod forces the journal to follow a circular path, rotating the crankshaft.

Simple Engine in Operation

When the engine is cranked, gasoline is atomized (reduced to minute particles) and mixed with air. This mixture is forced through an intake port and into the cylinder, where it is compressed by the piston on the upstroke and ignited by an electrical spark.

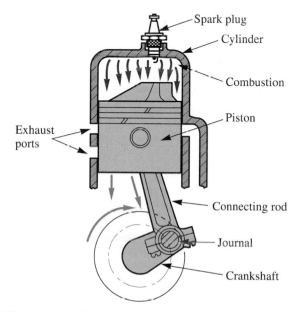

Figure 4-1. *Combustion forces the piston down to rotate crankshaft.*

Burning rapidly, the heated gases trapped within the cylinder (combustion chamber) expand and apply pressure to the walls of the cylinder and to the top of the piston. This pressure drives the piston downward on the power stroke, causing the crankshaft to turn. See **Figure 4-1.**

As the piston and connecting rod push the crankshaft journal to the bottom of the stroke, the pressure of the burned gases is released through an exhaust port. Meanwhile, a fresh air-fuel charge enters the cylinder and the momentum of the power stroke turns the crankshaft journal through bottom dead center (BDC) and into the upstroke on another power cycle.

Gasoline

Gasoline is a hydrocarbon fuel (mixture of hydrogen and carbon), refined from petroleum. *Petroleum* is a dark, thick liquid that is extracted from the earth by oil wells. Luckily, petroleum is the second most plentiful liquid in the world; only water is available in greater quantity. Gasoline, however, cannot be recycled as water can. Therefore, it is imperative that we conserve gasoline and use it wisely.

Gasoline contains a great amount of energy. For engine use, it should:
- Ignite readily, burn cleanly, and resist detonation (violent explosion).
- Vaporize easily, without being subject to vapor lock (vaporizing in fuel lines, impeding flow of liquid fuel to carburetor).
- Be free of dirt, water, and abrasives.

Gasoline is assigned an *octane number* that corresponds to its ability to resist detonation. Premium grade gasoline burns slower than regular gasoline. It has a high octane number and is used in engines with high compression. Regular grade gasoline has a lower octane number and burns relatively fast. Generally, regular gasoline is used in small, low compression, one-cylinder and two-cylinder, gasoline engines.

Gasoline was once available in both leaded and unleaded varieties. The use of lead compounds was the most economical way to increase gasoline's octane number. For many years, most gasolines contained *tetraethyl lead.*

Since the mid 1970s unleaded gasoline has replaced leaded gasoline. Instead of lead compounds, *oxygenates (alcohols and ethers)* are commonly added to these fuels to increase octane

levels. The main reason that unleaded gas was introduced was to provide fuel for automobiles equipped with catalytic converters. These vehicles will not operate properly on leaded fuel.

Modern unleaded gasoline is a complex substance. Ongoing research is necessary to seek ways to produce fuels that offer efficient engine performance and meet air pollution standards.

In the 1970s, a 10% ethanol blend of gasoline, known as gasohol, was introduced. Today, this product is often sold as *super unleaded* or *premium unleaded* gasoline, depending on the octane level.

The main drawback to these gasoline blends is their ability to absorb moisture, which can pass through the fuel filter and into the combustion chamber. These fuels should never be stored in high humidity areas or used in engines that set idle for long periods of time. Gasoline containing alcohol can also corrode fuel tank linings, shrink carburetor floats and seals, increase carbon deposits, and pit metal parts. For maximum performance and engine life, only use the type of gasoline recommended by the engine manufacturer. See *Appendix* for gasoline recommendations.

Gasoline must burn quickly

Gasoline placed in a container and ignited will produce a hot flame, yet it will not burn fast enough to produce the rapid release of heat necessary to run an engine. Even though a considerable quantity of fuel may be involved, a large flame will not necessarily result. See **Figure 4-2.**

> ⚠ Under no circumstances should experiments illustrated in this chapter be performed. Gasoline can be a very dangerous fuel and must be handled with caution. Illustrations and examples discussed here are meant to demonstrate how gasoline is prepared and used in an engine.

In **Figure 4-2,** the surface area of the wick in the lighter is small. Vapor from the surface of the liquid, combined with oxygen, is what burns readily. If the surface of the liquid is small, relatively little vapor will be given off to provide combustion. Since the liquid must change to vapor before it is burned, it would take considerable time to use up the fuel at this rate.

By placing the same amount of fuel in a shallow, wide container, more surface area will contact air and the fuel will burn rapidly. See **Figure 4-3.**

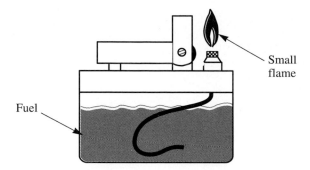

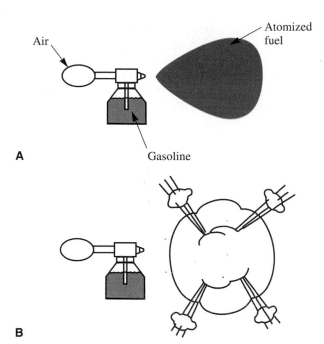

Air

Atomized fuel

A

Gasoline

Figure 4-2. *A small flame is produced, due to small area of exposed fuel.*

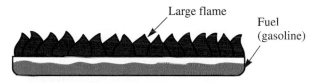

Large flame

Fuel (gasoline)

Figure 4-3. *A large flame is produced by a large area of exposed fuel.*

Fuel is atomized

The more surface area of gasoline exposed to the air, the faster a given amount will burn. To produce the rapid burning required in an engine, gasoline must be broken up into tiny droplets and mixed with air. This is called **atomizing**.

Once the entire surface of each droplet of the air-fuel mixture is exposed to the surrounding air, a huge burning area becomes available. Given a spark, the entire amount of gasoline will flash into flame almost instantly. In effect, atomization causes a sudden, explosive release of heat energy. See **Figure 4-4.**

B

Figure 4-4. *A—When atomized fuel is exposed to the surrounding air, a large burning area is available. B—When the atomized fuel is ignited, heat energy is released with an explosive force.*

Explosion must be contained

To perform useful work, the explosive force caused by the burning gas must be contained and controlled. To illustrate this point, imagine that a metal lid is suspended on a string and held several inches from the ground. If a mixture of gasoline and air (atomized) were sprayed under it and ignited, the lid would be raised a short distance by the force of the explosion. See **Figure 4-5.**

The reason the lid hardly moved is because the explosion was not confined and directed

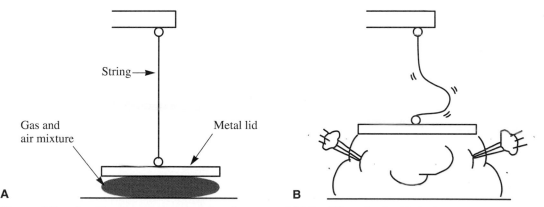

String

Gas and air mixture

Metal lid

A

B

Figure 4-5. *A—Atomized fuel sprayed under a metal lid. B—When the fuel is ignited, the lid moves a short distance.*

toward the lid. Instead, the explosion exerted force in all directions, and much of the force was lost. If the gasoline and air mixture is sprayed inside a metal container with a lid, the full force of the explosion will be directed against the lid when the mixture is ignited. This will blow the lid high into the air. See **Figure 4-6.**

Further improvement

Even though the burning air-fuel mixture is confined by the container, once the lid starts to lift, a large amount of the force escapes to the sides. To eliminate this loss, a long, cylindrical container may be used with the lid having a close, sliding fit. See **Figure 4-7.** With the fuel mixture slightly compressed in the bottom of the container by the weight of the lid, the fuel will burn and direct most of the pressure against the lid as it travels up through the container. When the lid reaches the top, it will be traveling at a high rate of speed. The expansion of the gas will be nearly complete and little force will be lost, even after the lid clears the container.

Basis for an Engine

An elementary engine can be formed by attaching a crankshaft and a connecting rod to the setup illustrated in **Figure 4-7.** The lid will serve as a piston and the container will act as a cylinder. See **Figure 4-8.** When the air-fuel mixture in the cylinder is ignited, it will drive the piston upward, causing the crankshaft to turn.

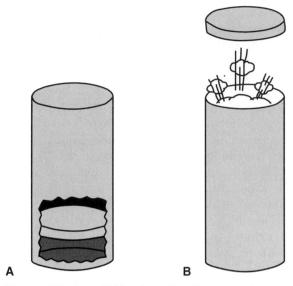

Figure 4-7. *A—Lid is placed in a long container. B—Most of energy of burning fuel is absorbed by lid, imparting greater speed to lid when explosion occurs.*

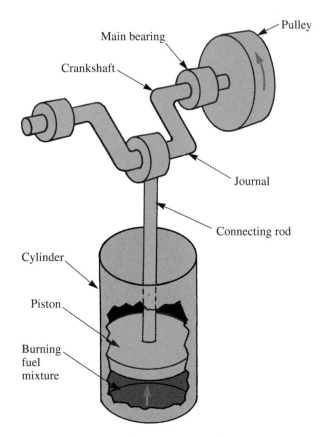

Figure 4-8. *Principles of operation illustrated here are the same as used in a modern gasoline engine. Note how burning fuel mixture forces lid (piston) upward to turn crankshaft and pulley.*

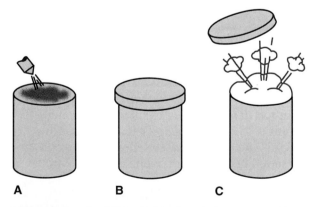

Figure 4-6. *A—Mixture of fuel and air is sprayed into a container. B—Lid is placed on top. C—Full force of explosion is directed toward base of lid when mixture is ignited, and lid is driven high into air.*

Although it is crude, this elementary engine illustrates the operating principles of a modern gasoline engine. Study the names of the various parts shown in **Figure 4-8.** Become acquainted with the parts and their application to engine design.

There are many faults with the engine pictured in **Figure 4-8.** These faults are explored with the following questions:

- How will a fresh air-fuel charge be admitted to the cylinder?
- How will the charge be ignited?
- What holds the various parts in alignment?
- How will the engine be cooled and lubricated?
- What will *time* the firing of the air-fuel mixture so that the piston will push on the crankshaft when the journal is in the correct position?
- How will the burned charge be removed (exhausted) from the cylinder?
- What will keep the crankshaft rotating after the charge is fired, and until another charge can be admitted and fired?

The previous questions can be categorized into five basic areas:

1. **Mechanical** (engine design and construction)
2. **Carburetion** (mixing gasoline and air, and admitting it to the cylinder)
3. **Ignition** (firing the fuel charge)
4. **Cooling** (heat dissipation)
5. **Lubrication** (oiling of moving parts)

In this chapter, emphasis will be placed on the mechanical aspects of engine design and construction. It will provide you with an opportunity to develop a workable engine. We will assume that the gasoline and air are being mixed correctly, the fuel charge is being fired at the right time, and the engine is properly cooled and lubricated.

Cylinder block

The *cylinder block* keeps all engine parts in alignment. See **Figure 4-9.** This critical engine component is usually a casting of iron or an aluminum alloy. The cylinder formed in the block can be produced accurately by modern methods. It may be bored directly into the casting, or a steel sleeve may be inserted into an oversize hole bored in the block.

Aluminum cylinder blocks are cast around a steel sleeve. Aluminum, being a soft metal, would wear out quickly due to the friction of the piston. Advantages of aluminum are its light weight and ability to dissipate heat rapidly.

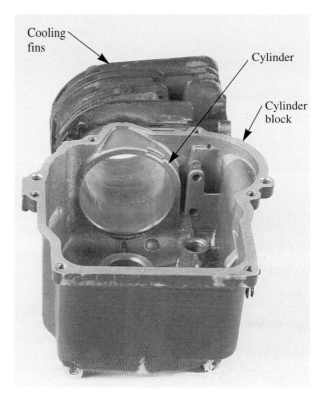

Figure 4-9. *Cylinder block is important because it keeps all moving parts in alignment.*

All air-cooled engines have **cooling fins** on the outside of the cylinder and cylinder head. The size, thickness, spacing, and direction of the cooling fins is carefully engineered for efficient air circulation and heat control.

The cylinder block must be rigid and strong enough to contain the power developed by the expanding gases. In some cases, the cylinder is a separate unit; in others, it is cast as part of the crankcase. Similarly, the cylinder head may be bolted to the block or it may be cast as one complete unit. The method employed depends on the intended application of the engine and the manufacturer's preference.

Figure 4-10 shows a combined cylinder block and crankcase with a separate, bolted cylinder head. Note the gasket that seals the unit. A sleeved, aluminum, die-cast cylinder is shown in **Figure 4-11.**

Crankshaft and crankcase

The **crankshaft** is the major rotating part of the engine. See **Figure 4-12.** Generally, it is forged or cast steel, with all bearing surfaces carefully

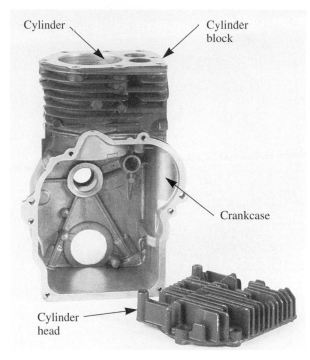

Figure 4-10. *A combined cylinder block and crankcase. Cylinder head and sealing gasket are bolted to cylinder block.*

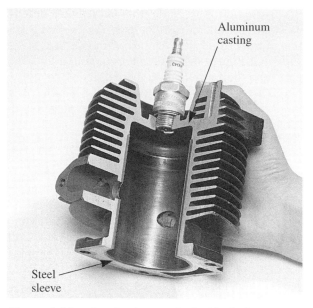

Figure 4-11. *An aluminum cylinder block die cast around a steel sleeve. Note fins for air cooling.*

machined and precision ground. Counterweights are used to balance the weight of the connecting rod, which is fastened to the journal. Since connecting rods are cast or forged from different weight materials, holes are often drilled in counterweights to balance the crankshaft and prevent vibration.

Figure 4-13 shows a crankshaft being installed in a crankcase. Note the tapered roller bearings. The flywheel is keyed to the end of the shaft with

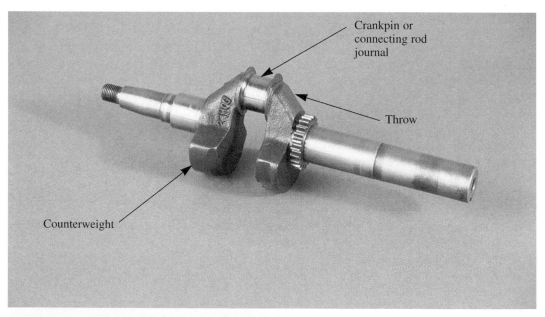

Figure 4-12. *Crankshaft for a single cylinder engine. Large counterweights opposite the crank journal balance rotational forces.*

Figure 4-13. *Crankshaft must be clean and carefully installed in crankcase. Tapered end fits into flywheel, which is secured with a key, lock washer, and nut. (Tecumseh)*

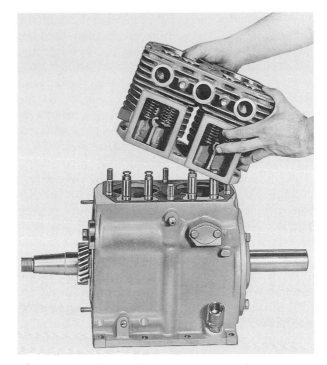

Figure 4-14. *Two cylinder crankcase and cylinder block being assembled. Stud bolts in crankcase will hold cylinder block in place when nuts are tightened.*

a Woodruff key. This type of key cannot slip out during operation. A lock washer and nut hold the flywheel in place.

The end of the crankshaft and the hole through the flywheel have matched tapers that provide good holding power. When roller bearings are used to support the crankshaft, highly polished, hardened alloy steel bearing races are pressed into the crankcase to reduce friction and provide good wearability.

The *crankcase* must be rigid and strong enough to withstand the rotational forces of the crankshaft, while keeping all parts in proper alignment. Oil for lubrication is contained in the crankcase on some engines. On others, a valve system is used that allows a fuel, air, and oil mixture to enter. The crankcase must be designed to protect the internal parts. Gaskets and oil seals are used to keep out dirt and keep in the clean oil.

The crankcase and the cylinder block may be cast as a unit or fastened together by bolts. *Casting* metal is pouring molten metal into a form of a desired shape. **Figure 4-14** shows a two cylinder engine with the cylinder block being placed on the crankcase with the crankshaft already installed. Note the tapered end on the crankshaft, which receives the flywheel.

In certain engine applications, the crankcase is not only an important part of the engine but also an integral part of the apparatus being driven. See **Figure 4-15**. A chain saw is a good example of this.

Pistons

The *piston* is the straight line driving member of the engine. It is subjected to the direct heat of combustion and must have adequate clearance in

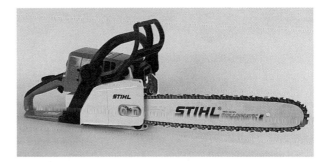

Figure 4-15. *The assembled engine components become the driving members of this lightweight chain saw.*

the cylinder to allow for expansion. The piston provides a seal between the combustion chamber and the crankcase. This is accomplished by cutting grooves near the top of the piston and installing *piston rings.* The piston rings fit the grooves with a slight side clearance and exert tension on the cylinder wall. Properly installed, piston rings prevent blowby of exhaust gases into the crankcase and leakage of oil into the combustion chamber.

The number of piston rings per piston depends upon the type of engine and its design. Note the two piston rings in **Figure 4-16.** The piston is hollow to reduce weight. The top may be flat, domed, or contoured to provide efficient flow of gases entering and leaving the combustion chamber.

There is a hole in each side of the piston through which a *piston pin*, or *wrist pin*, is placed. This pin acts as a hinge between the connecting rod and piston and holds the two together. Generally, spring retainers hold the piston pin in place.

Connecting piston to crankshaft

The sliding piston is connected to the rotating crankshaft with a metal link called a *connecting rod.* The big end of the connecting rod encircles the crankshaft journal and contains a bearing to permit free movement. The upper, or small end, of the connecting rod also must be movable. Note how the metal piston pin is passed through the connecting rod and piston. See **Figure 4-17.**

Also note the needle-type roller bearings, bearing race shells, and retainers, which must be installed in the large end of the connecting rod when it is placed on the crank journal. The bearing cap holds the assembly together with connecting rod bolts or screws. **Figure 4-18** shows the relative position of the connecting rod and cap.

Expanding gases push the piston toward the crankshaft, causing the connecting rod to turn the shaft. **Figure 4-19** shows how the reciprocating (up and down) movement of the piston is changed to rotary (revolving) motion by the crankshaft. Notice that the upper connecting rod bearing allows the connecting rod to swing back and forth while the lower bearing permits the crankshaft journal to rotate within the rod.

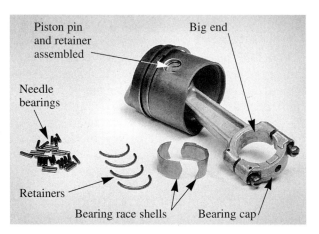

Figure 4-17. *Big end of connecting rod is* capped *to fit around crankshaft journal. Needle bearings are inserted to help reduce friction. Piston pin and retainers are shown assembled. (Jacobsen Mfg. Co.)*

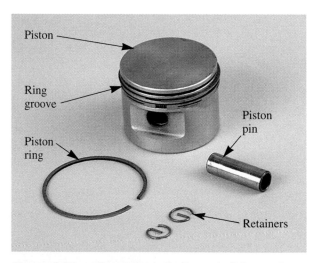

Figure 4-16. *The piston is the largest sliding-reciprocating part in engine. Piston rings seal combustion chamber from crankcase and must fit properly.*

Figure 4-18. *Shown is the relative positions of connecting rod parts.*

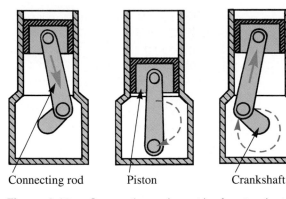

Connecting rod   Piston   Crankshaft

Figure 4-19. *Connecting rod must be free to pivot on piston pin, while crank journal follows a rotary path. Connecting rod must withstand severe stress in operation.*

Intake and exhaust ports

In developing an engine, we need to provide a way in which a fresh air-fuel mixture can be admitted to the engine and, once burned, the waste products exhausted. This can be done by using *ports* (openings) that are alternately covered and exposed by the piston (two-stroke cycle design) or by using *poppet valves* to open and close the port openings (four-stroke cycle design). Both two-stroke and four-stroke designs are commonly used. Each has definite advantages and disadvantages. The four-stroke cycle engine will be discussed here.

Poppet valves

For a four-stroke cycle engine (see Chapter 5 for additional information on fundamentals), passages leading to and from the cylinder area must be constructed. **Figure 4-20** shows two ports cast into the cylinder block, one intake and one exhaust. The cylinder head is recessed to provide a passage from the ports to the cylinder.

By installing a valve in each port, it is possible to control the flow of fresh fuel mixture into the cylinder and provide a means of exhausting the burned gases. During the period of expansion of the burning gases that drive the piston downward, both valves are tightly closed. See **Figure 4-21.**

The angled face of each valve will close tightly against a smooth seat cut around each port opening. To align the valve and assure accurate raising and lowering in relation to the seat, the valve stem passes through a machined hole in block. This hole is called a *valve guide*.

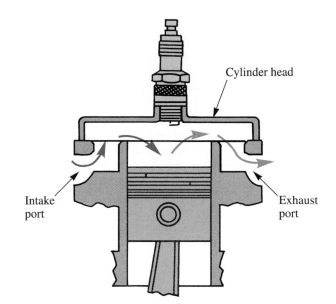

Figure 4-20. *Intake port permits air-fuel mixture to enter. Exhaust port allows burned gases to escape.*

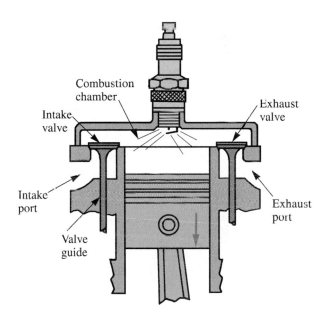

Figure 4-21. *Poppet valves seal intake and exhaust ports during power stroke. Valve guides keep valves aligned with valve seats.*

Valve spring assembly

A *valve spring* must be used on each valve to hold it firmly against the seat. Placed over the valve stem, the spring is compressed to provide tension. It is connected to the valve stem by means of a washer and keeper (lock).

The spring allows the valve to be opened when necessary and will close it when pressure is removed from the valve stem. **Figure 4-22** shows the location of the spring and keeper assembled on the valve. An enlarged view of the *horseshoe* valve lock system is shown in **Figure 4-23.**

A valve in the open position is illustrated in **Figure 4-24.** When pressure is removed from the end of the valve stem, the spring will draw the valve down against the seat and seal off the port from the combustion chamber. For the engine to function properly, the valves must be opened the right amount at the right time. They must remain open for a specific period and close at the correct instant.

By using a shaft with two thick sections spaced to align with the valve stems, a basic device for opening and closing the valves is provided. By grinding the thick sections into a cam shape, the **camshaft** is formed. When the shaft is revolved, the cam lobe will cause the valve to rise and fall, opening and closing the ports. This procedure is shown in **Figure 4-25.**

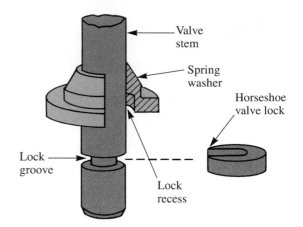

Figure 4-23. *Typical method of retaining valve spring on valve stem. Special tool generally is used to compress spring prior to removing horseshoe valve lock.*

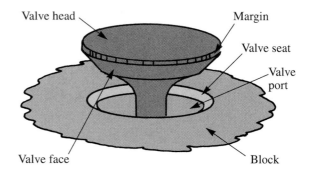

Figure 4-24. *Valve face and valve seat must be ground to correct angles, and concentric to centerline of guide, to seal properly.*

Valve lifter or tappet

In actual practice, the cam lobe does not contact the valve stem directly. By locating the camshaft some distance below the valve stem end, it is possible to insert a **valve lifter** between the lobe and stem. The valve lifter may have an adjustment screw in the upper end to provide a means of adjusting valve stem-to-lifter clearance. Without this adjustment, proper clearance must be obtained by grinding the end of the lifter or valve stem. The base of the lifter may be made wider than the body to provide a larger cam lobe-to-lifter contact area. See **Figure 4-26.**

By drilling a hole in the block above the camshaft, a guide is formed in which the lifter can operate. See **Figure 4-27.** The base of the lifter rides on the cam and the adjusting screw almost

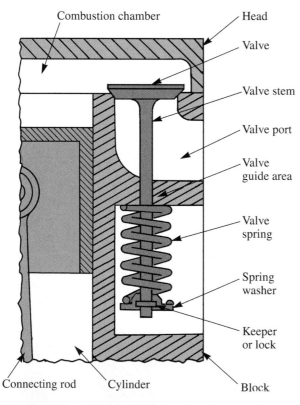

Figure 4-22. *Valve spring keeps tension on valve to ensure proper seating. Valve spring keeper and washer hold spring in place and permit removal when necessary.*

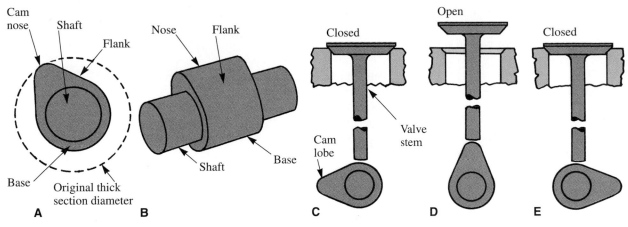

Figure 4-25. *A, B—By grinding a round shaft into a cam shape, a camshaft is formed. C, D, E—When camshaft is revolved, cam lobe will open valve.*

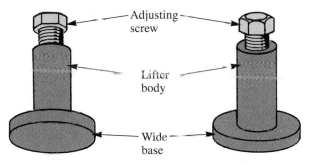

Figure 4-26. *Valve lifter may be called a tappet or cam follower. Adjustment screw allows setting of proper valve clearance. Wide base provides a larger contact area.*

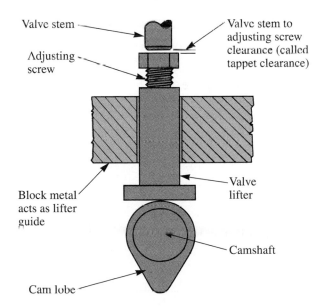

Figure 4-27. *As camshaft turns, cam lobe will operate valve lifter to open valve, then allow it to close.*

touches the end of the valve stem. As the camshaft revolves, the lifter will rise and fall, opening and closing the valve.

Camshaft

Generally, the camshaft is located in the crankcase, directly below the valve stems and valve lifters. The ends of the camshaft are supported in bearings in the block. See **Figure 4-28.** One type of

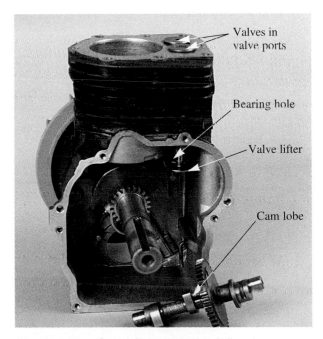

Figure 4-28. *Cam lobes are located directly under valve lifters. Camshaft turns in lubricated bearing holes.*

valve assembly is illustrated in **Figure 4-29.** Study the relationship of the parts. **Figure 4-30** shows the location of each valve part in relation to the rest of the engine block.

The camshaft is driven by the crankshaft through gears. **Figure 4-31** shows the large camshaft gear meshed with the smaller crankshaft gear. The camshaft gear is always twice as large as the crankshaft gear. This gear ratio will be explained in Chapter 5 under four-stroke cycle engine.

Flywheel

Even though the crankshaft moves fast during the power stroke, it is relatively light and tends to slow down or stop before the next power stroke. This periodic application of power, followed by coasting, would cause the engine to speed up, slow down, and/or run roughly.

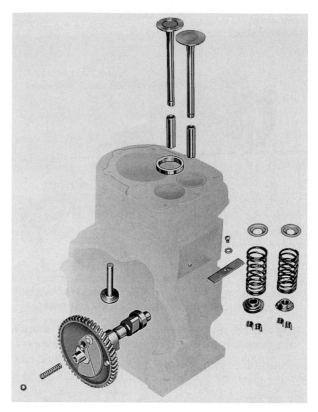

Figure 4-30. *Valve parts and their positions relative to cylinder block and crankcase.*

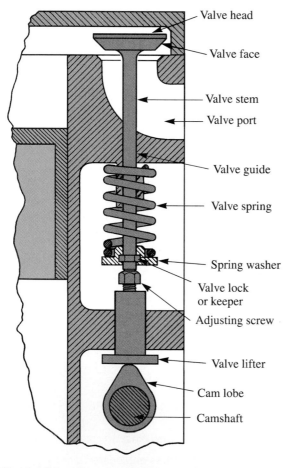

Figure 4-29. *Complete valve train. Study part names and their relationship to each other.*

Valve head
Valve face
Valve stem
Valve port
Valve guide
Valve spring
Spring washer
Valve lock or keeper
Adjusting screw
Valve lifter
Cam lobe
Camshaft

Crankshaft gear

Cam gear

Timing mark

Figure 4-31. *Camshaft is gear driven from crankshaft. Camshaft gear is always twice as large as crankshaft gear for proper timing. During assembly, timing marks must be matched.*

To improve the running quality of the engine, an additional weight in the form of a round *flywheel* is fastened to one end of the crankshaft. See **Figure 4-32.** During the nonpower strokes, the inertia of the heavy flywheel keeps the crankshaft spinning and smoothes engine operation. Metal fins on the flywheel act as a fan that forces air over the cylinder to cool the engine. Magnets cast into the flywheel produce electrical current for the ignition system.

Figure 4-32. *Flywheel is fastened to crankshaft. When rotating, its weight smoothes engine operation.*

Summary

A gasoline engine is designed to transform the chemical energy of burning fuel into mechanical energy. For engine use, gasoline should ignite readily, burn cleanly, and vaporize easily. It should also be free from dirt and oil and resist detonation.

For efficient small engine use, gasoline must be broken into small particles and mixed with air. This process is called atomizing.

To perform useful work, the explosive force caused by burning gasoline must be contained and controlled by a piston and a cylinder.

The cylinder block keeps all engine parts in alignment. Air-cooled engines have cooling fins on the outside of the cylinder block. The crankshaft is the major rotating part in the engine. The crankcase is designed to protect internal engine parts and must be rigid enough to withstand the rotational forces of the crankshaft.

The piston is the straight line driving member of the engine. It provides a seal between the combustion chamber and the crankcase. The piston is connected to the crankshaft by the connecting rod.

Air-fuel mixture is admitted and burned gases are exhausted through ports in the engine. In some engines, valves are installed in the ports to help control intake and exhaust. A camshaft and lifters are used to open and close the valves.

Know These Terms

gasoline
petroleum
octane number
atomizing
cylinder block
cooling fins
crankshaft
crankcase
casting
piston
piston rings

piston pin
wrist pin
connecting rod
ports
poppet valves
valve guide
valve spring
camshaft
valve lifter
flywheel

Chapter 4 Review Questions

Answer the following questions on a separate sheet of paper.
1. Fuel must be atomized in the engine for the purpose of _____.
2. Force on the piston is transmitted to the crankshaft by the _____.
3. Name five desirable characteristics of gasoline for use in small engines.

4. Small gasoline engines generally use
   _____.
   a. high octane fuel
   b. low octane fuel
5. What effect did lead (tetraethyl lead) have when introduced into gasoline?
6. What was the main reason that unleaded gasoline was introduced?
7. What adverse effects can result from using gasoline with alcohol blends?
8. The bearings that support the crankshaft are called _____ bearings.
9. The cylinder block is generally made of _____ or _____.
10. When are sleeved cylinders used?
11. Why are sleeved cylinders used?
12. Why are metal fins made as a part of the cylinder?
13. _____ are designed into the crankshaft to provide for engine balancing.
14. Why do some pistons have a contoured face?
15. The _____ will cause the valve to rise and fall, opening and closing the ports.
16. The angled face of each valve will close tightly against a smooth _____ cut around each port opening.
17. Which is larger, the camshaft gear or the crankshaft gear? How much larger?
18. What are three tasks which are performed by the flywheel?

Suggested Activities

1. Visit a local gasoline station and find out the following:
   a. Various fuel prices.
   b. Octane ratings of the fuels sold.
   c. Kind of containers fuel may be sold in.
   d. Quantities that legally may be stored at home.
2. Disassemble an engine and identify the parts discussed in this chapter. Carefully analyze the function of each part as it relates to the others.
3. If the engine used for disassembly is a used one, look for possible defects such as worn bearings, burned valves, broken or worn piston rings, a scored cylinder, or a loose piston pin.
4. Write to manufacturers of small gasoline engines requesting specifications for the models they produce. Write a report on the types of pistons, connecting rods, and crankshafts they use.
5. Prepare a display of the major components of a small gasoline engine. Use actual parts, photos, drawings, and cutaways to show the principal use of each part.

5

Two-Cycle and Four-Cycle Engines

After studying this chapter, you will be able to:
- ▼ Describe four-stroke cycle engine operation and explain the purpose of each stroke.
- ▼ Explain the concept of valve timing.
- ▼ Compare the lubrication system in a four-cycle engine to the system in a two-cycle engine.
- ▼ Describe two-stroke cycle engine operation and explain the principles of two-cycle operation.
- ▼ List the advantages and disadvantages of two-cycle and four-cycle engines.

Small Engine Identification

A basic design feature that aids in small engine identification is the number of piston strokes required to complete one operating (power) cycle. A four-stroke cycle engine, for example, requires four strokes per cycle; a two-stroke cycle engine requires two.

A *stroke* of the piston is its movement in the cylinder from one end of its travel to the other. Each stroke of the piston, then, is either toward the rotating crankshaft or away from it. Each stroke is identified by the job it performs (intake, exhaust, etc.).

Four-Stroke Cycle Engine

In a *four-stroke cycle engine* (called a *four-cycle*), four strokes are needed to complete the operating cycle. The four strokes are as follows:
- *intake stroke*
- *compression stroke*
- *power stroke*
- *exhaust stroke*

Two strokes occur during each revolution of the crankshaft. Therefore, a four-stroke cycle requires two revolutions of the crankshaft. **Figure 5-1** illustrates each of the four strokes taking place in proper sequence.

Intake stroke

Figure 5-1A shows the piston traveling downward in the cylinder on the *intake stroke.* As piston moves down, the volume of space above it is increased. This creates a partial vacuum that draws the air-fuel mixture through the intake valve port and into the cylinder.

With the intake valve open during the intake stroke, atmospheric pressure outside the engine forces air through the carburetor. This gives a large boost to the air-fuel induction process. With nature balancing unequal pressures in this manner, it follows that the larger the diameter of the cylinder and the longer the stroke of the piston, the greater the volume of air entering the cylinder on the intake stroke.

Bear in mind that the intake valve, **Figure 5-2,** performs several key functions. These key functions are as follows:
1. It must open at the correct instant to permit intake of air-fuel mixture.
2. It must close at the correct time and seal during compression.
3. Its shape must be streamlined, so the flow of gases into combustion chamber will not be obstructed.

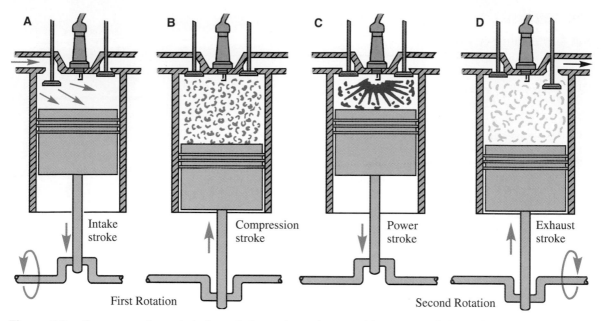

Figure 5-1. *Sequence of events in four-stroke cycle engine, requiring two revolutions of crankshaft and one power stroke out of four.*

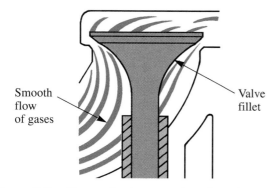

Figure 5-2. *Shape of valve smoothes flow of gases around it. Note how flow follows fillet, speeding entry or expulsion. (Cedar Rapids Engineering Co.)*

The intake valves are not subjected to as high temperatures as the exhaust valve. The incoming air-fuel mixture tends to cool the intake valve during operation.

Compression stroke

The *compression stroke* is created by the piston moving upward in the cylinder. See **Figure 5-1B.** Compression is a squeezing action while both valves are closed. On this stroke, the valves are tightly sealed and the piston rings prevent leakage past the piston.

As the piston moves upward, the air-fuel mixture is compressed into a smaller space. This increases the force of combustion for two reasons:
1. When atoms that make up tiny molecules of air and fuel are squeezed closer together, heat energy is created. Each molecule of fuel is heated very close to its flash point (point at which fuel will ignite spontaneously). When combustion does occur, it is practically instantaneous and complete for the entire air-fuel mixture.
2. The force of combustion is increased because tightly packed molecules are highly activated and are striving to move apart. This energy, combined with expanding energy of combustion, provides tremendous force against the piston.

It is possible to run an engine on uncompressed mixtures, but power loss produces a very inefficient engine.

Power stroke

During the *power stroke,* both valves remain in the closed position. See **Figure 5-1C.** As the piston compresses the charge and reaches the top of the cylinder, an electrical spark jumps the gap between the electrodes of the spark plug. This ignites the air-fuel mixture, and the force of the explosion (violent burning action) forces the piston downward.

Actually, the full charge does not burn at once. The flame progresses outward from the spark plug, spreading combustion and providing even pressure over the piston face throughout the power stroke.

The entire fuel charge must ignite and expand in an incredibly short period of time. Most engines have the spark timed to ignite the fuel slightly before the piston reaches *top dead center (TDC)* of the compression stroke. This provides a little more time for the mixture to burn and accumulate its expanding force.

Basically, the amount of power produced by the power stroke depends on the volume of the air-fuel mixture in the cylinder and the compression ratio of the engine. The *compression ratio* is the proportionate difference in volume of cylinder and combustion chamber at bottom dead center and at top dead center. If the compression ratio is too high, the fuel may be heated to its flash point and ignite too early.

Exhaust stroke

After the piston has completed the power stroke, the burned gases must be removed from the cylinder before introducing a fresh charge. This takes place during the *exhaust stroke.* The exhaust valve opens and the rising piston pushes the exhaust gases from the cylinder. See **Figure 5-1D.**

The exhaust valve has to function much like the intake valve. When closed, the valve must seal. When open, it must allow a streamlined flow of exhaust gases out through the port. See *Figure 5-2.* The removal of gases from the cylinder is called *scavenging.*

The passageway that carries away exhaust gases is referred to as the exhaust manifold or exhaust port. Like the intake manifold, the exhaust manifold must be designed for smooth flow of gases.

The heat absorbed by the exhaust valve must be controlled or the valve will deteriorate rapidly. See **Figure 5-3.** Some valve heat is carried away by conduction through the valve stem to the guide. However, the hottest part of the valve, the valve head, transfers heat through the valve seat to the cylinder block. See **Figure 5-4.**

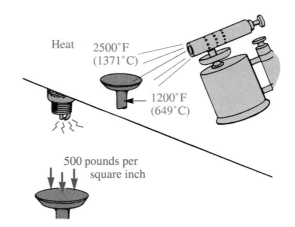

Figure 5-3. *Exhaust valve temperature may range from 1200°F (649°C) to 2500°F (1371°C) due to the hot gases surrounding it. Pressure of combustion may be as high as 500 pounds per square inch. (Briggs & Stratton Corp.)*

Valve timing

The degree at which the valves open or close before or after the piston is at *top dead center (TDC)* or *bottom dead center (BDC)* varies with

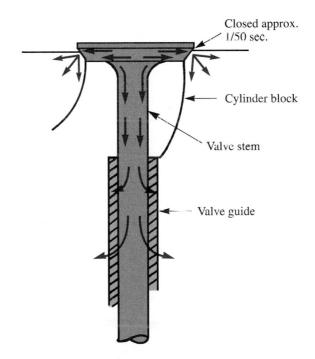

Figure 5-4. *Exhaust valve must cool during an incredibly short period (1/50 sec. at 3600 rpm). Heat is conducted from valve through seat to cylinder block. Some heat conducts down stem and to valve guide.*

different engines. However, if the timing marks on the crankshaft and camshaft gears are aligned, the valve timing will take care of itself.

 Engineers also specify the point at which the spark must occur. Chapter 9 of this text explores this in more detail.

Figure 5-5 shows one complete operating cycle of a four-cycle engine. Beginning at point A, the intake valve opens 10° before TDC and stays open through 235°. The exhaust valve closes 30° after TDC. *Valve overlap* occurs when both valves are open at the same time.

During the compression stroke, the intake valve closes and ignition occurs 30° before TDC. The power stroke continues through 120° past TDC. The exhaust valve opens 60° before BDC and stays open through 270°. During the last 40°, the intake valve is also open and the second cycle has begun.

Lubrication

Lubrication of the four-cycle engine is provided by placing the correct quantity and grade of engine oil in the crankcase. Several methods are used to feed the oil to the correct locations. The two most common methods are the splash system and the pump system. Some engines employ one or the other; others use a combination of both.

The multiple vee cylinder engine utilizes a combination splash and pressure lubrication system. See **Figure 5-6.** The pump picks up the oil from the crankcase and circulates some oil through the filter and directly back to the crankcase. This keeps a clean supply available.

Oil is also pumped through a spray nozzle aimed at the crankshaft. As the shaft rotates, it deflects the oil toward other moving parts. In addition, the splash finger on the bearing cap dips into the crankcase oil and splashes it on various internal surfaces.

Part of the engine oil is pumped through a tube to lubricate the governor assembly above the engine. Oil holes are provided in the connecting rod for lubricating the bearings and piston pin.

Obviously, the oil in a four-cycle engine must be drained periodically and replaced with clean oil. Also worth noting, four-cycle engines must be operated in an upright position or the oil will flow away from the pump or splash finger, preventing lubrication.

Two-Stroke Cycle Engine

The *two-stroke cycle engine* (commonly called *two-cycle*) performs the same cycle of events as the four-cycle engine. The main difference is that intake, compression, power, and exhaust functions take place during only two strokes of the piston. The two strokes occur during each revolution of the crankshaft. Therefore, it takes only one revolution of the shaft to complete one two-stroke cycle.

A two-cycle engine has several advantages over a four-cycle unit. It is much simpler in design than the four-cycle engine because the conventional camshaft, valves, and tappets are unnecessary. See **Figure 5-7.**

Additionally, a two-cycle engine is smaller and lighter than a four-cycle engine of equivalent horsepower. Unlike the four-stroke cycle engine, the two-cycle engine will get adequate lubrication when operated at extreme angles. It receives its lubrication as fuel mixed with oil is passed through the engine.

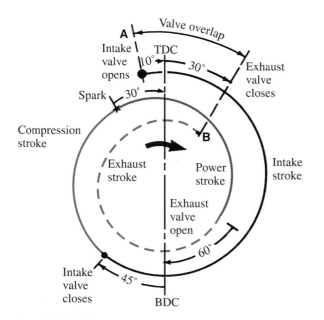

Figure 5-5. *Four-stroke cycle diagram shows exact number of degrees each valve is open or closed and time spark ignition occurs. Note that both valves are open (overlap) through an arc of 40°, permitting exhausting gases to create a partial vacuum in cylinder and help initiate a mixture of fuel in cylinder.*

Installing the correct mixture of fuel and oil is a critical factor in maintaining a two-cycle engine in good working condition. The prescribed type and grade of engine oil must be mixed with the fuel in proper proportion before being placed in the fuel tank.

In this way, there is a continuously new, clean supply of oil to all moving parts while the engine is

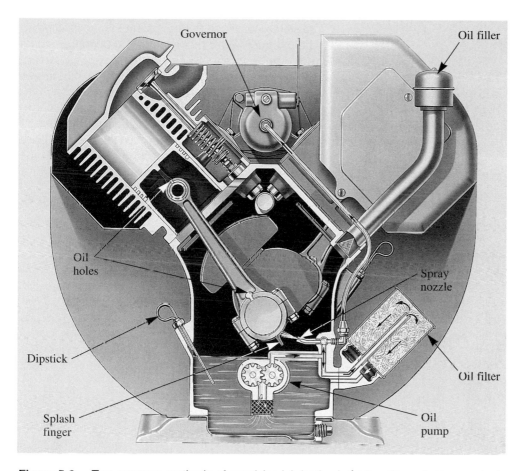

Figure 5-6. *Two common methods of supplying lubrication in four-cycle engines are splash system and pressurized system. Engine shown employs both methods. Splash finger churns oil into a mist that makes its way into oil holes and other parts. Gear pump directs oil to remote parts and sprays some on critical parts. (Wisconsin Motors Corp.)*

Moving Parts—Four-Cycle Engine

Moving Parts—Two-Cycle Engine

Figure 5-7. *The number of moving parts in a four-cycle engine is greater than in a two-cycle engine. Other differences are listed in chart at the end of this chapter. (Lawn-Boy Power Equipment, Gale Products)*

running. The oil eventually burns in the combustion chamber and is exhausted with other gases.

Two-cycle engines are popular in lawn mowers, snowmobiles, chain saws, string trimmers, leaf blowers, and other high-rpm applications.

Variations in design

Two basic types of two-cycle engines are in general use. They are the cross-scavenged and loop-scavenged designs. See **Figure 5-8.**

The *cross-scavenged* engine has a special contour on the piston head, which acts as a baffle to deflect the air-fuel charge upward in the cylinder. See **Figure 5-8A.** This prevents the charge from going straight out the exhaust port, which is located directly across from the intake port.

Cross-scavenged engines usually employ reed valves or a rotary valve, which is attached to the flywheel. See **Figure 5-8B.** These valves hold the incoming charge in the crankcase so it can be compressed while the piston moves downward in the cylinder. With this design, the piston acts as a valve in opening and closing intake, exhaust, and transfer ports. The transfer port permits passage of the fuel from the crankcase to the cylinder.

The *loop-scavenged* engine does not have to deflect the incoming gases, so it has a relatively flat or slightly domed piston, as shown in **Figure 5-8C.** The fuel transfer ports in loop-scavenged engines are shaped and located so that the incom-

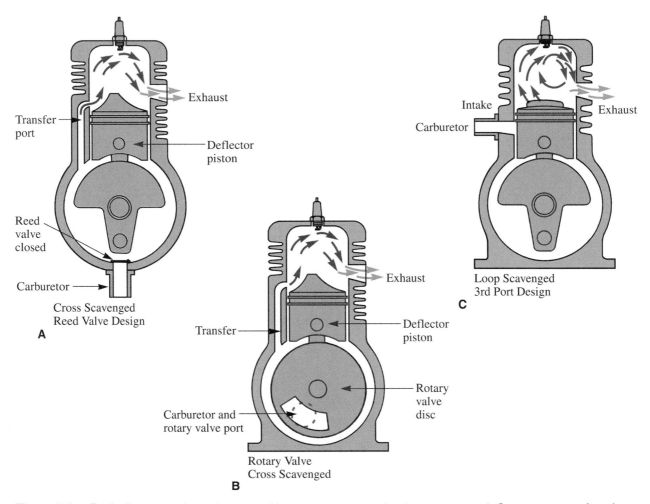

Figure 5-8. *Basically, two-cycle engines are either cross scavenged or loop scavenged. Cross-scavenged engines have a contoured baffle on top of piston to direct air-fuel mixture upward into cylinder while exhaust gases are being expelled. Loop-scavenged engines have flat or domed pistons with more than one transfer port. Note three styles of crankcase intake valves. (Kohler Co.)*

ing air-fuel mixture swirls. This controlled flow of gas helps force exhaust gases out and permits a new charge of air and fuel to enter.

Principles of operation

The location of the ports in a two-cycle engine is essential to correct timing of the intake, transfer, and exhaust functions. The cutaway cylinder in **Figure 5-9A** shows the exhaust port at the highest point, the transfer port next, and the intake port at the lowest point. Some engines, particularly loop-scavenged engines, have more than one transfer port. See **Figure 5-9B**.

Intake into crankcase

As the piston moves upward in the cylinder of a two-cycle engine, crankcase pressure drops and the intake port is exposed. Because atmospheric pressure is greater than the crankcase pressure, air rushes through the carburetor and into the crankcase to equalize the pressures. See **Figure 5-10A**.

While passing through the carburetor, the intake air pulls a charge of fuel and oil along with it. This charge remains in the crankcase to lubricate ball and needle bearings until the piston opens the transfer port on the downstroke.

Fuel transfer

Figure 5-10C and **Figure 5-10D** show the piston moving downward, compressing the air-fuel charge in the crankcase. When the piston travels far enough on the downstroke, the transfer port is opened and the compressed air-fuel charge rushes through the port and into the cylinder. The new charge cools the combustion area and pushes (scavenges) the exhaust gases out of the cylinder.

Ignition-power

As the piston travels upward it compresses the air-fuel charge in the cylinder to about one tenth of its original volume. See **Figure 5-10A**. The spark is timed to ignite the air-fuel mixture when the piston reaches TDC. See **Figure 5-10B**.

On some small engines, spark occurs almost at TDC during starting, then automatically advances so that it occurs earlier. This is done to get better efficiency from the force of combustion at higher speeds.

Peak combustion pressure is applied against the piston top immediately after TDC. Driving downward with maximum force, the piston transmits straight line motion through the connecting rod to create rotary motion of the crankshaft. See **Figure 5-10C**.

Exhaust

Several things happen during the exhaust phase. See **Figure 5-10C**. As the piston moves to expose the exhaust port, most of the burned gases are expelled. Complete exhausting of gases from the cylinder and combustion chamber takes place when the transfer ports are opened and the new air-fuel charge rushes in. See **Figure 5-10D**. This completes one cycle of operation.

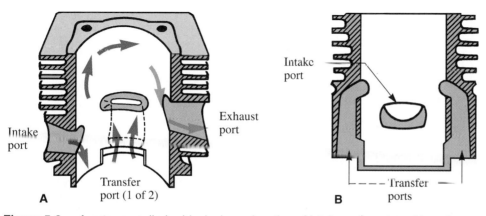

Figure 5-9. *A cutaway cylinder block shows location of intake, exhaust, and transfer ports of a loop-scavenged engine. A—Due to cutaway, only one of two transfer ports is shown. B—Section is revolved 90° to show both ports.*

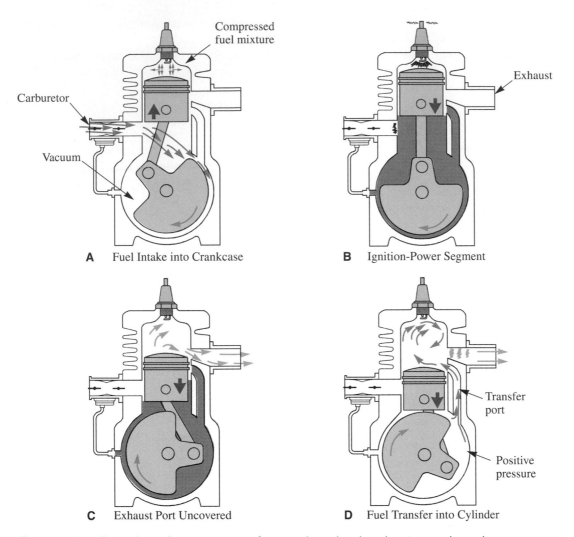

Figure 5-10. *Illustrations show sequence of events that take place in a two-cycle engine. Compression and intake occur simultaneously, then ignition occurs. Exhaust precedes transfer of fuel during lower portion of power stroke. Piston functions as only valve. (Rupp Industries, Inc.)*

Scavenging and tuning

When properly designed, the exhaust system scavenges all exhaust gases from the combustion chamber. The system allows the new fuel charge to move in more rapidly for cleaner and more complete combustion.

For best efficiency, the fuel charge should be held in the cylinder momentarily while the exhaust port is open. This helps prevent fuel from being drawn out of the cylinder with exhaust gases.

Some well engineered exhaust systems use the energy of sound waves from the exhaust gases for proper tuning. **Figure 5-11** shows a megaphone-like device, which amplifies the sound to speed up scavenging. The sound waves are reflected back into the megaphone to develop back pressure, which prevents the incoming air-fuel mixture from leaving with the exhaust gases. Compare this device with straight pipe operation shown in **Figure 5-12.**

Rotary disc valve engine

Figure 5-13 illustrates a two-cycle engine equipped with a *rotary disc valve.* The intake port is located directly in the crankcase, allowing room for additional transfer ports that promote better fuel transfer and scavenging.

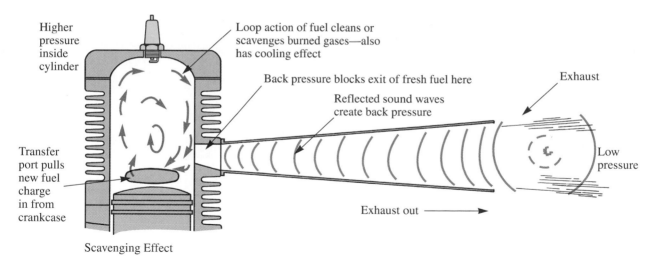

Scavenging Effect

Figure 5-11. *Pressure pulse exhaust tuning is an effective way of increasing power and efficiency in two-cycle engines. Exhaust sound waves reflected back into manifold create a back pressure that stops fuel mixture from leaving cylinder before piston closes port. This system requires precise engineering. (Kohler Co.)*

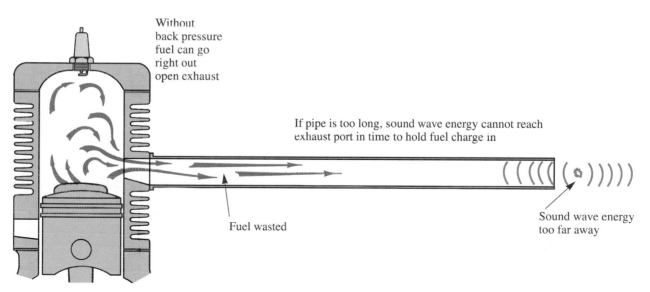

Overscavenged Effect (Pipe Too Long)

Figure 5-12. *A straight pipe may sound louder and more powerful than tuned exhaust, but actually is far less efficient. In this illustration, center of sound is too far and lacks amplification to have any beneficial effect on engine.*

Reed valve engine

The **reed valve engine** permits fuel intake directly into the crankcase. See **Figure 5-14.** The reed is made of thin, flexible spring steel, which is fastened at one end. See **Figure 5-15.** The opposite end covers the intake port. The **reed stop** is thick and inflexible. It prevents the reed from opening too far and becoming permanently bent.

In operation, the reed is opened by atmospheric pressure during the intake stroke. It is closed by the springiness of the metal and the compression in the crankcase on the power stroke. **Figure 5-16A** illustrates the air-fuel mixture entering the crankcase. **Figure 5-16B** shows how the reed valve is closed by crankcase pressure.

There are many reed valve designs. Some typical configurations are illustrated in **Figure 5-17.**

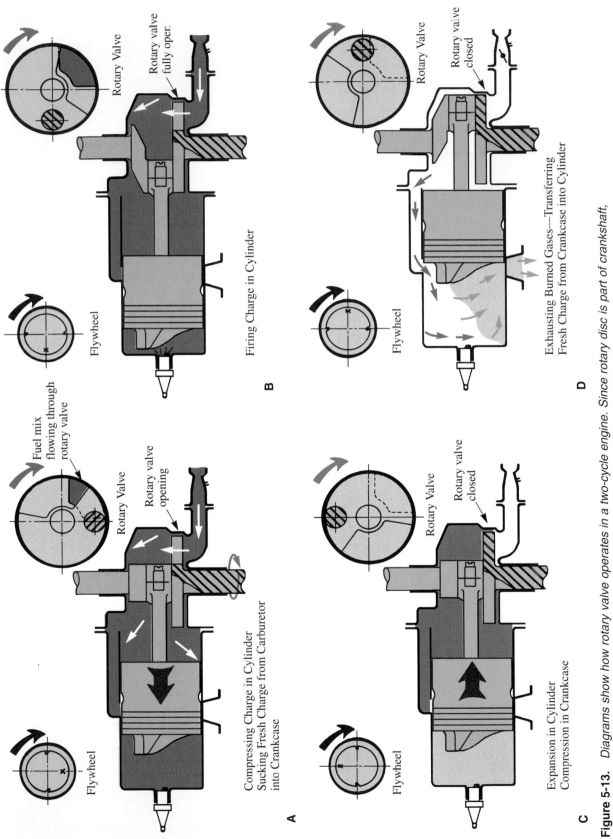

Figure 5-13. *Diagrams show how rotary valve operates in a two-cycle engine. Since rotary disc is part of crankshaft, port is open only when hole in disc and crankcase port are in alignment. (Evinrude Motors)*

Fuel mix flowing through rotary valve

Rotary Valve

Rotary valve opening

Flywheel

A Compressing Charge in Cylinder Sucking Fresh Charge from Carburetor into Crankcase

Rotary Valve

Rotary valve fully open

Flywheel

B Firing Charge in Cylinder

Rotary Valve

Rotary valve closed

Flywheel

C Expansion in Cylinder Compression in Crankcase

Rotary Valve

Rotary valve closed

Flywheel

D Exhausting Burned Gases—Transferring Fresh Charge from Crankcase into Cylinder

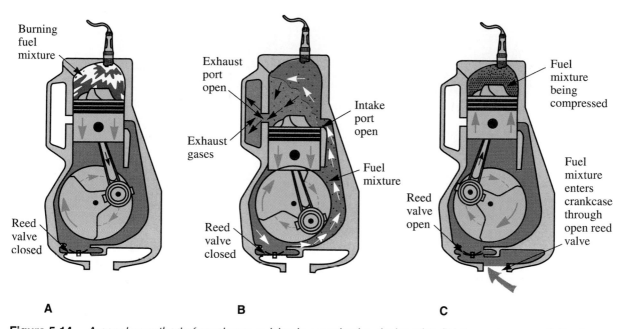

Figure 5-14. *A popular method of crankcase valving is a reed valve designed to fit into crankcase wall. It relies upon difference between atmospheric pressure and crankcase pressure to be opened. At rest position is closed position.*

A　　　　　　　　　　B　　　　　　　　　　C

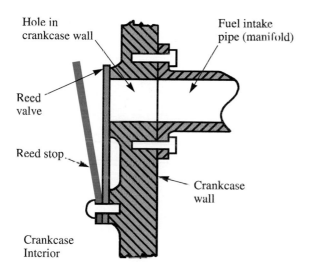

Figure 5-15. *Side view of a reed valve shows spring steel reed covering entry hole. Reed stop controls distance reed may open. This prevents permanent distortion and failure of reed to return snugly against port during time of crankcase compression.*

Four-Cycle Engine vs. Two-Cycle Engine

The advantages and disadvantages of any engine are directly related to the purpose for which the engine is intended. It cannot be said that one type of engine is better than another without considering every aspect of its application.

The chart in **Figure 5-18** lists the differences between two- and four-cycle engines.

Summary

The stroke of a piston is its movement in the cylinder from one end of its travel to another. Four-stroke cycle engines need four strokes to complete the operating cycle: intake, compression, power, and exhaust. Lubrication of four-cycle engines is generally provided by a splash system or a pump system.

In a two-stroke cycle engine, the intake, compression, power, and exhaust functions take place during two strokes of the piston. Two-cycle engines have many advantages over four-cycle units. They do not have conventional valves, tappets, or a camshaft. Two-cycle engines are smaller and lighter than four-cycle engines of equivalent horsepower.

The two-cycle engine receives its lubrication as a fuel-oil mixture is passed through the engine. Therefore, it will receive adequate lubrication when operated at extreme angles.

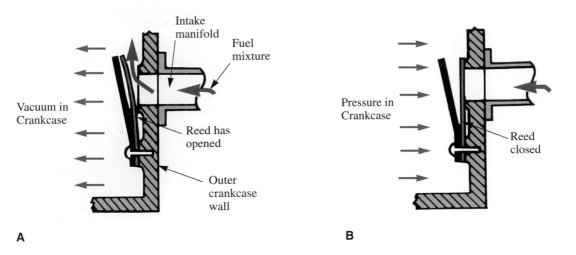

Figure 5-16. *Reed valve action. A—Vacuum in crankcase formed by upward moving piston causes atmospheric pressure to force air-fuel mixture through port opening. B—Downward piston movement compresses fuel mixture in the crankcase to a pressure greater than atmospheric pressure. Springy reed and crankcase pressure act together to close port.*

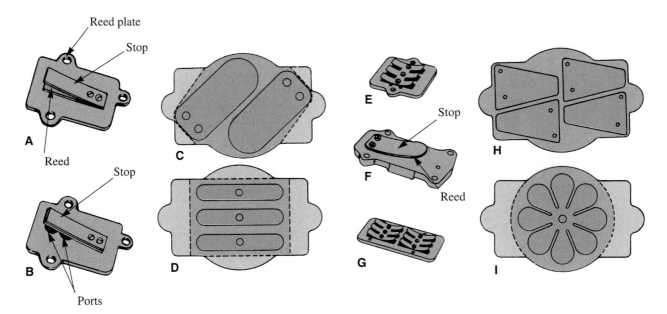

Figure 5-17. *Several forms of reed valves. A—Single reed, closed position. B—Single reed, open position. Note how the reed opening distance is controlled by the stop. C—Twin reed. D—Triple reed. E—Another form of triple reed. F—Single reed. G—Multiple reed. H—Four reed. I—Multiple reed.*

Characteristics	Four-Cycle Engine (equal hp) One Cylinder	Two-Cycle Engine (equal hp) One Cylinder
1. Number of major moving parts	Nine	Three
2. Power strokes	One every two revolutions of crankshaft	One every revolution of crankshaft
3. Running temperature	Cooler running	Hotter running
4. Overall engine size	Larger	Smaller
5. Engine weight	Heavier construction	Lighter in weight
6. Bore size equal hp	Larger	Smaller
7. Fuel and oil	No mixture required	Must be premixed
8. Fuel consumption	Fewer gallons per hour	More gallons per hour
9. Oil consumption	Oil recirculates and stays in engine	Oil is burned with fuel
10. Sound	Generally quiet	Louder in operation
11. Operation	Smoother	More erratic
12. Acceleration	Slower	Very quick
13. General maintenance	Greater	Less
14. Initial cost	Greater	Less
15. Versatility of operation	Limited slope operation (Receives less lubrication when tilted)	Lubrication not affected at any angle of operation
16. General operating efficiency (hp/wt. ratio)	Less efficient	More efficient
17. Pull starting	Two crankshaft rotations required to produce one ignition phase	One revolution produces an ignition phase
18. Flywheel	Requires heavier flywheel to carry engine through three nonpower strokes	Lighter flywheel

Figure 5-18. *Chart lists the differences between two-stroke and four-stroke cycle engines.*

Know These Terms

stroke
four-stroke cycle
intake stroke
compression stroke
power stroke
exhaust stroke
scavenging
valve timing
top dead center

bottom dead center
valve overlap
two-stroke engine
cross scavenged
loop scavenged
rotary disc valve
reed valve engine
reed stop

Chapter 5
Review Questions

Answer the following questions on a separate sheet of paper.

1. Name the four strokes of a four-cycle engine in proper order.
2. Name three important intake valve functions.
3. Explain why a four-cycle engine runs cooler than a two-cycle engine.
4. Why is there a difference in temperature between the intake and exhaust valves?
5. The exhaust valve is cooled mainly by _____.
   a. radiation
   b. conduction
   c. convection
   d. air-fuel circulation
6. How does compression increase engine power?
7. The compression ratio must be limited in gasoline spark ignition engines, because _____.
   a. there is no power advantage after compressing the fuel to a certain point
   b. the engine becomes too difficult to start
   c. mechanically it is not possible to increase the compression ratio
   d. the heat of compression will ignite the air-fuel mixture too soon
8. What are the two methods employed for lubricating four-cycle engines?
9. What are the two types of scavenging systems used in two-cycle engines?
10. Why can two-cycle engines be run in any position?
11. Name three valving systems employed in two-cycle engines.

12. The baffle on the contoured piston is for _____.
    a. creating turbulent flow of gases
    b. slowing the air-fuel mixture entering the combustion chamber
    c. directing the flow of air-fuel mixture upward in the cylinder
    d. directing oil evenly to the cylinder walls
13. The _____ _____ type of two-cycle engine requires a contoured piston.
14. In a properly tuned exhaust system, _____ _____ prevent the air-fuel mixture from leaving with the exhaust.
15. What advantage is there in having the intake port lead directly into the crankcase?
16. Time during the four-stroke cycle when both valves are open is called _____ _____.
17. A four-cycle engine accelerates slower than a two-cycle engine, because _____.
    a. there is only one power stroke in four
    b. the flywheel is heavier to carry the engine through three nonpower strokes
    c. there are more moving parts to be driven by the engine
    d. All of the above.

Suggested Activities

1. Look up additional information about internal combustion engines development. Names to look up: Christian Huygens, Philip Lebon, Samuel Brown, William Barnett, Pierre Lenoir, Beau DeRochas, Dr. N. A. Otto, Atkinson, Gottlieb Daimler, Priestman and Hall, Herbert Akroyd Stuart, Rudolph Diesel.
2. Begin a collection of engine repair and service manuals.
3. Using a worn out engine, cut away portions that will make the working parts visible while still enabling them to move. Report on the operation and timing of each part. After further study, replace the spark plug of the cutaway engine with a small light bulb switched on and off by the breaker points to simulate ignition.
4. Make a bulletin board display that illustrating the principles of two- and four-cycle engines.

Measuring Engine Performance

After studying this chapter, you will be able to:

▼ Define engine performance.

▼ Define and compute bore, stroke, displacement, compression ratio, force, work, power, energy, and horsepower.

▼ Differentiate between the various types of horsepower.

▼ Explain the function of a Prony brake and a dynamometer.

▼ Define and calculate torque.

▼ Explain volumetric efficiency, practical efficiency, mechanical efficiency, and thermal efficiency.

The Engine

The small gasoline engine belongs in the *heat engine* category. Other heat engines include automotive reciprocating piston engines, gas or steam turbines, steam engines, diesel engines, rotary combustion engines, rocket engines, and jet engines. Only the small, one-cylinder and two-cylinder piston engines will be discussed here.

The small gasoline engine is called an internal combustion engine because an air/fuel mixture is ignited (fired) and burned inside the engine. See **Figure 6-1.** The heat from the burning mixture

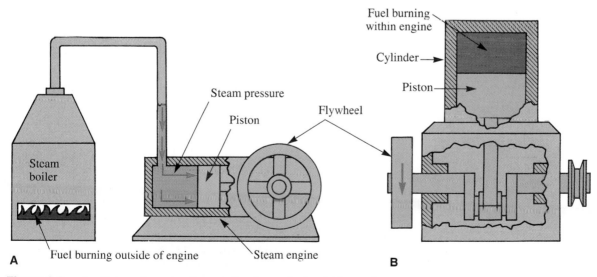

Figure 6-1. *A—External combustion engine burns fuel outside engine. B—Internal combustion engine burns fuel within engine.*

causes the gases to expand rapidly within the closed cylinder. The expanding gases apply strong force and push out in all directions within the cylinder, but only the piston can move.

The piston is pushed away from the center of combustion. If it were not fastened to the crankshaft, it would come out of the cylinder the way a bullet comes out of a gun. The crankshaft and connecting rod keep the piston under control and allow it to travel only a short distance. See **Figure 6-2.**

When the piston has moved downward as far as it can go, a port is opened. This allows burned gases to escape as the piston returns upward. The burned, escaping gases are called exhaust and come out through the exhaust pipe or manifold. A manifold is a chamber that collects the exhaust and directs it to the exhaust pipe.

The piston does not pause long at the end of its stroke before the movement of the crankshaft and flywheel carry it back to the top of the cylinder. As you may have learned from your science classes, bodies in motion tend to continue in motion. This tendency is called *inertia*. It is what keeps the crankshaft and flywheel moving in the engine.

When the piston reaches the top of its stroke, it is ready to be forced back down again. Because the piston continues this back and forth or up and down motion, piston engines are often called *reciprocating engines*.

Basic Terminology

To better understand how a gasoline engine works and to appreciate the power it provides, you must learn certain basic terms. These terms will be defined here only as far as necessary to provide a background for further discussion of measuring engine performance. *Performance* can be defined as the work engines do and how well they do it.

Engine bore and stroke

Engine *bore* is the diameter or width across the top of the cylinder. *Stroke* is the up or down movement of the piston. Length of stroke is determined by the distance the piston moves from its uppermost position (top dead center or TDC) to its lowest position (bottom dead center or BDC), or in reverse order.

The amount of crank offset determines the length of the stroke. *Crank offset* is the distance from the centerline of the connecting rod journal to the centerline of the crankshaft. A 2″ offset would produce a 4″ stroke. See **Figure 6-3.**

When the bore diameter is the same as the stroke, the engine is referred to as *square.* When the bore diameter is greater than the stroke, it is termed *over square.* Where bore diameter is less than the stroke, the engine is called *under square.*

Engine displacement

In a single-cylinder engine, engine size, or *displacement,* refers to the total volume of space increase in the cylinder as the piston moves from the top to the bottom of its stroke.

To work out a given engine's displacement, first determine the circular area of the cylinder (0.7854 by diameter squared, or D^2). Then multiply that answer by the total length of the stroke (piston travel).

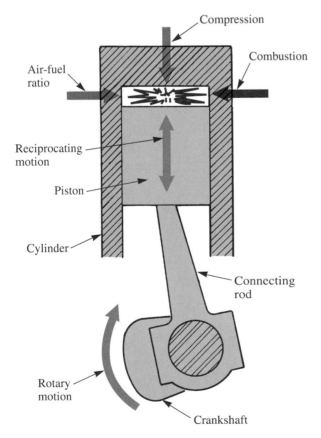

Figure 6-2. *Piston travel is controlled by connecting rod and crankshaft. Notice change from reciprocating (up and down) motion to rotary motion. (Deere & Co.)*

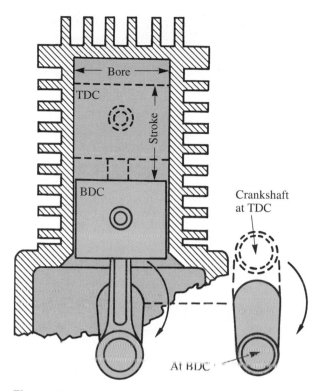

Figure 6-3. *Engine* bore *refers to diameter of cylinder.* Stroke *indicates length piston travels as it moves from TDC to BDC.*

The formula is as follows:

$$\text{Engine Displacement} = 0.7854 \times D^2 \times \text{Length of Stroke}$$

If the engine has more than one cylinder, multiply the answer to the above formula by the number of cylinders. For example, say that a two-cylinder engine has a bore of 3 1/4″ and a stroke of 3 1/4″. Using the displacement formula, you would have the following:

$$0.7854 \times D^2 \ (10.563) \times \text{Length of Stroke } (3\ 1/4″) \times \text{Number of Cylinders } (2) = 53.9 \text{ cubic inches (cu in)}$$

Figure 6-4 illustrates piston displacement. In drawing A, the piston is at TDC (top dead center). Green liquid has been added to fill up the space left between the piston and the cylinder head. The piston in drawing B is at the bottom of its stroke. Note the increased space above the piston now. In drawing C, orange liquid is added until the cylinder is once again filled. The amount of orange liquid (in cu in) would represent the piston displacement for this cylinder.

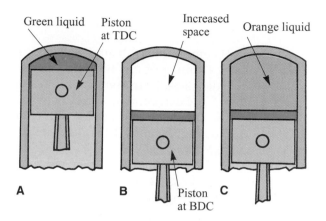

Figure 6-4. *Engine displacement is difference in volume of the cylinder and combustion chamber above piston when it is at TDC and when it is at BDC. Red area shows displacement.*

Compression ratio

The ***compression ratio*** of an engine is a measurement of the relationship between the total cylinder volume when the piston is at the bottom of its stroke (BDC) and the volume remaining when the piston is at the top of its stroke (TDC).

For example, if cylinder volume measures 6 cu in when the piston is at BDC (view A in **Figure 6-5**) and 1 cu in when at TDC, (view B in **Figure 6-5),** the compression ratio of the engine is 6 to 1. Many small gasoline engines have 5 or 6 to 1 compression ratios. Certain motorcycle engines have 9 or 10 to 1 compression ratios.

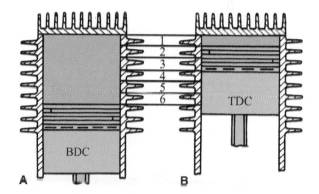

Figure 6-5. *Compression ratio is relationship between cylinder volume with piston A at BDC and piston B at TDC. Volume has been compressed to one-sixth of its original size, which indicates a 6 to 1 compression ratio. (Briggs and Stratton Corp.)*

Force

Forces are being applied all around us. *Force* is the pushing or pulling of one body on another. Usually two bodies must be in contact for force to be transmitted. For example, as you read this you are applying a force to a chair if you are sitting or to the floor if you are standing. The force is equal to the weight of your body.

You can easily measure it with a scale. See **Figure 6-6.** This is known as gravitational force and it acts upon all materials on and around the earth.

Some forces are stationary (motionless); others are moving. For example, if you push against a wall, force is applied but the wall does not move. The use of force may or may not cause motion. Force itself cannot be seen, but there are many ways of using it.

Centrifugal force acts upon a body whenever it follows a circular or curved path. The body tries to move outward from the center of its path. Modern examples are the man-made satellites that orbit the earth. The circular path and speed of the satellite produces a centrifugal force outward that is equal to the earth's gravitational force inward so that each is opposed and balanced. Therefore, the satellite neither goes up nor comes down. This is one case where a force is applied without one body touching another body. A ball swung on a string applies centrifugal force as shown in **Figure 6-7.**

Many forces interact when a gasoline engine is operating. The rotational speed of the crankshaft and flywheel create centrifugal force, which causes *tensile stress* (tension or pull) within the materials making up these parts. If the outward pulling force becomes greater than the strength of the material, the engine could fly apart. The rapid reciprocation (backward and forward or up and down motion) of the piston may put high forces on the connecting rod, crank journal, and piston pin. **Figure 6-2** shows how these parts work together.

One of the forces used efficiently in the gasoline engine is that which is applied to the top of the piston by the rapidly expanding gases. This force is produced by burning gasoline mixed with air. The greater the force applied to the piston, the greater the amount of power and work that can be done by the engine.

Force is measured in units of some standard weight such as pounds, ounces, or grams. For example, to support a shop vise weighing 16 lb, a person would have to apply a lifting force of 16 lb. Obviously, only half the lifting force would be needed to support an 8 lb vise. See **Figure 6-8.**

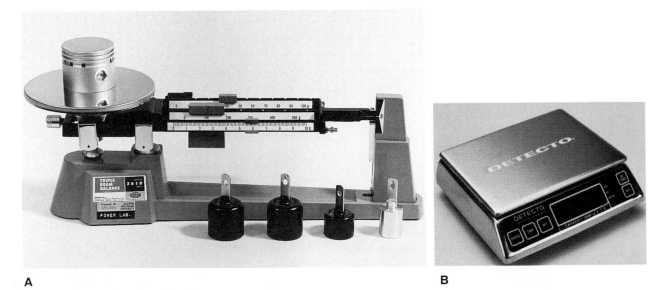

A   **B**

Figure 6-6. *A—A triple beam balance is used to measure the force of gravity applied to an engine piston. Weight is read from three scales. The top scale is graduated to ten gram increments, the center scale is graduated in 100 gram increments, and the bottom scale graduated in single grams. B—Digital scales are becoming more common because of accuracy and ease of reading. (Detecto; Cardinal Scale Mfg. Co.)*

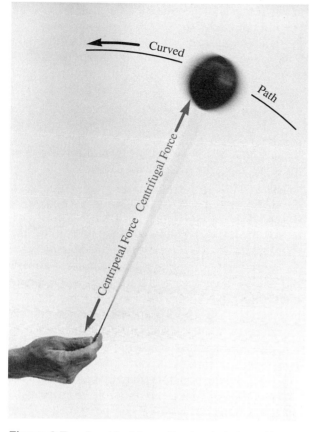

Figure 6-7. *An object in motion tends to travel in a straight line. The additional force required to cause deviation (or change) is called centripetal force. Reaction to centripetal force is called centrifugal force.*

Figure 6-8. *A small shop vise being lifted with a force equal to, or slightly greater than, its own weight.*

Force and pressure are often confused. These terms should be understood in the way they are applied. ***Pressure*** is a force per given unit of area.

For example, a piston with a face area of 5 square inches (sq in) may have a total force of 500 lb applied to it by the expanding gases. However, the pressure being applied is 500 lb divided by 5 sq in, which equals 100 pounds per square inch (psi). This means that every square inch on the piston face has the equivalent of a 100 lb weight pushing on it.

When we speak of pressures in mathematical calculations, we use letters to represent several words. For example, psi means *pounds per square inch*. The psi formulas are as follows:

$$psi = \frac{Force}{Area}$$

or $Force = psi \times Area$

or $Area = \frac{Force}{psi}$

The area of a circle can be found by multiplying pi, or π ($\pi = 3.1416$), by the radius squared. The written formula is as follows:

$$Area = \pi r^2$$

Another method is to multiply the constant .7854 by the diameter squared. The written formula is as follows:

$$Area = .7854D^2$$

For example, we shall calculate a force applied to a 3″ diameter piston, **Figure 6-9,** if the cylinder pressure is 125 psi.

$Area = \pi r^2$
$Area = 3.1416 \times (1.5 \times 1.5)$
$Area = 3.1416 \times 2.25$
$Area = \mathbf{7.0686}$

Then the piston area equals: 7.0686 sq in

$Force = psi \times Area$
$Force = 125 \times 7.0686$
$Force = 883.575$

This results in a total force of 883.575 lb.

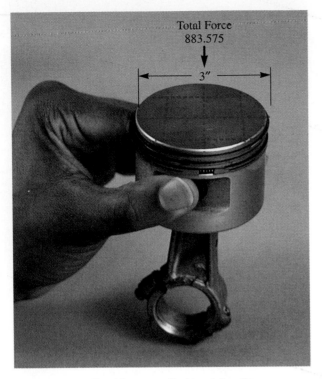

Figure 6-9. *Total force applied to piston face is equal to its area in square inches multiplied by psi (pounds per square inch).*

Work

Work is accomplished only when a force is applied through some distance. If a given weight is held so that it neither rises nor falls, no work is done even though the person holding the weight may become very tired. If the weight is raised some distance, then work is being done. The amount of work performed is the product or result of the force and the distance through which the weight is moved.

If a weight of 20 lb is lifted 3′, then 60 ft-lb of work is accomplished. The distance must always be measured in the same direction as the applied force. This results in the formula as follows:

$$Work = Force \times Distance$$

Because the formula calls for multiplying feet times pounds, the answer is expressed in foot-pounds (ft-lb).

The gasoline engine utilizes the principles of a number of simple machines. These machines include the lever, the inclined plane, the pulley, the wheel and axle, the screw, and the wedge.

With each of these simple machines, you will find that to increase the output force with a given input force, the input distance will have to be increased in the same proportion or percentage. Therefore, without considering loss through friction, the foot-pounds of output are equal to the foot-pounds of input.

By using the principle of the lever, for example, a mechanical advantage is possible. **Figure 6-10** illustrates how a heavy load can be moved a short distance with a small force exerted through a relatively great distance. The formula for computing leverage, as it applies to **Figure 6-10,** is as follows:

$$MA \text{ (mechanical advantage)} = \frac{ED \text{ (effort distance)}}{RD \text{ (resistance distance)}}$$

$$MA = \frac{ED}{RD} = \frac{6}{2} = 3$$

$$E \text{ (effort)} = \frac{R(resistance)}{MA} = \frac{600}{3} = 200 \text{ lb}$$

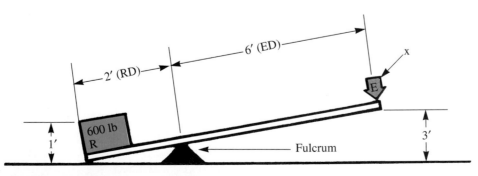

Figure 6-10. *The principle of the lever is the same as those applied to other forms of simple machines.*

or
$$X:6 = 2:600$$
$$X = \frac{2 \times 600}{6}$$
$$X = 200 \text{ lb}$$

Power

In studying the formula for work, note that it does not consider the time required to do the work. For example, if a small gasoline engine weighing 50 lb is lifted 3′ from the floor to the workbench, 150 ft-lb of work is done. The same amount of work would be performed whether it took 50 seconds (sec) to lift the engine or only 5 sec.

Because it is important to know the rate at which work is done, the word power enters the picture. *Power* is the rate at which work is performed. The rate (amount of time) is given in seconds. Power can then be considered as foot-pounds per second. The formula for power is as follows:

$$\text{Power} = \frac{\text{Work}}{\text{Time}}$$

$$\text{or} \quad \text{Power} = \frac{\text{Feet} \times \text{Pounds}}{\text{Seconds}}$$

or Power = ft-lb per second (ft-lb/sec)

When the engine is lifted in 5 sec, 150 ft-lb of work is performed. Using the power formula, it can be seen that:

$$\text{Power} = \frac{\text{Work}}{\text{Time}}$$

$$\text{Power} = \frac{150 \text{ ft-lb}}{5 \text{ sec}}$$

$$\text{Power} = 30 \text{ ft-lb/sec}$$

When lifting the engine in 50 sec, the formula shows the following:

$$\text{Power} = \frac{\text{Work}}{\text{Time}}$$

$$\text{Power} = \frac{150 \text{ ft-lb}}{50 \text{ sec}}$$

$$\text{Power} = 3 \text{ ft-lb/sec}$$

Work is a force applied to an object that causes the object to move, and power is the rate at which the work is done. The standard unit of power is termed *horsepower*.

Energy

Energy is the capacity to perform work. It is grouped into various types, including potential energy (PE), kinetic energy (KE), mechanical energy (ME), chemical energy (CE), and thermal energy (TE). Typical sources of the various types of energy are shown in **Figure 6-11.**

Many of the words just used will be unfamiliar to you. We will explain them as we go along. To begin with, energy is something we cannot define. It puts *life* into matter, giving it warmth, light, and motion. Energy cannot be seen, weighed, or measured. It does not take up space. However, we know it is there. The warmth and light of a bonfire, the electrical spark that jumps the gap of a spark plug, or the turning of a wheel are things we can sense. Thus, they must exist.

Matter and energy cannot be destroyed. Only the nature of matter and the forms of energy change. For example, when a piece of charcoal burns and disappears, we may think it is completely gone. But the charcoal material has combined with air and formed a like quantity of ash, water, and gases. The energy that did this existed as flame (light energy) and heat (heat energy). All of this will continue until another change takes place.

Whenever a form of matter can be separated from other forms of matter so that a part or all of its energy can be released, it is said to contain

	Types of Energy				
	PE	*KE*	*ME*	*CE*	*TE*
Coal, oil, gasoline				X	
Clock, spring-wound	X				
Moving auto		X			
Water behind a dam	X				
Jacking up an auto	X				
Cooking fire					X
Dynamite, stored			X		
Dynamite, exploding			X		X
Crankshaft, turning		X			
Man, running		X			

Figure 6-11. *Potential, kinetic, mechanical, chemical, and thermal energy and some of their sources. (Go-Power Corp.)*

potential energy. Examples of such matter include crude oil and the gasoline taken from it. We have learned different ways to release a part of the energy stored in these substances.

Engines are designed to release and change the potential energy of gasoline into mechanical power. Mechanical power does the work at hand.

Horsepower

For hundreds of years, men used horses to perform work. It was only natural that when machines were invented, their ability to perform work would be compared to the horse.

In his work with early steam engines, James Watt wanted some simple way to measure their power output. In measuring the power or rate of work performed by a horse, he found that most work horses could lift 100 lb a distance of 330' in 1 min.

If 1 lb is lifted 1' in 1 min, 1 ft-lb of work is done. The horse lifted 100 lb a distance of 330 ft in 1 min. Using the work formula (Work = Distance × Force), Watt found that the horse performed 33,000 ft-lb of work.

In determining the *rate of power* developed by the horse, we use the following formula:

$$\text{Power} = \frac{\text{Work}}{\text{Time}} = \frac{33{,}000 \text{ ft-lb (Work)}}{1 \text{ min}}$$

$$\text{or} \quad \frac{550 \text{ ft-lb}}{1 \text{ sec}} = 1 \text{ Horsepower (hp)}$$

The 550 ft-lb/sec (ability to lift 550 lb a distance of 1' in 1 sec) was then established as 1 hp. This standard is still in use today.

Horsepower formula

Engine horsepower can be calculated by dividing the total rate of work (ft-lb/sec) by 550 ft-lb/sec. For example, if an engine lifted 330 lb a distance of 100' in 6 sec, its total rate of work would be as follows:

$$\frac{100 \text{ ft} \times 330 \text{ lb}}{6 \text{ sec}} = 5500 \text{ ft-lb/sec}$$

Dividing this by 550 ft-lb/sec (1 hp) as follows:

$$\frac{5500 \text{ ft-lb/sec}}{550 \text{ ft-lb/sec}} = 10$$

You find that the engine rated at 10 hp. The horsepower formula would then be as follows:

$$1 \text{ Horsepower} = \frac{\text{Rate of Work in ft-lb/sec}}{50 \text{ ft-lb/sec}}$$

This formula may be used also to determine the exact horsepower needed for other tasks.

Kinds of horsepower

The word horsepower is used in more than one way. Some of the common terms include: brake horsepower, indicated horsepower, frictional horsepower, and rated horsepower.

Brake horsepower

Brake horsepower (bhp) indicates the actual usable horsepower delivered at the engine crankshaft. Brake horsepower is not always the same. It increases with engine speed. At very high and generally unusable engine speeds (depending on engine design), the horsepower output will drop off somewhat.

Figure 6-12 shows how horsepower increases with speed for two different engine models. The top speeds in this chart do not run high enough to show a drop in horsepower.

	Horsepower	
	Engine Models	
RPM	**ACN**	**BKN**
1600	2.5	3.5
1800	2.9	4.0
2000	3.5	4.4
2200	3.7	4.9
2400	4.2	5.4
2600	4.5	5.8
2800	4.8	6.2
3000	5.2	6.5
3200	5.6	6.7
3400	5.8	6.9
3600	6.0	7.0

Figure 6-12. *Brake horsepower increases with engine speed. Note that bhp at 3600 rpm is about twice that developed at 1600 rpm. (Wisconsin Motors Corp.)*

Measuring engine brake horsepower

There are two common methods of measuring brake horsepower. It can be measured by using a Prony brake or an engine dynamometer.

The *Prony brake* is a friction device that grips an engine-driven flywheel and transfers the force to a measuring scale. See **Figure 6-13.** One end of the Prony brake pressure arm rests on the scale and the other wraps around a spinning flywheel driven by the engine under test. A clamp is used to change the frictional grip on the spinning flywheel.

To check brake horsepower, the engine under test is operated with the throttle wide open. Then engine speed is reduced to a specific number of revolutions per minute by tightening the pressure arm on the flywheel. At exactly the right speed, the arm pressure on the scale is read. By using the scale reading (W) from the Prony brake, the fly-wheel rpm (R), and the distance in feet from the center of the flywheel to the arm support (L), brake horsepower can be computed. The formula used to determine brake horsepower on the Prony brake is as follows:

$$bhp = \frac{2\pi \times R \times L \times W}{33,000}$$

$$or \quad bhp = \frac{R \times L \times W}{5252}$$

where:
R = Engine rpm or speed.
L = Length from center of flywheel to the point where beam presses on scale in feet.
W = Weight as registered on scale in pounds.
1 hp = 550 ft-lb/sec

Since engine rpm is on a per-minute basis, it is necessary to multiply the 550 by 60 (60 sec per min), giving the figure 33,000.

As with the Prony brake, the *dynamometer* loads the engine and transfers the loading to a measuring device. Instead of using a dry friction loading technique (clamping pressure arm to a spinning wheel), the dynamometer utilizes either hydraulic or electric loading. Several different types are illustrated in **Figure 6-14.**

In **Figure 6-14A,** the engine drives an electric generator that is attached to a spring scale. When an electrical load is placed in the circuit, the generator housing (an enclosure holding the moving parts) attempts to spin, applying a force to the scale.

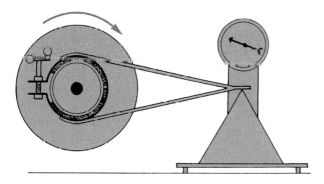

Figure 6-13. *When using a Prony brake, one end of the pressure arm surrounds a spinning flywheel driven by the engine. By tightening the friction device, torque is transmitted to and measured at the scale.*

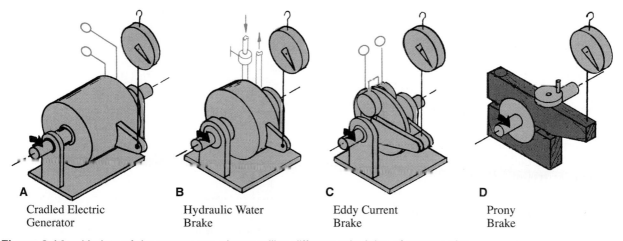

A	B	C	D
Cradled Electric Generator	Hydraulic Water Brake	Eddy Current Brake	Prony Brake

Figure 6-14. *Variety of dynamometers shown utilize different principles of construction.* (Go-Power Corp.)

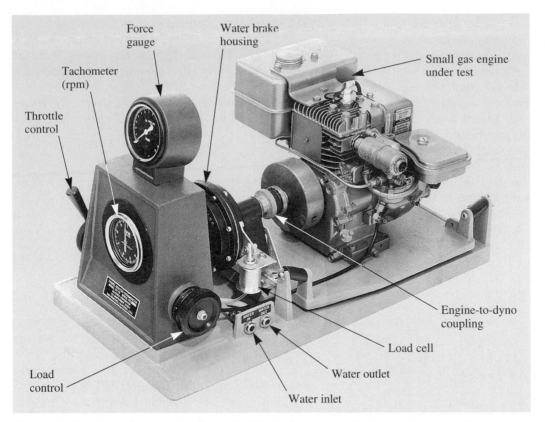

Figure 6-15. *A small gas engine is readied for testing on a water brake dynamometer. (Go-Power Corp.)*

In **Figure 6-14B,** a hydraulic water brake is attached to the scale. (Hydraulic means moved or operated by a liquid.)

The engine is loaded by admitting more and more water into the brake, causing the housing to try to rotate. This exerts a force on the scale. **Figure 6-14C** shows an Eddy Current brake, and **Figure 6-14D** illustrates the Prony brake.

A hydraulic load cell is used with one type of water brake dynamometer. When the housing tries to rotate, it creates a force on the hydraulic load cell. This cell is a type of piston, which pushes against the water. The *push* is transmitted through tubing to the gauge, where it is read in pounds. **Figure 6-15** shows a small gas engine setup for testing on such a dynamometer. Note the load cell. See how it is attached to the housing. **Figure 6-16** pictures a dynamometer that was built to test model airplane engines.

Indicated horsepower

Indicated horsepower (ihp) measures the power developed by the burning fuel mixture

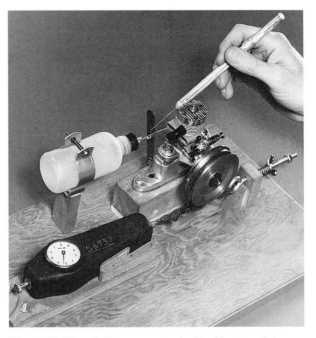

Figure 6-16. *A dynamometer built with a precision spring scale for testing a model airplane engine. Notice vibration tachometer being used to record rpm.*

inside the cylinder. To measure *ihp*, you must determine the pressure inside the cylinder during the intake, compression, power, and exhaust strokes. A special measuring tool is used to provide constant checking of cylinder pressure. This pressure information is placed on an indicator graph, like the one in **Figure 6-17.**

At this point, the ***mean effective pressure (mep)*** must be determined. To do this, we subtract the average pressure during the intake, compression, and exhaust strokes from the average pressure developed during the power stroke. The mean effective pressure changes according to engine type and design. After finding the *mep*, the following formula is used to determine indicated horsepower:

$$\text{Indicated horsepower (ihp)} = \frac{\text{PLANK}}{33,000}$$

where:

P = mep in pounds per square inch.
L = length of piston stroke in feet.
A = cylinder area in square inches.
N = power strokes per minute: rpm/2 (four-cycle engine).
K = number of cylinders.

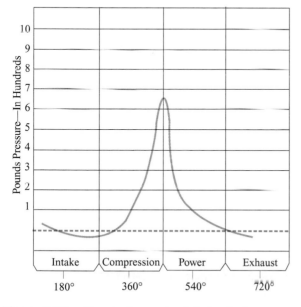

Figure 6-17. *Graph shows simulated cylinder pressure developed in cylinder of a specific four-cycle engine. Atmospheric pressure is shown by dotted line. Graph makes it possible to establish mean effective pressure (mep).*

Frictional horsepower

Frictional horsepower (fhp) represents that part of the potential or indicated horsepower lost because of the drag of engine parts rubbing together. See **Figure 6-18.**

Despite smooth contact surfaces and proper lubrication, a certain amount of friction (resistance to movement between two objects that are rubbing together) is always present and represents a sizable horsepower loss. Actual loss will vary with engine design and use, but will generally run about 10%. Friction loss is not always the same. It increases with engine speed.

Frictional horsepower is determined by subtracting brake horsepower from indicated horsepower or by the following formula:

$$\text{fhp} = \text{ihp} - \text{bhp}$$

Rated horsepower

An engine used under a brake horsepower load that is as great as the engine's highest brake horsepower rating will overheat. Excessive pressure on the bearings (loading) will seriously shorten the engine's service life. In some cases, complete engine failure can occur in a very short period of time.

As a general rule, never load an engine to more than 80% of its highest brake horsepower rating. For example, if a job requires a horsepower loading of 8 hp, you would use an engine with at least a 10 hp rating. Then, the load would be no more than 80% of the engine's maximum hp.

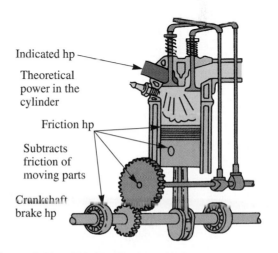

Figure 6-18. *Frictional horsepower is determined by subtracting crankshaft brake horsepower from indicated horsepower. (Deere & Co.)*

An engine's *rated horsepower* generally will be 80% of its maximum brake horsepower. Note in **Figure 6-19** how the rated horsepower (recommended maximum operating bhp) is less than engine maximum bhp.

Corrected horsepower

Standard brake horsepower ratings are based on engine test conditions with the air dry, temperature at 60°F, and a barometric pressure of 29.92 inches of mercury, or Hg (sea level). Horsepower, however, can be greatly affected by changes in atmospheric pressure, temperature, and humidity (amount of moisture in the air).

Corrected horsepower is a *guess* at horsepower of a given engine under specific operating conditions that are not the same as those present during actual dynamometer testing. Facts to consider are:

- For each 1000′ of elevation above sea level, horsepower will drop around 3 1/2%.
- For each 1″ drop in barometric pressure, horsepower will drop another 3 1/2%.

- Each 10°F of temperature increase results in a horsepower loss of 1%.
- New engines will develop somewhat less horsepower (due to increased friction) until they have been operated a number of hours.
- An increase of 200°F-400°F in head operating temperature can lower horsepower by 10%.
- Quality of fuel, mechanical conditions, and *state of tune* can also affect horsepower.

When horsepower tests are conducted under conditions varying from standard, corrections must be applied to establish true horsepower.

Correction factor

The correction factor (a factor is a condition which would change an answer) is determined by using the following formula:

$$\text{Correction Factor} = \text{Temperature Correction} \times \text{Pressure Correction} \times \text{Humidity Correction}$$

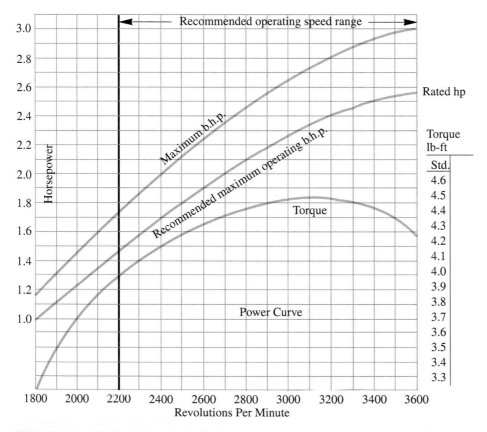

Figure 6-19. *Maximum operating brake horsepower loading is charted for a specific engine. At all speeds,* rated *hp is about 80% of maximum bhp. (Briggs and Stratton Corp.)*

For example, suppose that the dynamometer tests were carried out at a temperature of 90°F with an atmospheric pressure of 28″ Hg and a wet bulb temperature (determines humidity) of 73.5°F. First, see the chart in **Figure 6-20.** Follow dotted line A from top of chart (90°F temperature) down until it crosses the temperature line B. See that by now moving left along the chart, the dotted line shows a temperature correction factor of 1.028.

Now, follow dotted line C up from the 28″ Hg marking at the bottom of the chart until it crosses pressure line D. By moving left at this point, a pressure correction factor of 1.068 is shown.

To find the humidity correction factor, use the chart in **Figure 6-21.** Follow the dotted line up from the 90°F dry bulb temperature mark until it meets the 73.5°F dotted wet bulb temperature line. Move across to the right and note that the humidity correction factor is 1.0084.

Using the correction factor formula as follows:

Correction Factor = Temperature Correction ×
                    Pressure Correction ×
                    Humidity Correction

Correction Factor = 1.028 × 1.068 × 1.0084

Correction Factor = 1.1071

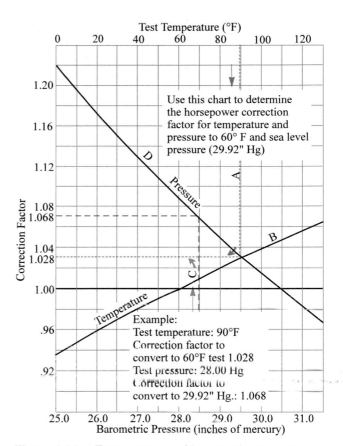

Figure 6-20. *Temperature and barometric pressure correction chart. (Go-Power Corp.)*

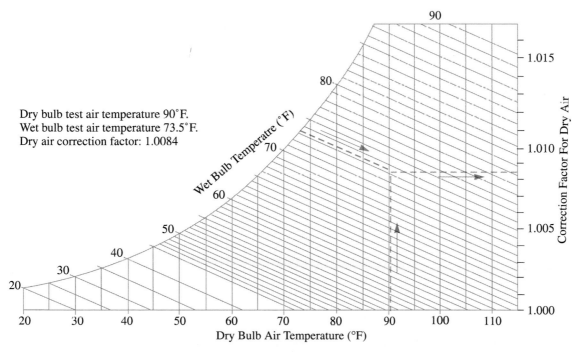

Figure 6-21. *Chart for determining humidity correction factor. (Go-Power Corp.)*

If the dynamometer test had shown 3.15 horsepower, this reading could be changed to standard test conditions by applying the correction factor of 1.1071 as determined above. Thus:

Corrected hp = Correction Factor × Test hp
Corrected hp = 1.1071 × 3.15
Corrected hp = 3.4873 hp

Torque

Torque is a twisting or turning force. Therefore, any reference to engine torque means the turning force developed by the rotating crankshaft.

In order to find torque, we must know the force (in pounds) and the radius (distance, in feet, from the center of the turning shaft to the exact point at which the force is measured). The formula would read as follows:

Torque = Force × Distance (Radius)

or

Torque = Pounds × Feet

or

Torque = lb-ft

Figure 6-22 shows a torque wrench attached to a rotating crankshaft. Imagine that it is attached with a friction device much like the Prony brake pressure arm shown in **Figure 6-13.** This will allow the crankshaft to turn while still applying turning force to the stationary torque wrench.

Suppose a scale placed exactly two feet from the center of the crankshaft indicates a force of 100 lb. By using the formula for determining torque (Torque = lb-ft or Torque = 100 lb × 2′), we find that this engine is developing 200 lb-ft of torque.

If the scale is 3′ from the shaft center and the force is 50 lb, the torque will be 150 lb-ft. When measuring torque, the reading is given in lb-ft. When measuring work, the reading is given in ft-lb.

Torque is not constant

Engine torque, for any engine and set of test conditions, will change according to engine speed. The pressure of the burning air-fuel mixture against the piston is transferred to the crankshaft by the connecting rod. The greater the pressure, the more torque the crankshaft will develop.

The point where gas pressure will be highest is the speed at which the engine takes in the largest volume of air-fuel mixture. This point will vary according to engine design but will always be at a lower speed than that at which the greatest horsepower is reached. Horsepower generally increases to quite a high rpm before it finally begins to drop off. Torque, on the other hand, decreases at a much lower rpm.

As engine speed is increased beyond idle, its torque increases. As it continues to speed up, a point will be reached where the natural restriction to airflow through the carburetor, intake manifold, and valve ports begins to limit the speed at which the air-fuel mixture can enter the cylinder. At this point, the highest torque is developed.

When engine speed goes higher than this point, the intake valve will open but the piston

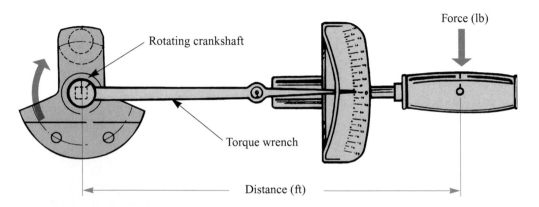

Figure 6-22. *Torque is determined by multiplying turning effort in pounds by the distance from shaft center to point at which force is read. (Dresser Industries Inc.)*

moves far down on the intake stroke before the mixture can get into the cylinder. This cuts down the amount of air-fuel mixture entering the cylinder. As a result, burning pressure is lowered as well as the torque. Beyond this point, torque will decrease as speed increases.

Torque and horsepower

Unlike torque (which drops off when engine revolutions per minute exceed point of maximum volumetric efficiency), horsepower continues to increase until engine speed is very high. Beyond a certain speed, however, horsepower will actually decrease.

Keep in mind that torque measures the twisting force generated by the crankshaft while horsepower measures the engine's ability to perform work. Even though torque may decline at higher speeds, the shaft is turning much faster. Therefore, it is able to perform work at a greater rate.

Figure 6-23 shows the relationship between torque and horsepower curves for one specific engine. Note the arrow indicating the rated horsepower. This is the horsepower at which the engine can be operated continuously without damage.

Volumetric efficiency

How well an engine *breathes,* or draws the air-fuel mixture into the cylinder, is referred to as its *volumetric efficiency.* It is measured by comparing the air-fuel mixture actually drawn in to the amount that could be drawn in if the cylinder were completely filled. See **Figure 6-24.**

Volumetric efficiency changes with speed. At high engine speeds, it can be very low. The reason for this is simple. As engine revolutions increase beyond a certain point, the piston moves down (intake stroke) so rapidly that it travels far down the cylinder before the air-fuel mixture begins to flow into the cylinder. The intake cycle can be complete (piston moving upward on compression stroke) before the cylinder is much more than half full.

Other factors can change the volumetric efficiency, including atmospheric pressure; air temperature; air cleaner, carburetor and intake manifold design; size of intake valve; engine temperature; throttle position; valve timing; and camshaft design.

Efficiency can be increased by using a larger intake valve, altering cam profiles (shapes) or cam

timing, increasing the size of the carburetor air horn, straightening and increasing the diameter of the intake manifold, improving exhaust flow, and/or adding a supercharger.

Figure 6-25 illustrates how volumetric efficiency varies with speed. There is a gradual buildup to a certain rpm, followed by a rapid decline as

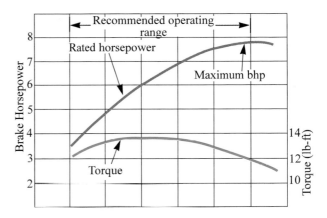

Figure 6-23. *An illustration of the relationship between torque an horsepower for one particular engine. (Clinton Engine Corp.)*

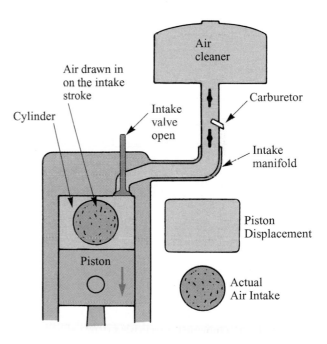

Figure 6-24. *Volumetric efficiency is the measurement of an engine's breathing ability. It compares intake of air-fuel mixture with piston displacement. Note that actual air intake is considerably less than piston displacement. (Deere & Co.)*

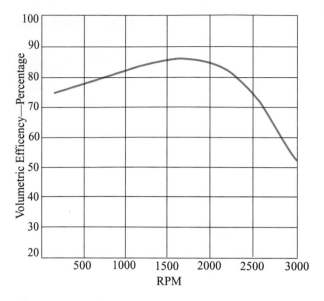

Figure 6-25. *Graph shows close relationship between volumetric efficiency and engine rpm. As engine speed reaches a certain point, efficiency declines rapidly. Torque is also greatest at the point of highest volumetric efficiency.*

speed is increased. Remember that the speed at which top volumetric efficiency is reached will vary with engine design.

Practical efficiency

In theory, each gallon of gasoline contains enough energy to do a certain amount of work. This may be thought of as potential energy. Unfortunately, engines are not efficient enough to use all the potential energy in the fuel. Practical efficiency takes into consideration power losses caused by friction, incomplete burning of the air-fuel mixture, heat loss, etc. *Practical efficiency* is simply an overall measurement of how efficiently an engine uses the fuel supply.

Mechanical efficiency

Mechanical efficiency is the percentage of power developed in the cylinder (indicated horsepower) compared to the power that is actually delivered at the crankshaft (brake horsepower). Brake horsepower is always less than indicated horsepower. The difference is due to friction losses within the engine. Mechanical efficiency runs

about 90%, indicating an internal friction loss of about 10%. The formula for mechanical efficiency is as follows:

$$\text{Mechanical Efficiency} = \frac{\text{Brake Horsepower}}{\text{Indicated Horsepower}} = \frac{\text{bhp}}{\text{ihp}}$$

Thermal efficiency

Thermal efficiency (heat efficiency) indicates how much of the power produced by the burning air-fuel mixture is actually used to drive the piston downward.

Much of the heat developed by the burning gas is lost to such areas as the cooling, exhaust, and lubricating systems. Thermal efficiency will run about 20% to 25%. Keep in mind that the percentages are only *about right* and will vary depending upon engine design and operation. See **Figure 6-26.** The exhaust system siphons off about 35% of the heat. The cooling and lubricating systems combine to absorb a like amount. The rest

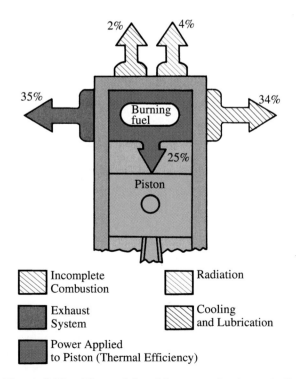

Figure 6-26. *Thermal (heat) losses make thermal efficiency rather poor. Note that it is generally about 25%.*

is lost through radiation and incomplete combustion. Use the following formula for computing brake thermal efficiency:

$$\text{Brake Thermal Efficiency} =$$

$$\frac{\text{Brake Horsepower (bhp} \times 33{,}000)}{778 \text{ Fuel Heat Value} \times \text{Weight of}}$$
$$\text{Fuel Burned per Minute}$$

The 778 in the above formula is Joule's equivalent (a number which is equal to another). The fuel heat (calorific) value is based on the Btu (British thermal unit) per pound.

Summary

Engine bore is the diameter across the top of the cylinder. Stroke is the up or down movement of the piston. Length of stroke is determined by the distance the piston moves from its uppermost position to its lowest position.

Engine displacement refers to the total volume of space increase in the cylinder as the piston moves from the top of its stroke to the bottom of its stroke.

An engine's compression ratio is a measurement of the relationship between total cylinder volume at BDC compared to the volume remaining when the piston is at TDC.

Force is the pushing or pulling of one body on another. The greater the force applied to the piston, the greater the amount of power and work that can be done by the engine. Pressure is a force per given unit of area.

Work is accomplished only when a force is applied through some distance. Power is the rate at which work is performed. The standard unit of power is horsepower.

Energy is the capacity to perform work. Types of energy include potential energy, kinetic energy, mechanical energy, chemical energy, and thermal energy.

Engine horsepower is calculated by dividing the total rate of work (ft-lb/sec) by 550 (ft-lb/sec).

Brake horsepower indicates the actual usable horsepower delivered at the crankshaft. Brake horsepower increases with engine speed.

Indicated horsepower refers to the power developed by the burning fuel mixture inside the cylinder. To measure indicated horsepower, you must determine the pressure inside the cylinder

during the intake, compression, power, and exhaust strokes.

Frictional horsepower represents the part of the indicated horsepower lost because of engine parts rubbing together. Frictional horsepower is determined by subtracting brake horsepower from indicated horsepower.

An engine's rated horsepower is generally 80% of its maximum brake horsepower.

Standard brake horsepower ratings are based on ideal engine test conditions. Horsepower can be greatly affected by changes in atmospheric pressure, temperature, and humidity. Corrected horsepower is an estimation of the horsepower of a given engine under specific operating conditions.

Engine torque refers to the turning force developed by the crankshaft. Engine torque will change according to speed.

Volumetric efficiency is the measurement of how well an engine draws the air-fuel mixture into the cylinder.

Practical efficiency is an overall measurement of how efficiently an engine uses the fuel supply.

Mechanical efficiency is the percentage of power developed in the cylinder (indicated horsepower) compared to the power that is actually delivered at the crankshaft (brake horsepower).

Thermal efficiency indicates how much of the power produced by the burning air-fuel mixture is actually used to drive the piston.

Know These Terms

inertia	horsepower
reciprocating engines	energy
performance	brake horsepower
bore	Prony brake
stroke	dynamometer
crank offset	indicated horsepower
square	mean effective pressure
over square	frictional horsepower
under square	rated horsepower
displacement	corrected horsepower
compression ratio	torque
force	volumetric efficiency
tensile stress	practical efficiency
pressure	mechanical efficiency
work	thermal efficiency
power	

Chapter 6
Review Questions

Answer the following questions on a separate sheet of paper.

1. What type of small gasoline engines are discussed in this text?
2. How do you compute engine displacement?
3. What force acts opposite to the direction of centrifugal force?
4. The force applied opposite to the way a compressive force is applied is called _____ force.
5. If an object is raised 3′ and is then moved 5′ sideways, at the same level, work is being done over a distance of _____.
   a. 8′
   b. 5′
   c. 3′
   d. 2′
6. Give the formula for *work*.
7. If a force of 35 lb is applied to an area of 5 sq in, what is the pressure in psi?
8. The top area of a 2 1/2″ diameter piston would be how many sq in?
9. If 60 lb is lifted 5′ in 6 sec, what amount of power is exerted in ft-lb/sec?
10. What is the definition of *horsepower*? How much is 1 hp?
11. A measure of the horsepower delivered at the engine crankshaft is called _____.
12. A Prony brake with a 12″ arm is applied to an engine flywheel. At 1200 rpm, the scale registers 20 lb. Calculate the horsepower to two decimal places.
13. What is indicated horsepower?
14. When testing an engine on the dynamometer, what are the standard test conditions?

15. What is the engine horsepower reduction for each 1000′ of elevation?
16. Explain the horsepower correction factor.
17. Volumetric efficiency is greatest at _____ speed.
18. Horsepower of an engine is greatest at maximum rpm. True or False?
19. On the average, what percentage of the energy from fuel is used to produce power?
20. An engine's rated horsepower is approximately _____% of its maximum brake horsepower.

Suggested Activities

1. Using the principles studied in this chapter, determine the horsepower required of several individuals to walk up one flight of stairs. The following items will be needed: a stopwatch, tape measure and bath scale.
2. Design and build a Prony brake for a small gasoline engine.
3. Use a dynamometer to develop a graph like the one in **Figure 6-23.** Compare the graph with the manufacturer's graph for the same model engine.
4. On an engine with the head removed, measure the position of the top of the piston (in relation to the top of the block) at TDC and again at BDC. Determine the stroke of the engine.
5. On the same engine, measure the diameter of the piston and determine if the engine is square, oversquare or undersquare.
6. Determine the cubic inch displacement of the engine from the facts learned in activities above.

Fuel and Emission Control Systems

After studying this chapter, you will be able to:
▼ Name various types of fuel that can be used in a small engine and list practical applications for each.
▼ Explain the importance of proper fuel-oil mixture in a two-cycle engine.
▼ Describe the purpose of fuel filters.
▼ Explain fuel pump operation.
▼ Describe the operation of a pressurized fuel system.
▼ Explain the importance of emission control.

Engine Fuels

Small gas engines can be designed to operate efficiently on gasoline, liquefied petroleum gas, natural gas, kerosene, or diesel fuel. See **Figure 7-1.** Gasoline is the most popular small engine fuel. Refer to *Fuel Recommendations* in the *Appendix* section.

In addition to its power potential, gasoline is readily available and easily transported for refueling. See **Figure 7-2.**

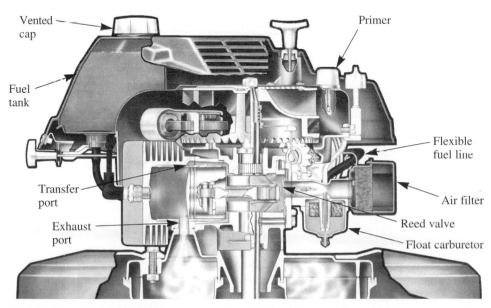

Figure 7-1. *Callouts identify components of a typical fuel system on a two-cycle engine used in a power mower application.*

Vented cap
Primer
Fuel tank
Flexible fuel line
Transfer port
Air filter
Exhaust port
Reed valve
Float carburetor

109

Figure 7-2. *Refueling a four-cycle engine requires use of an approved fuel can and a fresh fill of regular gasoline.*

Gasoline

Most small engine manufacturers specify the use of regular grade, unleaded gasoline with an octane rating around 90. Occasionally, premium fuels are recommended for use in hot climates. This practice may prevent detonation or *dieseling* (after-run). However, a heavier buildup of solid materials in the combustion chamber can be expected from premium fuels because they contain more additives than regular grade fuels.

 Gasoline should be clean, free from moisture, and reasonably *fresh*. After prolonged storage, especially in small quantities, gasoline tends to become *stale*. This is caused by oxidation that forms a sticky, gum-like material. This gum can clog small passageways in the carburetor and cause poor engine performance or hard starting.

Liquefied petroleum gas and natural gas

Liquefied petroleum gas (LPG) may be propane, butane, or a mixture of both. Properly designed fuel systems will allow the use of LPG with no appreciable loss of horsepower, as compared to a similar engine burning gasoline. LPG burns cleanly and leaves few combustion chamber deposits. Because they emit fewer noxious fumes, these engines are often used in warehouses, factories, etc. LPG also has a high anti-knock rating.

Natural gas generally causes a horsepower loss of around 20% when compared with gasoline. Both LPG and natural gas require a different fuel system setup than the conventional type used for supplying gasoline. **Figure 7-3** shows the components of one type of LPG system.

Combustion of LPG

LPG burns slower than gasoline because it has higher ignition temperatures. For this reason, the timing is often advanced on LPG engines.

Due to the higher ignition temperatures, greater voltage at the spark plugs may be needed for LPG combustion. *Colder* plugs or smaller spark plug gaps may solve this problem. Check the engine manual for recommendations.

Less heat is required at the intake manifold to vaporize LPG than gasoline. LPG vaporizes at much lower temperatures than gasoline. It vaporizes at room temperature. This results in less wasted heat and more heat being converted to engine power.

Advantages of LPG
- Cheaper, especially when close to the source (refinery).
- Less oil consumption due to engine wear.
- Reduced maintenance costs—longer engine life between overhauls.
- Smoother power from the slow, even burning of LPG.
- Fewer noxious or poisonous exhaust gases, such as deadly carbon monoxide gas.

Disadvantages of LPG
- Initial equipment costs are high. Bulk fuel storage and carburetion equipment are costly.
- Fewer accessible fuel points (gas stations).
- Harder to start LPG engines in cold weather—0°F (−18°C) or below.

Kerosene and diesel fuels

Some non-diesel type small gas engines can be converted to operate successfully on **kerosene** or fuel oil through the installation of a low compression cylinder head and a special carburetor. These engines are started and operated on gasoline

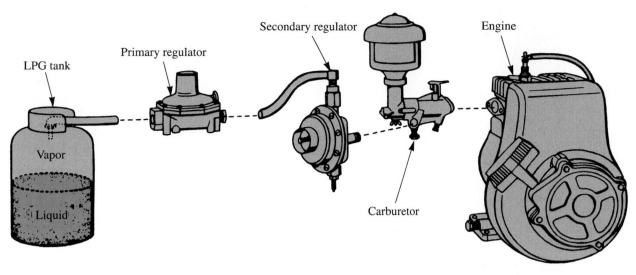

Figure 7-3. *Typical LPG fuel system using vapor withdrawal. (Clinton Engine Corp.)*

until fully warm, then switched over to the kerosene or fuel oil. These fuel installations are generally limited to heavy-duty, industrial engines.

A true diesel engine uses *diesel fuel* injected into the cylinder where it is ignited by the heat of compression. It is not unusual to have compression ratios as high as 20 to 1. Currently, however, diesel application in the small engine field is somewhat limited. It is only practical on applications where continuous use for long periods of time are common.

Two-cycle fuel mixtures

Most two-cycle engines receive lubrication only from the oil mixed with the gasoline. Because of this, it is important that the correct quantity and proper quality of oil is thoroughly mixed with a specific amount of gasoline.

 Always follow the manufacturer's recommended specifications as to the type and quantity of oil to use.

Too little oil can cause the engine to overheat. Overheating, in turn, causes expansion of parts and possible scoring of machined surfaces. Eventually, the pistons may seize (bind, then stick) in the cylinders. Excessive oil, on the other hand, will cause incomplete combustion and rapid buildup of carbon, fouling the spark plugs and adding weight to the pistons.

Tanks, lines, and fittings

Small engine fuel tanks are made of metal or plastic. Some are mounted away from the engine. See **Figure 7-4.** Others are contoured to fit snugly around the engine. See **Figure 7-5.**

The tank filler cap is vented. If the vent becomes clogged, the engine will create enough vacuum in the tank to cause fuel starvation. Most filler caps have baffles and filters. See **Figure 7-6.**

The purpose of a fuel tank filler cap with a screw vent is to prevent fuel evaporation when the

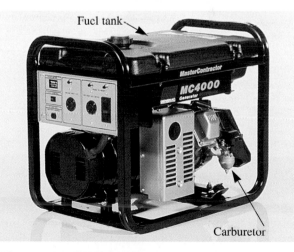

Figure 7-4. *Portable engine-driven generator with the fuel tank mounted on top side opposite carburetor. (Generac Corp.)*

Figure 7-5. *On this lawn mower the plastic fuel tank is contoured to fit snugly around engine.*

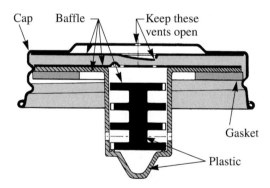

Figure 7-6. *Vented fuel filler caps are baffled to prevent dirt and dust from entering fuel tank. (Clinton Engine Corp.)*

vent is closed. The vent should be opened before starting the engine. A variety of cap styles are shown in **Figure 7-7.**

Fuel tanks used in all terrain vehicles (ATVs) and snowmobiles often have the *fuel pick-up line* inserted from the top of the tank. The pick-up line usually is very flexible and weighted at the bottom, so the line will always be where the fuel is deepest in the tank when the vehicle is at a steep angle.

Figure 7-7. *Fuel filler caps: A—Plastic cap with plastic fabric as a filter and a perforated fiber disc. B—Cap with a threaded screw vent, which will seal tank. C—Standard cap with a single vent hole. D—Three-piece plastic cap showing maze-type baffle with fiber gaskets.*

Fuel filters

Some small engines have a fuel line fitting in the bottom of the tank. A filter screen is placed in the tank fitting or at the end of the pick-up line. See **Figure 7-8.** A top mounted pick-up line with the filter element at the bottom end is shown in **Figure 7-9.**

Other engines have a bottom mounted fuel fitting with a shutoff valve threaded into the tank. See **Figure 7-10.**

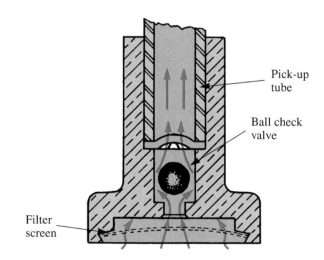

Figure 7-8. *A filter and ball check valve attached to end of a fuel pick-up tube. Ball check valve prevents fuel from draining back into tank when the engine is running. (Clinton Engine Corp.)*

Older small engines have a more elaborate filter incorporated in a glass sediment bowl. See **Figure 7-11.** The gasket, screen, and bowl can be removed for inspection and cleaning. A similar filter is shown in **Figure 7-12.** It is mounted directly on the fuel tank and has a shut-off valve.

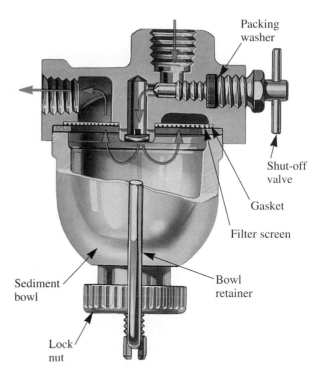

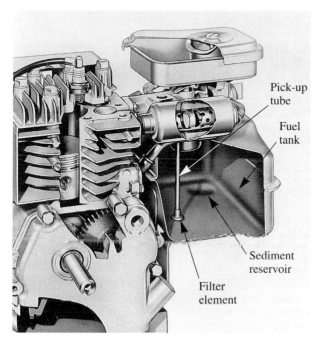

Figure 7-9. *Fuel tank cutaway shows pick-up tube in tank with ball check valve and filter element attached. (Briggs and Stratton Corp.)*

Figure 7-11. *Some small engines have remote fuel strainers located somewhere along fuel line. Glass bowl permits visual inspection without dismantling. (Wisconsin Motors Corp.)*

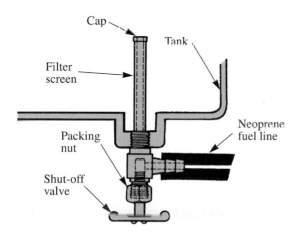

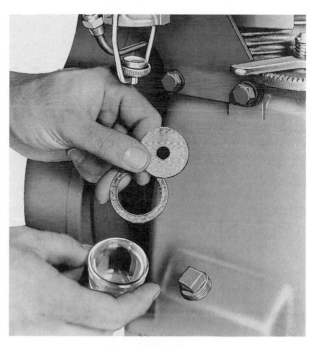

Figure 7-10. *Shutoff valve in tank is necessary to stop loss of fuel whenever a part of fuel system is undergoing work. Filter screen is permanent part of valve. Care during installation is necessary. (Lawn-Boy Power Equipment, Gale Products)*

Figure 7-12. *When moisture or dirt is found in sediment bowl, fuel strainer can easily and quickly be taken apart for cleaning. Fuel shutoff valve is closed before removing bowl.*

Fuel Pumps

Fuel pumps are used on engines that have the fuel tank mounted in such a way that a gravity fuel supply system will not operate. In these applications, the tank and fuel level is lower than the carburetor, or the fuel level may be above the carburetor at times and below the rest of the time. For example, an ATV may have the fuel tank mounted away from the engine and the angle of the vehicle may change constantly.

Fuel pumps provide constant, pressurized fuel flow to the carburetor under changing conditions. They help ensure that the engine can always provide quick acceleration and constant full power.

Mechanical fuel pumps

The mechanical fuel pump used on small engines is basically the same as the type used on automobile engines. It may include a filter as part of the pump design. **Figure 7-13** is a cutaway view of a combination fuel pump and filter system. Trace the arrows to follow the flow of fuel.

Fuel pump operation

The typical mechanical fuel pump shown in **Figure 7-13** operates by means of a diaphragm and atmospheric pressure on the surface of the fuel in the tank. As the engine camshaft revolves, an eccentric (7) actuates the fuel pump rocker arm (6) pivoted at (8). This pulls the pull rod (11) and diaphragm (13) down against spring pressure (12), creating a depression in the pump chamber (15). Fuel drawn from the tank enters the glass bowl from the pump intake (3). After passing through the filter screen (17) and the inlet valve (1), the fuel enters the pump chamber (15).

On the return stroke, pressure of the spring (12) pushes the diaphragm (13) upward, forcing fuel from the chamber (15) through the outlet valve (16) and outlet (14) to the carburetor. When the carburetor bowl is full, the carburetor float will seat the needle valve, preventing any flow from the pump chamber (15). This will hold the diaphragm (13) down against the spring pressure (12). It will remain in this position until the carburetor requires additional fuel and the needle valve opens. The rocker arm (6) operates the connecting link (9) by making contact at (5). This construction allows idling movement of the rocker arm without moving the fuel pump diaphragm. The

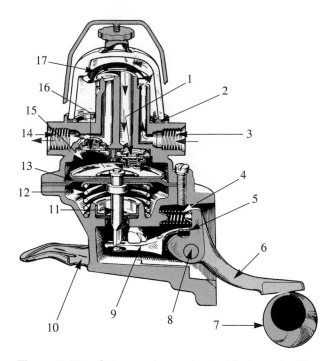

Figure 7-13. *Cutaway of a mechanical fuel pump with a combined fuel strainer on top. Pumps are activated by camshaft (7) as shown. Diaphragm (13) pulsates and forces fuel through check valves (1) and (16).*

spring (4) keeps the rocker arm in constant contact with the eccentric (7) to eliminate noise.

Fuel pump hand primer

The *hand primer* shown as (10) in **Figure 7-13** is used when the carburetor float bowl or pump bowl has become empty. By pulling the hand primer upward, the float bowl will fill and ensure easy starting without prolonged use of the starter.

Because of the special construction of the pump, it is impossible to overprime the carburetor. After several strokes of the hand primer, its handle will become free acting. This indicates that the float bowl is full.

Fuel pump without a filter system

The fuel pump in **Figure 7-14A** is a mechanical pump without a filter chamber. It is mounted on the crankcase, **Figure 7-14B.** A separate filter may be installed somewhere along the fuel line. In some applications, the system may rely solely on the pick-up line filter, a filter in the carburetor, or a combination of both. **Figure 7-15** shows a combination fuel pump and filter system. Note shut off needle valve and primer lever.

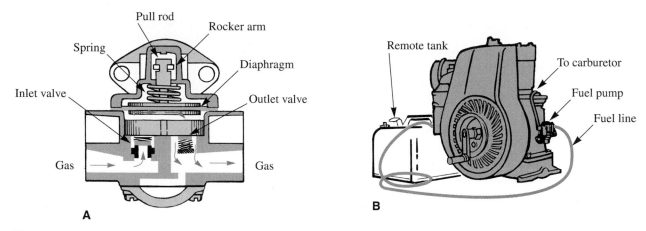

Figure 7-14. *A—Diaphragm fuel pump without a combined filter. B—Pump is cam operated and can pump fuel from portable fuel tanks if necessary. (Clinton Engine Corp.)*

Impulse diaphragm fuel pumps

One type of diaphragm fuel pump sometimes used on small gas engines is activated by the pulsing vacuum in the intake manifold or crankcase. Four-cycle engines use the intake manifold vacuum; two-cycle engines use crankcase vacuum.

A typical impulse diaphragm pump is illustrated in **Figure 7-16.** When vacuum draws the diaphragm upward against spring tension, the inlet check valve opens to allow fuel to flow in. When vacuum is relieved, the spring pushes the diaphragm downward to force fuel through the outlet check valve. This process is repeated as long as the engine is running.

It is common practice today to design the impulse diaphragm fuel pump directly into the carburetor. This design principle will be explained under the section on carburetors.

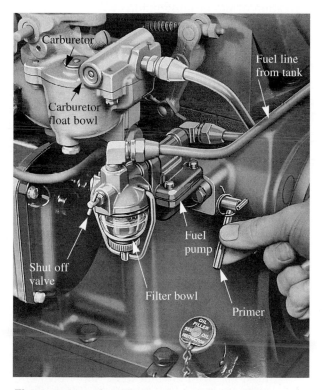

Figure 7-15. *Combination fuel pump and filter mounted in tandem. Fuel pump has manual primer used to initially get fuel to carburetor. (Wisconsin Motors Corp.)*

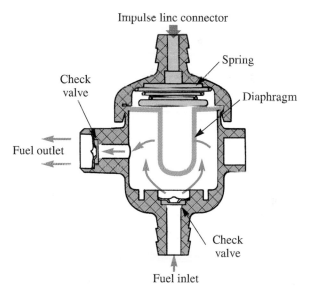

Figure 7-16. *This fuel pump is operated by vacuum pulses transferred from engine crankcase. It can be mounted in any convenient location on engine. (Clinton Engine Corp.)*

Pressurized fuel system

A *pressurized fuel system* is used when the fuel tank is located a considerable distance below the carburetor. See **Figure 7-17.** Outboard engines, for example, often operate from portable tanks resting in the bottom of the boat. The pressurized fuel system shown in **Figure 7-17** operates as follows:

1. The carburetor (1) is connected to the fuel tank (15) through the fuel filter by fuel lines (2 and 5).

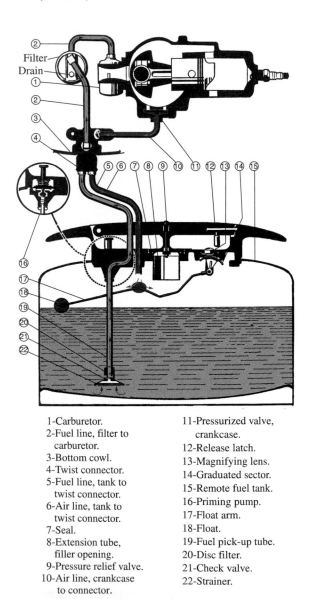

2. Fuel flow from the tank to the carburetor is induced by pressure transmitted from the crankcase to the air space above the fuel line level via the air line (10), which runs from pressurized valve (11) to the twist connector (4) and into the top of the fuel tank (15).
3. For starting, initial flow to the carburetor is induced by the hand-operated priming pump (16).
4. A disc filter (20) is incorporated in the bottom of the fuel pick-up tube (19).
5. The fuel level is indicated by a graduated sector (14), actuated by a float (18) attached to arm (17).
6. The pressure relief valve (9) in the center of the carrying handle permits relieving pressure when necessary.
7. The check valve in the twist connector (4) permits disconnecting the air-fuel line without loss of tank pressure.
8. The tank pressure forces fuel up through the pick-up line (19), through the filter, and into the carburetor (1).
9. The check valve (21) is essential to the operation of the priming pump (16).

Vapor return fuel systems

If the temperature of the air around or inside a carburetor becomes high enough to vaporize the gasoline, pockets of vapor will stop all flow of fuel. When this occurs, the engine will become *vapor locked.* It will not run until the temperature drops low enough for the vapor to condense (return to a liquid).

One of the best ways to prevent vapor lock is to use a carburetor with a vapor return line. In these systems, any vapor that forms is directed back into the fuel tank where the built-up pressure is vented to the atmosphere.

A diagram of a typical vapor return fuel system is shown in **Figure 7-18.** The carburetor in this system has a built-in diaphragm fuel pump. The impulse tube operates the pump.

Small Engine Emissions

Emissions from lawnmowers, snow blowers, chain saws, leaf vacuums, and similar outdoor power equipment are significant sources of pollution. In the United States alone, walk-behind lawnmowers, chain saws, string trimmers, garden

1-Carburetor.
2-Fuel line, filter to carburetor.
3-Bottom cowl.
4-Twist connector.
5-Fuel line, tank to twist connector.
6-Air line, tank to twist connector.
7-Seal.
8-Extension tube, filler opening.
9-Pressure relief valve.
10-Air line, crankcase to connector.
11-Pressurized valve, crankcase.
12-Release latch.
13-Magnifying lens.
14-Graduated sector.
15-Remote fuel tank.
16-Priming pump.
17-Float arm.
18-Float.
19-Fuel pick-up tube.
20-Disc filter.
21-Check valve.
22-Strainer.

Figure 7-17. *Pressurized fuel system uses crankcase pressure transferred to fuel tank. Pressure on surface of fuel forces fuel into engine. (Evinrude Motors)*

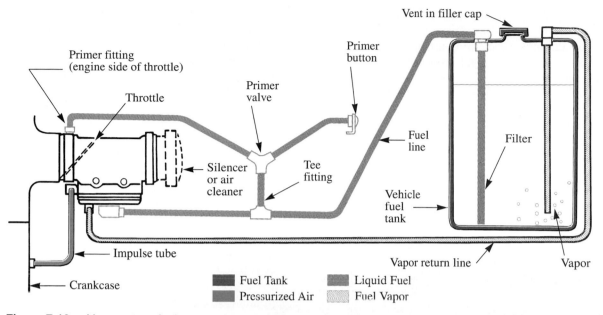

Figure 7-18. *Vapor return fuel system is one of the best methods for preventing a vapor lock. Vapors formed by heat are directed back to fuel tank where they are cooled and condensed to liquid form. (Kohler Co.)*

tractors, Rototillers®, and leaf blowers generate an estimated 6.8 million tons of air pollution (the combined hydrocarbon, oxides of nitrogen, and carbon monoxide contribution) each year. Mowing a lawn for a half-hour with a typical mower can produce as much smog as driving a car 172 miles.

Engine exhaust emissions have come under considerable examination. Today's small engines emit oxides of nitrogen (NO_x) and high levels of carbon monoxide (CO). CO is an odorless, colorless, and poisonous gas. These engines also emit hydrocarbons (HC), which contribute to the formation of ground-level ozone. Ground-level ozone, a component of smog, is a noxious irritant that impairs lung function and inhibits plant growth. Small engines contribute approximately 50% of the national nonroad engine hydrocarbon emissions, representing 5% of the national hydrocarbon inventory. See **Figure 7-19.**

Although exhaust by-products are the major source of harmful small engine emissions, evaporative emissions also contribute to environmental and health problems. Lawn-and-garden equipment refueling losses account for 72% of the evaporative emissions from nonroad engines. Americans spill over 17,000,000 gallons of gasoline every summer while refueling their lawn-and-garden equipment. This is enough gasoline to fill

approximately 24 Olympic-size swimming pools. All this gasoline evaporates into the air, where it contributes to the formation of ground-level ozone, irritating the eyes, damaging the lungs, and aggravating respiratory problems. **Figure 7-20** shows a new type of pouring device designed to eliminate gasoline spillage. The Sure Pour Nozzle® stops pouring automatically when the fuel tank is full. After use, the nozzle seals the container to prevent evaporation of fuel into the atmosphere.

Emission control regulations

In 1970, the U.S. government made a significant effort to combat the steadily increasing level of air pollutants. At this time, the *Federal Clean Air Act* was passed. This act was aimed at ridding the atmosphere of harmful road vehicle emissions.

Urban Summertime Hydrocarbon Nonroad Sources	
Small spark-ignition engines	50%
Recreational boats	30%
Other nonroad engines	20%

Figure 7-19. *Nonroad sources contribute to the national hydrocarbon inventory.*

Since that time, there has been an increased awareness of the harmful emissions generated by small gas engines. The 1990 amendment to the Clean Air Act initiated legal authority to regulate small engine emissions.

Small gas engine manufacturers have dedicated great amounts of time, money, and engineering expertise to make their engines perform more efficiently to meet demanding requirements. The *California Air Resources Board (CARB)* established a public hearing in 1990 to consider regulations regarding the California exhaust emission standards and test procedures for utility and lawn-and-garden equipment engines. This resulted in very detailed and extensive regulations for the testing and monitoring of manufactured engines.

The new regulations first took effect in California in 1995. The restrictions were met by engines manufactured during the 1995 model year. Most engine manufacturers claim to have accomplished this through combinations of better oil control, design of combustion chambers, carburetion, ignition systems, overhead valves, and valve

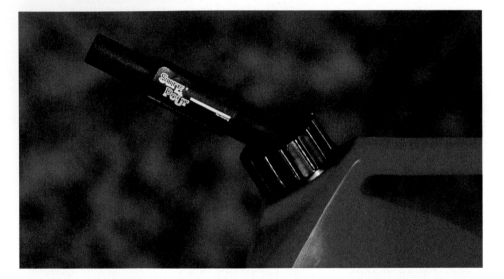

Rest Nozzle on Tank | Push Down to Pour | Stops Automatically | Closes When Removed

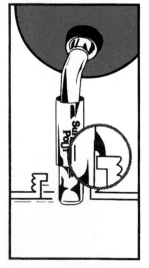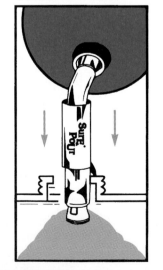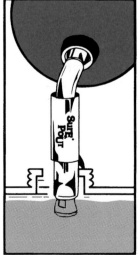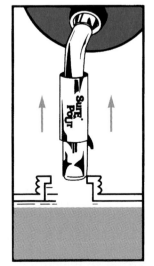

Figure 7-20. *New products, such as this pouring device, are produced to help prevent environmental and health hazards. (VEMCO, Inc.)*

Engines Subject to Proposed Phase 1	
Non-hand-held	*Hand-held*
lawnmowers	trimmers
tillers	edgers
chippers	brush cutters
generator sets	leaf blowers
pumps	leaf vacuums
air compressors	chain saws
aerial lifts	shredders
lawn and garden tractors	augers
wood splitters	
commercial turf	
equipment	
pressure washers	
golf carts	
forklifts	
sweepers	

Figure 7-21. *Several types of utility, and lawn and garden equipment are regulated under the Phase 1: Small Spark-Ignition Engine Rule.*

timing. So far, catalytic converters similar to those used on road vehicles have not been needed. When stricter standards are required, catalytic converters, fuel injection, and overhead valve arrangements may be necessary.

In the mid-1990s, more legislation and amendments were passed that established exhaust emission standards and test procedures for engines used in lawn-and-garden equipment, as well as those used in other utility equipment. Stricter regulations are expected in the near future.

The *Environmental Protection Agency (EPA)* has also become involved in reducing air pollution generated by small gasoline engines. In April, 1994, the EPA proposed the Federal Limits. In July of 1995, the EPA published the Phase 1 Standards for Small Spark-Ignition Engines to regulate air emissions from small engines used in residential and commercial equipment. Phase 2 Standards for Small Spark-Ignition Engines are currently being developed. In addition to pursuing regulatory emission controls, the EPA works with manufacturers, dealers and retailers, environmental and health groups, and the consumer.

With expected growth in the use of small engine–powered equipment, air pollution from these sources will increase over the next twenty years. When complete fleet turnover occurs, the Phase 1 Standards should reduce hydrocarbon and carbon monoxide emissions from small engines by 32% and 14%, respectively.

Phase 1 Standards for Small Spark-Ignition Engines

Controlling exhaust emissions is the first step in preventing pollution from engines used for lawn, garden, and other small engine–driven equipment. The EPA, with input from engine and equipment manufacturers, has developed *Phase 1 Standards for Small Spark-Ignition Engines.*

Phase 1 Standards set emission regulations for hydrocarbons, carbon monoxide, and oxides of nitrogen from certain nonroad spark-ignition engines that have a gross power output at or below 25 hp (19 kW). These standards took effect for most new engines produced in the 1997 model year.

Refer to the Appendix, to find Phase 1 Standards, which were published July 3, 1995.

The engines subject to these standards are typically used to power lawn-and-garden equipment, small farm and construction equipment, commercial turf equipment, and utility equipment. **Figure 7-21** provides a sampling of equipment that is regulated. The emission standards selected were chosen because they have been demonstrated to be achievable and cost effective using currently available technology. Manufacturers can use any method to meet the standards, as long as it does not compromise safety.

The limits on the output emissions are set according to five displacement classes—two for non-hand-held equipment and three for hand-held equipment. See **Figure 7-22.** Non-hand-held limits deal primarily with four-stroke engines; hand-held limits deal with two-stroke engines. Hand-held limits are higher than those for non-hand-held equipment. The EPA has decided that two-stroke lawnmowers can follow less stringent hand-held limits, because at this time it is impossible for a two-stroke engine to meet the non-hand-held limits. Two-stroke lawnmowers must comply with the non-hand-held limits by 2003.

EPA Phase 1 Emission Limits for nonroad engines rated at or below 25 hp					
Class	*Engine Displacement*	*HC + NO$_x$*	*HC*	*CO*	*NO$_x$*
I	under 225cc	16.1	—	519	—
II	225cc and up	13.4	—	519	—
III	hand-held; under 20cc	—	295	805	5.36
IV	hand-held; 20 to 49.9cc	—	241	805	5.36
V	hand-held; 50cc and up	—	161	603	5.36

Figure 7-22. *EPA Phase 1 emission limits for nonroad engines rated at or below 25 hp.*

The goal of the Phase 1 standards is to create low-emissions engines that are both user friendly and environmentally friendly. Low-emissions engines must be set precisely, which results in a finer margin of error when doing repairs. Repairs that put the engine out of compliance are prohibited. Service technicians must be precise in their repair and adjustment of ignition and fuel system components.

Proposed Phase 2 Standards for Small Spark-Ignition Engines

The Environmental Protection Agency is currently proposing *Phase 2 Standards for Small Spark-Ignition Engines.* These standards, which would be phased in between 2001 and 2005, will reduce hydrocarbons plus oxides of nitrogen by an additional 30%.

Phase 2 standards will require that most larger non-hand-held engines be equipped with overhead valves. Smaller engines will require improvements to existing designs. Catalysts will not be required to meet the proposed standards. The cost of meeting the Phase 2 standards is expected to range from $1 to $18 per unit, depending on the size and complexity of the engine. These costs will be offset by increased durability and fuel efficiency.

Phase 2 Standards also institute a program of certification, assembly line testing, and in-use testing to ensure that an engine meets the new emission requirements throughout its active service life.

Role of the consumer

Both residential and commercial consumers must take an active role in preventing pollution from lawn-and-garden equipment. The type of equipment chosen and the way the equipment is used can have an impact on preventing pollution.

New technology is beginning to appear in the marketplace in the form of changes to traditional gasoline powered engines, as well as alternative power sources, such as electricity and solar energy. The full impact of cleaner equipment has not been established, since the Phase 1 standards did not become effective until August 1, 1996 and were not implemented until the 1997 model year.

As the new technology emerges it will be increasingly important for the consumer to follow manufacturer's suggested maintenance procedures. This will result in reduced pollution, longer lasting and better performing engines, and a healthier environment.

In an attempt to involve the consumer, the EPA is beginning a pilot pollution prevention communication strategy. The focus of this program is to educate residential and commercial small engine users about spillage during refueling. This is a problem only the consumer can fix.

Summary

Small gas engines can be designed to operate on gasoline, LPG, natural gas, kerosene, or diesel fuel. Most manufacturers specify the use of unleaded gasoline with an octane rating around 90. Gas should be clean, free from moisture, and reasonably fresh. Two-cycle engines receive lubrication from oil that is mixed with fuel. Always follow the manufacturer's specifications for the type and quantity of oil to use. Small engine fuel tanks are made of metal or plastic. Various types of fuel filters are used in small engines.

Fuel pumps are used on engines that do not have a gravity fed fuel supply system. Fuel pumps provide constant, pressurized fuel flow to the carburetor under changing conditions. Mechanical fuel pumps are usually driven by the camshaft. Diaphragm fuel pumps are activated by the pulsing vacuum in the intake manifold or the crankcase. Pressurized fuel systems are used when fuel tanks are located a considerable distance below the carburetor.

Emissions from outdoor power equipment are significant sources of pollution. Today's small engines emit oxides of nitrogen (NO_x), carbon monoxide (CO), and hydrocarbons (HC). Although exhaust by-products are the major source of harmful small engine emissions, evaporative emissions also contribute to environmental and health problems.

In 1970, the Federal Clean Air Act was passed. This act was aimed at ridding the atmosphere of harmful road vehicle emissions. The 1990 amendment to the Clean Air Act initiated legal authority to regulate small engine emissions.

The California Air Resources Board (CARB) established a public hearing in 1990 to consider regulations regarding the California exhaust emission standards and test procedures for utility and lawn-and-garden equipment engines. This resulted in very detailed and extensive regulations for the testing and monitoring of manufactured engines. The new regulations took effect in 1995.

In the mid-1990s, more legislation was passed that established exhaust emission standards and test procedures for engines used in lawn-and-garden equipment, as well as those used in other utility equipment.

In July of 1995, the EPA published the Phase 1 Standards for Small Spark-Ignition Engines to regulate emissions from small engines used in residential and commercial equipment. Phase 2 Standards, which are currently being developed, will reduce hydrocarbons plus oxides of nitrogen by an additional 30%. Phase 2 Standards are expected to be implemented between 2001 and 2005.

Know These Terms

gasoline
LPG
kerosene
diesel fuel
two-cycle mixture
filler cap vent
fuel pick-up line
fuel filter
fuel pump
hand primer
impulse diaphragm
   fuel pump
pressurized fuel system
vapor return fuel system

vapor lock
Federal Clean Air Act
California Air
   Resources Board
   (CARB)
Environmental
   Protection Agency
   (EPA)
Phase 1 Standards for
   Small Spark-
   Ignition Engines
Phase 2 Standards for
   Small Spark-
   Ignition Engines

Chapter 7 Review Questions

Answer the following questions on a separate sheet of paper.

1. In addition to the power available from gasoline, give three other reasons for its wide acceptance for engine use.
2. Most manufacturers specify regular grade, unleaded gasoline for small engines. True or False?
3. Premium fuels are sometimes recommended for use in hot climates. True or False?
4. A greater build-up of solid materials in the combustion chamber could be expected from using regular grade fuel. Yes or No?
5. Premium fuels contain more additives than regular grade fuels. True or False?
6. LPG is either _____ or _____ or a mixture of both.

7. Natural gas used as a small engine fuel is generally accompanied by a horsepower loss of _____ percent.
8. If excessive oil is mixed with the fuel for two-cycle engine _____.
   a. overheating may result
   b. spark plugs may become overheated
   c. incomplete combustion may occur
   d. seizing will result
9. Filler caps with screw vents are for the purpose of _____.
   a. preventing fuel evaporation when closed
   b. preventing fuel starvation when open
   c. preventing contamination in the tank
   d. All of the above
10. The two types of fuel pumps discussed in this chapter are _____.
    a. atmospheric pressure and gravity vacuum
    b. impulse diaphragm and mechanical
    c. gravity vacuum and mechanical
    d. gravity vacuum and impulse diaphragm
11. When a carburetor has been removed and replaced, the engine will be slow starting because of lack of fuel. This problem can be overcome if the engine has a fuel pump with a(n) _____.
12. One satisfactory fuel system that prevents vapor lock is the _____ _____ system.
13. CARB stands for _____ .
14. What year did the public hearing established by the CARB take place?
15. Newly manufactured engines were required to meet the California emissions standards by the year _____.
16. Stricter requirements needed in the future may cause manufacturers to resort to equipping their engines with _____ converters and fuel _____.
17. What are the three major pollutants from exhaust fumes?
18. Spillage of gasoline during refueling operations produces what is called _____ pollution.

19. Small spark-ignition engines produce _____% of national nonroad emissions.
20. Estimated refueling losses due to spillage equate to about _____ million gallons of gasoline.
21. When did the EPA *Phase 1* rule go into effect?
22. Spillage of gasoline contributes to ground level _____, a component of smog.
23. What two categories do regulated equipment fall into?
24. The additional cost of cleaner running engines meeting *Phase 2 Standards* is estimated to range from _____ to _____ per unit.
25. The new cleaner running engines will require consumers to follow manufacturer's suggested _____ to reduce pollution and retain good performance.

Suggested Activities

1. Collect a variety of tank filler caps. Either cut them in half or disassemble them. Make a display board showing the baffle and the filter system.
2. Make a display board of cutaway drawings of fuel tanks with gravity feed fuel lines and top mounted pick-up lines.
3. Obtain and cut away some old fuel pumps so that internal parts can be seen and worked. Note the function and location of each internal component.
4. Cut away parts of an old fuel filter so that the fuel circuit can be traced.
5. Demonstrate proper methods of engine fueling that will minimize spillage of gasoline. Use a standard fuel can and filler nozzle. List equipment and/or methods of improving the procedure, such as those shown in **Figure 7-21.**

Carburetion

After studying this chapter, you will be able to:
▼ List and explain the principles of carburetion.
▼ Identify the three basic types of carburetors.
▼ Explain float-type carburetor operation.
▼ Explain the operation of the diaphragm-type carburetors.
▼ Define manual throttle controls.
▼ List the basic functions of a governor.
▼ Adjust and maintain common governors.
▼ Describe the purpose of an air cleaner.

Principles of Carburetion

A carburetor's primary purpose is to produce a mixture of fuel and air to operate the engine. This function, in itself, is not difficult. It can be done with a simple mixing valve.

The mixing valve, however, is limited in efficiency. It cannot, for example, provide economical fuel consumption and smooth engine operation over a wide range of speeds. Meeting these performance goals requires a much more complex mechanism. This is the main reason why there are so many styles and designs of carburetors.

Gasoline engines cannot run on *liquid* gasoline. The carburetor must vaporize the fuel and mix it with air in the proper proportion for varying conditions:
- Cold or hot starting.
- Idling.
- Part throttle.
- Acceleration.
- High speed operation.

Basically, air enters the top of the carburetor and is mixed with liquid fuel, which is fed through carburetor passages and sprayed into the airstream. The *air-fuel mixture* that results is forced into the intake manifold by atmospheric pressure and burned in the combustion chamber of the engine.

Figure 8-1 shows how a typical carburetor operates. In this particular small gasoline engine

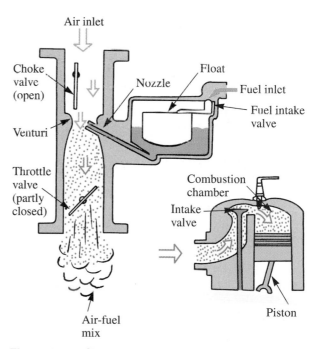

Figure 8-1. *Air entering carburetor mixes with fuel in proper proportion, and mixture flows into combustion chamber. (Deere & Co.)*

application, the engine is at part throttle operation. Note that the choke valve is open, while the throttle valve is partly closed.

Air-fuel mixture

The amount of air needed for combustion is far greater than the amount of fuel required. The usual weight ratio is 15 parts of air to 1 part of fuel. One pound of air would take up a much greater space than one pound of fuel. Therefore, by volume, one cubic foot of gasoline would have to be mixed with 9000 cu ft of air to establish a 15 to 1 weight ratio.

Small gasoline engines use varying air-fuel ratios, depending on engine speed and load. The chart in **Figure 8-2** shows how the mixture changes for various operating conditions.

Carburetor pressure differences

A *carburetor* is a device that is operated by pressure differences. When discussing pressure differences, several terms are commonly used. They are vacuum, atmospheric pressure, and venturi principle.

Vacuum

An *absolute vacuum* is any area completely free of air or atmospheric pressure. This condition is difficult to obtain and is never reached in a small gasoline engine. Therefore, any pressure less than atmospheric pressure generally is referred to as a *vacuum.*

Atmospheric pressure

The pressure produced by the weight of air molecules above the earth is called *atmospheric pressure.* The amount of atmospheric pressure varies with altitude. A person standing on a beach at sea level, for example, would be under a higher vertical column of air than a person standing on a mountaintop. Therefore, the total weight of air molecules would be greater at sea level. See **Figure 8-3.**

Furthermore, any time air molecules are removed from a particular space, a vacuum is created. If conditions permit, this space immediately fills with air under atmospheric pressure.

The effect of atmospheric pressure can be related to small gasoline engines. The downward

movement of the piston creates a partial vacuum in the cylinder. As soon as the intake valve opens or the intake port is uncovered, atmospheric pressure forces air through the carburetor and manifold to fill that vacuum.

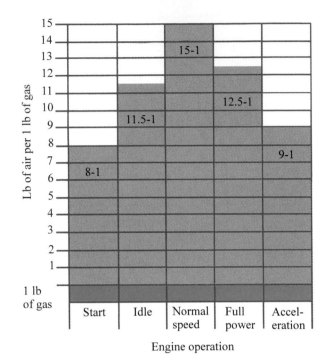

Figure 8-2. *Air-fuel mixture requirements vary depending upon operating conditions. Chart shows approximate air-fuel ratios for various operating conditions.*

Mean Atmospheric Pressure at 68° F (20°C)	
Altitude (ft)	*Atmospheric Pressure (inches Hg.)*
9000	20.92
8000	21.92
7000	22.92
6000	23.92
5000	24.92
4000	25.92
3000	26.92
2000	27.92
1000	28.92
Sea level	29.92

Figure 8-3. *Weight of air exerted on a given object is determined by height and density of a column of air above object. Air is less dense at higher altitudes. For every 1000' above sea level, mercury column pressure is reduced by 1.0".*

Venturi principle

The carburetor creates a partial vacuum of its own by means of a venturi for the purpose of drawing fuel into the airstream. A *venturi* is a restriction in a passage, which causes air to move faster (increase velocity). The gauge shown at the top in **Figure 8-4** indicates no change in velocity. Therefore there is no change in pressure. The area in which the air is moving faster (middle gauge) develops a lower pressure.

Figure 8-5 shows a simple carburetor with fuel being drawn from the float bowl through the main discharge nozzle. This nozzle is located so that its outer end is in the low pressure area of the venturi section. Fuel coming from the discharge nozzle is still in relatively large liquid droplets that do not burn well. To further atomize the fuel, an air bleed passage is built into the air horn. See **Figure 8-6.** A small portion of the air rushing through the carburetor is forced through the air bleed passage to the main discharge tube. This air mixes with the stream of fuel, breaking it into small particles before it reaches the venturi. The small particles of fuel are broken into even finer particles by the air rushing through the venturi.

When the fuel moves into the intake manifold (which is under partial vacuum), the boiling point of the gasoline is lowered. This causes many of the atomized particles to *boil* or *flash* into a vapor. See **Figure 8-7A.** As the partially vaporized fuel moves through the manifold, it is warmed by the heat of the manifold walls. This causes further vaporization. See **Figure 8-7B.** When the mixture enters the combustion chamber, the swirling motion and the sudden increase in temperature due to the compression stroke complete the vaporization of the fuel.

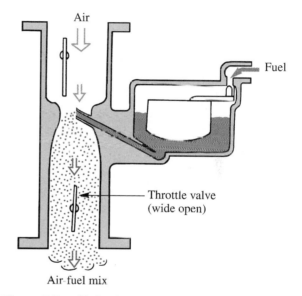

Air

Fuel

Throttle valve (wide open)

Air-fuel mix

Figure 8-5. *Air flowing through venturi has reduced pressure around nozzle. Fuel is drawn up nozzle by vacuum and mixes in airstream. (Deere & Co.)*

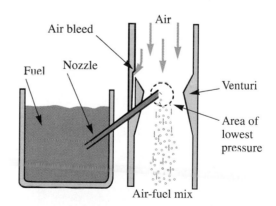

Air bleed

Air

Fuel

Nozzle

Venturi

Area of lowest pressure

Air-fuel mix

Figure 8-6. *To atomize fuel into finer particles, an air bleed is used. Higher pressure of air horn forces some air to enter at a port midway in nozzle, so fuel is partly atomized before leaving nozzle.*

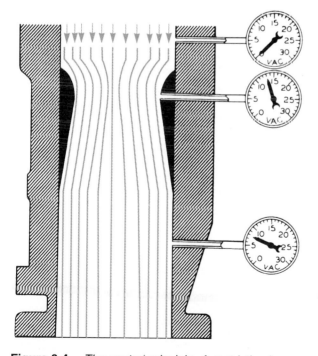

Figure 8-4. *The venturi principle. A restriction in a passage will cause incoming air to increase its velocity, while pressure will be reduced. Reduction in pressure draws fuel into airstream.*

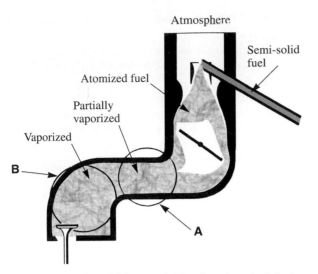

Figure 8-7. *In addition to air bleed and venturi, fuel is further vaporized. A—By vacuum in manifold. B—Engine heat.*

Types of Carburetors

The three basic types of carburetors are named according to the direction that air flows from their outlets to the engine manifold. These types are the natural draft or side draft, the updraft, and the downdraft.

The *natural draft carburetor* is used when there is little space on top of the engine. The air flows horizontally into the manifold. See **Figure 8-8.**

Updraft carburetors are placed low on the engine and use a gravity-fed fuel supply. See **Figure 8-9.** However, the air-fuel mixture must be forced upward into the engine. The air velocity must be high, so small passages must be used in the carburetor and manifold.

Downdraft carburetors operate with lower air velocities and larger passages. See **Figure 8-10.** This is because gravity assists the air-fuel mixture flow to the cylinder. The downdraft carburetor can provide large volumes of fuel when needed for high speed and high power output.

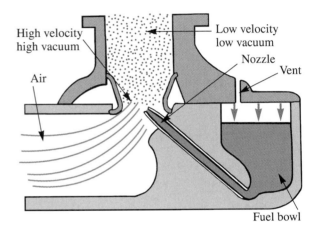

Figure 8-9. *Air flowing through updraft carburetor moves vertically upward into venturi. Passages must be comparatively smaller than those in the downdraft carburetor to increase air velocity so it will carry fuel upward.*

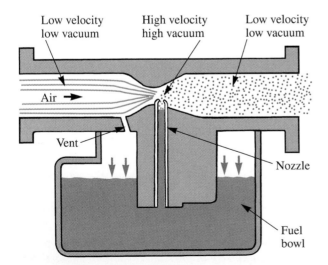

Figure 8-8. *Natural draft carburetor has horizontal air flow through it. (Deere & Co.)*

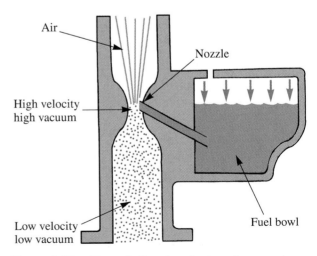

Figure 8-10. *Downdraft carburetor has downward flow of air through venturi. Since it can operate with lower velocities, it has larger passages.*

Float-type carburetors

The carburetor *float* is a small sealed vessel made of brass or plastic. Some floats are made of solid flotation materials that eliminate the possibility of leakage. See **Figure 8-11.**

The purpose of the carburetor float is to maintain a constant level of fuel in the float bowl. The float rises and falls with the fuel level. As fuel is used from the float bowl, the float lowers and unseats a needle valve, which lets fuel enter the bowl. This, in turn, raises the float, seating the needle and shutting off fuel supply to the bowl.

The closed position of the needle valve is illustrated in **Figure 8-12.** The needle valve illustrated in **Figure 8-13** shows valve action in greater detail. The neoprene needle point is soft and seats well in the valve. Also, it is not as likely to wear out as a brass needle point.

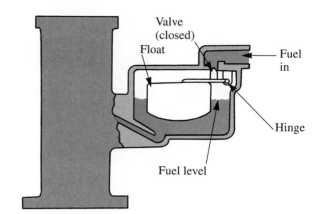

Figure 8-12. *Float in float bowl maintains a constant fuel level. When fuel level rises, float closes needle valve, stopping incoming fuel. When fuel level lowers, float unseats needle and lets more fuel in. (Deere & Co.)*

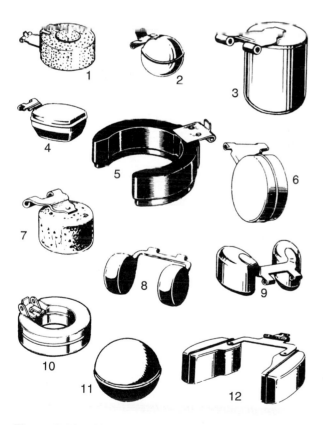

Figure 8-11. *Various float designs. 1—Doughnut-shaped cork. 2—Ball-shaped metal. 3—Cylindrical metal. 4—Rectangular metal. 5—Horseshoe-shaped plastic. 6—Cylindrical metal. 7—Round cork. 8 and 9—Twin-type metal. 10—Doughnut-shaped metal. 11—Ball-shaped metal. 12—Twin-type plastic.*

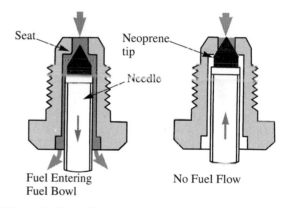

Figure 8-13. *Needle in float bowl opens and closes fuel passage into chamber. Needle is operated by hinged arm of float. (Evinrude Motors)*

Float bowl ventilation

Most carburetors are sealed and balanced to maintain equal air pressure. The air pressure above the fuel in the bowl and the air pressure entering the carburetor are equalized by a vent in the float bowl. See **Figure 8-9.** This vent assures a continuous, free flow of fuel.

Choke system

The carburetor *choke* is a round disc mounted on a shaft located at the intake end of the carburetor. See **Figure 8-14.** When closed, the choke provides a rich air-fuel mixture, which is necessary when starting a cold engine. It allows less air to enter the carburetor. The manifold vacuum draws harder on the fuel nozzle. Therefore, more fuel and less air enters the combustion chamber.

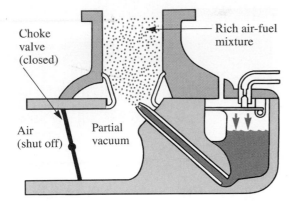

Figure 8-14. *Choke valve is closed and vacuum is high in carburetor. Fuel mixture entering intake manifold is extremely rich.*

Throttle system

Like the choke, the *throttle* is a round disc mounted on a shaft. This valve, however, is located beyond the main fuel nozzle. See **Figure 8-15.**

The main purpose of the throttle valve is to regulate the amount of air-fuel mixture entering the cylinders. It also permits the operator to vary engine speed to suit conditions or to maintain a uniform speed when the load varies.

On many engines, a linkage connects the throttle valve to a governor. The governor, in turn, is connected to a speed control lever. When the speed control lever is set for a given speed, the governor will maintain that speed until the engine reaches its limit of power.

When the load on the engine increases, the governor automatically opens the throttle valve. This permits more air-fuel mixture to enter the engine,

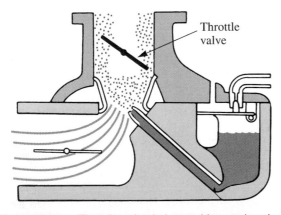

Figure 8-15. *Throttle valve is located beyond main fuel nozzle. Throttle regulates amount of air-fuel mixture entering engine.*

providing increased power to maintain a uniform speed. When the load decreases, the governor closes the throttle to reduce engine power. More details on governors is presented later in this chapter.

Load adjustment

The amount of fuel entering the main discharge nozzle is sometimes regulated by a *load adjusting* needle. See **Figure 8-16.** Many carburetors have a fixed jet or orifice, which is preset to allow proper fuel flow for maximum power and economy. Carburetors equipped with a fixed jet are nonadjustable.

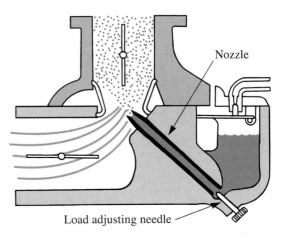

Figure 8-16. *A load adjusting needle, located as shown, regulates amount of fuel entering main nozzle.*

Acceleration system

When the throttle valve is opened quickly for acceleration, a large amount of air is allowed to enter. Unless some method is used to provide additional fuel to maintain a satisfactory air-fuel ratio, the engine will slow down and possibly stop. On larger engines and multi-cylinder engines, a mechanical plunger-type pump is connected to the throttle linkage. When the throttle valve is opened on acceleration, the pump automatically depresses and forces fuel into the carburetor.

Acceleration well

An *acceleration well* is a reservoir of fuel. During idling (when load nozzle is inactive), fuel rises inside the nozzle. The fuel flows through holes in the side of the nozzle and into the acceleration well. See **Figure 8-17.**

When the throttle valve is opened quickly, the stored fuel rushes through the holes in the nozzle

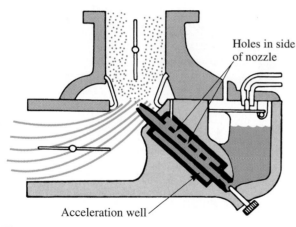

Figure 8-17. *Acceleration well stores fuel for use during rapid acceleration. When fuel has been used from acceleration well, nozzle holes act as air bleeds. (Deere & Co.)*

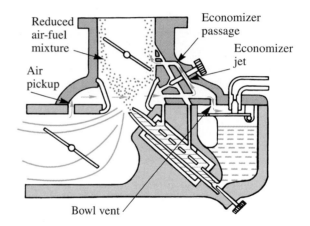

Figure 8-18. *Economizer system creates a reduced pressure in float bowl during part throttle operation, which retards amount of fuel discharged from main nozzle.*

without being metered by the adjusting needle. This fuel combines with the fuel in the nozzle, and the double charge enters the airstream. This provides a much richer air-fuel mixture when there is a sudden need for more power. As the fuel supply decreases in the accelerating well and the holes are uncovered, they become air bleeds for the main nozzle.

Economizer circuit

During operation at part throttle, the full capacity of the main nozzle is not required. To reduce capacity, some carburetors are equipped with economizer circuits. The *economizer circuit* is designed to retard fuel flow to the engine at part throttle.

The basic *economizing* process is the same for all carburetors. **Figure 8-18** shows an updraft carburetor with the bowl vent passage extended to a point near the throttle valve. When the throttle valve is partially open, the economizer passage is on the engine side of the plate. This permits the engine to draw air through the passage, reducing air pressure in the bowl and cutting down on fuel flow from the nozzle.

Idling circuit

During idling operation, the throttle valve is closed. In this condition, the idling system of any type of carburetor supplies just enough air-fuel mixture to keep the engine running. However, actual idling system operation varies in updraft, downdraft, and natural draft carburetors.

The *updraft carburetor* in **Figure 8-19** is in the idling mode of operation. The choke is partially

closed, directing airflow through the pickup. Since the throttle valve is closed, the air moves through a passage outside of the venturi to the idle orifice. At this point, the idle adjusting needle regulates the amount of air mixing with the fuel in the idle orifice. Less air provides a richer mixture, more air produces a leaner mixture.

At slow idle, the throttle valve is closed. Only the primary orifice is exposed to allow fuel to

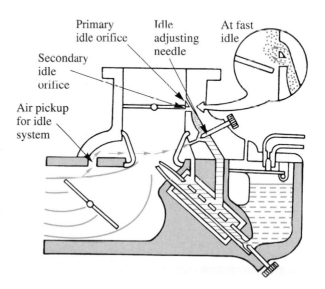

Figure 8-19. *During idling, some incoming air is directed through a passage around venturi. This air mixes with fuel and is drawn out primary and secondary idle orifices. Throttle valve is closed for idle and slightly opened for fast idle.*

enter into the manifold. At fast idle, the throttle valve opens slightly to expose both primary and secondary orifices. Remember, the speed and power of the engine is directly related to the amount of air-fuel mixture allowed to enter the cylinder. Note that at idling speed, the main discharge nozzle is inoperative due to lack of airflow through the venturi.

The *downdraft carburetor* in **Figure 8-20** is in the idling mode. The air bleed is located above the venturi and serves both the idling ports and main discharge nozzle.

Main discharge nozzle is not shown in **Figure 8-20** for purpose of clarity. It would be located as shown in **Figure 8-10.** The idle adjustment screw in this carburetor regulates flow of air-fuel mixture.

The *natural draft carburetor* in **Figure 8-21** is in the idling mode. The throttle valve is closed, and the engine is running from the primary idle discharge hole. The choke valve is wide open. The engine is idling.

Part-throttle, full-throttle sequence

Beyond idling speed, the carburetor has other circuits for part-throttle and full-throttle operation. In **Figure 8-22,** the throttle valve in this natural draft carburetor is partly open. The primary and

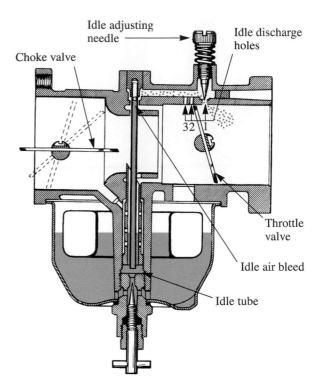

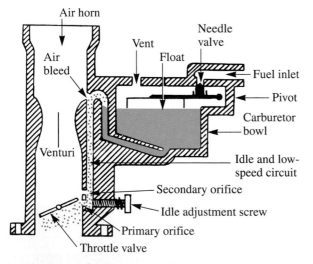

Figure 8-20. *In downdraft carburetor, incoming air enters in air bleed above venturi and travels with fuel to idle orifice. Carburetor is in idling state, since throttle valve has uncovered primary orifice only. (Deere & Co.)*

Figure 8-21. *Idling. Throttle valve is closed, and engine is operating from primary idle orifice. (Zenith Div., Bendix Corp.)*

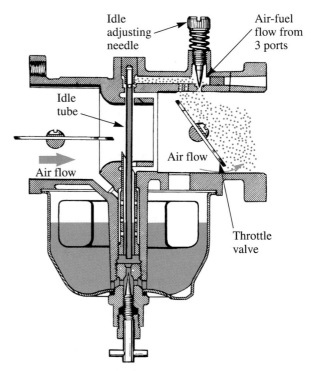

Figure 8-22. *Part throttle. Engine is running from primary and secondary orifices.*

secondary discharge holes are open, allowing more air-fuel mixture to enter. The engine is running at part throttle.

See **Figure 8-23** for full-throttle mode of operation. The throttle is wide open, and the maximum amount of air is flowing through the venturi. The main discharge nozzle is operating because of high vacuum in the nozzle area. The maximum air-fuel mixture is entering the cylinders, and the engine is developing full speed and power.

Figure 8-24 shows an exploded view of the natural draft carburetor shown in **Figure 8-21** through **Figure 8-23**.

Primer

Some carburetors are equipped with primers. The *primer* is a hand-operated plunger, which, when depressed, forces additional fuel through the main nozzle prior to starting a cold engine.

In operation, the primer pumps air pressure into the float bowl, forcing fuel up the nozzle. A primer mounted on a float-type carburetor is shown in **Figure 8-25.**

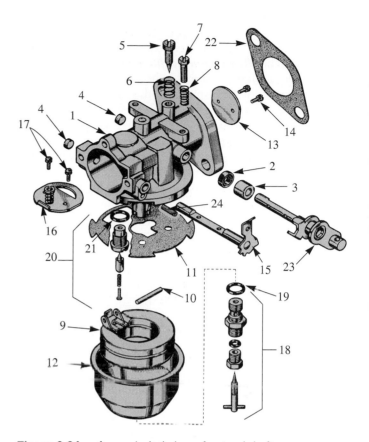

Figure 8-24. *An exploded view of natural draft carburetor: 1—Throttle body. 2—Seal. 3—Retainer. 4—Cup rings. 5—Idle adjustment needle. 6—Spring. 7—Throttle stop screw. 8—Spring. 9—Float and hinge assembly. 10—Float pin. 11—Gasket. 12—Fuel bowl. 13—Throttle valve. 14—Screw. 15—Lever and shaft assemble for choke. 16—Choke valve. 17—Screw. 18—Main jet and adjustment assembly. 19—Washer. 20—Fuel valve and seat assembly. 21—Gasket. 22—Flange gasket. 23—Throttle shaft and lever assembly. 24—Spring. (Zenith Div., Bendix Corp.)*

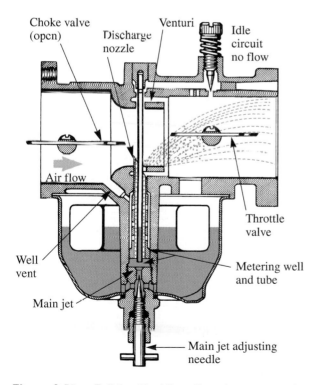

Figure 8-23. *Full throttle. Idle orifices have stopped feeding fuel due to reduced vacuum in that part of carburetor. A full flow of fuel is being drawn from main nozzle.*

Figure 8-25. *A primer plunger mounted on the carburetor.*

Diaphragm-type carburetors

The **diaphragm carburetor** does not have a float system. Instead, the difference between atmospheric pressure and the vacuum created in the engine pulsates a flexible diaphragm. A diaphragm, control needle, and needle seat are shown in **Figure 8-26.** The diaphragm draws fuel into a chamber of the carburetor from which it is readily drawn into the venturi.

The carburetor shown in **Figure 8-27A** is a diaphragm-type, natural draft carburetor. Views **B** and **C** illustrate the operating system of the carburetor.

In **Figure 8-27B,** vacuum created in the manifold draws fuel from the upper chamber through the check valve into the venturi. Then, reduced pressure in the upper chamber allows atmospheric pressure to lift the diaphragm, compressing the inlet tension spring. Finally, movement of the diaphragm opens the fuel valve, permitting fuel to flow into the upper chamber. Remember, this action takes place on the intake stroke of the piston.

In **Figure 8-27C,** manifold pressure increases to equal atmospheric pressure when the piston rises on the compression stroke. Since there is no difference in pressure between the upper chamber and the lower chamber, the inlet tension spring closes the fuel valve and returns the diaphragm to a neutral position. The check valve closes immediately when the pressure is equalized. Therefore, the retracting diaphragm draws fuel into the upper chamber before the fuel valve closes completely.

The pulsation of the diaphragm takes place on every intake and compression stroke, regardless of the number of engine cylinders. On four-cycle engines, fuel is drawn into the cylinder on the downstroke of the piston. On two-cycle engines, fuel is drawn into the crankcase during the upstroke of the piston.

In some applications, the diaphragm spring is adjustable (adjustment screw not shown) to balance the force of the inlet tension spring.

Diaphragm carburetor operation

A study of the various circuits of a typical diaphragm-type carburetor will help clarify operating principles. The carburetor shown in **Figure 8-28** is in starting condition with the choke valve closed. Follow the arrows that indicate direction of fuel flow.

Fuel is drawn from the idle discharge ports and main nozzle because manifold vacuum is high. The carburetor diaphragm is drawn upward during the intake stroke of the engine piston, unseating the fuel inlet needle to allow fuel to flow. Note that the natural draft carburetor in

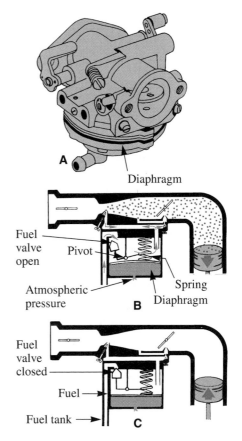

Figure 8-27. A—A diaphragm-type, natural draft carburetor. B—Diaphragm is lifted by manifold vacuum while fuel is being drawn from jets. C—When vacuum is reduced, diaphragm returns to normal, drawing new fuel into upper fuel chamber.

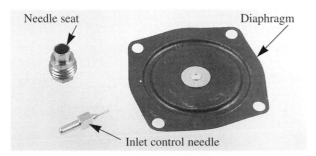

Figure 8-26. A diaphragm, control needle, and needle seat used in a diaphragm carburetor system. (Deere & Co.)

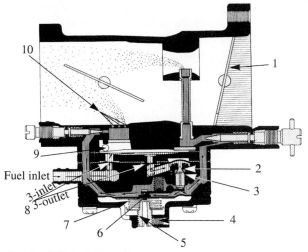

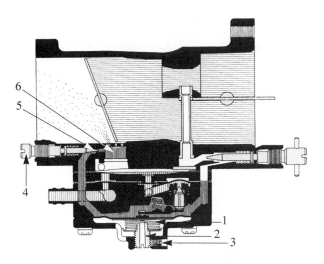

Starting (Choke) Operation

1-Choke valve.
2-Inlet control valve.
3-Valve seat.
4-Lock screw.
5-Adjustment screw

6-Inlet control lever.
7-Diaphragm spring.
8-Check valve.
9-Impulse channel.
10-Idle discharge port.

Figure 8-28. *With choke plate closed, a very strong vacuum is formed in air horn. A large quantity of gasoline is sucked out of idle jets and main nozzle. A rich mixture results, which can support cold engine operation.*

Idling Operation

1-Diaphragm spring.
2-Adjustment screw.
3-Lock screw.
4-Idle mixture screw.

5-Idle mixture screw orifice.
6-Idle fuel supply channel.

Figure 8-29. *During idling or very slow speed operation, only the primary orifice (passage) is feeding fuel to engine. Air velocity is not high enough to draw fuel out of the high speed circuit. Remember, this circuit controls fuel mixture when an engine is idling.*

Figure 8-30 also incorporates a fuel pump diaphragm in its body.

Figure 8-29 shows idling operation with the choke valve open and the throttle valve closed. Vacuum is in effect on the engine side of the throttle valve. Since only one idle discharge port is exposed, a small quantity of fuel is being used, and the engine runs slowly.

Figure 8-30 shows the throttle partially open for intermediate speed. Airflow through primary venturi is still not great enough to draw fuel up the main nozzle. Three idle discharge ports are feeding fuel for medium speed. These extra idle discharge ports are termed off-idle ports. They must supply more fuel than the single idle port, yet not as much as the main discharge port. The intermediate circuit must provide fuel for transition from idle to high-speed operation.

Figure 8-31 illustrates high-speed operation with maximum air and fuel flowing through the carburetor. All idle ports and the main nozzle are feeding fuel. Notice that the choke and throttle valves are open.

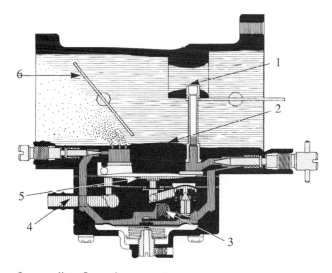

Intermediate Operation

1-Primary venturi.
2-Secondary venturi.
3-Inlet tension spring.

4-Fuel inlet.
5-Fuel pump diaphragm.
6-Throttle valve.

Figure 8-30. *Fuel feeding from idling discharge ports provides intermediate speed operation. (Rupp Industries, Inc.)*

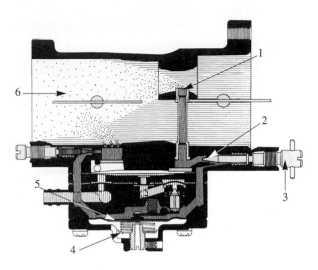

High-Speed Operation

1-Main fuel discharge port.
2-High-speed mixture
  screw orifice.
3-High-speed mixture screw.
4-Spring seat.
5-Main diaphragm.
6-Maximum airflow.

Figure 8-31. *During high-speed operation, fuel flow from main nozzle and idle jets combines with maximum airflow.*

Manual throttle controls

A basic *manual throttle control* consists of either mechanical linkage or a flexible cable. One end of the control is attached to the throttle shaft lever. The other end is connected to a lever, slide, or dial that is operated manually to open and close the throttle valve.

The manual throttle can be used as the sole control for positioning the throttle valve. Typical applications of this type of throttle are chain saws, motorcycles, snowmobiles, and outboard engines. In some installations, the manual control is used in conjunction with a governor. This setup permits governed speed to be changed when desired.

Figure 8-32 shows a manual remote control using a flexible cable to transmit motion from the speed control lever to the throttle valve. In this case: A—Motion changes tension on the governor spring. B—Tension on governor spring changes the throttle valve position.

Figure 8-33 shows a throttle control that varies governor spring tension, positions the throttle valve, and actuates the choke for starting. In **Figure 8-33A,** the control knob is turned to *Start*,

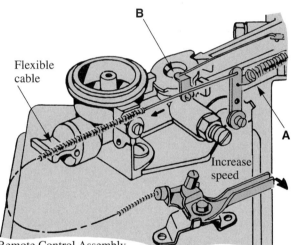

Remote Control Assembly

Figure 8-32. *This manual throttle control uses a flexible cable to transmit motion from hand lever to governor spring lever. (Briggs and Stratton Corp.)*

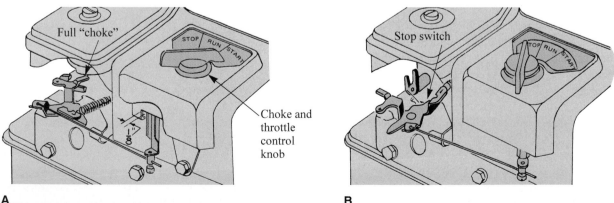

A **B**

Figure 8-33. *Combined manual throttle and choke control. A—Control knob turned to* Start *activates choke. B—Knob turned to* Stop *closes stop switch.*

which rotates the choke valve shaft to the choked position. When the engine is started, the control knob is turned to *Run*.

To stop the engine, the knob is turned to *Stop*, as shown in **Figure 8-33B.** The stop switch grounds the ignition system, cutting off the flow of electricity to the engine.

 This switch is illustrated in detail in Chapter 9 of this text.

Governor throttle controls

In many small gasoline engine applications (lawn mowers, generators, and garden tractors), the load on the engine can change instantly. The change in load would require constant throttle changes on the part of the operator. Instead, *governors* are used to provide a smooth, constant speed, regardless of engine loading.

What an engine governor does

Governors can be designed to serve three basic functions:

- Maintain a speed selected by operator that is within range of governor.
- Prevent overspeeding that may cause engine damage.
- Limit both high and low speeds.

In **Figure 8-34,** observe how tractor speed varies without a governor, but stays constant with a governor.

Small engine governors are generally used to maintain a fixed speed not readily adjustable by the operator or to maintain a speed selected by means of a throttle control lever. In either case, the governor protects against overspeeding. If the load is removed, the governor immediately closes the throttle. If the engine load is increased, the throttle will be opened to prevent engine speed from being reduced.

For example, a lawn mower normally has a governor. When mowing through a large clump of grass, engine load increases suddenly. This tends to reduce engine speed. The governor reacts by opening the carburetor throttle valve. Engine power output increases to maintain cutting blade speed. When the mower is pushed over a sidewalk (no grass or engine load), engine speed tends to go up. The governor reacts by closing the carburetor

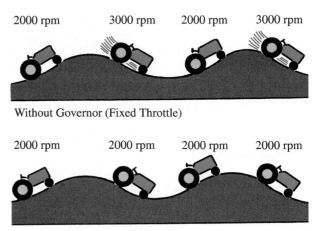

Without Governor (Fixed Throttle)

With Governor (Keeps Same Speed)

Figure 8-34. *Notice what the governor does for the engine. (Deere & Co.)*

throttle valve. This limits maximum cutting blade speed. As a result, mower engine and cutting blade speeds stay relatively constant.

Types of governors

There are several types of engine governors: *air vane* (also called *pneumatic*), *centrifugal* (also called *mechanical* and shown in **Figure 8-35**), and *vacuum*. Most modern governors are air vane types, which adjust fuel intake according to engine demands. See **Figure 8-36.** However, centrifugal governors are also fairly common. Vacuum governors are usually found on farm and industrial engines. Basically, all governors accomplish the same purpose—to protect the engine from overspeeding and to maintain a constant speed, independent of load. However, different speed sensing devices are used.

Air vane governor

The *air vane governor* is operated by the stream of air created by the flywheel cooling fins. The force developed by the airstream is in direct proportion to the speed of the engine.

A lightweight, thin strip of metal called an air vane is placed in the direct path of the airstream. It is pivoted on a pin or shaft set near one end. The vane is connected, with linkage, to the throttle shaft lever. When the engine is running, the airstream pivots the vane and attempts to close the throttle valve. See **Figure 8-36.**

The governor spring is attached to the throttle lever or to linkage from the vane. This spring is

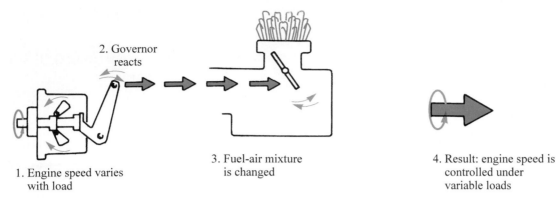

2. Governor reacts

1. Engine speed varies with load

3. Fuel-air mixture is changed

4. Result: engine speed is controlled under variable loads

Figure 8-35. *Centrifugal governor controls engine speed by varying fuel mixture. (Deere & Co.)*

designed to pull the throttle valve to wide open position.

Note how the airflow pivots the vane, causing it to exert rotary pressure on the upper throttle shaft lever while the governor spring tries to pull the throttle valve open. The ratio of pressure

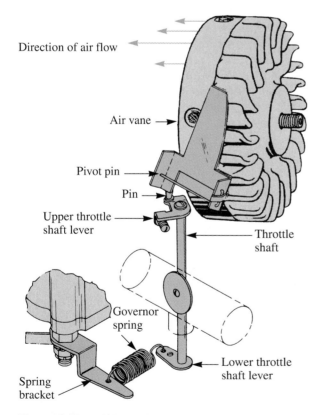

Direction of air flow

Air vane

Pivot pin

Pin

Upper throttle shaft lever

Throttle shaft

Governor spring

Lower throttle shaft lever

Spring bracket

Figure 8-36. *Schematic illustrates operation of an air vane governor. Vane tries to close throttle valve, while governor spring tries to open it. Balance between these two forces determines throttle position. (Tecumseh Products Co.)*

developed by the vane, as opposed to the tension of the governor spring, determines throttle valve position. When the engine is stopped, the airstream ceases and the throttle valve is pulled to wide open position by the governor spring. The governor spring bracket can be moved to vary the amount of tension exerted by the spring. This will alter vane spring pressure balance and establish a new throttle setting.

Governor spring tension is carefully calibrated by the manufacturer. If the spring is stretched or altered in any way, it should be replaced with a new spring designed for that make and model engine. If the linkage is bent, worn, or damaged, it should be straightened, repaired, or replaced. See **Figure 8-37.**

If it is necessary to replace any component on an air vane governor, the top no-load rpm should be checked with an accurate tachometer. The top speed *must not exceed* the maximum recommended rpm for the implement being driven.

In the case of lawn mowers, blade tip speed should not exceed 19,000 feet per minute in a no-load condition. If necessary, change the governor spring or adjust the top speed limit device so the engine stops accelerating at the recommended rpm, which is based on blade length. See **Figure 8-38.**

Since blade tip speed is a function of blade length and engine rpm, longer blades require lower engine speeds. It is suggested that top governed engine speed be adjusted at least 200 rpm lower than the speeds shown in **Figure 8-38** to account for tachometer inaccuracy.

Fixed speed. If the engine is designed to run at only one specific rpm setting, the tension of the governor spring is carefully adjusted until the speed is correct. Then it is left at this setting,

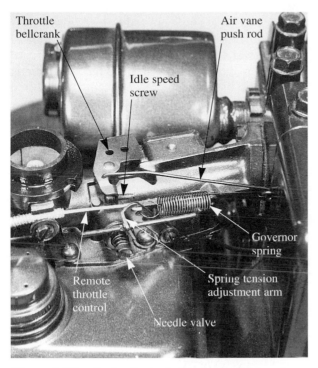

Figure 8-37. *Parts of a typical governor system for actual engine. Note location of idle speed screw and needle valve screw.*

Figure 8-39. Fixed speed engines of this type have a limited range of governor spring adjustment. When the engine is started, the force of the airstream on the vane closes the throttle until that force is equal to spring tension.

Variable speed. It is often desirable to have an engine operate at many different speeds that may be quickly and easily set by the operator. In this case, a variable speed air vane governor is used. See **Figure 8-40.** Engine rpm is changed by pivoting the governor spring bracket. Remember, the throttle control adjusts governor spring tension. The spring is not connected directly to the throttle lever.

Blade Length in Inches (millimeters)	Maximum Rotational (rpm)
18 (460)	4032
19 (485)	3820
20 (510)	3629
21 (535)	3456
22 (560)	3299
23 (585)	3155
24 (610)	3024
25 (635)	2903
26 (660)	2791

Figure 8-38. *Chart lists various lengths of lawn mower blades and maximum rotational speeds, which produce blade tip speeds of 19,000 feet per minute. It is recommended that top speeds be set 200 rpm less than shown.*

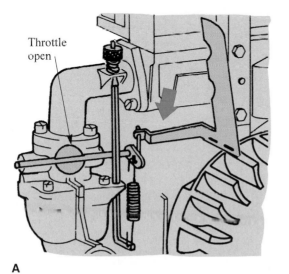

A

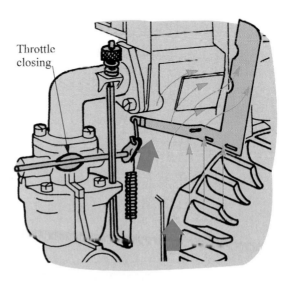

B

Figure 8-39. *Air vane governor. A—Engine stopped. Spring holds throttle open. B—Engine running. Air pressure pivots vane of fixed speed governor and shuts throttle valve until spring pressure and vane pressure are balanced. Knurled nut alters spring tension and adjusts speed. (Briggs and Stratton Corp.)*

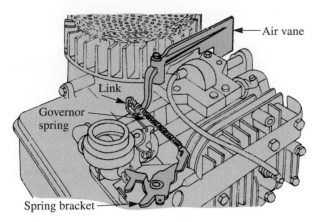

Figure 8-40. *Variable speed air vane governor. To change rpm, operator alters governor spring tension by rotating spring bracket.*

Centrifugal or mechanical governor

Like the air vane governor, the centrifugal (mechanical) governor also controls engine speed. The centrifugal governor, however, utilizes pivoted flyweights that are attached to a revolving shaft or gear driven by the engine. With this setup, governor rpm is always directly proportional to engine rpm.

Figure 8-41 shows how centrifugal governors operate. When the engine is stopped, the heavy ends of the flyweights are held close to the shaft by the governor spring. The throttle valve is held fully open as illustrated in **Figure 8-41A.**

When the engine is started, the governor is rotated. As its speed increases, centrifugal force increases and causes the flyweights to pivot outward. This forces the spool upward, raising the governor lever until spring tension equals the centrifugal force on the weight. This action partially closes the throttle valve shown in **Figure 8-41B.**

If the engine is subjected to a sudden load that reduces rpm, the reduction in speed lessens centrifugal force on the flyweights. The weights move inward, lowering the spool and governor lever. This series of actions opens the throttle valve and lost rpm is regained.

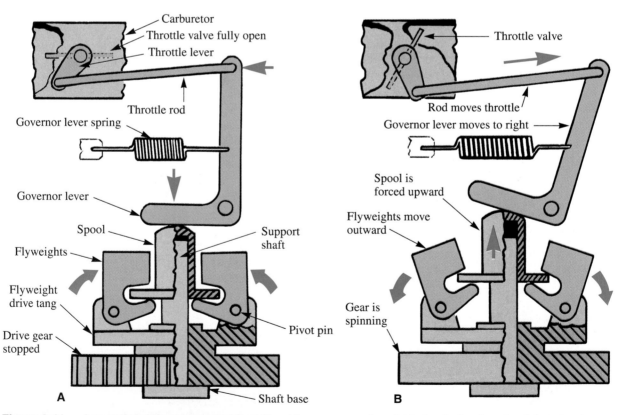

Figure 8-41. *A centrifugal-type governor. Centrifugal force causes flyweights to pivot outward, raising spool. Spool rotates governor lever which closes throttle valve. Balance between centrifugal force and governor spring tension determines throttle valve setting.*

Changing governor speed setting. The centrifugal governor speed setting can be changed by turning a knurled adjusting nut on the end of the tension rod. See **Figure 8-42.** This system is used when the engine is expected to run at a constant speed setting for long periods of time.

A remote, hand-controlled cable that alters spring tension also can be used to set governor speed. See **Figure 8-43.** Movement of the control handle increases or decreases spring tension, which speeds up or slows down the engine. The operator can quickly select any speed within the range of the governor.

Another type of governor speed adjusting arrangement is shown in **Figure 8-44.** Movement of the governor adjusting lever changes spring tension and engine rpm.

Hunting of centrifugal governors. Frequently, when an engine is first started or is working under load, its speed becomes erratic or oscillates. The engine speeds up rapidly; the governor responds and engine speed drops quickly. The governor stops functioning and engine speed again increases. The governor responds and this action is repeated over and over. This condition is known as *hunting*.

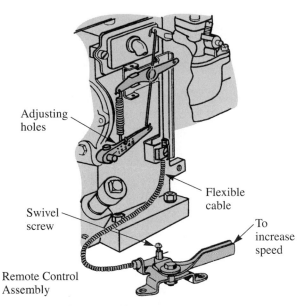

Figure 8-43. *Operator-controlled governor speed setting device allows quick, wide, governed speed changes. (Briggs and Stratton Corp.)*

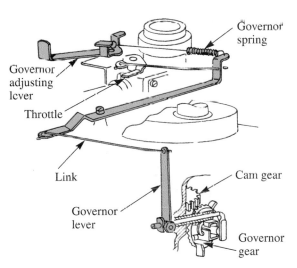

Figure 8-44. *Method of setting governed speed by using an adjusting lever to control governor spring tension.*

Hunting is usually a result of improper carburetor adjustment. Leaning or richening the fuel mixture can often correct the problem. Also, the governor may cause hunting if it is too stiff or binds at some point. It must work freely

Vacuum governors

Farm and industrial engines are often equipped with a vacuum governor for regulating maximum engine speed. The vacuum governor is located

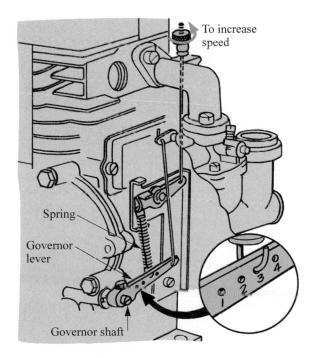

Figure 8-42. *Single speed setting, using knurled nut to provide a limited governor speed range. (Briggs and Stratton Corp.)*

between the carburetor and the intake manifold. See **Figure 8-45.** It senses changes in intake manifold pressure (vacuum). There is no other mechanical connection between the governor and other parts of the engine.

As the engine speed and the suction (vacuum) increase, the governor unit closes the throttle butterfly valve. This causes a decrease in fuel flow and engine speed.

When engine speed and vacuum decrease, the spring opens the throttle valve. This action causes the fuel flow and engine speed to increase. An adjustment of spring tension is used to set the desired speed range.

Governor features

The operating principles of the governor mechanism are quite simple and reliable. Governors provide accuracy and efficiency of operation combined with convenience and comfort for the operator. Two of the most important operating features of engine speed and power output governors are stability and sensitivity. *Stability* is the ability to maintain a desired engine speed without fluctuating. Instability results in hunting or oscillating due to over-correction. Excessive stability results in a dead-beat governor (one that does not correct sufficiently for load changes). *Sensitivity* is the percent of speed change required to produce a corrective movement of the fuel control mechanism. High governor sensitivity will help keep the engine operating at a constant speed.

Air cleaners and air filters

An engine breathes a tremendous quantity of air during its normal service life. If the incoming air is not thoroughly cleaned by passing it through a filtering device, dirt and grit entering the cylinder would cause rapid wear and scoring of machined parts throughout the engine. Engine life, under severe dust conditions, actually could be reduced to minutes. Refer to the *Appendix* for engine failure analysis information.

Three types of air cleaners widely used in small gasoline engines are the *oil-wetted, dry types,* and *dual-element.*

Oil-wetted air cleaner

The oil-wetted air cleaner utilizes a filtering element (crushed aluminum, polyurethane foam, etc.) dampened with engine oil. **Figure 8-46** shows a polyurethane foam oil-wetted element in place and cover open. In operation, the air is drawn directly through the oil-wetted element where the damp material effectively filters out contaminants. This type of element can be reused by rinsing in soapy water, drying, and re-oiling.

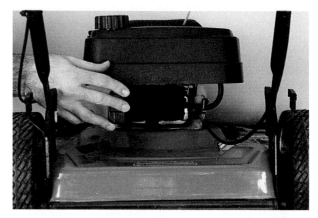

Figure 8-46. *An oil-wetted air cleaner. Polyurethane foam is dampened with oil and contained in a vented case attached to carburetor.*

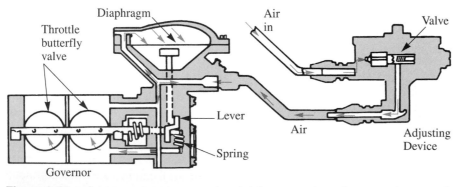

Figure 8-45. *A vacuum governor must maintain a preset maximum engine speed, independent of engine load.*

The air cleaner functions to keep dirt, dust and gritty substances from entering the engine during running. In doing so, the dirt, dust and grit becomes trapped on the external side of the filter and even within the filter material. As the filter pores become filled with debris, less air can get through the filter which has the effect of changing the air-fuel mixture ratio. The engine will run as if the choke is partially closed and a very rich mixture of fuel with less air enters the combustion chamber. The engine begins to run poorly, use more fuel, generate less power, and carbon accumulation on internal parts increases. Also, lubricating oil becomes dirty, diluted, and internal engine wear increases. When the air cleaner gets very dirty the engine may become difficult to start. When the engine does start, black smoke from the rich mixture may show up in the exhaust. Removing an air filter element is usually quite easy so that it can be maintained regularly by the operator. However, this simple task is sometimes neglected until engine damage is already done.

When to service the air cleaner

There are many different designs of air cleaners as illustrated by a few examples in **Figure 8-47.** Because engines are used in many different environments the length of running time before cleaning may vary considerably. However, for average applications such as lawn mowing, cleaning every 25 hours of operating time is recommended, or once a season, whichever comes first. In dustier conditions cleaning should be done more often. Regular inspections of the filter element is the best way to determine whether it should be cleaned or even replaced.

Removing air cleaner foam element

Remove the screw, wing nut, or other fastening device to uncover the air cleaner plastic foam element. **Figure 5-48** shows the filter receptacle with the cover open and foam element removed. The plastic foam filter material can be pulled from the receptacle. Care should be taken not to drop any dirt into the carburetor throat during this procedure. **Figure 5-49** illustrates an air cleaner and its parts with some instructions about assembly.

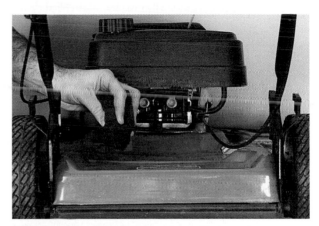

Figure 8-48. *Polyurethane foam filter can be removed, cleaned, and reoiled.*

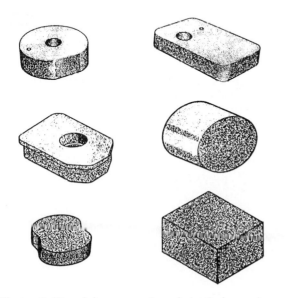

Figure 8-47. *A few examples of plastic foam air cleaners. Many special shapes are available for specific engines.*

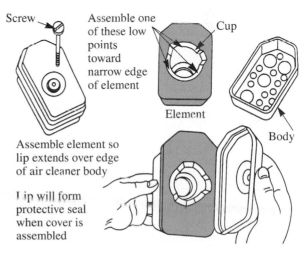

Figure 8-49. *Removal of the air cleaner is relatively easy so servicing can be done regularly. (Briggs and Stratton Corp.)*

Cleaning the plastic foam air cleaner element

To clean the plastic foam air cleaner element, do the following:

1. Wash the plastic foam element in kerosene or liquid detergent and water.

 Kerosene is a flammable liquid and should be used with extreme care and away from heat or flame. Detergent and water are safer and recommended here. See **Figure 8-50A.**

2. Wrap the plastic foam element in dry cloth and squeeze the element dry. Absorbent toweling works well for this procedure. See **Figure 8-50B.**
3. The plastic foam should be saturated with clean engine oil. See **Figure 8-50C**.
4. Squeeze excess oil out of the foam as shown in **Figure 8-50D.**

 Reassemble the air filter unit. Follow any special instructions found in the owners manual for the specific engine and filter.

Dry-Type Air Cleaner

Dry-type air cleaners pass the airstream through treated paper, felt, fiber, or flocked screen. Some filter elements (flocked screen) can be cleaned, but most are designed to be thrown away when they become dirty.

A typical treated paper air cleaner element is shown in **Figure 8-51.** You can clean this filter by tapping it on a flat surface to dislodge light accumulations of dirt. However, when it will not tap clean, a treated paper filter must be replaced with a new element designed for the given engine application.

A new style of dry-type filter cartridge is the pleated paper design shown in **Figure 8-52.** The

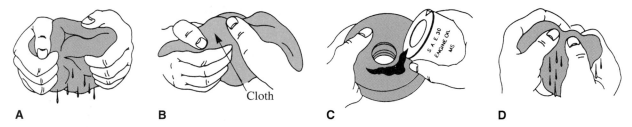

Figure 8-50. *Follow these steps when cleaning an oil-wetted foam element: A—Wash the element in a detergent and water solution. B—Wrap the element in a clean cloth and squeeze dry. C—Saturate the element in clean engine oil. D—Squeeze the excess oil out of the element. (Briggs and Stratton Corp.)*

Figure 8-51. *Dry-type filter element can be partially cleaned by tapping gently. New element should be installed as needed.*

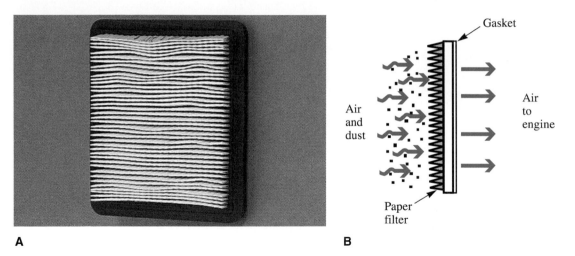

A   B

Figure 8-52.   *Air and dust particles travel through pleated paper filter element trapping dust. Clean air continues to engine. Gasket seals element.*

paper filter material is a special treated paper with porosity of extremely small size to let air flow through while preventing fine particles of dirt and dust from penetrating. The pleated design provides great surface area to collect particulates. This design is similar to the filters used in today's automobiles. A flexible plastic gasket material is molded around the edges of the paper filter to establish the overall dimensions and seal the unit in the receptacle. Cleaning is accomplished by tapping the cartridge with the external side down to shake off dirt. Care must be taken to prevent distortion of the pleated paper. If the dirt will not separate from the filter, throw it away and replace it with a new filter.

> ⚠ Never use compressed air to clean the paper filter because the porosity enlarges and the filter will not prevent fine dirt from penetrating and entering the engine.

When installing the pleated paper type air filter cartridge the pleated paper should face the external side of the receptacle. See **Figure 8-53.**

Dual element air cleaners

Engines to be used in greater than normal dusty conditions may have dual element air cleaners that provide more protection. Dual element air cleaners contain a dry or oiled plastic foam filter pre-cleaner followed by a pleated paper type cartridge. See **Figure 8-54.** Some pre-cleaners are oiled and some are non-oiled type. The dual ele-

Figure 8-53.   *When installing the pleated paper air cleaner cartridge place the paper element toward the external cover side of the receptacle.*

ment filters are contained in specially designed receptacles. See **Figure 8-55.** The cartridges and receptacles are found in a variety of designs such as the cylindrical vertical mount, the front mount, and the horizontal mount. See **Figure 8-56.** The pre-cleaner can be washed clean with detergent and water and squeezed dry. The paper filter cartridge should be cleaned by shaking.

Every engine to be serviced or repaired should have the air cleaner examined. If the air cleaner element and/or cartridge is damaged or shows signs of restriction replace them. Worn or damaged mounting gaskets and air cleaner gaskets

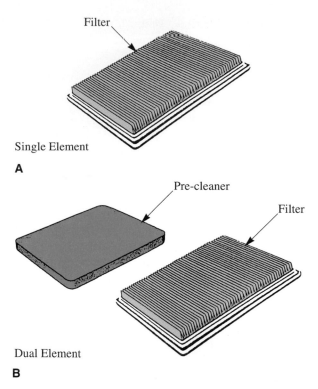

Single Element

A

Dual Element

B

Figure 8-54. *Single filters used for mild conditions. Dual element filters used for extremely dirty conditions.*

should be replaced to prevent dirt and dust from entering the engine through improper sealing around the filter.

Summary

The main purpose of the carburetor is to produce a mixture of fuel and air to operate the engine. The gasoline engine cannot run on *liquid* gasoline. The carburetor must vaporize the fuel and mix it with air.

The amount of air needed for combustion is far greater than the amount of fuel required. The average weight ratio is 15 parts air to 1 part fuel.

The carburetor is operated by pressure differences. It creates partial vacuum by means of a venturi. A venturi is a restriction in a passage that causes air velocity to increase and pressure to decrease. Reduction in pressure draws fuel into the airstream.

Three basic types of carburetors include the natural draft, the updraft, and the downdraft. These carburetors are named according to the direction that the air flows from their outlets to the engine manifold.

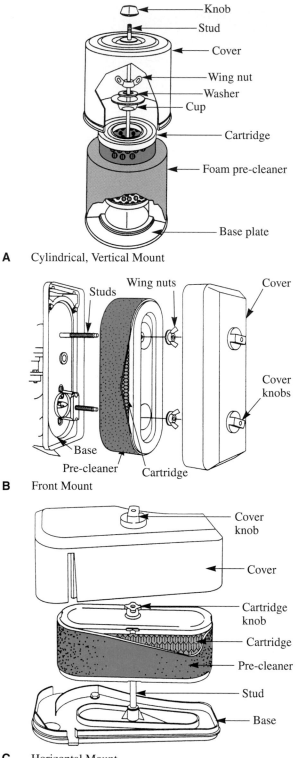

A Cylindrical, Vertical Mount

B Front Mount

C Horizontal Mount

Figure 8-55. *A—A dual element air cleaner of the cylindrical vertical mount. B—Front mounted dual element air cleaner. C—Horizontal mounted dual element air cleaner. (Briggs and Stratton Corp.)*

Some fuel systems are equipped with float-type carburetors. The purpose of a carburetor float is to maintain a constant level of fuel in the float bowl.

A diaphragm carburetor does not have a float. Instead, the engine vacuum pulsates a flexible diaphragm. The diaphragm draws fuel into a chamber of the carburetor from which it is readily drawn into the venturi.

A manual throttle control consists of mechanical linkage or a flexible cable that is operated manually to open and close the throttle valve.

In many small engine applications, the load on the engine changes constantly. This change would require constant throttle changes on the part of the operator. Governors work to maintain speed selected by the operator, prevent overspeeding, and limit high and low speeds.

A filtering device is used to clean incoming air. If air is not properly filtered, dirt entering the cylinder will cause rapid wear and scoring of machined parts throughout the engine.

Know These Terms

carburetor	load adjustment
air-fuel mixture	acceleration well
vacuum	economizer circuit
atmospheric pressure	idling circuit
venturi	primer
venturi principle	diaphragm carburetor
natural draft carburetor	manual throttle control
updraft carburetor	governor throttle control
downdraft carburetor	hunting
float	stability
choke	sensitivity
throttle	air cleaner

Chapter 8 Review Questions

Answer the following questions on a separate sheet of paper.

1. Give five different engine running conditions that must be met by the carburetor.

2. Normal air-fuel mixture by weight is _____.
   a. 12 to 1
   b. 13 to 1
   c. 14 to 1
   d. 15 to 1

3. If barometric pressure on a standard day at 1500′ MSL was 29.95, then at 3500′ MSL the barometric pressure would be _____. (See **Figure 8-3**.)
   a. 30.95″ Hg
   b. 31.95″ Hg
   c. 29.95″ Hg
   d. 27.95″ Hg

4. In a venturi, _____.
   a. air pressure is greatest where velocity is greatest
   b. air pressure is least where velocity is greatest
   c. velocity is least where air pressure is least
   d. the volume of air entering is slightly greater than the air leaving due to the restriction to flow

5. Name the three basic types of carburetors. (Consider direction of airflow.)

6. Which type of carburetor would normally require a smaller air passage than the other two types?

7. Needle valve points in the carburetor float chamber are usually made from one of two materials. The two materials are _____ and _____.

8. The choke valve in the carburetor is always located _____.
   a. nearest the intake end of carburetor
   b. nearest the manifold end of carburetor
   c. in the center of the carburetor
   d. above the float chamber level

9. Richest air-fuel mixture takes place during _____.
   a. full throttle
   b. half throttle
   c. idle
   d. starting

10. On some carburetors, the amount of fuel that enters the main discharge nozzle is regulated by _____.
    a. the float level
    b. a load adjusting needle
    c. the idle adjustment needle
    d. a spray bar needle valve

11. The acceleration well fills when the engine is _____.
    a. running at steady high speed
    b. running at half throttle
    c. under heavy load
    d. idling
12. The primary purpose of air bleeds is to _____.
    a. increase the air-fuel ratio
    b. improve atomization of the fuel
    c. remove air bubbles that may be mixed with the fuel
    d. prevent vapor lock
13. During idle and fast idle conditions, the main discharge nozzle is _____.
    a. discharging a small amount of fuel
    b. inoperative
    c. acting as an air bleed
    d. providing most of the fuel
14. The carburetor economizer _____.
    a. reduces float bowl pressure
    b. reduces the amount of fuel discharged into the venturi
    c. operates only after the engine reaches part throttle
    d. All of the above.
15. One main advantage of a diaphragm carburetor as compared to a float-type carburetor is its ability to _____.
16. The diaphragm in a carburetor _____.
    a. forces fuel through the main discharge nozzle
    b. operates only at high speed
    c. draws fuel under spring pressure
    d. draws fuel during a vacuum pulse from the manifold or crankcase
17. Governors serve three basic functions. What are they?
18. Name two basic types of small engine governors.
19. On a governor installation, the governor spring is attached to the throttle lever. The governor spring is intended to _____.
    a. have no effect on the throttle valve
    b. close the throttle valve
    c. open the throttle valve
    d. return the throttle lever to the off position
20. Name three types of air cleaners.

21. The *safest* solution in which to wash an air cleaner plastic foam element is _____.
    a. low lead gasoline
    b. #1 kerosene
    c. detergent and water
22. The proper way to clean an air filter cartridge of the pleated paper type is to _____.
    a. soak it in kerosene
    b. shake it external side down
    c. blow it clean with compressed air
23. The newest type of air cleaner for small gasoline engines is the _____.
    a. flat pleated paper cartridge filter
    b. oil bath air cleaner
    c. oil-wetted foam filter

Suggested Activities

1. Make a venturi tube. Provide a connection so that air can be forced through the venturi. Install one pressure gauge before the restriction and one gauge in the restriction. Demonstrate what happens when the air is applied to the venturi.
2. Working with the same venturi used in activity number one, remove the pressure gauge in the restriction. Place a pickup tube in the restriction and draw water out of a beaker. Demonstrate the atomization of the water particles.
3. Make a cutaway of a float-type carburetor so that the float, needle, throttle valve, choke, and internal passages can be seen.
4. Make a working mock-up model of a variable speed centrifugal governor system that demonstrates governor principles.
5. Make a large cross section of a float-type carburetor mounted on a board. Make the choke, throttle valve, float, and needle movable from the back. Paint the various parts, passages, and ports with bright colors. Give a demonstration to the class of choked, idle, part throttle, and full throttle carburetor functions.
6. Remove the air cleaner from an engine and identify the type. Examine it, properly service it, and if needed, replace it.

Ignition Systems

After studying this chapter, you will be able to:

▼ List the primary purposes of the ignition system.
▼ Identify the components in a typical magneto system and describe the function of each part.
▼ Describe small engine ignition advance systems.
▼ List the advantages of a solid state ignition system.
▼ Identify the three general classifications of magneto ignition systems and explain the operation of each.
▼ Describe the operation of a battery ignition system.

Basic Ignition System Operation

The primary purpose of the ignition system, **Figure 9-1,** of a small gasoline engine is to provide sufficient electrical voltage to discharge a spark between the electrodes of the spark plug. See **Figure 9-2.** The spark must occur at exactly the right time to ignite the highly compressed air-fuel mixture in the combustion chamber of the engine.

The ignition system must be capable of producing as many as 30,000 volts to force electrical current (electrons) across the spark plug gap. The intense heat created by the electrons jumping the gap ignites the air-fuel mixture surrounding the electrodes.

The rate, or times per minute, at which the spark must be delivered is very high. For example, a single cylinder, four-cycle engine operating at 3600 rpm requires 1800 ignition sparks per minute. A two-cycle engine running at the same speed

requires 3600 sparks per minute. In multi-cylinder engines, the number of sparks per minute is multiplied by the number of cylinders.

Every spark must take place when the piston is at exactly the right place in the cylinder and during

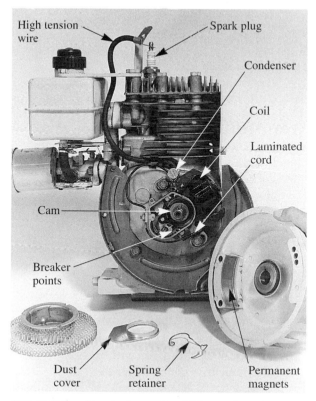

Figure 9-1. *The major parts of this small engine magneto system are the breaker points, condenser, coil, flywheel magnets, and the spark plug.*

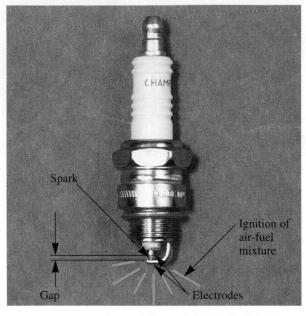

Figure 9-2. *Ignition system of small engine works hard to produce enough voltage to force electrons to jump spark plug gap.*

the correct stroke of the power cycle. Refer to *Chapter 5* of this text. Considering the high voltage required, the precise degree of timing, and the high rate of discharges, the ignition system has a remarkable job to do.

Most small gasoline engines use magneto systems to supply ignition spark. See **Figure 9-1.** *Magneto systems* produce electrical current for ignition without any outside primary source of electricity. They serve as simple and reliable ignition systems. Basic parts of a magneto system include:

- Permanent magnets.
- High tension coil.
- Mechanical or electronic switching device.
- Condenser (used only with mechanical switching device).
- High tension spark plug wire.
- Spark plug. See **Figures 9-1** and **9-2.**

Today, there are several types of magneto systems used on small engines. The mechanical breaker ignition (MBI) system uses mechanical breaker points to control current in the ignition coil. This type system was used exclusively until the development of the solid state ignition system. Solid state systems use electronic devices (transistors, capacitors, diodes) to control various ignition system functions. Several types of magneto systems will be discussed in detail later in this chapter.

Fundamental electrical principles

To make it easier to understand how various magneto parts function, we must first understand the fundamental electrical principles. These principles cover the electron theory, electrical units of measure, Ohm's law, magnetism, and magnets and electricity. The following five sections of this chapter cover these basic electrical principles.

The electron theory

All matter is composed of atoms. An atom is extremely small, so small that it cannot be seen with the most powerful microscope. It is the smallest particle of an element that can exist, alone or in combination; yet it consists of electrons, protons, and neutrons.

Atoms can be broken down into types, determined by the number and arrangement of the electrons, protons, and neutrons. A few types of atoms are hydrogen, oxygen, carbon, iron, copper, and lead. There are many others, about 100 in all.

The structure of the atom determines the weight, color, density, and other properties of an element. The electrons, though varying in number, are identical in all elements. An electron from silver would be the same as an electron from copper, tin, or any other substance.

Electrons travel in orbits around the center of the atom. See **Figure 9-3.** They are very light and their number per atom varies from one element to another. Electrons have negative (−) electrical charges. Protons are large, heavy particles when compared with the electrons. One or more protons help form the nucleus (center) of the atom and are positively (+) charged. A nucleus is made up of protons and neutrons bound tightly together. Neutrons are electrically neutral and are located in the nucleus of the atom. The number of electrons is equal to the number of protons in any atom.

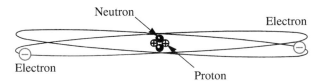

Figure 9-3. *All atoms consist of electrons, neutrons, and protons. Neutrons and protons form the nucleus. Neutrons have no electrical charge, but each proton carries a positive (+) charge. Electrons orbit the nucleus and carry negative (−) charges.*

Normally, atoms are electrically neutral because the negative electrons cancel the positive force of the protons.

Actually, an atom is held together because unlike electrical charges attract each other. The positively charged protons hold the negatively charged electrons in their orbits. Since like electrical charges repel each other, the negative electrons will not collide as they spin.

The ease with which an electron from one atom can move to another atom determines whether a material is an electrical conductor or nonconductor. In order to have electric current, electrons must move from atom to atom. Materials allowing electrons to move in this way are called conductors. Examples are copper, aluminum, and silver.

In nonconductors, it is difficult, if not impossible, for electrons to leave their orbits. Nonconductors are called insulators. Some examples are glass, mica, rubber, plastic, and paper.

Electron flow in a conductor follows a path similar to that shown in **Figure 9-4.** The flow of electrons will take place only when there is a complete circuit and a difference in electrical potential. A difference in potential exists when the source of electricity lacks electrons, or is positively (+) charged. Since electrons are negatively (–)

charged and unlike charges attract, the electrons move toward the positive source. See **Figure 9-5.**

Electrical potential is produced in three ways:
- Mechanically.
- Chemically.
- Statically.

The electrical generator is a mechanical producer of electricity and can be run by water power, steam turbines, or internal combustion engines. A magneto is a type of generator. Mechanical energy from the crankshaft is used to rotate a permanent magnet.

Electricity used in homes and factories is produced mechanically. Batteries are chemical producers of electricity. Lightning is a result of static electricity.

Electrical units of measurement

Three basic units of electrical measurement are as follows:
- Amperes (rate of electron flow).
- Volts (force that causes electrons to flow).
- Ohms (resistance to electron flow).

An *ampere (A)* is a measurement of the number of electrons flowing past any given point in a specific length of time. One ampere of current is equal to 6,240,000,000,000,000,000 (6.24 × 10^{18})

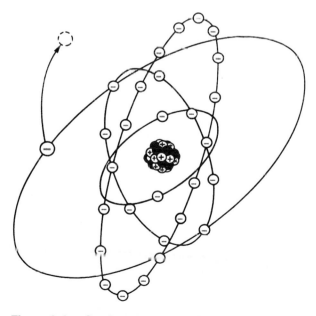

Figure 9-4. *Conductors are materials having electrons that can easily leave orbit of one atom and move to orbit of another atom. When many electrons do this, electricity is produced.*

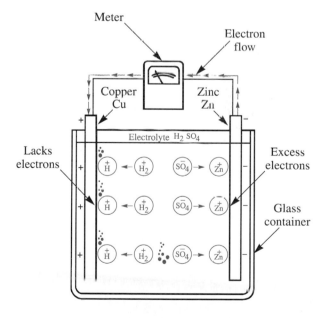

Figure 9-5. *A difference in potential exists if source of electricity lacks electrons and, therefore, is positively (+) charged. Electrons, being negatively (–) charged, are attracted to positive source.*

electrons per second. Since electricity is generally transmitted through wires, the greater the number of electrons flowing, the larger the wire size must be.

The difference in electrical potential between two points in a circuit is measured in *volts (V)*. Voltage is the force, or potential, that causes the electrons to flow.

Resistance to electron flow is measured in *ohms (Ω)*. Some materials produce a strong resistance to electron flow; others produce little resistance. If a wire is too small for the amount of current produced at the source, the wire will create excessive resistance and will get hot.

The air gap between spark plug electrodes is highly resistant to electron flow, creating the need for high voltage to cause the electrons to jump the gap. This high resistance also creates heat, which ignites the fuel in the cylinder.

Ohm's law

Every electrical circuit operates with an exact relationship of volts, amps, and ohms. It is possible to work out their mathematical relationship through the application of *Ohm's law.* The formula for Ohm's law is as follows:

$$I = \frac{E}{R}$$

where:

I = amperes
E = volts
R = ohms

A helpful visual aid for remembering the Ohm's law is shown in **Figure 9-6.** To use this aid, simply cover the unknown value. For example, to find amperes cover the I to get E over (divided by) R. From this aid we can also find ohms (unknown), if volts and amperes are known. Cover the R, to

get E over (divided by) I. We can also find volts (unknown), if amperes and ohms are known. Cover the E, to get I next to (multiplied by) R. The following are examples of using Ohm's law.

If circuit voltage is 12V and resistance is 8Ω, the current would be found as follows:

$$I = \frac{E}{R} \qquad I = \frac{12}{8} \qquad I = \textbf{1.5A}$$

If amperage is 15Ω and voltage is 6V, resistance would be found as follows:

$$R = \frac{E}{I} \qquad R = \frac{6}{15} \qquad R = \textbf{.4Ω}$$

If amperage is 3A and resistance is 10Ω, the voltage would be found as follows:

$$E = I \times R \qquad E = 3 \times 10 \qquad E = \textbf{30V}$$

Magnetism

The molecular theory of magnetism is the one most widely accepted by scientists. Molecules are the smallest particles of matter that are recognizable as being that matter. For example, a molecule of aluminum oxide will contain atoms of aluminum and atoms of oxygen.

In most materials, the magnetic poles of adjoining molecules are arranged in a random pattern, so there is no magnetic force. See **Figure 9-7.** Iron, nickel, and cobalt molecules, however, are able to align themselves so that all their north poles point in one direction and south poles in the opposite direction. See **Figure 9-8.** The individual magnetic forces of each molecule combine to produce one strong magnetic force. In magnets, opposite poles attract each other and like poles repel each other, much in the same way that like and unlike electrical charges react.

Figure 9-6. *This learning tool is used to find formulas for Ohm's law.*

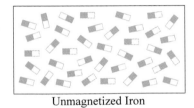

Unmagnetized Iron

Figure 9-7. *An unmagnetized substance is made up of molecules whose poles are not aligned. Molecules have north and south poles, like bar magnets.*

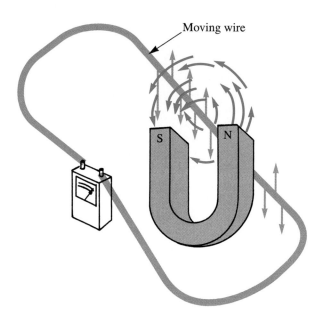

Figure 9-8. A magnetized substance has all molecules in alignment, north to south. Individual molecules combine magnetic forces to produce a strong overall magnetic force.

Certain materials have good magnetic retention. That is, they retain their molecular alignment. These materials are suitable as permanent magnets. Some materials maintain their molecular alignment only when they are located within a magnetic field. When the field is removed, the molecules disarrange themselves into random patterns and the magnetism is lost. When permanent magnets are cut into pieces, each piece takes on the polarity of the parent magnet, as illustrated in **Figure 9-9.**

Magnets and electricity

The fact that there is a close relationship between electricity and magnetism serves as the basis for making a workable magneto system. Over 100 years ago, Michael Faraday discovered that electricity could be produced from magnetism. One of his experiments showed that if a wire is moved past a magnet, the magnetic field is cut by the wire and current will flow. See **Figure 9-10.** When movement of the wire is stopped, the current flow also stops. See **Figure 9-11.** Therefore, elec-

Figure 9-10. If a conductor, such as copper wire, is moved so that it cuts magnetic lines of force, an electron flow is induced in conductor. Flow of electrons (electricity) can be measured with a sensitive meter.

tricity will flow when the magnetic lines of force are being cut by the wire.

An important principle used in magneto construction is that a magnetic field is developed when electrons flow through a coil of wire. **Figure 9-12** shows a simple coil of wire with current flowing through it. In this illustration, the magnetic field is indicated by lines around the coil. The lines of force come from the north pole and return to the south pole. If the direction of current is changed, the polarity also changes. In magneto

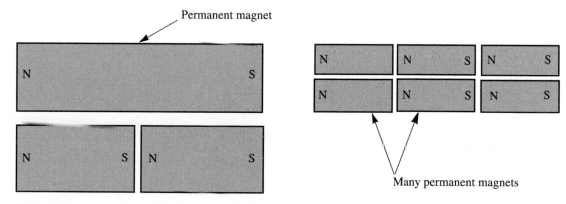

Figure 9-9. If a permanent bar magnet is broken into subparts, each subpart has a north and south pole, like the parent magnet. If parts could be further broken into individual molecules, each molecule would be an individual magnet.

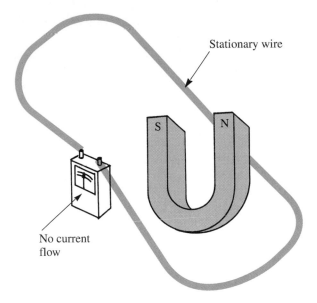

Figure 9-11. *A conductor that is not moving and not cutting magnetic lines of force will not induce electrical current.*

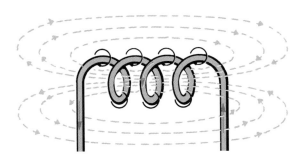

Figure 9-12. *A coil of wire with current flowing through it will produce a magnetic field around itself and around each turn of wire in coil.*

operation, an electric current is passed through a coil, which develops a magnetic field.

Ignition coil

The *ignition coil* used in a magneto system operates like a transformer. The coil contains two separate windings of wire insulated from each other and wound around a common laminated iron core. See **Figure 9-13.** The primary winding is heavy gage wire with fewer turns than the secondary winding, which has many turns of light gage wire.

When electrical current is passed through the primary winding, a magnetic field is created around the iron core. When the current is stopped, the magnetic field collapses rapidly, cutting through the secondary windings. This rapid cutting of the field by the wire in the coil induces high voltage in the secondary circuit. The high secondary voltage, in turn, causes a spark to jump the spark plug gap and ignite the air-fuel mixture.

Spark plugs

At first glance, an assortment of spark plugs may look very much alike. Actually, there are many variations. Using the correct spark plug for a given engine application can greatly increase the efficiency, economy, and service life of the engine.

Figure 9-14 shows major parts of a typical spark plug. The terminal nut is the external contact with the high tension coil. Some terminal nuts are removable, others are not. Many of the major parts of the spark plug are used to identify the actual type of the plug. Other considerations include

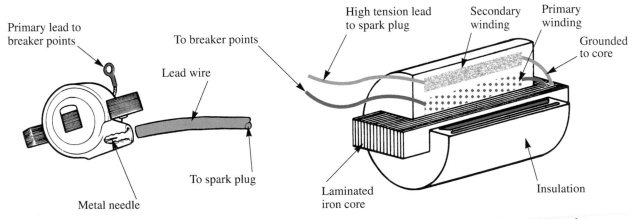

Figure 9-13. *Ignition coil consists of two windings, one inside the other. Coil functions as a step-up transformer to produce high voltage and low amperage from low voltage and high amperage.*

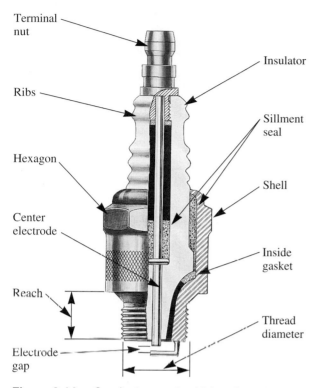

Figure 9-14. *Spark plug carries high voltage current produced by ignition system. It also must withstand the high temperatures and shock of combustion, insulate center electrode against current loss, and seal against compression leakage. (Deere & Co.)*

Terminal nut
Ribs
Hexagon
Center electrode
Reach
Electrode gap
Insulator
Sillment seal
Shell
Inside gasket
Thread diameter

construction, heat rating number, and firing end construction. To learn how to identify spark plug types, use the spark plug symbols chart in **Figure 9-15**.

Two common methods of **high tension lead** connections are shown in **Figure 9-16**. Application A uses the exposed clip, which is satisfactory in uses where moisture, oil, or dirt will not get on the plug or can easily be wiped off. The boot type, shown at B, provides better plug protection.

The spark plug **insulator** is usually an aluminum oxide ceramic material, which has excellent insulating properties. The insulator must have high mechanical strength, good heat conducting quality, and resistance to heat shock. Generally, ribs on the insulator extend from the terminal nut to the shell of the plug to prevent *flashover.* Flashover is the tendency for current to travel down the outside of the spark plug instead of through the center electrode. See **Figure 9-17**.

The **center electrode** carries the high voltage current to the spark gap. If the electrical potential is great enough to cause the current to jump the

plug gap, the grounded electrode will complete the circuit to ground.

> Always refer to manufacturer's specifications for correct electrode gap. See **Figure 9-18**.

The sillment seal is a compacted powder that helps ensure permanent assembly and eliminates compression leakage under all operating conditions. The inside gasket also acts as a seal between the insulator and the steel shell.

Spark plug **reach** varies with type of spark plug. Some are long, others quite short. See **Figure 9-19**.

> Never use a spark plug that has a longer reach than specified. See **Figure 9-20**. Serious engine damage can result if the piston hits the plug.

Several standard thread sizes are commonly used. Threads on some spark plugs are metric sizes, usually 14mm.

Spark plug heat transfer

Heat transfer in spark plugs is an important consideration. The heat of combustion is conducted through the plug as shown in **Figure 9-21**. Spark plugs are manufactured in various **heat ranges** from *hot* to *cold*. See **Figure 9-22**. Cold running spark plugs are those which transfer heat readily from the firing end. They are used to avoid overheating in engines having high combustion temperatures.

In figuring spark plug heat range, the length of the insulator nose determines how well and how far the heat travels. The spark plug at A in **Figure 9-22**, for example, is a hot plug because the heat must travel a greater distance to the cylinder head. Spark plug at D is comparatively colder than A. A cold plug installed in a cool running engine would tend to foul. Cool running usually occurs at low power levels, continuous idling, or in start/stop operation.

The tip of the insulator is the hottest part of the spark plug and its temperature can be related to preignition (firing of fuel charge prior to normal ignition) or plug fouling. Experiments show that if combustion chamber temperature exceeds 1750°F (954°C) in a four-cycle engine, preignition is likely to occur. If insulator tip temperature drops below 700°F (371°C), fouling or shorting of the plug due to carbon is likely to occur.

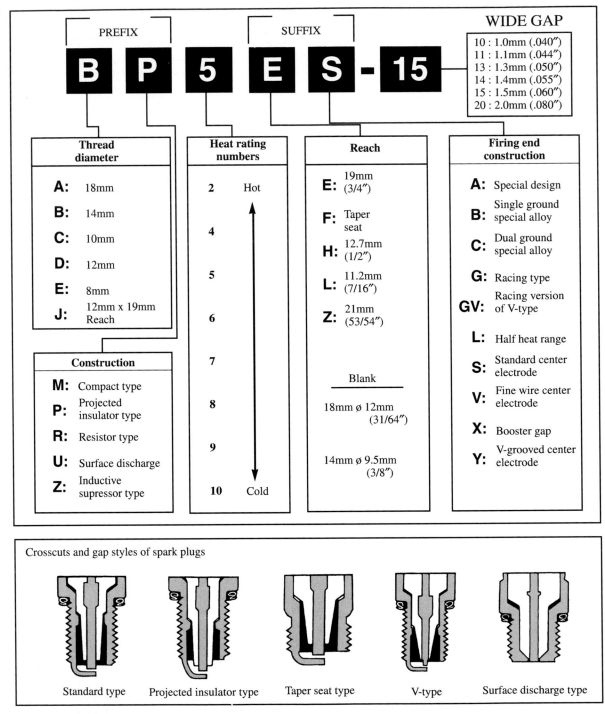

Spark Plugs
Spark Plug Symbol Explanation

PREFIX SUFFIX

B P 5 E S - 15

WIDE GAP

10 :	1.0mm (.040″)
11 :	1.1mm (.044″)
13 :	1.3mm (.050″)
14 :	1.4mm (.055″)
15 :	1.5mm (.060″)
20 :	2.0mm (.080″)

Thread diameter

- **A:** 18mm
- **B:** 14mm
- **C:** 10mm
- **D:** 12mm
- **E:** 8mm
- **J:** 12mm x 19mm Reach

Construction

- **M:** Compact type
- **P:** Projected insulator type
- **R:** Resistor type
- **U:** Surface discharge
- **Z:** Inductive supressor type

Heat rating numbers

- 2 — Hot
- 4
- 5
- 6
- 7
- 8
- 9
- 10 — Cold

Reach

- **E:** 19mm (3/4″)
- **F:** Taper seat
- **H:** 12.7mm (1/2″)
- **L:** 11.2mm (7/16″)
- **Z:** 21mm (53/54″)

Blank

18mm ø 12mm (31/64″)

14mm ø 9.5mm (3/8″)

Firing end construction

- **A:** Special design
- **B:** Single ground special alloy
- **C:** Dual ground special alloy
- **G:** Racing type
- **GV:** Racing version of V-type
- **L:** Half heat range
- **S:** Standard center electrode
- **V:** Fine wire center electrode
- **X:** Booster gap
- **Y:** V-grooved center electrode

Crosscuts and gap styles of spark plugs

Standard type   Projected insulator type   Taper seat type   V-type   Surface discharge type

Figure 9-15. *This chart explains how to identify spark plugs.*

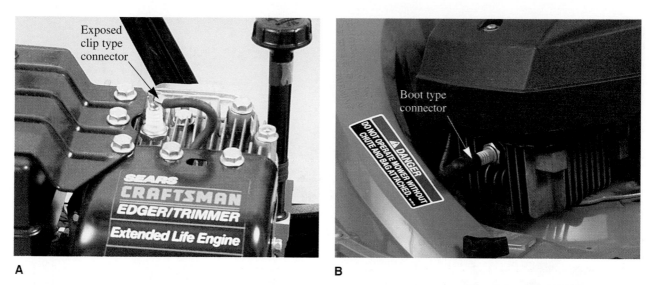

Figure 9-16. *Two common high tension lead connectors. A—Exposed type. B—Neoprene boot type. Exposed clip connector is used in conjunction with a metal strip stop switch.*

Flashover

Figure 9-17. *Flashover is caused by moisture or dirt. It can also be caused by a worn out terminal boot, which allows voltage to short across outside of ceramic insulator.*

Figure 9-19. *Spark plug reach (length of thread) can vary considerably from one plug to another. Too long a reach can damage a piston. Too short a reach provides poor combustion. (AC Spark Plug Div., GMC)*

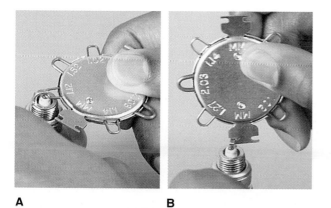

A   **B**

Figure 9-18. *A—Technician is checking the correct electrode gap. B—Technician is setting the correct electrode gap from manufacturer's specifications.*

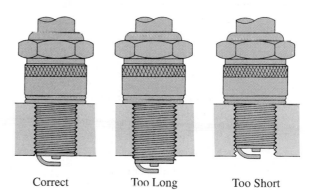

Correct   Too Long   Too Short

Figure 9-20. *Spark plug reach is determined by thickness of cylinder head.*

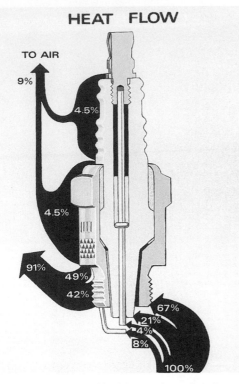

Figure 9-21. *Heat of ignition and combustion must be conducted away from critical parts of spark plug to prevent preignition and burning of electrodes.*

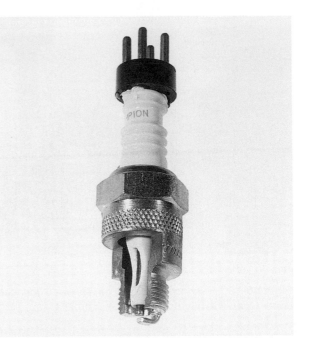

Figure 9-23. *Operating temperature of spark plug can be studied with a special spark plug having a thermocouple (heat sensor) installed in it. (Champion Spark Plug Co.)*

Measuring spark plug temperature

Specially assembled thermocouple spark plugs are used to accurately determine spark plug temperatures during actual engine operation. See **Figure 9-23.** These plugs have a small temperature sensing element embedded in the insulator tip and serve as a valuable aid to engineers in gathering spark plug and combustion data.

Types of Electrodes

Electrode configurations vary considerably. In **Figure 9-24,** plug A is a retracted gap type used in some engines where clearance is a problem or protection of the firing tip is desirable. Plug B is a

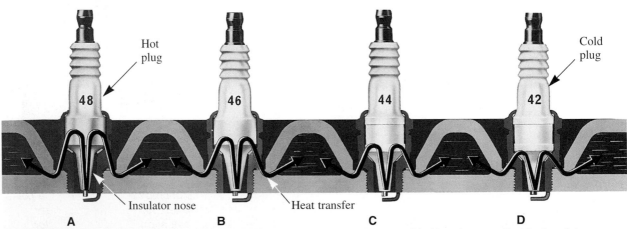

Figure 9-22. *Spark plug heat transfer determines whether plug is* hot *or* cold. *Heat is controlled by insulator nose. (AC Spark Plug Div., GMC)*

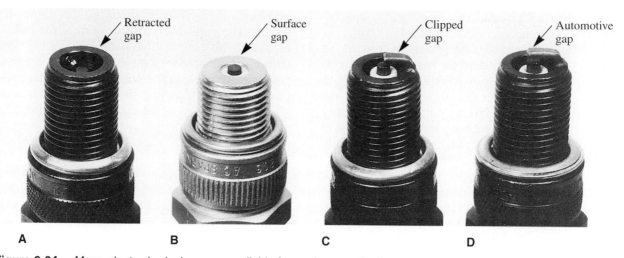

Retracted gap   Surface gap   Clipped gap   Automotive gap

A   B   C   D

Figure 9-24. *Many electrode designs are available for engine use. Surface gap spark plug is extremely cold and is finding application with capacitor discharge ignition systems. (AC Spark Plug Div., GMC)*

surface gap type, which is extremely cold. Surface gap spark plugs are sometimes used in engines with capacitive discharge ignition systems. Plug C is a clipped gap type, in which the side electrode extends only part way across the center electrode. Plug D is a standard gap automotive spark plug.

Switching devices

Switching devices are used in the ignition system to control the primary current to the ignition coil. The switching devices are either mechanically or electronically controlled. In the mechanically controlled type, the primary current flows when breaker points physically make contact. In the electronically controlled type, the primary current flows when an electrical circuit is completed and the switch is closed.

Breaker points

Some types of ignition systems use *mechanical breaker points* to control primary current to the coil. The breaker points generally consist of two tungsten contacts. One contact point is stationary, the other is movable. Each contact is fastened to a bracket.

Tungsten is a hard metal with a high melting temperature. These characteristics are needed to withstand the continual opening and closing that takes place and the eroding effect of the arc that occurs when the points *break* (start to open). **Figure 9-25** shows the breaker point assembly with the dust cover removed.

The breaker point assembly is an electrical switch. When the points are closed, the magnetic field created by the flywheel magnets is being cut by the primary coil winding. This induces current

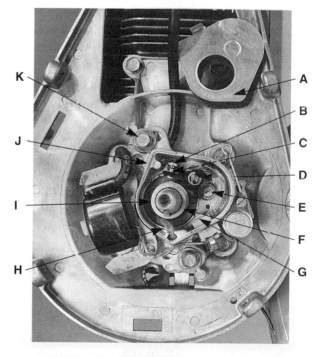

Figure 9-25. *Magneto breaker point assembly. A—Dust cover. B—Stationary point. C—Movable point. D—Pivot pin. E—Gap adjustment screw. F—Wear block. G—Breaker point spring. H—Felt lubricator. I—Cam block. J—Point housing. K—Advance adjustment screw.*

in the primary circuit, and the primary winding builds its own magnetic field around the coil.

When the breaker opens, current flow in the primary circuit stops and the magnetic field collapses. Immediately upon collapse, a surge of high voltage induced in the secondary winding of the coil forces the electrons to jump the spark plug gap. This process is repeated every time the breaker points close and open. The spark plug sparks only at the instant the breaker points open.

Electronic switching devices

Some ignition systems use *electronic switching devices* instead of the mechanical breaker points to control the primary current to the ignition coil. An electronic switching device is more dependable than a mechanical type. There are no breaker points to wear or burn out.

The electronic switching device is part of an electrical circuit that controls ignition. The electronic switch device is simply an on/off switch. When the switch is closed it is in the *on position*, and the primary current flows to the ignition coil. When the switch is opened it is in the *off position*, and current cannot flow.

The switch is activated by a positive electrical charge. When a positive charge is present, the switch is closed. The primary current can flow to the ignition coil. When there is not a positive charge present, the switch is open. The primary current cannot flow to the ignition coil.

The MBI magneto system

The *mechanical breaker point ignition (MBI)* magneto system supplies the ignition spark on most small engines. A magneto system will produce current for ignition without any outside primary source of electricity. Major components of a typical flywheel magneto ignition system are illustrated in **Figure 9-26.** Except for the spark plug; the coil, condenser, and breaker points may be found inside or outside of the flywheel. This varies with engine type, but the principles of operation remain basically the same.

The *condenser* plays an important part in MBI operation. Its primary purpose is to prevent current from arcing across the breaker point gap as the points open. If arcing would occur, it would burn the points and absorb most of the magnetic energy stored in the ignition coil. Not enough energy would be left in the coil to produce the

necessary high voltage surge in the secondary circuit. The condenser absorbs current the instant breaker points begin to separate. Since the condenser absorbs most of the current, little is left to form an arc between the points. A condenser must be selected that has just the right capacitance to absorb the amount of energy required to produce an arc. The construction of the condenser is quite simple. **Figure 9-27** shows a condenser partially opened up to expose laminations of aluminum foil and an insulating strip. Two strips of aluminum foil of specific length are wound together with the insulator strip between them. One foil strip is grounded, while the other is connected across the breaker points.

Magnets are usually cast into the flywheel and cannot be removed. They are strong permanent

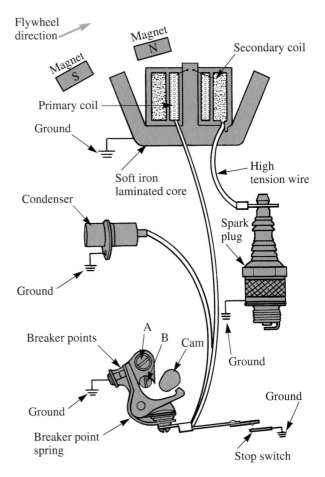

Figure 9-26. *Coil, condenser, breaker points, and spark plug make up primary and secondary circuits of this magneto system. Magnets create current flow in primary winding of coil, which induces current in secondary winding.*

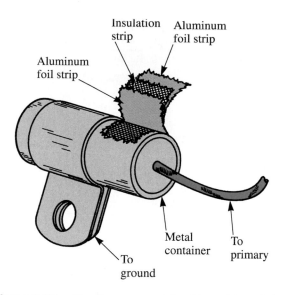

Figure 9-27. *Condenser consists of two strips of aluminum foil separated by an insulating material (dielectric). One foil strip is connected to metal container, other is attached to primary lead. (Kohler Co.)*

magnets made of **Alnico** (aluminum, nickel, cobalt alloy) or of a newer ceramic magnetic material.

The coil in **Figure 9-26** is cut away to show primary and secondary windings. The primary winding usually has about 150 turns of relatively heavy copper wire. The secondary winding has approximately 20,000 turns of very fine copper wire. One end of primary and one end of secondary are grounded to the soft iron laminated core which, in turn, is grounded to the engine.

The odd shape of the core of the coil is designed to efficiently direct the magnetic lines of force. The spacing (air gap) between the magnets and the core ends is critical and can greatly affect the whole system. This gap can be checked with special gauges or standard feeler gauges. The high tension wire is heavily insulated because it carries high voltage. If the insulation deteriorates, much of the voltage can be lost by arcing to nearby metallic parts of the engine.

In the system illustrated in **Figure 9-26,** the breaker points are mechanically actuated, opened by the cam and closed by the breaker point spring. The breaker point gap is adjusted by loosening clamp screw A and turning eccentric screw B to move the stationary point. Remember that some magneto systems use solid state switching devices instead of mechanical breaker points.

The MBI magneto cycle

As the flywheel turns, the magnets pass over the legs of the laminated core of the coil. When the north pole of the magnet is over the center leg of the coil, the magnetic lines of force move down the center leg through the coil, across the bottom of the lamination, and up the side leg to the south pole. See **Figure 9-28.**

As the flywheel continues to turn, **Figure 9-29,** the north pole of the magnet comes over the side leg and the south pole is over the center leg of the core. Now the lines of force move from the north pole down through the side leg, up through the center leg and coil, and to the south pole. At this point, the lines of force have reversed direction.

Figure 9-30 shows the field reversal taking place in the center leg of the core and coil. The reversal induces low voltage current in the primary

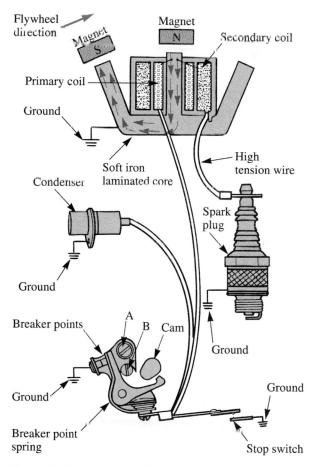

Figure 9-28. *As flywheel continues to turn, magnets realign with center and outside leg of core, causing magnetic field to reverse and induce low voltage in primary coil.*

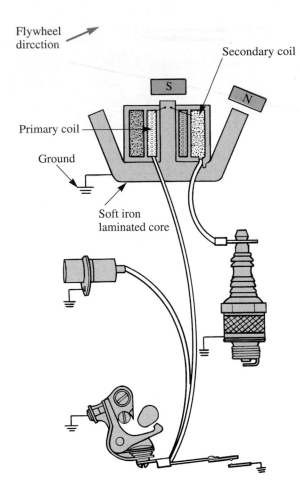

Figure 9-29. *As flywheel turns and magnets align with legs of laminated core of coil, a magnetic field is conducted through the primary winding.*

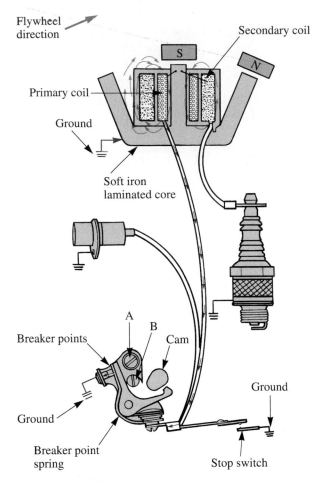

Figure 9-30. *Magnetic field reversal takes place as magnets pass from left to right. Breaker points are closed.*

circuit through the breaker points. Current flowing in the primary winding of the coil creates a primary magnetic field of its own, which reinforces and helps maintain the direction of the lines of force in the center leg of the lamination. It does this until the magnets' poles move into a position where they can force the existing lines of force to change direction in the center leg of the lamination. Just before this happens, the breaker points are opened by the cam.

Opening of the points breaks the primary circuit and the primary magnetic field collapses through the turns of the secondary winding. See **Figure 9-31.** The condenser makes the breaking of the primary current as instantaneous as possible by absorbing the surge of primary current to prevent arcing between the breaker points.

As the magnetic field collapses through the secondary winding of the coil, high voltage

current is induced in the secondary winding. At exactly the same time, the charge stored in the condenser surges back into the primary winding, **Figure 9-32,** and reverses the direction of current in the primary windings. This change in direction sets up a reversal in direction of the magnetic field cutting through the secondary and helps increase the voltage in the secondary circuit. The potential of the high voltage causes secondary current to arc across the spark plug gap.

The stop switch

The spark plug can only fire when the ignition points open the primary circuit. Using this as a basis for a stop switch, the switch is designed to ground the movable breaker point so that, in effect, the points never open. See **Figure 9-33.** Therefore, the engine stops running.

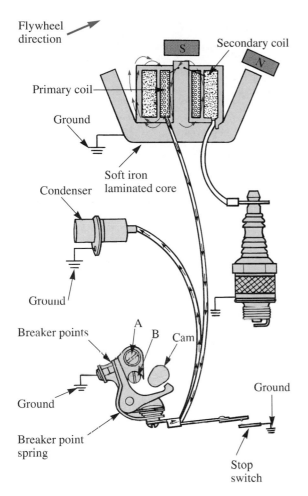

Figure 9-31. *Breaker points open, causing primary magnetic field to collapse at an extremely high rate through secondary winding. This induces high voltage required to fire spark plug. Field collapse also cuts through windings of primary coil, inducing moderate voltage that is absorbed by condenser.*

Another common method of stopping single cylinder engines is by means of a strip of metal fastened to one of the cylinder head bolts. When the engine is running, the strip is suspended about 1/2″ from the spark plug wire terminal. By depressing the strip against the plug wire, the current flows down the strip to the cylinder head to prevent a spark at the plug. There is no danger of shock to the operator.

 When stopping single cylinder engines by means of a strip of metal fastened to one of the cylinder head bolts, do not touch the spark plug directly.

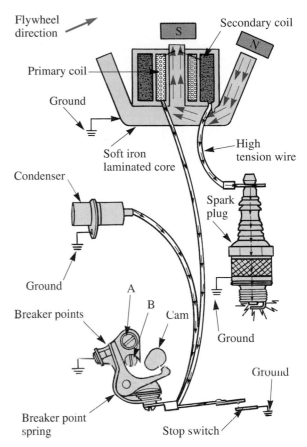

Figure 9-32. *Spark plug fires and condenser discharges voltage back into primary circuit.*

Ignition advance systems

Some small engines have mechanical systems that retard occurrence of spark for starting. For intermediate and high speed operation, the *ignition advance system* causes spark to occur earlier in the cycle.

One type of ignition advance system is illustrated in **Figure 9-34.** Two different spark timings are provided, one for starting and one for running. For starting, the spark-advance flyweight holds the cam in a position so that the ignition spark occurs at 6° of crankshaft rotation before the piston reaches top dead center (TDC). See **Figure 9-34A.**

When the engine reaches a speed of nearly 1000 rpm, centrifugal force moves the flyweight out, forcing the cam to rotate. This position of the cam causes the points to open and a spark to occur at 26° before top dead center. See **Figure 9-34B.**

Figure 9-35 shows the flyweight and cam assembled on the crankshaft in starting (solid lines) and running (dashed lines) positions.

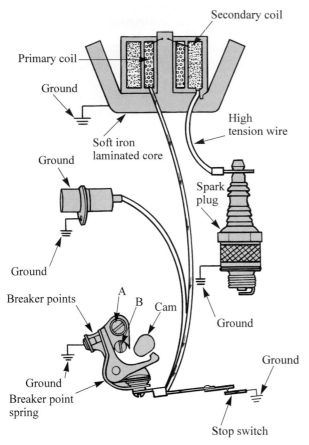

Figure 9-33. *When stop switch is closed, breaker point system is grounded and engine is stopped.*

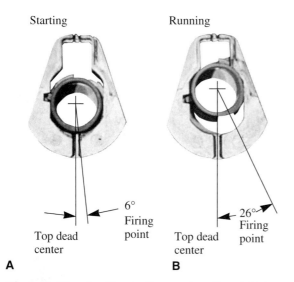

Figure 9-34. *A—For starting engine, flyweight is held in retarded position by a spring. B—When engine reaches 1000 rpm, flyweight overcomes spring and rotates cam to advanced position. (Lawn-Boy Power Equipment, Gale Products)*

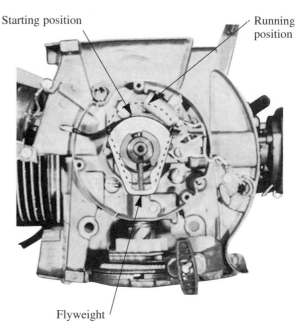

Figure 9-35. *Flyweight, spring, and cam rotate together as a single assembly. Outer position of flyweight is for advanced operation.*

Dwell and cam angle

Dwell (cam angle) is the time the breaker points stay closed during one revolution of the cam. Dwell is the number of degrees measured around the cam from the point of closing to the point of opening. **Figure 9-36** shows the direct relationship between the breaker point gap setting and dwell time.

Figure 9-36A indicates what might be considered normal dwell time. **Figure 9-36B** shows a large point gap with a correspondingly short dwell. Note that the wear block must travel quite a distance to a lower position on the lobe before the points make contact. Then, they open with just the slightest rise in the lobe.

The narrow point gap illustrated in **Figure 9-36C** causes the breaker points to close after only a slight travel down the lobe, but more distance is required up the lobe to open the points. Remember that the cam is driven directly from the crankshaft. When the breaker points open, the spark plug fires.

Obviously, then, changing the point setting can also change spark timing. The engine manufacturer specifies which gap setting is best (usually between .020″ to .030″) and the number of degrees before top dead center (BTDC) that the spark should occur.

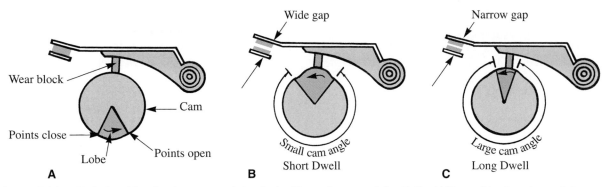

Figure 9-36. *Relationship of point gap and dwell. A—Normal gap and dwell. B—With a wider breaker point gap setting, dwell decreases. C—A narrower gap increases dwell.*

Electronic ignition

Solid state is a broad term applied to any ignition system that uses electronic semi-conductors (diodes, transistors, silicon controlled rectifiers, etc.) in place of one or more mechanical ignition components.

Electronic components in solid state systems are extremely small. Since there are no moving parts, mechanical adjustments are not required. Solid state ignition systems provide many advantages:

- Eliminate ignition system maintenance.
- No breaker points to burn, pit, or replace.
- Increase spark plug life.
- Easy starting, even with fouled plugs.
- A flooded engine will start easily.
- Higher spark output and faster voltage rise.
- Spark advance is electronic and automatic. It never needs adjusting.
- Electronic unit is hermetically sealed and unaffected by dust, dirt, oil, or moisture.
- System delivers uniform performance throughout component life and under adverse operating conditions.
- Improve idling and provide smoother power under load.

Operation of capacitive discharge ignition (CDI) system

The *capacitive discharge ignition (CDI) system* is a solid state ignition system. It is one of the newest systems used in an internal combustion engine. It is standard equipment in many applications and has improved the reliability of modern small gasoline engines.

The CDI system is breakerless. The mechanical points and accessories are replaced with elec-

tronic components. The only moving parts are the permanent magnets in the flywheel. **Figure 9-37** shows the CDI module installed on a small gasoline engine.

The CDI module can be tested to see if it is producing a spark by grounding a known good spark plug to the engine. Turn the flywheel and watch the plug electrode gap for a spark. If there is no spark, the ignition switch or switch lead may be faulty; or the air gap may be incorrect (spacing between flywheel and CDI module). No other troubleshooting is necessary.

The ignition switch is the most vulnerable part of the solid state ignition system. The switch must be kept dry and clean.

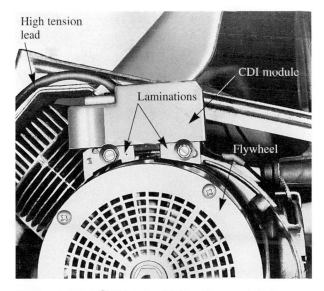

Figure 9-37. *Solid state, CDI ignition module is compact and maintenance free. Only moving parts are flywheel magnets.*

Refer to the following illustrations to progressively trace current flow through the various electronic components in a typical CDI system.

In **Figure 9-38,** flywheel magnets rotate across the CDI module laminations, inducing a low voltage alternating current (ac) in charge coil. The ac passes through a rectifier and changes to direct current (dc), which travels to the capacitor where it is stored.

In **Figure 9-39,** the flywheel magnets rotate approximately 351° before passing the CDI module laminations and induce a small electrical charge in the trigger coil. At starting speeds, this electrical charge is just great enough to turn on the silicon controlled rectifier (SCR) in a retarded firing position (9° BTDC). This provides for easy starting.

In **Figure 9-40,** when the engine reaches approximately 800 rpm, advanced firing begins. The flywheel magnets travel approximately 331°, at which time enough voltage is induced in the trigger coil to energize the silicon controlled rectifier in the advanced firing position (29° BTDC).

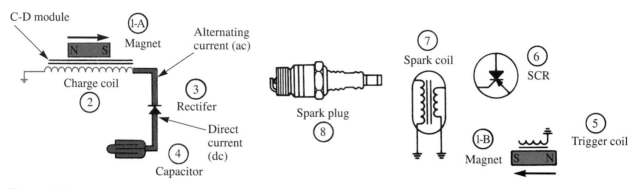

Figure 9-38. *Flywheel operation. 1A—Magnets induce low voltage alternating current into charge coil at 2. 3—Rectifier changes alternating current to direct current. 4—Direct current from rectifier is stored in capacitor (condenser). See Figure 9-40.*

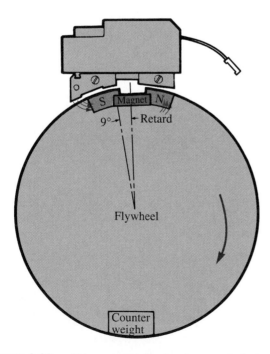

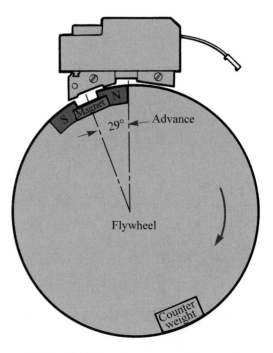

Figure 9-39. *At low speed, flywheel magnets induce a small current in trigger coil, which turns on silicon rectifier at 9° BTDC for easy starting.*

Figure 9-40. *At 800 rpm, stronger trigger coil current turns on silicon rectifier at 29° BTDC for satisfactory ignition during normal engine operation.*

In **Figure 9-41,** when the silicon controlled rectifier is triggered, the 300V dc stored in the capacitor travels to the spark coil. At the coil, the voltage is stepped up instantly to a maximum of 30,000V. This high voltage current is discharged across the spark plug gap.

Operation of transistor controlled ignition (TCI) system

The individual components that make up the *transistor controlled ignition system* are given in a chart in **Figure 9-42.** Study the function of each part carefully.

There are a variety of transistor controlled circuits. Each has its own unique characteristics and modifications. **Figure 9-43** illustrates a typical circuit for a transistor controlled ignition. Refer to this circuit as its principles are described in the following section.

As the engine flywheel rotates, the magnets on the flywheel pass by the ignition coil. The magnetic field around the magnets induces current in the primary windings of the ignition coil.

The base circuit of the ignition system has current flow from the coil primary windings, common grounds, resistor (R1), base of the transistor (T1), emitter of the transistor (T1), and back to the primary windings of the ignition coil.

Current flow for the collector circuit in **Figure 9-43** is from the primary windings of the coil, common grounds, collector of transistor (T1), emitter of transistor (T1), and back to the primary windings.

When the flywheel rotates further, the induced current in the coil primary increases. When the current is high enough, the control circuit turns on and begins to conduct current. This causes transistor (T2) to turn on and conduct. A strong magnetic field forms around the primary winding of the ignition coil. See **Figure 9-43.**

The trigger circuit for this ignition system consists of the primary windings, common grounds, control circuit, base of transistor (T2), and emitter of transistor (T2).

When transistor (T2) begins to conduct current, the base current flow is cut. This causes the collector circuit to shut off and transistor (T1) stops conducting current.

When transistor (T1) stops conducting, current stops flowing through the primary of the ignition coil. This causes the primary magnetic field to collapse across the secondary windings of the ignition coil. High voltage is then induced into the secondary to *fire* the spark plug.

The secondary circuit includes the coil secondary windings, high tension lead, spark plug, and common grounds returning to the coil secondary.

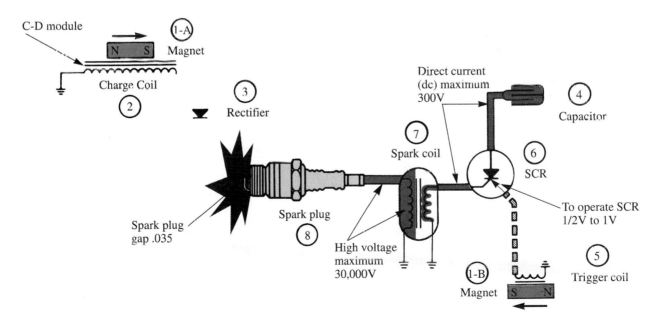

Figure 9-41. *1B—Magnet induces a small current in trigger coil. 5—Trigger coil switches on silicon controlled rectifier. 6—Rectifier permits capacitor to discharge 300V into primary winding of spark coil. 7—Spark coil steps up voltage in secondary winding. 8—Spark plug fires.*

Diode (D1, D2)	A ——▷|—— K	Allows one way current from Anode "A" to Cathode "K" as rectifier.
Flywheel	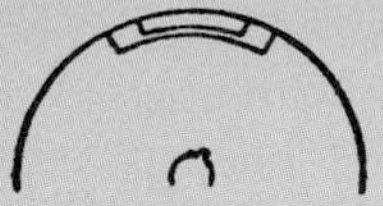	Provides magnetic flux to primary windings of ignition coil.
High-tension lead		Conducts high voltage current in secondary windings to spark plug
Ignition coil	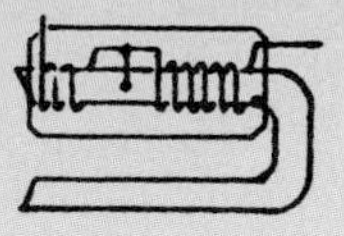	Generates primary current, and transforms primary low voltage to secondary high voltage.
Ignition switch	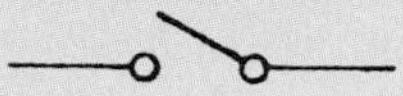	No spark across gap of spark plug when switch is at "STOP" position.
Resistor (R1, R2)	 —/\/\/—	Resists current flow.
Spark plug		Ignites fuel-air mixture in cylinder.
Thyristor (S)	A ——▷|—— K G	Switches from blocking state to conducting state when trigger current/voltage is on gate "G".
Transistor (T, T1, T2)	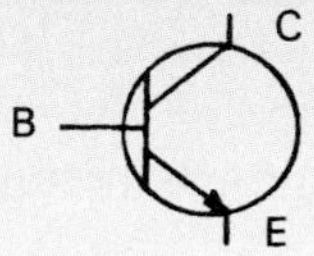 B —— C E	Very small current in the base circuit (B to E) controls and amplifies very large current in the collector circuit (C to E). When the base current is cut, the collector current is also cut completely.

Figure 9-42. *Study components of transistor controlled ignition system.*

When the ignition switch is off, **Figure 9-43,** the primary circuit is grounded to prevent the plug from firing. Diode (D1) is installed in the circuit to protect the TCI module from damage.

The ESG circuit in **Figure 9-43** is used to retard the ignition timing. At high engine rpm, the ESG circuit conducts. This bypasses the trigger circuit and delays when the current reaches the base of transistor (T2).

Magneto ignition systems compared

The following are definitions for the three general classifications of magneto ignition systems.

1. A mechanical breaker ignition (MBI) system is a flywheel magneto inductive system commonly used for internal combustion engines. It employs mechanical breaker contacts to time or trigger the system.

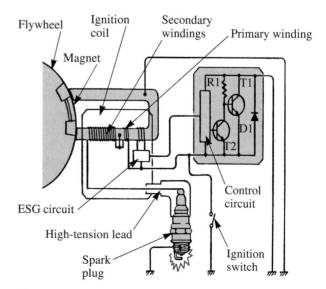

Figure 9-43. *Study how this transistor circuit is used to operate the ignition coil.*

2. A capacitor discharge ignition (CDI) system is a solid state (no moving parts) system that stores its primary energy in a capacitor and uses semiconductors for timing or triggering the system.

3. A transistor controlled ignition (TCI) system is an inductive system that *does not* use mechanical breaker contacts. It utilizes semiconductors (transistors, diodes, etc.) for switching purposes.

Figure 9-44 compares these three types of magneto ignition systems. Study them!

Battery ignition systems

The *battery ignition system* has a low voltage primary circuit and a high voltage secondary circuit. Like the magneto system, it consists of a coil, condenser, breaker points (or solid state switching device), and spark plug. The basic difference is that the source of current for the primary circuit is supplied by a lead-acid battery. See **Figure 9-45.**

When the ignition switch is turned on, current flows from the positive post of the battery to the ignition coil. Current traveling through the primary windings of the coil builds up a magnetic field. See **Figure 9-46.** During this time, the breaker

Comparisons	Mechanical breaker Ignition System	Transistor Controlled Ignition System	Capacitor Discharge Ignition System
Abbreviation	MBI	TCI	CDI
Circuit type	Conventional	Solid state	Solid state
Energy source	Primary current of ignition coil	Primary current of ignition coil	Stored in capacitor
Trigger switch	Breaker contacts	Power transistor	Thyristor
Secondary voltage	Standard	Standard	Higher
Spark duration	Standard	Standard	Shorter
Rise time*	Standard	Standard	Shorter
Maximum operating speed	Standard	Higher	Higherspeed
Maintenance	Regap and retime	none	none

*Rise time–time required for maximum voltage to occur.

Figure 9-44. *Chart compares breaker point, transistor controlled, and capacitor discharge systems. Note differences and similarities.*

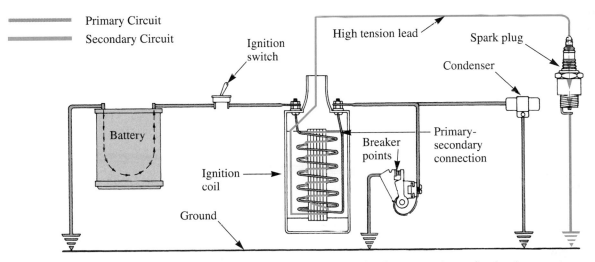

Figure 9-45. *Battery system is similar to a magneto system, except that battery replaces flywheel magnets.*

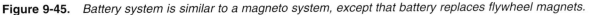

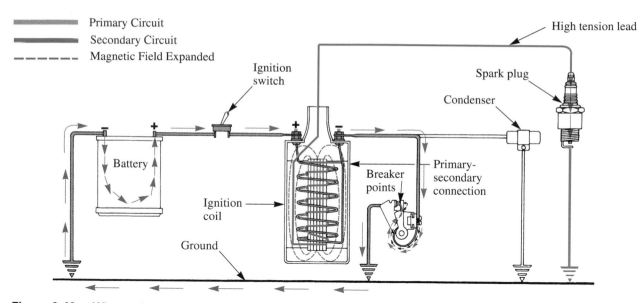

Figure 9-46. *When points close in battery ignition system, primary current builds a magnetic field around coil. (Kohler Co.)*

points are closed. Ignition at the plug is not required, so the current returns to the battery through the common ground.

Then, at the exact time when ignition at the plug is required, the breaker points are opened by the cam. Current flow stops abruptly, causing the magnetic field surrounding the coil to collapse. See **Figure 9-47.** This rapid change of magnetic flux causes voltage to be induced in every turn of

the primary and secondary windings. Voltage in the primary winding (about 250V) is quickly absorbed by the condenser.

Without the condenser, the current would arc at the breaker point gap, burning the points. The condenser acts as a reservoir for the sudden surge of power in the primary windings of the coil. The condenser holds the current for an instant, then releases it to the primary circuit, as shown in **Figure 9-48.**

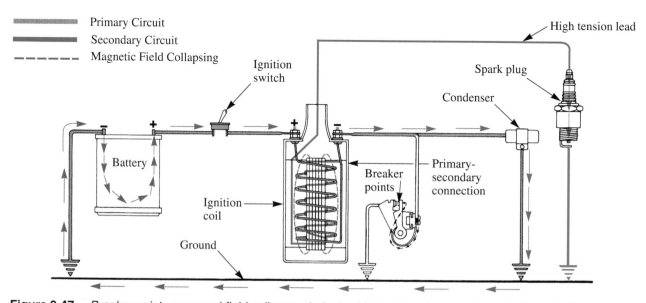

Figure 9-47. *Breaker points open and field collapses, inducing high voltage in secondary winding of coil. Condenser absorbs voltage surge in primary winding.*

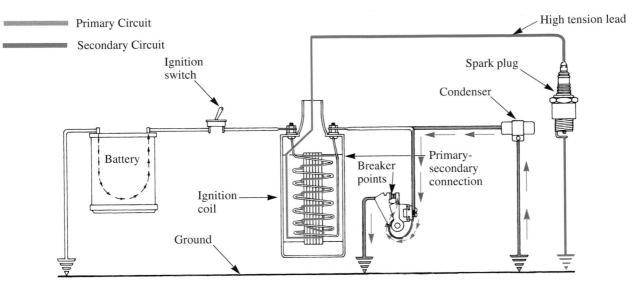

Primary Circuit

Secondary Circuit

Ignition switch

High tension lead

Spark plug

Condenser

Battery

Ignition coil

Breaker points

Primary-secondary connection

Ground

Figure 9-48. *Condenser discharges back into primary circuit.*

Note that many battery ignition systems employ solid state ignition components instead of mechanical breaker points.

High voltage secondary current

The voltage built up in the secondary winding of the coil can become as high as 25,000V. It has approximately 100 times as many turns of wire as the primary. See **Figure 9-49.** Normally, the voltage does not reach this value. Once it becomes great enough to jump the spark plug gap, the voltage drops. Usually, the amount required to jump the gap is between 6000V and 20,000V. The actual amount of voltage required depends upon variables such as compression, engine speed, shape and condition of electrodes, spark plug gap, etc.

Auto-transformer type ignition coil

The *auto-transformer type ignition coil* used on some small gasoline engines serves as a step-up transformer. It increases low voltage primary current to the high voltage required to bridge the spark plug gap. The primary and secondary windings are connected, and the common ground of the battery and primary circuit is used to complete the secondary circuit.

With this type of coil, very little primary current can flow into the secondary circuit because the secondary circuit is normally open at the spark plug gap. Primary current is just not great enough to jump the gap. Therefore, the two circuits function separately.

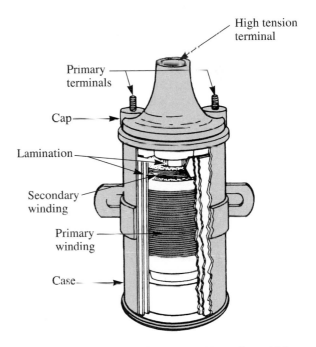

High tension terminal

Primary terminals

Cap

Lamination

Secondary winding

Primary winding

Case

Figure 9-49. *Auto-type ignition coil has about 100 times as many turns in secondary winding as in primary winding. Coil can produce 25,000V, if needed.*

The primary winding of the coil consists of about 200 turns of heavy copper wire. The secondary winding has approximately 20,000 turns of very fine copper wire. Because the magnetic field collapses through such a great number of conductors in the secondary, a very high voltage (electrical potential) is developed. The amperage (rate of electron flow), however, is proportionately low. This is a typical characteristic of any transformer.

Laminated iron is used as the center core of the coil. See **Figure 9-49.** It also forms the outer shell of the inner assembly, providing maximum concentration of the magnetic field. The inner assembly is sealed in a coil case, and the remaining space inside is filled with a special oil to minimize the effects of heat, moisture, and vibration.

The top of the coil is provided with two primary terminals. They are marked positive (+) and negative (−). The positive terminal must be connected to the positive side of the battery. The negative terminal connects to the breaker points. See **Figure 9-48.** The center tower of the coil contains the high tension terminal.

The lead-acid battery

The battery is the sole source of energy for the battery ignition system of a small gasoline engine. A generator is used to replenish energy in the battery. However, the generator does not supply energy directly to the ignition system.

Lead-acid type batteries are used in battery ignition systems. The cell plates are made of lead, and a sulfuric acid and water solution serves as the electrolyte. *Wet charged* or *dry charged* types are available. Wet batteries are supplied with the electrolyte in them, ready for use if the charge has been kept up. Dry charged batteries must have electrolyte installed after purchase. Both types of batteries function in the same way.

Battery construction

The typical 12V battery is constructed with a hard rubber case and six separate compartments called cells. See **Figure 9-50.** There is a specific number of negative and positive plates in each cell. The greater the number of plates per cell, the higher the ampere-hour rating (capacity to provide current for a specific length of time) of the battery. The positive plates have a lead oxide covering. The negative plates have a porous or spongy surface.

Battery voltage

A chemical reaction causes each negative plate to lose electrons and each positive plate to gain electrons when surrounded by electrolyte. The plates, therefore, develop an electrical potential between them. All plates of a like charge are electrically connected, causing accumulative charges to be present at the positive and negative battery terminals.

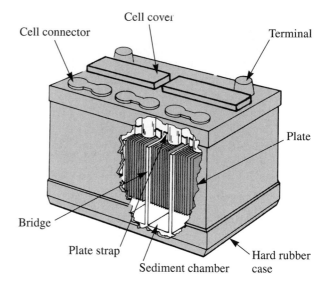

Figure 9-50. *Battery provides all current for battery ignition system. A 12V battery has six cells each producing 2V.*

Each cell of a battery in good condition contributes approximately 1.95V to 2.08V. Six fully charged cells will produce at least 12V. If not, the battery must be recharged or replaced.

A discharging battery

When a battery discharges without replacement of energy, the sulfuric acid is chemically withdrawn from the electrolyte, specific gravity goes down, and lead sulfate deposits accumulate on the plates. If sulfate deposits become too great or the level of the electrolyte falls lower than the top of the plates, permanent damage may be done to the battery.

When a battery is recharged, a controlled direct current is passed through the battery in reverse direction from normal operation. This causes a reversal in chemical action and restores the plates and electrolyte to active condition.

Summary

The primary purpose of the ignition system is to provide sufficient voltage to discharge a spark between the electrodes of a spark plug. Many small engines use magneto systems to supply ignition spark. Magneto systems are self-contained systems that produce electrical current for ignition without an outside primary source of electricity. Basic magneto system parts include: permanent magnets; high tension coil; breaker points, transistor, or capacitor; high tension spark plug wire; and spark plugs.

Using the correct spark plug can greatly increase engine efficiency and service life. Reach, heat range, and electrode consideration must all be considered.

There are three general types of magneto ignition systems. The mechanical breaker ignition system is a flywheel magneto inductive system. It employs mechanical breaker points to time triggering the ignition system. Some small engines are equipped with mechanical systems that retard and advance timing. Dwell (cam angle) is the amount of time that the breaker points stay closed during one revolution of the cam.

Many small engines are equipped with solid state ignition systems that use electronic devices in place of one or more mechanical ignition components. Most solid state systems have no moving parts and do not require mechanical adjustments. The two most common solid state systems are the capacitive discharge ignition (CDI) system and the transistor controlled ignition (TCI) system.

The CDI system stores primary energy in a capacitor and uses semiconductor devices to trigger the ignition system. The TCI system is an inductive system that utilizes semiconductor devices (transistors, diodes, etc.) for switching purposes.

Instead of a magneto, some ignition systems use a lead-acid battery to supply primary current. These systems generally employ an auto-type ignition coil. Because the battery is the only source of energy for battery ignition systems, a generator is used to replenish energy in the battery.

Know These Terms

magneto system	heat ranges
amperes (A)	switching devices
volts (V)	breaker points
ohms (Ω)	tungsten
Ohm's law	electronic switching
ignition coil	devices
high tension lead	mechanical breaker
insulator	point ignition (MBI)
center electrode	condenser
reach	Alnico

ignition advance system	transistor controlled
dwell (cam angle)	ignition (TDI)
solid state	battery ignition system
capacitive discharge	auto-transformer type
ignition (CDI)	coil

Chapter 9
Review Questions

Answer the following questions on a separate sheet of paper.

1. The two major tasks performed by an ignition system are to _____ and _____.
2. If a four-cycle engine runs at 3600 rpm, the number of sparks per minute required at the spark plug would be _____ per minute.
3. Name the six main electrical components that make up the magneto ignition system.
4. Regarding atoms, _____.
   a. electrons and protons are tightly bonded together
   b. electrons are positively charged particles
   c. neutrons are negatively charged, protons are positively charged
   d. electrons do not collide because they are negatively charged
5. The electron theory states that _____.
   a. like charges attract each other
   b. like charges repel each other
   c. only negative charges repel each other
   d. electrons orbit the nucleus because the protons repel the electrons
6. Substances that have electrons that can move freely from atom to atom are said to be good _____.
   a. conductors
   b. dielectrics
   c. insulators
   d. nonconductors
7. List three good electrical current conductors.
8. List three good insulating materials.
9. A source of electricity _____.
   a. has an excess of electrons and is positively charged
   b. has an excess of electrons and is negatively charged
   c. lacks electrons and is negatively charged
   d. lacks electrons and is positively charged

10. Electrical current in a battery flows from _____.
    a. negative to positive
    b. positive to negative
11. Electrical current in the primary circuit of a battery ignition system flows from _____.
    a. negative to positive
    b. positive to negative
12. What are the three ways in which an electrical potential can be produced?
13. Match the units of electrical measurement:
    a. _____ Rate of electron flow.   Ohms
    b. _____ Resistance to   Volts
              electron flow.   Amperes
    c. _____ Electrical potential.
14. In a circuit of 5 amperes and 2 ohms, the voltage would be _____.
15. In a circuit with 12 volts and 5 amperes, the resistance would be _____ ohms.
16. Soft iron used in the laminations of a coil or transformer has _____.
    a. good magnetic retention
    b. poor magnetic retention
    c. nonmagnetic properties
17. What happens when a coil of wire is passed through a magnetic field?
18. What happens when electric current is passed through a coil of wire?
19. In the ignition coil, the primary winding has _____.
    a. many turns of fine wire
    b. few turns of fine wire
    c. few turns of heavy wire
    d. many turns of heavy wire
20. The coil acts as a transformer that _____.
    a. steps down the voltage and increases the output amperage
    b. steps up the voltage and amperage
    c. steps down the voltage and amperage
    d. steps up the voltage and decreases the output amperage
21. The greatest resistance to the highest voltage in the ignition circuit occurs at _____.
    a. spark plug
    b. breaker points
    c. condenser
    d. secondary of the coil
22. Would a *cool* spark plug have a short or long insulator nose?
23. Spark plugs of the *cool* type should always be matched with cool running engines. True or False?

24. The device used by engineers and technicians to measure spark plug temperatures under running conditions is a(n) _____ _____.
25. Breaker point contacts are made of a very hard material called _____.
26. When the breaker points in the magneto are closed, _____.
    a. current is induced in the primary circuit by the flywheel magnets and coil
    b. the spark plug fires
    c. a high voltage is induced in the secondary circuit
    d. the condenser absorbs current from the primary circuit for later use in the secondary
    e. All of the above.
27. The spark plug fires when _____.
    a. magnetic field collapses through the secondary windings of the coil
    b. piston is nearing top of compression stroke
    c. breaker points open
    d. voltage potential is high enough to bridge the electrodes
    e. All of the above.
28. The condenser acts as a voltage reservoir or shock absorber for the primary circuit. Its primary purpose is to prevent _____.
29. When the breaker points open, voltage in the primary circuit may surge to as much as _____.
30. The stop switch grounds the _____.
31. When connecting the auto-type ignition coil in the circuit, the positive terminal of the battery must be connected to _____.
    a. the positive terminal of the coil
    b. the negative terminal of the coil
    c. either terminal of the coil
32. When breaker points are set with a wider gap, the dwell _____.
    a. becomes greater
    b. becomes less
    c. does not change
33. Name five advantages of a solid state ignition system.

Suggested Activities

1. Experiment with a coil of wire and a magnet to induce a current. Use a galvanometer to show the current flow.

2. With a dry cell, iron filings, copper wire, and a piece of paper, demonstrate the magnetic flux produced when the current flows through the wire.
3. Using copper wire, form it into a loose coil. Use a dry cell to pass a current through the wire and determine the polarity with a permanent magnet. Reverse the polarity.
4. Using the coil of wire from Suggested Activity number 3, wrap insulating tape around an iron bar and place it in the coil. Demonstrate how this can improve the magnetic strength of the coil when current is flowing.
5. Make a visible magneto mounted on a display board or built into a clear acrylic box so that it can be manually turned with a crank. Old, but usable, engine parts can be used.
6. A workable battery ignition system can be built and mounted as a display board. Demonstrate the operation and principles involved in this system.
7. Make a large Ohm's law pie (visual aid discussed in this chapter) and display it until it has been learned. Make up some problems that require the use of formulas.
8. Make a collection of various kinds of spark plugs.
9. Section a spark plug, so that the inner construction can be studied.
10. Cut away a *dry* type battery that has not had electrolyte added to it and discuss how it works.
11. Section an old ignition coil to show the primary and secondary windings around the core.
12. Carefully open a condenser to display the lamination of aluminum foil and insulation.
13. Collect several engines with different types of ignition advance mechanisms. Demonstrate them to the class.
14. Disassemble a magneto and demonstrate how it works.

This walk-behind greens mower is used for golf course maintenance. It is driven by a four-cycle, overhead valve 3.7 horsepower engine. (Deere & Co.)

Lubrication Systems

After studying this chapter, you will be able to:
▼ Define friction and explain how it affects the internal engine components.
▼ List the functions of lubricating oil.
▼ Differentiate between the lubrication systems in two-cycle engines and four-cycle engines.
▼ Explain the operation of ejection pumps, barrel pumps, and positive displacement pumps.
▼ Explain the function of oil filter systems and differentiate between the three main types.

Principles of Lubrication

Lubrication is the process of reducing friction between sliding surfaces by introducing a slippery or smooth substance between them. See **Figure 10-1.** Lubricants come in dry (powdered), semi-dry (grease), and liquid (oil) forms. Oil is the most important lubricant for small engine use, simply

because it is often the only lubrication the engine needs. See **Figure 10-2.**

Friction

Friction is the resistance to motion created when one dry surface rubs against another. Even highly polished metal surfaces have irregularities (when studied under a microscope) that will create a great deal of friction if rubbed together. The microscopic roughness will resist movement and create heat.

As the relatively rough projections on the contact surfaces rub across each other, they eventually break off and become loose particles. These loose particles, in turn, work between the contact

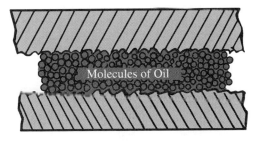

Figure 10-1. *A film of engine oil serves to separate and lubricate machined surfaces. When parts are in motion, oil molecules roll over one another like microscopic ball bearings.*

Figure 10-2. *Oil is the most important lubricant used for engine lubrication. (Briggs and Stratton Corp.)*

surfaces and gouge grooves in the metal. Then, as friction and heat increase, the metal parts expand, causing greater pressure between the surfaces and creating even greater friction. This condition of wear exists until the parts either weld themselves together or seize (expand so much that mating parts cannot move).

In some cases, the excessively worn parts lose so much material from their contact surfaces that they become too loose to function properly. When this happens, the scored part should be replaced.

Preventing wear due to friction

In designing a small gasoline engine, the manufacturer selects suitable materials for parts that will be in moving contact with one another. For example, precision insert bearing shells are used in connecting rods and caps, and in main bearing saddles and caps. See **Figure 10-3.** The steel backing has a cast babbitt surface. *Babbitt*, an alloy of tin, copper and antimony, has good antifriction qualities.

Rod and main bearing inserts must withstand reciprocating and rotational forces while fitting closely to the crank journal or throw. Notice that the oil hole in the insert in **Figure 10-3** permits oil to enter the oil groove. With the crankshaft in motion, the oil groove distributes oil all the way around the insert. Then, oil travels outward to bearing edges and back into crankcase.

Some bearing inserts have laminations of other metals to back up the babbitt. See **Figure 10-4.**

This construction enhances the properties of the insert, making it more heat resistant and durable.

The lock tab illustrated in **Figure 10-4** is used to prevent the bearing insert from rotating in the connecting rod or the main bearing cap. An insert with the tab in place in a rod assembly is shown in **Figure 10-5.**

Regardless of the quality of the material used, all bearing surfaces in small gasoline engines must have oil separating moving parts that are in close contact. See **Figure 10-6.** A thin film of oil must coat the area between: piston/piston rings and cylinder wall, piston pin and piston or connecting rod, valve stem and guide, valve tappet and guide,

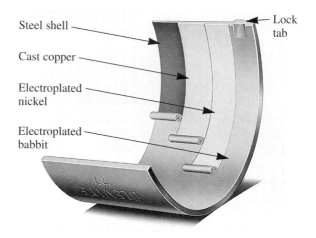

Figure 10-4. *Some precision inserts have several layers of material between steel shell and babbitt. (Clevite Corp.)*

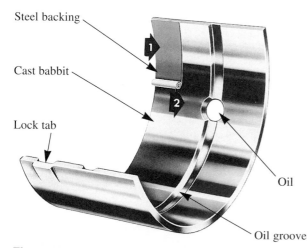

Figure 10-3. *Precision insert type of engine bearing has a steel shell with a cast babbitt inner surface. Oil hole and groove carries pressurized oil to bearing surfaces.*

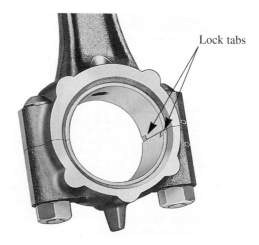

Figure 10-5. *Lock tabs prevent bearing insert from rotating in connecting rod. Tabs must be seated in recesses provided in rod. (Wisconsin Motors Corp.)*

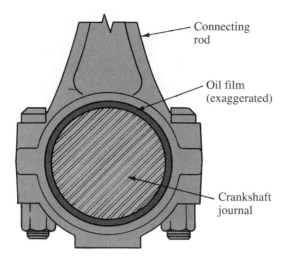

Figure 10-6. *Oil film between close fitting metallic parts prevents actual contact. Oil provides a relatively frictionless movement of parts.*

main and rod bearing inserts and crankshaft journals, etc.

The thin film of oil between the close fitting parts may be only a few molecules thick, yet this is enough to prevent the two metal surfaces from actually touching. The molecules of oil then roll over one another, acting like microscopic ball bearings between the surfaces. See **Figure 10-1.**

Lubricating oil

Modern motor oil is a highly specialized product, which has been developed by engineers and chemists to perform many essential functions in an engine. In order to operate efficiently, engines depend on motor oil to do the following:

- Permit easy starting.
- Lubricate engine parts.
- Protect against rust and corrosion.
- Keep engine parts clean.
- Cool engine parts.
- Seal combustion pressures.
- Prevent foaming.
- Aid fuel economy.

Permits easy starting

The proper oil must be used if an engine is to start easily. If the oil is too thick, it will create so much drag between moving parts that the engine will not crank fast enough to start quickly and keep running.

Cold temperatures thicken oil. Oil for winter use must be thin enough to permit adequate cranking speeds at the lowest anticipated temperature. Once the engine is started, the oil must be fluid enough to flow quickly to the bearings to prevent wear. On the other hand, the oil must also be thick enough to provide adequate protection when the engine reaches normal operating temperatures.

Lubricates and prevents wear

Once an engine is started, oil must circulate quickly to prevent the metal-to-metal contact that can cause wear, scoring, or seizure of engine parts. See **Figure 10-1.** Bearings and cylinder walls are particularly sensitive to movement, pressure, and oil supply. Oil supplies to these components must be continually replenished by adequate flow and distribution. Refer to *Engine Failure Analysis* in *Appendix.*

Once the oil reaches the moving parts, it must lubricate and prevent wear of the moving surfaces. The oil is expected to establish a complete, unbroken film between surfaces. Lubrication engineers call this full-film or ***hydrodynamic lubrication.***

Under some conditions, it is impossible to maintain a continuous oil film between moving parts, and there is intermittent metal-to-metal contact between the high spots on sliding surfaces. Lubrication engineers refer to this as ***boundary lubrication.*** When this occurs, the friction generated by the contact can produce enough heat to cause the metals to melt and weld together. Unless counteracted by proper additive treatment, the result is either part roughening or tearing.

Boundary lubrication always exists during engine starting and often exists during the operation of a new or rebuilt engine. Boundary lubrication is also found around the top piston ring where oil supply is limited, temperatures are high, and piston motion is reversed.

Protect against rust and corrosion

Normally, burning fuel forms carbon dioxide and water. Gasoline engines, however, do not burn all of their fuel completely. Some of the partially burned gasoline undergoes complex chemical changes during combustion and, under some conditions, forms soot or carbon. Some of this soot and partially burned fuel escapes through the exhaust in the form of black smoke. Part of the soot and fuel escapes past the rings and into the crankcase. They tend to combine with water to form sludge and varnish deposits on critical engine parts. Sludge buildup can clog oil passages, reducing oil flow. Varnish buildup interferes with proper clearances, restricts oil circulation, and causes vital engine parts to stick, resulting in rapid engine failure.

Water causes a considerable problem in the engine. For each gallon of fuel burned, more than one gallon of water is formed. Although most of this water is in vapor form and escapes with the exhaust, some condenses on the cylinder walls or escapes past the piston rings and is trapped temporarily in the crankcase. This occurs most frequently in cold weather, before the engine reaches its normal operating temperature.

In addition to water, other corrosive combustion gasses get past the rings and are condensed or dissolved in the crankcase oil. Add to this the acids formed by the normal oxidation of oil, and the potential for rust and corrosive engine deposits becomes significant.

Engine life depends in part on the ability of motor oil to neutralize the effects of these corrosive materials. Due to extensive research by oil chemists, effective, oil-soluble, chemical compounds have been developed and are added to motor oil to provide protection for engine parts.

Keeps engine parts clean

Engines are unable to tolerate excessive amounts of sludge and varnish on critical parts. Sludge can collect on oil pump screens and limit the flow of oil to vital engine parts. Accumulation of varnish can cause piston rings to drag, preventing the engine from developing full power. Plugged oil-control rings prevent the removal of excess oil from the cylinder walls, resulting in excessive oil consumption.

Straight mineral oils have very limited ability to keep contaminants from forming masses of sludge in the engine. Therefore, various *detergent/dispersant additives* have been blended into modern motor oils. These additives keep engine parts clean by suspending fine particles of the oil contaminants until they can be trapped by the oil filter. See **Figure 10-7.**

Detergent/dispersant additives are also effective at preventing varnish deposits within an engine. Varnish-forming materials react chemically with the oxygen in the crankcase to form complex chemical compounds. These compounds, which react with each other and with oxygen, are baked into a hard coating on hot engine parts. Piston rings

Oil filter

Figure 10-7. *An oil filter traps contaminants picked up by engine oil. Most disposable filters are located on side of engine. (Briggs and Stratton Corp.)*

and bearings are particularly sensitive to varnish deposits. When deposits are formed on these components, engine operation can be severely impaired.

Cools engine parts

Crankshaft, camshaft, timing gears, bearings, pistons, and other components in the lower part of the engine are directly dependent on the motor oil for cooling. Each of these parts has definite temperature limits that must not be exceeded. Some can tolerate fairly high temperatures, while others, such as the main and connecting rod bearings, must run relatively cool to avoid failure. These parts must get an ample supply of cool oil to absorb heat and transfer it back to the crankcase.

Seals combustion pressures

The surfaces of the piston rings, ring grooves, and cylinder walls are not completely smooth. When examined under a microscope, these surfaces consist of minute hills and valleys. Therefore, the rings cannot completely prevent high combustion and compression pressures from escaping into the low pressure area of the crankcase. This results in a reduction in engine power and efficiency. Motor oil fills in the hills and valleys between ring surfaces and cylinder walls and helps seal compression and combustion pressures. Because the oil film at these points is quite thin, it cannot compensate for excessively worn rings, ring grooves, or cylinder walls. When such conditions exist, oil consumption may be high. New and rebuilt engines may also consume excessive amounts of oil until the hills and valleys on the surfaces have smoothed out enough to allow the oil to form a good seal.

Prevents foaming

Modern engine oil has antifoaming additives to prevent it from whipping into bubbles that do not readily collapse. Oil foam does not cool well and does not provide adequate lubrication.

Aids fuel economy

Oils are available that have been formulated to improve fuel economy in gasoline-fueled engines. The fuel economy benefits are achieved by various means, including the use of friction-modifier additives.

Oil selection

The cost of oil for a small gasoline engine is relatively low. However, the particular oil selected for use in a small engine is extremely important to the life of the engine. See **Figure 10-8.**

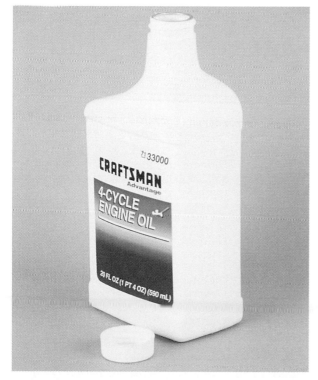

Figure 10-8. *When a full quart of engine oil is not used, seal with the cap to keep oil clean during storage.*

The oil recommended for use in a given engine may be shown on the engine nameplate or on a special label attached to the engine. The operator's manual also carries the manufacturer's recommendation, as does the lubrication guide provided by major oil companies.

Specifications for engine oils are given in two ratings:

- Society of Automotive Engineers (SAE) Viscosity, referred to as *viscosity grade.*
- American Petroleum Institute (API) Engine Oil Service Classification, often referred to as the *type of oil.*

SAE viscosity grade

Viscosity must be considered when selecting engine oil. *Viscosity* is a measure of the oil's resistance to flow. This resistance keeps the oil from being squeezed out from between engine surfaces as they move under load or pressure. The resistance to flow is a function of the molecular structure of the oil. Because it is this resistance that causes most of the drag during starting, it is important to use an oil with viscosity characteristics that ensure satisfactory cold cranking, good oil circulation, and adequate temperature protection.

The SAE has established a viscosity range classification system for engine lubricating oils. All motor oils are classified according to this system, which is used worldwide. Each oil is assigned an SAE grade (or grades) that signifies the range into which it falls. Single grade motor oils commonly used today are SAE 5W, 10W, 15W, 20W, 20, 30, 40, and 50. Thick, slow-flowing oils have high numbers. Thin, free-flowing oils have low numbers. The *W* denotes oils suitable for use at low ambient (encompassing) temperatures. SAE numbered oils that do not have *W* designations are measured for viscosity at 212°F (100°C) to ensure adequate viscosity at normal engine operating temperatures.

Figure 10-9 compares the viscosity recommendations of five manufacturers for their four-cycle engines at various operating temperatures. Note that the higher viscosity oils (more resistant to flow) are recommended for higher temperatures. A thick oil in low temperature operation makes a cold engine very difficult to start, and may deprive critical parts from adequate lubrication while the oil is gaining heat from combustion. *Cold running* can result in scored cylinder walls and engine bearings.

Four-Cycle Crankcase Lubrication (Viscosity-grade) Recommendations of Manufacturers					
Manufacturer	**Above 40°F**	**Above 32°F**	**Below 5°F**	**Below 0°F**	**Below -10°F**
Briggs and Stratton	SAE 30 or 10W-30	5W-20 or 10W			
Clinton	SAE 30		10W		5W
Kohler	SAE 30		SAE 10W	5W or 5W-20	
Wisconsin	SAE 30	SAE 20 or SAE 20W	10W		
Tecumseh	SAE 30		10W-30		

Figure 10-9. *A comparison of viscosity-grade recommendations by five engine manufacturers. Recommendations are for specific models only, not for full-line coverage.*

The temperature effect on viscosity varies widely with different types of oils. A standard has been developed for measuring the relationship between viscosity and temperature. This standard is called the viscosity index (V.I.). Oil with a high viscosity index shows little change in viscosity over a wide range of temperatures. Today, through the use of selective crude oil stocks, new refining methods, and special chemical additives, there are many high viscosity index oils that are light enough to provide easy cranking at low temperatures and heavy enough to perform satisfactorily at high temperatures. These oils, which meet the viscosity requirements of two or more SAE grades, are known as *multigrade* or *multiviscosity* oils. Examples are oils labeled as SAE 5W-20, SAE 5W-30, and SAE 10W-30.

Some of the oils listed in **Figure 10-9** are single viscosity grade oils, such as SAE 20. Others are multiviscosity grade oils such as SAE 5W-20. Although multiviscosity oils can be substituted for single viscosity grades in four-cycle engines, they should not be used in two-cycle engines.

API engine oil service classification

The *API engine service classification* is a dynamic arrangement that allows new categories to be added as engine designs change, placing more demands on motor oil. Currently, the classification system defines 18 categories of engine oil service. Of these categories, only SJ, CF, CF-2, CF-4, and CG-4 are considered suitable for any road vehicle in operation today. All other categories are considered obsolete for automotive and heavy-duty truck service. They are shown for historical purposes only. Category SH became obsolete after August 1, 1997. See **Figure 10-10.** The typical applications of the active categories include:

- **SJ.** (1997 gasoline engine warranty maintenance service) Service typical of gasoline engines in current and earlier passenger cars, sport utility vehicles, vans, and light trucks operating under the vehicle manufacturers' recommended maintenance procedures.

API Engine Service Categories	
I. Passenger Cars, Vans, and Light-Duty Trucks	
SA	Obsolete
SB	Obsolete
SC	Obsolete
SD	Obsolete
SE	Obsolete
SF	Obsolete
SG	Obsolete
SH	Obsolete (Can still be used in conjunction with an API C- service category)
SJ	Current
II. Heavy-Duty Diesel (Commercial and Fleet Service)	
CA	Obsolete
CB	Obsolete
CC	Obsolete
CD	Obsolete
CD-II	Obsolete
CE	Obsolete
CF	Current
CF-2	Current
CF-4	Current
CG-4	Current

Figure 10-10. *American Petroleum Institute (API) has published oil service classifications in which oils are recommended for specific service conditions.*

- **CF.** (Off-road indirect-injected diesel engine service) Service typical of off-road indirect-injected diesel engines and other diesel engines that use a broad range of fuels. CF oils may be used when CD oils are recommended.
- **CF-2.** (Two-stroke diesel engine service) Service typical of two-stroke engines requiring effective control over cylinder and ring face scuffing and deposits. CF-2 oils may be used when CD-II oils are recommended. These oils do not necessarily meet the requirements of CF or CF-4 oils.
- **CF-4.** (1990 diesel engine service) Service typical of high-speed four-stroke diesel engines. CF-4 oils exceed the requirements of the CE oils, providing improved control of oil consumption and piston deposits.
- **CG-4.** (Severe-duty diesel engine service) Service typical of high-speed four-stroke diesel engines used in highway and off-road applications. CG-4 oils control piston deposits, corrosion, wear, foaming, oxidation, and soot accumulation. They are especially effective in engines designed to meet 1994 exhaust emission standards and may be used when CD, CE, or CF-4 oils are recommended.

The API engine oil service classification symbol

The API engine oil service classification symbol provides the consumer with information about an oil's characteristics and applications. See **Figure 10-11.** The symbol typically appears on the label of an oil container. It may also be found on the engine's oil fill cap or in the owner's manual. The symbol is divided into three parts including:

- **Top.** Specifies the oil's service classification or recommended applications.
- **Center.** Describes the oil's viscosity.
- **Bottom.** Reserved for information on the oil's fuel saving properties. Oils labeled *energy conserving II* offer better fuel-saving properties than oils labeled *energy conserving.*

Engine lubrication

The way that moving parts are lubricated differs in two-cycle and four-cycle small gasoline engines. In preparing a two-cycle engine for use, a specified amount of two-cycle engine oil is mixed with each gallon of gasoline to provide fuel for the engine. Refer to manufacturer's recommended oil-to-fuel ratio for specific engine make and model. **Figure 10-12** shows the weight of oil to mix per gallon of gasoline to obtain a given ratio. The oil should be weighed on a scale that is calibrated in ounces. Before pouring the oil into the container the scale should be zeroed with the empty container. This fuel is thoroughly mixed in a separate container before being poured into the fuel tank. See **Figure 10-13.** The correct method to mix oil and fuel is to first empty about half the fuel out of the fuel container. Now, weigh the oil in a paper or plastic cup, and pour that oil into the fuel container. Then, place the cap on the fuel container and shake it so the oil mixes thoroughly with the fuel. Add the fuel that was removed and shake the container thoroughly once more. In operation, an oil mist is created that lubricates the cylinder wall and all internal engine parts.

Figure 10-11. *Typical API engine oil service classification symbol for one type of SAE 10W-30 oil.*

Two-Cycle Engine Fuel/Oil Ratio Chart						
Gallons of Gasoline	*16:1*	*20:1*	*24:1*	*32:1*	*40:1*	*50:1*
1	8	6	5	4	3	3
2	16	13	11	8	6	5
3	24	19	16	12	10	8
4	32	26	21	16	13	10
5	40	32	27	20	16	13
6	48	38	32	24	19	15

Figure 10-12. *This chart is used when mixing oil and gasoline. Note that for 1 gallon of gasoline mixed at a 40/1 or 50/1 ratio, 3 ounces is used for both.*

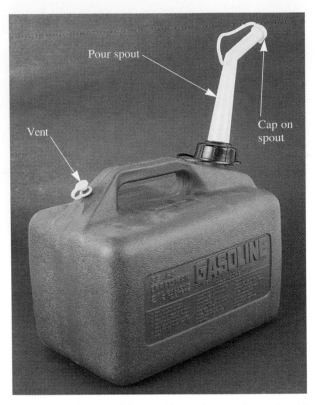

Figure 10-13. *Fuel for small gasoline engines should be stored in a clean, properly marked can with a vent cap and pouring spout.*

In readying a four-cycle engine for use, the fuel tank is filled with fresh gasoline. In addition, a specified amount of four-cycle engine oil (type and viscosity class recommended by the manufacturer) is poured into the crankcase of the engine. In operation, the air-gasoline mixture is *fired* in the combustion chamber. At the same time, the oil sump in the crankcase supplies lubrication for the cylinder wall and all of the internal engine parts.

Two-cycle engine lubrication systems

Air-cooled engine operation covers a wider range of varying speeds with much higher combustion chamber temperatures than water-cooled engines. Therefore, automotive engine oils *are not* suitable for two-cycle, air-cooled engines. Likewise, multiviscosity and detergent type oils may contain additives not intended for two-cycle engine operation and should not be used. Always use the type and viscosity grade of oil recommended by the engine manufacturer.

Certain manufacturers recommend the use of a specific SAE viscosity, diluted, two-cycle engine

oil. The diluent is added to make the oil pour more freely, particularly at low temperatures. As engine oil gets colder, one of its ingredients forms interlocking crystalline structures, which bind together until the oil finally solidifies. The diluent blocks the formation of these crystals, and the oil stays fluid.

Special additives are often recommended for two-cycle engine use. Since two-cycle engines are lubricated by mixing oil with the fuel, the oil eventually enters the combustion chamber and is burned. Some regular engine oils have additives that do not burn completely and leave a residue that fouls spark plugs and clogs exhaust ports.

The spark plugs shown in **Figure 10-14** were used in four test engines. Each engine ran for the same length of time, and identical preventive maintenance and adjustments were performed. The only exception in the test procedures was in the brand of engine oil used. Spark plug A in **Figure 10-14** was taken from an engine that used the oil recommended by the manufacturer. The other three spark plugs were removed from test engines operating with other brands of engine oil.

Not all engine oils will produce deposits of the type shown in **Figure 10-14,** but the importance of using the recommended oil should be quite clear from this comparison. Special additives for two-cycle oils must be selected to avoid or prevent unburned deposits. Oils containing these additives generally are sold under the brand name of the engine manufacturer.

Four-cycle engine lubrication systems

Four-cycle engines must be operated with the proper *type of oil*. For this reason, always use

Figure 10-14. *Spark plugs A, B, C, and D were used in identical test engines using different oils. Plug A was taken from engine using oil recommended by engine manufacturer.*

engine oil service classifications that have been established by the SAE and API. Most four-cycle engine manufacturers recommend oils supplemented by additives. These chemicals are added to improve the quality of the oil. They may prevent corrosion, provide a better cushioning effect between moving parts, help prevent scuffing, and/or reduce wear.

Detergent/dispersants, commonly called *detergents,* are added to some oils. Detergents suspend the dirt and sludge in the oil, where the contaminants can be trapped by a filter or readily drained before fresh oil is put in. Basically, if the recommended oil is used in an engine, special oil treatments should not be required.

Splash lubrication system

Small, four-cycle, gasoline engines generally use some type of *splash lubrication system* to lubricate internal machined surfaces. The splash lubrication system shown in **Figure 10-15** features an oil dipper arm on the connecting rod cap. The *dipper* is designed to pick up oil from the crankcase on every revolution of the crankshaft, splashing oil on the various moving parts as it is carried around by the crank throw.

With the splash system, the cylinder wall receives a generous amount of oil. To avoid oil-burning problems, the oil control ring on the piston removes excess oil, returning it to the crankcase as illustrated in **Figure 10-16.** The connecting rod bearings and piston pin receive lubrication through oil passage holes. See **Figure 10-17.**

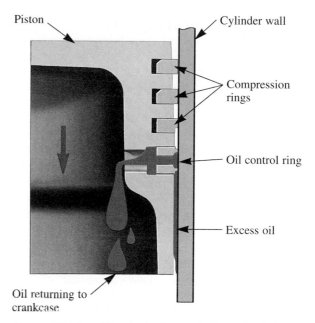

Figure 10-16. *Oil splashed on cylinder wall lubricates piston and piston rings. Excess oil is scraped from wall by oil control ring.*

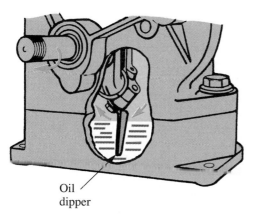

Figure 10-15. *With splash system, some oil dippers are cast onto connecting rod, others are bolted on. Oil level must be high enough for dipping action. (Briggs and Stratton Corp.)*

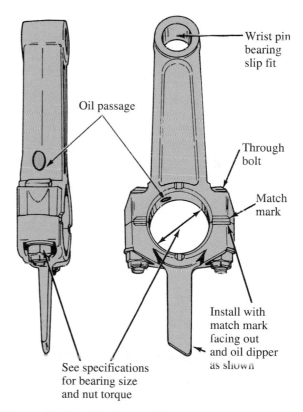

Figure 10-17. *The holes drilled in connecting rod, bearing insert, and piston pin boss provide passageways for lubricating oil.*

Constant level splash system

The *constant level splash system* provides three major improvements over the simple splash system. These improvements are as follows:

- A lubrication oil pump.
- A splash trough.
- A strainer.

The cam-operated pump supplies oil to the trough where the oil dipper picks it up and distributes it to the cylinder wall and moving parts. The term *constant level* is used because the pump can supply more oil than the dipper can remove. Therefore, the trough is always full. Oil returning to the crankcase must pass through the filter before it can be pumped back to the trough. This keeps large contaminants in the crankcase.

The constant level splash system will provide adequate lubrication as long as there is enough oil to supply the pump. However, if the oil level is low, the cooling effect of the oil is reduced.

Ejection and barrel pumps

The *ejection pump* forces oil under pressure against the rotating connecting rod. Some oil enters the connecting rod bearings, while the remaining oil is deflected to other parts in the crankcase. The ejection pump system is similar to the splash system, but it provides a more forceful spray of oil.

The *barrel pump* is a cylinder and plunger type of lubrication pump. See **Figure 10-18.** By design, an eccentric on the camshaft moves the plunger in and out of the pump cylinder. The camshaft is hollow and has holes from the center of the shaft to the eccentric.

In operation, the plunger is drawn out until a hole in the eccentric aligns with a hole in the plunger. This allows the cylinder to fill with oil. When the plunger is forced in, a different hole in the eccentric aligns with the plunger, and oil is forced through passages to the main bearings and crankshaft connecting rod journal.

Positive displacement oil pumps

Several types of *positive displacement oil pumps* are used in pressurized lubrication systems. See **Figure 10-19.** One common type is the gear pump shown in **Figure 10-20.** The end cover has been removed to expose two meshed gears. One gear is shaft-driven from the engine. It drives the second gear.

Note that the driving gear in **Figure 10-20** is keyed to the driving shaft. As the gears turn, oil

Figure 10-18. *Plunger of barrel pump is actuated by an eccentric shaft driven by valve camshaft. Ball-shaped plunger end is held in a socket so that it can pivot as shaft turns.*

fills the spaces between the teeth and is carried around to the oil outlet. No oil passes between the gears where the teeth are meshed, because of the tight fit.

If, for some reason, oil flow is restricted somewhere in the engine, the increase in pressure would raise the ball against the spring in the pressure relief valve. When this happens, oil will pass through the valve and recirculate through the pump. Recirculation of the engine oil continues until the restriction to flow ceases and pressure declines, allowing the ball to seat and the relief valve to close. Without a pressure relief valve in the system, pressures would become excessively high during high engine speeds.

Full-pressure lubrication systems

A *full-pressure lubrication system* is the type used in automobile engines. On some larger small engines, an almost completely pressurized system is used, including a positive displacement gear or rotor pump. Passages for oil flow are drilled to all critical points, such as camshaft bearings, main bearings, connecting rod bearings, and piston pins.

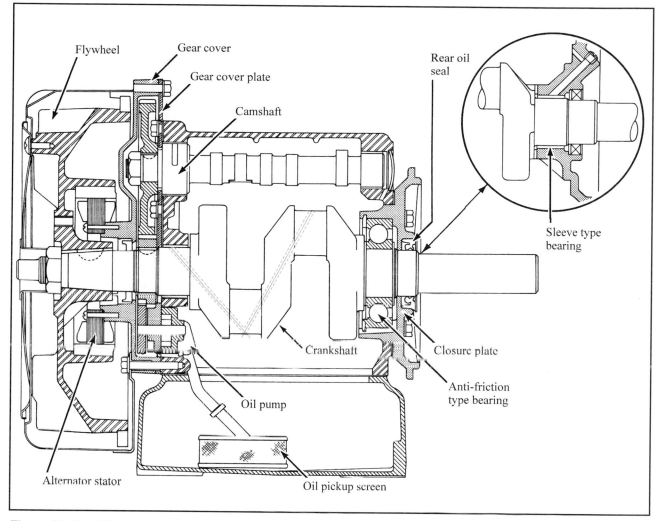

Figure 10-19. *Oil pump used in two cylinder engine supplies oil to moving internal engine parts. Note drilled crankshaft.*

A splash system is used in conjunction with the pressure system, particularly for lubricating cylinder walls.

Oil Filter Systems

Oil filters are used on some small engines. **Figure 10-21** shows one type of oil filter. Filters trap dirt, carbon, and other harmful materials, preventing them from circulating through the engine. The oil filter prevents very fine particles from circulating. The oil strainer, **Figure 10-22** (usually attached to intake side of oil pump), prevents large particles from entering the filter.

Three basic types of oil filter systems are in common use: bypass, shunt, and full-flow. Generally,

the filter element is replaceable and it can be discarded when dirty.

Bypass Systems

The bypass filter system shown in **Figure 10-22** pumps part of the oil through the filter, while the remaining oil is pumped to the engine bearings. Oil pumped through the filter is returned directly to the crankcase. The primary purpose of the filter in the bypass system is to keep a clean supply of oil in the crankcase.

The pressure relief valve (regulating valve) in the bypass system controls the maximum allowable pressure in the system. If there is a restriction to oil flow, pressure buildup will overcome relief valve spring tension and the valve will open. When

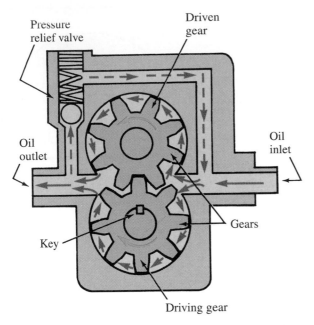

Figure 10-20. *In gear-type pump operation, the oil is carried between teeth of matching gears. If oil pressure is too high, relief valve recirculates oil through pump.*

Figure 10-21. *Filters, such as this one, trap unwanted materials and prevent those materials from circulating through the engine. (Deere & Co.)*

this occurs, oil pressure will be relieved and oil will flow through the valve, back to the crankcase.

Pressure relief valves are installed in all pressurized lubrication systems. Often they are an integral part of the oil pump.

Shunt Filter Systems

In the *shunt filter system,* part of the oil delivered by the pump is filtered and directed to the engine bearings. Some of the oil is shunted past the

filter. The remaining oil is circulated through the pressure relief valve and back into the crankcase.

Full-flow filter systems

The *full-flow filter system* directs the entire volume of pumped oil through the filter to the bearings. See **Figure 10-23.** If the filter element becomes clogged with dirt, oil pressure will increase. The added pressure will open the relief valve, permitting the oil to flow. If the filter did not have a relief valve and the filter became clogged, serious engine damage would result.

The filter cartridge must correspond to the filter system on the engine. For example, a full-flow cartridge used in a partial-flow system will give longer service life, but initial efficiency will be poor due to its high-flow rate. On the other hand, a partial-flow cartridge used in a full-flow system would drastically reduce oil pressure. Proper oil filtration in modern gasoline engines cannot be overemphasized.

Summary

Lubrication is the process of reducing friction by introducing a slippery substance between sliding surfaces. Friction is the resistance to motion created when one dry surface rubs against another.

All bearing surfaces in small gas engines must have oil separating moving parts that are in close contact. The oil film between close fitting parts may only be a few molecules thick, but it is enough to keep the metal surfaces from actually touching.

In addition to lubricating, oil cools and cleans the engine. The oil also provides a seal between the piston rings and the cylinder wall.

When selecting oil for use in a small engine, service classification and viscosity grade must be considered.

The way that moving parts are lubricated differs in two-cycle and four-cycle engines. Oil in a two-cycle engine is mixed with gasoline and poured into the fuel tank. In operation, an oil mist is created that lubricates all internal parts. In a four-cycle engine, oil is poured into the crankcase. The oil sump in the crankcase supplies lubrication to all internal engine parts.

Oil filters are used on many small engines. These filters trap dirt, carbon, and other harmful materials, preventing them from circulating through the engine. Three common types of oil filter systems include the bypass, shunt, and full-flow.

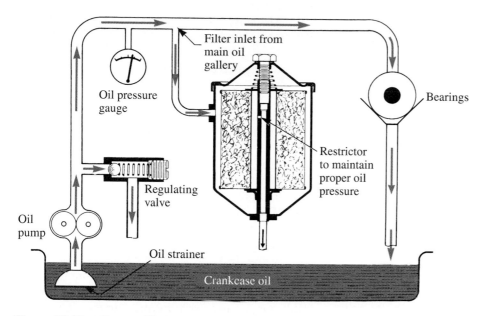

Figure 10-22. *Bypass filter system pumps some engine oil through filter. Remaining oil goes to engine bearings. (Wix Filters)*

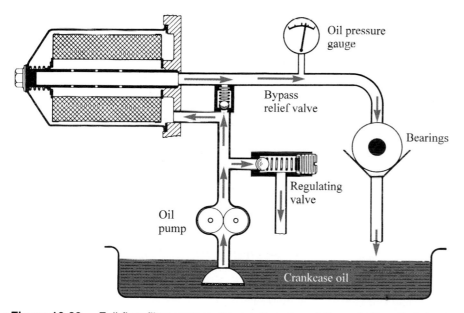

Figure 10-23. *Full-flow filter system directs all engine oil through filter. Relief valve opens if filter becomes clogged.*

Know These Terms

lubrication
friction
oil
API service classification
viscosity
viscosity grade
multigrade
multiviscosity
detergent/dispersants
splash lubrication
  system
dipper

constant level splash
  system
ejection pump
barrel pump
positive displacement
  pump
full pressure lubrication
  system
oil filter
bypass filter system
shunt filter system
full-flow filter system

Chapter 10 Review Questions

Answer the following questions on a separate sheet of paper.

1. Give three general types of lubricants.
2. Name four important jobs performed by a good engine lubricant.
3. Which engine lubrication system utilizes a trough in the crankcase?
   a. barrel type lubrication system
   b. ejection pump system
   c. constant level system
   d. splash system
4. Which small engine lubrication system is similar to the system used on automobiles?
   a. ejection pump system
   b. constant level splash system
   c. barrel type pump system
   d. full pressure system
5. Babbitt metal used as a bearing material is an alloy of three metals. Name them.
6. The type of bearing shell shown in **Figure 10-3** is referred to as a(n) _____ _____ bearing.
7. The lock tabs on the bearing in **Figure 10-3** are for the purpose of _____.
8. The lubricating oil recommendation for a particular make and model engine would be found on a special label on the engine. True or False?

9. API stands for _____.
10. SAE stands for _____.
11. *Type of oil* refers to its _____.
    a. service classification
    b. viscosity
    c. grade
    d. maximum operating temperature
12. Most four-cycle engine manufacturers recommend oils supplemented by additives. True or False?
13. What is the purpose of detergent/dispersants in oils?
14. Why are some oils with additives not suitable for two-cycle engine use?
15. A high viscosity oil would _____.
    a. be thin
    b. be suitable for high temperature use
16. Several types of positive displacement oil pumps are used in pressurized lubrication systems. Yes or No?
17. Name three types of oil filter systems in use.
18. Generally, the oil filter element is replaceable. True or False?

Suggested Activities

1. Demonstrate friction to the class by first rubbing two sheets of waste paper together, then rub two sheets of coarse abrasive together. Discuss how a lubricant could reduce friction in each case.
2. Make viscosity measurements of several engine oils with a viscosimeter.
3. Compare a multiviscosity oil with single viscosity oils, using a viscosimeter. Test the oils at various temperatures and make a chart of your results.
4. Compare a detergent oil with a non-detergent oil by placing equal amounts of carbon black in each. Shake them up, then let them set for several days. Explain any observed differences in the oils.
5. Produce condensation in a glass beaker partially filled with gasoline. Place the loosely covered beaker in a warm place or in direct sunlight. When the air in the beaker is warm, pack ice around the outside. Repeat the

process until a quantity of water can be seen in the bottom of the beaker. A hot humid day will produce best results.

6. Create heat with friction. Place a dull, unwanted drill in a drill press. Try to drill a bar of metal. Show what happens to drill as a result of friction.

7. Demonstrate to the class how oil cleans. With engine grease and grime on your hands, wipe them clean in a container of clean engine oil. Compare the cleaning power with soap or a detergent.

8. Remove the end cover from a gear-type oil pump and demonstrate how it works. Reinstall the cover, place the pump in oil, and operate it by hand.

9. Obtain an approved gasoline container. List the following information that should be found on the container:
   a. Who approved the container?
   b. What standard code does the container use?
   c. What is the capacity of the container?
   d. Is the container clearly marked *Gasoline?*
   e. Is the container metal or plastic? What color?
   f. What cautions are on the container?
   g. What health and safety instructions are printed on the container?

10. Demonstrate the proper and safe procedure of weighing and mixing oil in gasoline for two-cycle engine use.

This 20 ton log splitter is powered by a 5 horsepower engine. (Brave Industries, Inc.)

Oil for winter application, such as snowmobiling, must be thin enough to permit adequate cranking speeds at the lowest anticipated temperatures. (Polaris)

C H A P T E R

11

Cooling Systems

After studying this chapter, you will be able to:
▼ Explain how air cooling, exhaust cooling, and water cooling work to lower engine operating temperatures.
▼ Define the basic function of a water pump and give examples of several common types.
▼ Describe the basic operation of outboard water circulation systems.
▼ Explain the function of a thermostat and a radiator.

Principles of Engine Cooling

The efficiency and life of an engine depend upon how well it is cooled. The average temperature of burned gases in the combustion chamber of an air-cooled engine is about 3600°F (1982°C).

About a third of the heat is carried away by the cooling system. See **Figure 11-1.** Various parts surrounding the air-cooled engine direct the flow of cooling air. The exhaust system carries away another third of the heat. That which remains is used to produce engine power.

Loss of heat through the cylinder walls to the cooling system reduces the temperature from 1200°F (649°C) to approximately 350°F (177°C). The temperature drops to 100°F (38°C) by the time it reaches the outer edges of the cooling fins. **Figure 11-2** illustrates cylinder wall temperatures at various locations.

There is a reason for the comparatively low temperature along the inner cylinder wall. A

boundary layer of stagnant gas lies next to the wall and acts as an insulator. Therefore, less heat travels through the cylinder wall than you might expect from the 3600°F (1982°C) produced by the burning fuel.

Figure 11-1. *The air cooling system of a small gasoline engine consists of the shroud, screen, flywheel, baffles, and cooling fins. Airflow follows the path shown by arrows.*

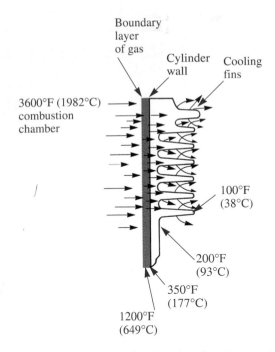

Figure 11-2. *High combustion chamber temperature is reduced by exhaust gases and cooling system. Large area of cooling fins controls heat dissipation from cylinder.*

How air cooling works

Operating temperature is lowered to about 200°F (93°C) as heat passes through the cylinder wall to the outer surfaces of the cylinder. Air forced over the cooling fins and directed by the sheet metal baffles rapidly dissipates the heat. The cooling system, then, carries heat away from the engine.

The heat of combustion (rapid burning and expansion of gas) travels from the cylinder through the cylinder walls by conduction. *Conduction* is heat transfer through a solid material.

When the heat reaches the outer surfaces of the cylinder, air forced over the surface carries it away by convection. *Convection* occurs when heat transfers through movement of a gas—in this case, air.

The flywheel has fins, which blow air around the cylinder wall and cooling fins. The flow of air is controlled and directed by a sheet metal shroud and baffles surrounding the flywheel and cylinder. An engine should never be run without the shroud in place or it will quickly overheat. For safety, the flywheel is covered with a screen or perforated plate, which allows air to be drawn through it. See

Figure 11-3. The screen should be kept clean to permit unrestricted airflow.

Thin *cooling fins* increase the surface area around the outside of the cylinder. See **Figure 11-4.** The greater the surface area in contact with cool air, the more rapidly the heat can be carried away. Cooling fins are necessary on air-cooled engines but not on water-cooled engines. Water is four times more effective than air for engine cooling.

Figure 11-3. *Cooling air intake screen must be kept clean for unrestricted airflow. Screen must be kept in place to prevent clogging of cooling fins. (Tecumseh Products Co.)*

How exhaust cooling works

As noted earlier, the exhaust system carries away approximately a third of the engine heat. If the exhaust system is restricted in any way, part of the 1200°F (649°C) temperature will remain in the metals of the engine. This heat buildup will increase friction. In turn, maintenance costs will increase on moving parts and engine life will be shortened. Dirt, grass clippings, leaves, straw, or other materials lodged between the cooling fins will tend to insulate the cylinder, causing *hot spots* and engine overheating. Keeping an engine clean, cool, and properly lubricated will significantly extend its life.

Figure 11-4. *Cooling fins are designed and placed to provide adequate cooling for each part of the cylinder. Thickness, surface area, and spacing are important considerations.*

How water cooling works

Water is an excellent medium for cooling engines. It is inexpensive, readily available, and absorbs heat well.

Some small engines are water cooled because they are used in or around a water source. The outboard engine in **Figure 11-5** is typical of a relatively small water-cooled engine. Notice the absence of cooling fins, which are found only on air-cooled engines.

Water-cooled engines are generally made with coolant passages surrounding the cylinder. See **Figure 11-6.** These passages are called *water jackets.* A small pump keeps water circulating through the jackets. The water absorbs the heat of combustion and carries it away from the engine. In cold weather, *antifreeze* solutions are added to the water to prevent it from freezing.

Water pumps

Water pumps of many different designs are used to move the water through the cooling system. In outboard engines, the pump is generally located in the lower unit near or below the water

Figure 11-5. *Outboard engines are frequently water cooled. They are identified by the absence of cooling fins. (Evinrude Motors)*

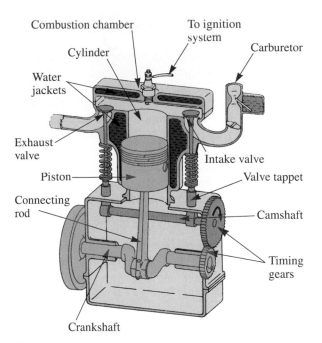

Figure 11-6. *Water is circulated around the cylinders through the water jacket, where it absorbs combustion heat.*

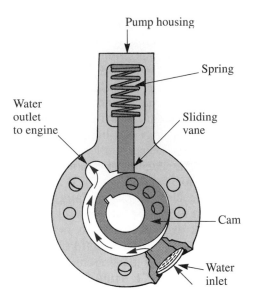

Figure 11-7. *Revolving cam in sliding vane water pump fills large volume of space, and water is pushed through outlet to engine. Sliding vane prevents water from revolving with cam.*

line. The main pump member is driven by the vertical drive shaft or horizontal propeller shaft. The pump must have an opening for water to enter and one for it to exit. These openings are called the inlet and outlet.

A *sliding vane pump* is illustrated in **Figure 11-7.** As the eccentric cam (an off-center enlargement on a shaft) is rotated in the sliding vane pump, the volume of space between the cam and pump housing is constantly changing. Water enters the inlet and fills this space. The cam rotates and closes the inlet, pushing the water ahead of it toward the pump outlet.

The outlet is connected to the water jacket by suitable tubing. A second tube carries the heated water out of the engine. The sliding vane is kept in close contact with the cam by a spring. The vane provides a seal so that water cannot continue past the outlet. However, during high-speed operation, water pressure may become high enough to lift the vane against the spring, allowing some water to recirculate through the pump. This action prevents too much pressure from building up in the cooling system.

The *rotor-type pump* operates much like the sliding vane pump. The vane and rotor are one piece, and the eccentric gyrates the rotor (rotates the rotor with a bobbing motion), causing a pumping action.

The *plunger pump* has a cylinder and plunger. See **Figure 11-8.** The plunger is raised and lowered in the cylinder by an eccentric on the propeller shaft. A spring keeps the plunger close to the eccentric.

When the spring forces the plunger down in the cylinder, water is drawn in through the inlet ball check valve while the outlet ball check valve

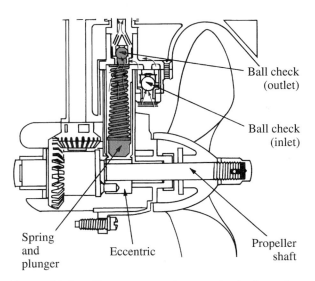

Figure 11-8. *Plunger pump draws water in through one ball check valve and forces it out through a second valve. Plunger is operated by an eccentric.*

is closed. As the eccentric lifts the plunger, the inlet check valve is closed by the increasing pressure in the cylinder and the outlet check valve is opened. Water in the cylinder is forced through tubing to the engine.

Vari-volume pumps use a synthetic rubber impeller. See **Figure 11-9A.** Since the impeller housing is off-center with the drive shaft, the impeller arms must flex as they revolve. The volume between the arms increases and decreases with rotation.

The inlet port or opening is located where the volume is increasing and the water is drawn into the housing. The outlet is on the side where volume is decreasing and the water is forced out into the water jacket.

When a vari-volume pump is driven at high speeds, the back pressure of the water in the system becomes great enough to force the impeller arms inward. See **Figure 11-9B.** The pump loses some of its effectiveness and moves only enough water through the engine to maintain proper operating temperature in the cylinders. Overcooling of the cylinders of an outboard engine can happen at high engine speeds unless some sort of flow control, such as this, is designed into the cooling system.

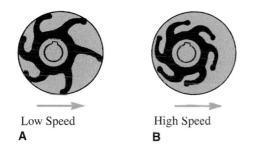

Low Speed
A

High Speed
B

Figure 11-9. *Vari-volume pump has a flexible synthetic rubber impeller that is offset from center of impeller housing. A —Space between impeller arms varies as they rotate. B—At high speed, increased water pressure forces impeller arms inward to prevent overcooling of engine.*

Outboard Water Circulation Systems

Some small outboard engines use a simple cooling device called a *pressure-vacuum water flow system*. See **Figure 11-10.** The water flow from the propeller tips creates pressure against the intake port and vacuum at the outlet port, which is located immediately forward of the propeller.

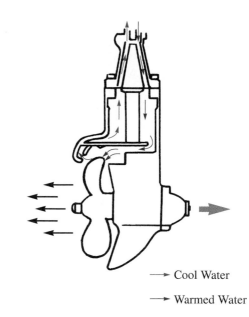

→ Cool Water

→ Warmed Water

Figure 11-10. *Pressure-vacuum cooling system utilizes thrust of propeller tips and venturi (narrowed section of a passage) vacuum principle created by forward motion of engine to circulate water.*

Propeller action and the forward motion of the boat provide water circulation.

At slow speeds, like those used for trolling, pressure of water from the propeller tips may not be enough to force water through the cooling system. Sufficient cooling is still maintained, however, by the siphon (sucking) effect of the discharge channels as the boat moves through the water. When starting an engine with a pressure-vacuum system, the engine should be given a short burst of speed to fill the channels and water jacket with water. See **Figure 11-10.**

Another type of pressure-vacuum water flow system has the discharge ports located in the propeller blades. The centrifugal force created by the turning propeller aids in discharging the water. *Centrifugal force* is the tendency of spinning matter to move away from the center of its path. Since there are no moving parts, except for the propeller, the system will function as long as the water channels and jackets remain unobstructed.

With this system, vacuum must be maintained, particularly at low speeds. Therefore, all water connections in the system are airtight. Air seepage into the cooling system would destroy the slow speed siphoning effect and cause overheating.

Worn propeller blades can also cause poor circulation. The propeller's reduced diameter puts the tips farther from the water scoop opening. This reduces water pressure. Salt water corrosion, marine growth, and mud-clogged water channels are all causes of faulty water circulation in this type of system.

Other outboard water circulation systems are in use. The basic principles are similar for most types. Some examples follow.

Figure 11-11 shows a pump-driven cooling system. Water is drawn into the intake, pumped through the water jacket surrounding the cylinders, and discharged. This is a relatively simple system. Notice that the pump is driven by the drive shaft.

In a similar system, the water pump is driven by the propeller shaft. When this pump location is used, the water is drawn in through ports in the propeller hub.

In some engines, water temperature is carefully controlled by a thermostat. The *thermostat* is simply a valve arrangement that stops the circulation of engine coolant until it reaches the proper operating temperature.

There are two types of thermostats, each opening and closing the valve at predetermined temperatures. Both work on the principle of the expansion of heated materials.

In the old bellows-type thermostat, the valve is attached to a gas-filled, sealed bellows made of thin copper. Heat causes the gas to expand, opening the valve.

The more popular pellet thermostat operates in the same way but depends upon a small, wax-filled, copper cylinder. When heated, the wax pushes against a rubber diaphragm (disc). This moves a piston that is attached to the valve. In some cases, the piston remains stationary and the cylinder moves away to open the valve.

A thermostatically controlled engine will be kept at a constant temperature regardless of speed or outside temperature. See **Figure 11-12.** Cold water enters the intake and passes through the pump to the water jacket. If the temperature of the water from the jacket is high enough to open the thermostat, the water is discharged from the engine. If the water is too cool to open the thermostat, it is recirculated through the pressure control valve and back to the pump. When water is recirculated, it retains some of its heat and eventually brings the cylinder head temperature up to thermostat temperature. This type of system can maintain a constant operating temperature and will automatically compensate for even slight

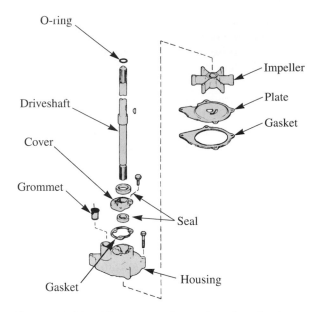

Figure 11-11.   *Water pump for outboard engine is driven by vertical drive shaft. Some pumps are mounted on propeller shaft.*

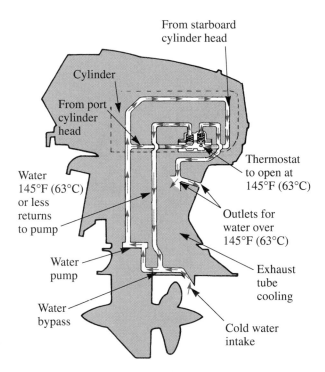

Figure 11-12.   *Some outboard engines have a thermostatically controlled cooling system. Water recirculates until thermostat opens, discharging heated water and allowing more cold water to enter intake. This thermostatically controlled cooling system employs a double feed bypass.*

changes in cooling water temperature. A thermostat in a cold engine will remain closed until it reaches thermostat opening temperature.

Observing thermostat operation

With the thermostat removed from the engine, it is easy to see how it works. Place thermostat in a beaker of cool water. Put the beaker on a hot plate and slowly heat the water. When the water has reached the proper temperature, the thermostat will open. By placing a thermometer in the beaker you can determine the temperature rating of the thermostat. See **Figure 11-13.**

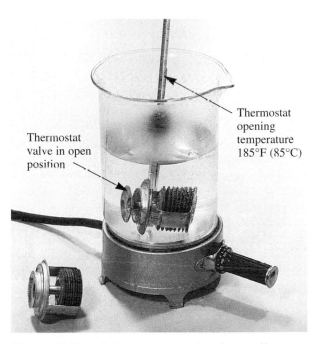

Figure 11-13. *A thermostat can be observed in action by heating it in a beaker of water. Thermometer registers temperature at which thermostat opens.*

Radiators

Radiators are water reservoirs, which are made from many thin copper or aluminum tubes. These tubes are connected at the top and bottom (or each side) to water tanks. The tubes are held in place by thin metal fins. The fins increase the cooling surface area of the tubes. The tube and fin assembly shown in **Figure 11-14** is called the *radiator core.*

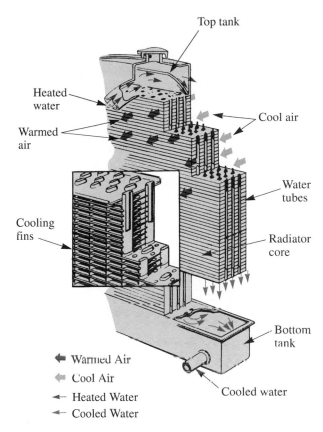

Figure 11-14. *Radiator carries water through tubes in core, where it is cooled by air forced across cooling fins.*

The water, which is heated by the engine, is pushed into the top or side of the radiator core by the water pump. The hot water travels downward or across through the tubes to the bottom tank. Cool air is forced between the fins in the core by an engine-driven fan. The air cools the water by conduction, radiation, and convection. The cooled water passes from the bottom tank to the engine water jacket, ready to absorb more heat. Engines cooled by radiators also have thermostats, which open when the proper operating temperature is reached. When engine temperature is below normal, the thermostat remains closed and the water is directed through a bypass channel. The water bypassing the radiator retains its heat and returns directly to the engine. This action retards forced cooling until the engine reaches its normal operating temperature.

The fine balance of cooling maintained by the radiator and thermostat assures that the engine will not overheat under loads in the hot summer or run too cold in the winter.

How oil cooling works

Most of this chapter has been devoted to engine cooling by air or water. Nevertheless, engine oil also plays an important role in the cooling process. Oil circulating through the engine also acts as a coolant. The oil absorbs heat as it contacts the various metal parts. It then draws off the heat as it leaves the engine (two-cycle engine) or transfers the heat through the crankcase to the outside air (four-cycle engine).

Summary

The efficiency and life of an engine depends on how well it is cooled. The heat from combustion is carried away from the engine by conduction and convection.

Approximately a third of engine heat is carried away by the cooling system, another third is carried away by the exhaust system, and the remaining heat is used to produce engine power.

The flywheel on a small engine has fins that blow air around the cylinder wall and cooling fins. The cooling fins increase the surface area around the outside of the cylinder. The greater the surface area in contact with air, the more rapidly heat can be carried away. In order to efficiently carry away heat, the exhaust system must be free from restrictions. Keeping an engine clean, cool, and properly lubricated will significantly extend its life.

Some small engines are water cooled. Water is an excellent medium for cooling engines.

Some of the outboard engines use a pressure-vacuum water flow system. Water flow from the propeller tips creates pressure against an intake port and vacuum at an outlet port. Another type of pressure-vacuum system has discharge ports located in the propeller blades. The centrifugal force created by the turning propeller aids in discharging the water.

In some engines, the temperature of the cooling water is controlled by a thermostat. The thermostat stops circulation of coolant until the engine reaches a specific operating temperature.

Some small engines are equipped with radiators. Radiators are water reservoirs made of many copper or aluminum tubes, which are connected at the top and bottom to water (coolant) tanks. The tubes are held in place by thin metal fins. The fins increase the cooling area of the tubes. If a radiator system is used, antifreeze must be added to water

to prevent freezing. Water pumps are used to move water through the cooling system.

Oil circulating through an engine also acts as a coolant. Oil absorbs heat as it contacts various metal parts.

Know These Terms

boundary layer
conduction
convection
cooling fins
water jacket
antifreeze
sliding vane pump
rotor-type pump

plunger pump
vari-volume pump
pressure-vacuum water
    flow system
centrifugal force
thermostat
radiator
radiator core

Chapter 11 Review Questions

Answer the following questions on a separate sheet of paper.

1. The average temperature of the burned gases in the combustion chamber is _____.
   a. 2400°F (1315.5°C)
   b. 3000°F (1649°C)
   c. 6200°F (3427°C)
   d. 3600°F (1982°C)
2. The heat of combustion is carried away in three ways. One third is removed by the cooling system and one third by the exhaust system. That which remains is used to produce _____.
3. Why is the temperature of the inside cylinder wall relatively cool compared to the combustion temperature?
4. The heat that reaches the cooling fins is carried away mainly by _____.
   a. convection
   b. conduction
   c. radiation
5. What are two ways to prevent overheating of air-cooled engines?

6. When used for engine cooling, water is about _____ times as effective as air.
7. Cylinders of water-cooled engines are surrounded by _____.
   a. insulation material
   b. a water reservoir
   c. a water jacket
   d. cooling water tubes
8. Name four types of water pumps discussed in this chapter.
9. The _____ water flow system in outboard engines utilizes the tips of the propeller to circulate water.
10. A heat sensing device that maintains a constant engine operating temperature is the _____.
    a. check valve
    b. thermostat
    c. rheostat
    d. flexible impeller
11. In cold weather, a(n) _____ should be added to water cooled engines.

12. What cooling system defect can cause *hot spots*?
13. What two fluids are used to cool engines?

Suggested Activities

1. Place a thermostat and thermometer into a beaker of water. Heat the water and check thermostat operation.
2. Operate an air-cooled engine and with a smoke generator observe the flow of air through the shroud and around the cooling fins.
3. Inspect a cutaway of a radiator and explain how it cools water.
4. Disassemble the lower unit of an outboard engine and study the pumping and circulation system.
5. Place the outboard engine in a tank and briefly operate it to observe the cooling system at work.

This personal watercraft is powered by a three-cylinder, two-stroke engine, which is water cooled. (Kawasaki Motors Corp., U.S.A.)

Preventive Maintenance and Troubleshooting

After studying this chapter, you will be able to:
- ▼ Keep engines clean.
- ▼ Change the oil in a four-cycle engine.
- ▼ Mix fuel and oil correctly for a two cycle engine.
- ▼ Perform preventive maintenance on various engine systems, including the crankcase breather, air cleaner, and muffler.
- ▼ Prepare a water cooling system for storage.
- ▼ Describe systematic troubleshooting.
- ▼ Use manufacturer's service manuals to determine engine specifications and explain why this information is necessary when servicing a small engine.

General Preventive Maintenance

There are certain maintenance tasks that must be performed regularly to keep an engine working properly. These tasks come under the heading of *preventive maintenance,* because they help prevent premature engine wear.

Keeping engines clean

Cleaning a small air-cooled engine periodically can prevent overheating. For proper cooling action, air must pass across the extended metal surfaces (cooling fins) of the cylinder block and cylinder head. If the *cooling fins* are insulated by dirt, leaves, and/or grass clippings, engine parts will retain most of the combustion heat. Parts will expand, probably distort, and possibly seize. Therefore, all finned surfaces should be cleaned to the bare metal.

Methods for cleaning small air-cooled engines vary. You can blow debris from the fins with compressed air and then use a cleaning solvent. See **Figure 12-1.** Or, you can scrape the dirty areas with a piece of wood and wipe them with a clean cloth. Various aerosol spray cleaners are suitable for use on small engines.

> ⚠️ When using compressed air, always be extremely careful where you direct the blast of air. Wear safety goggles. Never direct the air blast toward skin or clothing.

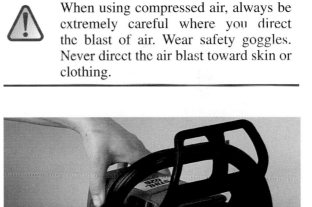

Figure 12-1. *This chain saw engine is being cleaned with compressed air.*

Checking oil level and condition

Crankcase oil in four-cycle engines should be checked periodically. Preferably, it should be checked each time fuel is added. The engine manufacturer provides a means of visually inspecting the level and condition of the oil. Use the type and viscosity grade of oil recommended by the manufacturer and maintain it at the proper operating level.

To check the oil level, withdraw the **dipstick** and wipe it dry. Reinsert the dipstick as far as it will go. Withdraw it a second time and observe the oil level. See **Figure 12-2.**

The markings on dipsticks may vary, but all will have a *Low* (*Add*) and a *Full* mark. Add oil if the level is below the *Low* mark. Do not run the engine with oil showing above the *Full* mark on the dipstick. If the crankcase oil level is high, drain some oil.

> ⚠ Overfilling the crankcase with oil can foul plugs and cause the engine to use too much oil.

Some small gasoline engines do not have dipsticks. Instead, they have a **filler plug** that seals out dirt and seals in the oil. **Figure 12-3** shows the proper method of loosening one type of filler plug. When the plug is removed, the oil level should be

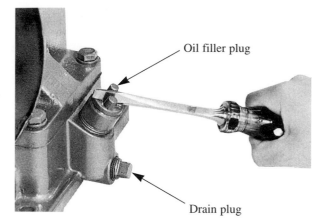

Figure 12-3. *Removing oil filler plug from engine crankcase. (Briggs and Stratton Corp.)*

at the top of the filler hole or to a mark just inside of the filler hole.

If the engine oil level drops at an excessive rate (requires addition of oil frequently), look for the cause. Refer to the troubleshooting chart for the particular engine at hand. Troubleshooting charts are covered later in this chapter. Typical causes are: external leaks, worn oil seals around the crankshaft, worn valve guides, worn piston rings, or a hot running engine.

The color of used oil is not always an accurate indication of its condition. Additives in the oil may cause it to change color, while not decreasing its lubricating qualities.

When to change oil

The small engine manufacturer will recommend oil changes at intervals based on hours of running time. A new engine should have the first oil drained after only a few hours of running to remove any metallic particles from the crankcase. After that, the time specified may vary from 10 hours to 50 hours.

Engine oil does not *wear out*. It always remains slippery. However, oil used for many hours of engine operation becomes contaminated with dirt particles, soot, sludge, varnish-forming materials, metal particles, water, corrosive acids, and gasoline. These contaminants finally render the oil *useless*. The harm they cause outweighs the lubricating quality of the oil.

The time interval for oil changes is selected so that the oil never reaches a *loaded* level of contamination. **Loaded oil** cannot absorb any more contaminants and still be an effective lubricant.

Figure 12-2. *Checking the oil is an important part of preventive maintenance.*

When oil reaches a loaded condition, varnish deposits begin to form on the piston and rings and sludge collects in the crankcase.

Changing oil

Draining engine oil is not difficult. First, run the engine until it is thoroughly warmed up. Warm oil will drain more completely, and more contaminants will be removed if the oil is agitated.

Stop the engine and disconnect the spark plug. The oil drain plug is located at a low point on the crankcase, usually along the outside edge of the base. See **Figure 12-4.** Some engines are drained through the filler cap and have no drain plug.

Clean the dirt from the drain plug area, then remove the plug with a proper wrench. Drain the oil for approximately five minutes to remove as much contaminated oil as possible. If possible, tilt the engine toward the drain hole if it is located on the side or top. When draining is complete, replace the drain plug.

If the engine is equipped with a disposable oil filter, it should be replaced each time the oil is changed. See **Figure 12-5.** To prevent oil leaks, always coat the O-ring seal with a light coat of clean oil before installing the filter.

Before putting fresh oil in the engine, clean the filler opening, funnel (or other container), and the top of the can of oil. Be sure to use the type, viscosity grade, and quantity of oil recommended by the manufacturer. Pour it in the engine's crankcase and check the level. Then replace the filler cap and connect the spark plug lead to the spark plug.

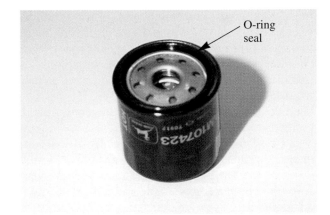

Figure 12-5. *Typical disposable oil filter. To prevent leaks, apply a light film of oil around O-ring seal before installation. (Deere & Co.)*

Start and run the engine for a few minutes. Stop the engine and recheck oil level. Check for oil leaks.

> ⚠️ After changing the oil, wash or destroy oily rags. Storing them may cause spontaneous combustion. Spontaneous combustion occurs when combustible material self-ignites.

If any oil is left in the container, keep it clean during storage. Keep the cap tightly sealed on the oil container to prevent condensation or contamination.

Mixing oil and fuel

A two-cycle engine can be seriously damaged by improperly mixing gasoline and oil, even though the recommended proportions are used. The proper way to mix gasoline and oil is to pour some of the gasoline into a clean container first. Add the oil to the gasoline and agitate (shake) the partial mixture. Add the rest of the gasoline and agitate again thoroughly.

> 📄 Some people may pour the oil into the container first, then allow the gas pump pressure to mix the gasoline and oil. However, this method *does not* always thoroughly mix the two elements.

Once the gasoline and oil are thoroughly mixed, the oil will remain in suspension indefinitely. If fuel is to be stored for several weeks (or

Figure 12-4. *Oil drain plug is located at a low point in the crankcase to permit complete drainage of used oil. (Briggs and Stratton Corp.)*

more), add a gasoline stabilizer to prevent the formation of oxidation, varnish and corrosive acids, which can ruin an engine.

For safety, always store fuel in containers manufactured for storage of gasoline and clearly marked *gasoline*. See **Figure 12-6.** If possible, keep the storage container full to prevent moisture condensation.

Condensation forms in partially filled containers because of temperature changes. During the day, the air above the fuel may become very warm and hold a considerable amount of moisture. At night, the warm air cools, contracts, and loses the water vapor.

As water droplets gather on the inside walls of the container, they run together and flow into the fuel. Each night, more water is added to fuel in this manner. Partially filled fuel tanks react to temperature changes in this way. For this reason, fuel tanks should be kept full of gasoline or gasoline/oil fuel mixture.

Air cleaner service

The carburetor *air cleaner* should be cleaned before each season of operation and at regular intervals thereafter. A plugged air filter can cause hard starting, loss of power, and spark plug fouling. Under severe dust conditions, air filters should be cleaned more often.

Three types of air cleaners widely used in small gasoline engines are the **oil-wetted, dry types,** and **dual-element.** Each has a different method of cleaning and servicing.

> For more information on these air cleaners, refer to the *Air cleaners and air filters* section in Chapter 8 of this text. Also, see the manufacturer's service manual for service information.

Crankcase breather service

If the small gasoline engine has a **crankcase breather,** it should be removed and cleaned periodically. The breather assembly is located over the valve stem chamber. It is held in place with two or more screws.

To service the breather, remove the screws and the cover. Under the cover is a filter element and a reed valve unit similar to the one in **Figure 12-7.** The breather allows outward airflow only. Inspect the reed valve to make sure it is not damaged or

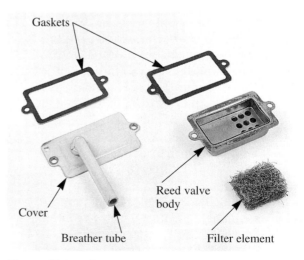

Figure 12-7. *Crankcase breather elements need periodic cleaning and inspection. Reed valve function is to only permit air to leave the crankcase. (Deere & Co.)*

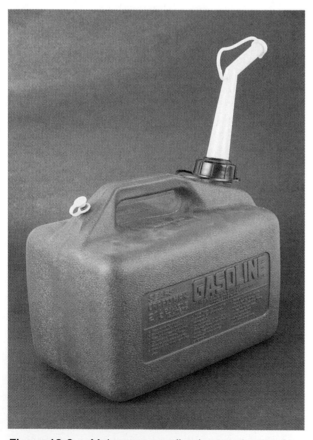

Figure 12-6. *Make sure gasoline is stored properly in a container that is clearly marked* gasoline.

distorted. Wash parts in a cleaning solvent and replace damaged gaskets. See that drain hole in reed valve body is open and located on the bottom when installed. It permits accumulated oil to return to engine. After all components have been cleaned and inspected, replace assembly and tighten screws.

Muffler service

An engine takes in large quantities of air mixed with fuel, and then burns the mixture. Unless the engine readily rids itself of the by-products of combustion, its efficiency will be greatly reduced. This is the task of the exhaust system, which, in small gasoline engines, mainly consists of exhaust port(s) and a small muffler.

Mufflers are designed to reduce noise and allow gases to escape. When it becomes clogged with carbon soot, gases cannot get out quickly enough to allow fresh air and fuel to enter. Then, a power loss occurs along with a tendency to overheat.

If it is a sealed muffler and clogging is suspected, install a new one and check for improved engine efficiency. See **Figure 12-8.** If a muffler is designed to be taken apart, it should be disassembled and cleaned in a solvent; never use gasoline.

Maintaining water cooling systems

The water cooling systems used in small engines require maintenance similar to that employed in the automobile engine. Because the combination of water and metal sometimes produces harmful chemical reactions that attack the water jacket, a chemical **rust inhibitor** should be added whenever a system is drained and refilled. If rust and scale are allowed to form and accumulate, the walls of the water jacket will become insulated.

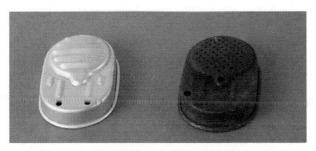

Figure 12-8. *Old clogged mufflers (right) should be replaced with new, more efficient ones (left).*

This will cause engine heat to be retained rather than removed.

Scale settling to the bottom may plug water passages in cylinder block and clog water tubes in the radiator. Without free circulation of water, the engine will run hot even when the thermostat is open. Local *hot spots* can occur in engine when the passages in the block are obstructed.

In severe cases, water may boil inside block, and steam will prevent water from contacting and cooling inner walls. Then, serious overheating and damage to parts of the engine are bound to occur.

The cooling fins that surround the tubes of the radiator should be kept clean for efficient heat transfer. Compressed air or pressurized water will remove any accumulations that might prevent air from passing through the fins and across the tubes. To remove the debris, direct the flow of air or water in the opposite direction of normal airflow.

Engine blocks and radiators may be cleaned periodically by **reverse flushing** the system with pressurized water. Disconnect the hoses from the radiator and the block. Force clean water in the opposite direction of normal circulation. This will push loose sediment out. Continue flushing until the water runs clear. Flushing should be done with the engine stopped and cool. To remove additional rust clinging to inner surfaces, use a commercial cooling system rust remover. Follow the manufacturer's instructions.

When adding coolant for warm temperatures, mix 50% water to 50% antifreeze (Ethylene glycol). Never use just water. For freezing temperatures, refer to chart on antifreeze container. Service should be done before the cold season begins. A frozen cooling system will damage engine block and radiator. Antifreeze also increases boiling point of water.

Systems cooled with saltwater

Outboard engines operated in salt water are exposed to extremely corrosive conditions. Exposed engine parts require careful maintenance.

Outboards used in saltwater should be removed from the water immediately after operation. If the engine cannot be removed, tilt the gearcase out of the water and rinse it with fresh water. (The gearcase must be removed from the water when not in use.)

Flushing the internal cooling system of an outboard engine is extremely important. Flushing is done by attaching a freshwater hose to the water scoop or by operating the engine in a barrel of freshwater for several minutes.

Rinse the engine with freshwater and wipe all lower unit parts with a clean, oily cloth. Ignition leads and spark plug insulators should be wiped frequently to prevent an accumulation of salt residue.

Storing water-cooled engines

Storing water-cooled engines for lengthy periods, particularly during winter, calls for special maintenance procedures. If the engine is *radiator cooled,* **antifreeze** must be added to the water to protect against freezing at the lowest possible temperatures. If rust inhibitor is not supplied in the antifreeze, it should be added.

If the engine will not be started at anytime during storage, drain the cooling system completely. Then, tag the engine to indicate its drained condition.

When storing outboard engines, remove all plugs from the gearcase and drive shaft housing. This allows accumulated water in the gearcase and cooling system to drain off.

Failure to take this precaution when winterizing may result in a cracked cylinder block and/or gearcase, plus possible damage to water channels and tubes.

Rock the engine from side to side to make certain all water has drained. Refill the gearcase with the type of grease specified by the engine manufacturer. Attend to all other lubrication recommendations made by the manufacturer for care of engines being stored.

Systematic Troubleshooting

Most small engine service and repair jobs can be done without taking the whole engine apart. If the engine will not start, is hard to start, runs rough, or lacks power, troubleshooting may be necessary. Troubleshooting is simply a number of tests and steps you go through to find the problem.

Sometimes the cause of an engine problem is easy to find. At other times, checking probable causes requires a certain amount of reasoning and the use of the process of elimination. Also, more than one fault can exist at the same time, making it harder to locate the trouble.

Always take a systematic approach when troubleshooting small engines. *Systematic troubleshooting* involves checking and/or testing one component after another component until the problem is located and corrected. There are two basic principles to keep in mind when trying to pinpoint small engine problems:

1. Check the easiest things first.
2. Verify the fundamental operating requirements.

Check the easiest things first

Always start troubleshooting with the simplest, most probable possibilities first. If an engine will not start, the problem could be something as simple as an empty fuel tank or a disconnected spark plug wire. Do not start working on the carburetor or ignition system until you have made a few basic checks to determine that a simple remedy will not cure the problem.

Verify the fundamental operating requirements

In order to start and run properly, an engine must meet five fundamental operating requirements. These requirements include:

- **Proper carburetion.** Clean, fresh fuel must be delivered in the correct proportion with combustion air.
- **Correct ignition system operation.** Strong ignition spark must be precisely timed for best performance and efficiency.
- **Adequate lubrication.** The proper amount of high-quality lubricating oil must reach critical engine components.
- **Sufficient cooling.** Ample supply of cooling air (no more than 20°F hotter than outside ambient air) must reach engine.
- **Proper compression.** 30–45 psi minimum for starting and 90 psi minimum for efficient operation and sufficient power.

Keep these operating requirements in mind when troubleshooting small gas engines. Through the process of elimination, you can easily isolate problems. For example, if an engine will not start but will spin normally, you can eliminate lubrication system problems because the engine is not locked-up. By spinning the engine, you can also determine whether or not it has sufficient compression. If the engine will not start after it has cooled down, the cooling system can be eliminated as a potential problem. In a matter of seconds, you have determined that your troubleshooting efforts should be concentrated in the areas of carburetion and ignition.

The engine's owner can also provide assistance with your troubleshooting efforts. Ask a few questions about the engine's performance before it stopped. Relate the answers to the operating requirements. For example, if an engine runs for 30 to 45 minutes and then stalls, you should ask if it restarts immediately after it stops. If the answer is yes, the problem is probably an ignition component that is intermittently experiencing heat-related breakdown. If the engine must cool before it will restart, vapor lock or sticking valves are possible problems. If a metallic snap is evident in the engine during the cooling period, the valves are likely to be the problem. The time you spend to ask a few pertinent questions can save a lot of time in the long run by eliminating additional problem possibilities.

Check rpm

When servicing small engines, it is often necessary to test or set maximum idle rpm or governor rpm. One way of doing this is by using a device that converts engine vibration from power pulses to rpm. See **Figure 12-9.**

Place the nose of the instrument against the running engine. A thin wire is moved in or out of the instrument barrel until it vibrates into a fan pattern. When the fan shape is at its widest point, take the rpm reading from the scale on the barrel of the instrument.

Another convenient and accurate method of measuring rpm is with a portable noncontact optical tachometer like the one in **Figure 12-10.** A small piece of reflective tape is placed on a rotating part of the engine such as the crankshaft, flywheel, or pulley. The tachometer is aimed at the rotating part. The rpm is read on the digital display in the window. This tachometer can measure from 5 to 100,000 rpm at a distance up to three feet and at an angle of 45°. These tachometers are small enough to fit in a pocket and weigh only 6 ounces. This is a safe method of measuring engine rpm, because contact with a hot engine is not necessary and a safe distance from moving parts can be maintained.

Many mechanics use an electronic unit called a *stroboscope* to check rpm. See **Figure 12-11.** This method is very accurate. To prepare for the test, make a chalk mark on the crankshaft, flywheel, or pulley of the engine.

The stroboscope produces a high intensity light, which flashes on and off at a controlled rate. Aim the light at the chalk mark on the rotating shaft and adjust the flashing frequency until the chalk mark appears to stand still. At this point, the

Figure 12-9. *Testing speed of engine through vibrations caused by power pulses of the piston. Note the adjustable slide. (Vibra-Tech)*

Figure 12-10. *This optical tachometer can measure rpm without contacting the engine by aiming it at a rotating part that has a reflective tape attached. (Monarch Instrument)*

time interval between each crankshaft rotation is equal to the time interval between each flash of the light. The scale on the stroboscope shows the number of light flashes per minute. This, in turn, is equal to the rpm of the crankshaft.

 When a stroboscope is used, parts that are actually moving at a high rate of speed appear to be stopped. Never attempt to touch the parts or the chalk mark.

Test compression

A cylinder *compression test* can be a first step toward determining the condition of the upper major mechanical parts of the engine. This test is especially valuable if an engine lacks power, runs poorly, and shows little or no improvement after fuel system and ignition adjustments. Use the following procedure for a compression test:

1. Run the engine until it is warm.
2. Disconnect all drives to the engine.
3. Open choke and throttle valves wide.
4. Remove the air cleaner.

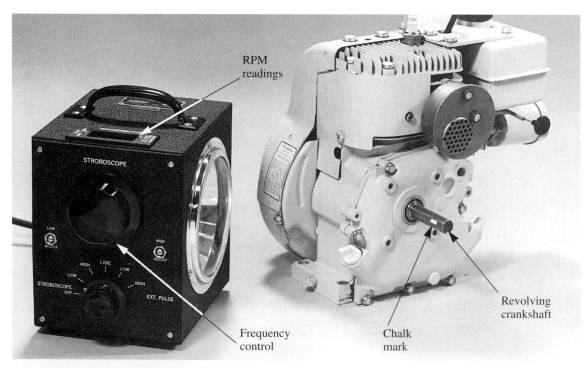

RPM readings

Frequency control

Chalk mark

Revolving crankshaft

Figure 12-11. *This flashing light of a stroboscope unit can be adjusted to match revolutions of the crankshaft, which gives engine rpm.*

5. Remove the spark plug and insert compression gauge. See **Figure 12-12.**

6. Crank the engine as fast as possible to obtain an accurate test. Repeat to ensure accuracy.

Engines equipped with compression release camshafts may have to be cranked in reverse rotation to obtain an accurate test. Most can be cranked forward.

An engine producing a compression less than the minimum suggested by the manufacturer usually has one or more of the following problems:

- Leaking cylinder head gasket.
- Warped cylinder head.
- Worn piston rings.
- Worn cylinder bore.
- Damaged piston.
- Burned or warped valves.
- Improper valve clearance.
- Broken or weak valve springs.

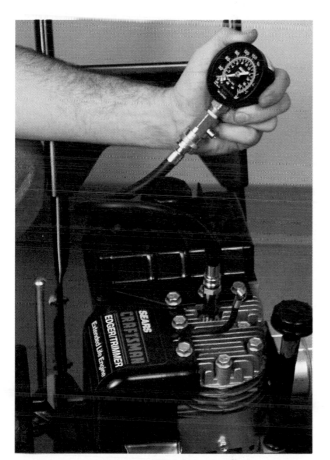

Figure 12-12. *A compression test can indicate the condition of various mechanical components of an engine.*

To determine whether the valves or rings are at fault, pour a tablespoonful of SAE 30 oil into the spark plug hole. Crank the engine several times to spread the oil and repeat the compression test. The heavy oil will temporarily seal leakage at the rings. If the compression does not improve, the rings are satisfactory and leakage is due to valves, cylinder head, or a damaged piston. If the compression is much higher than the original test, the leakage is due to defective piston rings.

Test differential pressure

A *differential pressure test* checks the compression of an engine by measuring leakage from the cylinder to other parts of the engine. This test device and procedure can identify a specific worn or damaged component in the engine that may, or may not, be directly related to the cylinder condition or rings. The device is designed so that specific leakages can be detected and isolated before disassembling parts of the engine. See **Figure 12-13.** The operation of the pressure tester is based on the principle that, for any given airflow through a fixed orifice, a constant pressure drop across the orifice will result. As the airflow varies, the pressure changes accordingly and in the same direction. If air is supplied under pressure to the

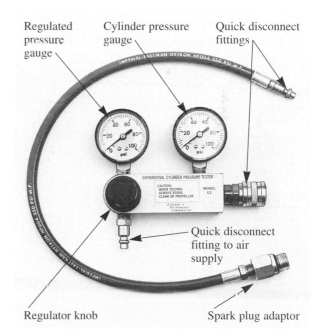

Figure 12-13. *A differential pressure tester can help identify specific worn or damaged components without disassembling the engine. (Eastern Technology Corp.)*

cylinder with both intake and exhaust valves closed, the amount of air that leaks by the valves or piston rings indicates their condition—the perfect cylinder would have no leakage.

The differential pressure tester requires the application of air pressure to cylinder being tested with the piston at top dead center on the compression stroke. The orifice size for engines up to 1000 cu in displacement should be .040″ diameter, .250″ long with a 60° air approach angle.

A schematic diagram of the differential pressure tester is shown in **Figure 12-14.** As the regulated air pressure is applied to one side of the restricter (metering) orifice with the air shutoff valve closed, there will be no leakage on the other side of the orifice and both gauges will read the same. However, when the air shutoff valve is opened and leakage through the cylinder increases, the cylinder pressure gauge will record a proportionally lower reading.

The differential pressure test can be applied to single, or multiple cylinder engines. Each cylinder is tested separately and in the same manner. Use the following procedure to test:

1. Run engine until it is warmed up to provide uniform lubrication to cylinder walls and rings.
2. Remove the spark plug wire(s) and spark plug(s).
3. Rotate the engine crankshaft until the piston of the cylinder being tested is at top dead center of the compression stroke. Both valves must be closed.

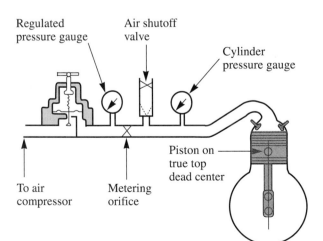

Figure 12-14. *Schematic diagram of differential pressure tester connected to engine cylinder. (Eastern Technology Corp.)*

4. Thread adaptor into spark plug hole.
5. Connect adaptor on rubber hose to quick disconnect fitting on tester.
6. Connect an air source of at least 90 psi to the tester adaptor.

 Make sure all personnel and objects are clear of the engine shaft, belts, pulleys, or other devices connected to the engine shaft.

7. Slowly adjust the regulated pressure gauge on the left with the regulator knob to read 80 psi.
8. The right-hand gauge indicates the relative pressure of the cylinder being tested.

 Due to standard engine clearances and normal wear, no cylinder is expected to maintain a perfect 80 psi. It is important only that all cylinders on multi-cylinder engines have a somewhat consistent reading. Good judgment should be used as to the allowable tolerance between cylinders. Always consult the manufacturer's engine specifications.

Recheck a cylinder that has a lower reading by approximately 15 psi of the other cylinder(s). If a second check is still 15 psi lower, it should be suspected of being defective.

By listening for escaping air at the carburetor intake, the exhaust system, and the crankcase breather you can determine the cause of low pressure readings. See **Figure 12-15.**

If the test indicates a valve problem, first check valve clearance to make sure the valve is not being held open due to inadequate clearance.

Service Information

Before starting engine work, look at the manufacturer's *service manual* or troubleshooting chart. Manuals generally include *service procedures.* These procedures list the steps to take in order to accomplish a task effectively. Manuals also contain *exploded views* of assemblies and systems. These detailed drawings can help you to disassemble and reassemble parts in the right order.

Air Escaping From:	Indication
carburetor intake	defective intake valve
exhaust system	defective exhaust valve
crankcase breather	defective piston rings

Figure 12-15. *The chart shows location of air leak and the indicated defect.*

Troubleshooting charts

Troubleshooting charts list the most common engine troubles along with possible causes and suggested remedies. See **Figure 12-16.** Experienced mechanics have learned what most symptoms mean and go about solving problems without using troubleshooting charts.

Engine Troubleshooting Chart	
Cause	**Remedy**
Engine fails to start or starts with difficulty	
No fuel in tank.	Fill tank with clean, fresh fuel.
Shut-off valve closed.	Open valve.
Obstructed fuel line.	Clean fuel screen and line. If necessary, remove and clean carburetor.
Tank cap vent obstructed	Open vent in fuel tank cap.
Water in fuel.	Drain tank. Clean carburetor and fuel lines. Dry spark plug and points. Fill tank with clean, fresh fuel.
Engine overchoked.	Close fuel shut-off and pull starter until engine starts. Reopen fuel shut-off for normal fuel flow.
Improper carburetor adjustment.	Adjust carburetor.
Loose or defective magneto wiring.	Check magneto wiring for shorts or grounds; repair if necessary.
Faulty magneto.	Check timing, point gap; if necessary, overhaul magneto.
Spark plug fouled.	Clean and regap spark plug.
Spark plug porcelain cracked.	Replace spark plug.
Poor compression.	Overhaul engine.
No spark at plug.	Disconnect ignition cut-off wire at the engine. Crank engine. If spark at spark plug, ignition switch, or safety switch interlock switch is inoperative. If no spark, check magneto.
Crankcase seals and/or gaskets leaking(two cycle only).	Replace seals and/or gaskets.
Exhaust ports plugged (two cycle only).	Clean exhaust ports.
Engine knocks	
Carbon in combustion chamber.	Remove cylinder head and clean carbon from head and piston.
Loose or worn connecting rod.	Replace connecting rod.
Loose flywheel.	Check flywheel key and keyway; replace parts if necessary. Tighten flywheel nut to proper torque.
Worn cylinder.	Replace cylinder.
Improper magneto timing.	Time magneto.

Figure 12-16. *Typical troublshooting chart for small gas engines.*

(Continued)

Engine Troubleshooting Chart

Cause	Remedy
Engine misses under load	
Spark plug fouled.	Clean and regap spark plug.
Spark plug porcelain cracked.	Replace spark plug.
Improper spark plug gap.	Regap spark plug.
Pitted magneto breaker points.	Replace pitted breaker points.
Magneto breaker arm sluggish.	Clean and lubricate breaker point arm.
Faulty condenser.	Check condenser on a tester; replace if defective.
Improper carburetor adjustment.	Adjust carburetor.
Improper valve clearance.	Adjust valve clearance to recommended specifications.
Weak valve spring.	Replace valve spring.
Reed fouled or sluggish (two cycle only).	Clean or replace reed.
Crankcase seals leak (two cycle only).	Replace worn crankcase seals.
Engine lacks power	
Choke partially closed.	Open choke.
Improper carburetor adjustment.	Adjust carburetor.
Magneto improrerly timed.	Time magneto.
Worn rings or piston.	Replace rings or piston.
Air cleaner fouled.	Fill crankcase to the proper level.
Lack of lubrication (four cycle only).	Clean air cleaner.
Valves leaking (four cycle only).	Grind valves and set to recommended specifications.
Reed fouled or sluggish (two cycle).	Clean or replace reed.
Improper amount of oil in fuel mixture (two cycle only).	Drain tank; fill with correct mixture..
Crankcase seals leak (two cycle only).	Replace worn crankcase seals.
Engine overheats	
Engine improperly timed.	Time engine.
Carburetor improperly adjusted.	Adjust carburetor.
Air flow obstructed.	Remove any obstructions from air passages in shrouds.
Cooling fins clogged.	Clean cooling fins.
Excessive load on the engine.	Check operation of associated equipment. reduce excessive load.
Carbon in combustion chamber.	Remove cylinder head and clean carbon from head and piston.
Lack of lubrication (four cycle only).	Fill crankcase to proper level.
Improper amount of oil in fuel mixture (two cycle only).	Drain tank; fill with correct mixture.

(Continued)

Figure 12-16. *Continued.*

Engine Troubleshooting Chart

Cause	Remedy
Engine surges or runs unevenly	
Fuel tank cap vent hole clogged.	Open vent hole.
Governor parts sticking or binding.	Clean and, if necessary, repair governor parts.
Carburetor throttle linkage or throttle shaft and/or butterfly binding or sticking.	Clean, lubricate, or adjust linkage and deburr throttle shaft or butterfly
Intermittent spark or spark plug.	Disconnect ignition cut-off wire at the engine. Crank engine. If spark, check ignition switch, safety switch, and interlock switch. If no spark, check magneto. Check wires for poor connections, cuts or breaks.
Improper carburetor adjustment.	Adjust carburetor.
Dirty carburetor.	Clean carburetor.
Engine vibrates excessively	
Engine not securely mounted.	Tighten loose mounting bolts.
Bent crankshaft.	Replace crankshaft.
Associated equipment out of balance.	Check associated equipment.
Engine uses excessive amount of oil (four cycle only)	
Engine speed too fast.	Using tachometer adjust engine RPM to specifications.
Oil level too high.	To check level, turn dipstick cap tightly into receptacle for accurate level reading.
Oil filler cap loose or gasket damaged causing spillage out of breather.	Replace ring gasket under cap and tighten cap securely.
Breather mechanism damaged or dirty causing leakage.	Replace breather assembly.
Drain hole in breather box clogged causing oil to spill out of breather.	Clean hole with wire to allow oil to return to crankcase.
Gaskets damaged or gasket surfaces nicked, causing oil to leak out.	Clean and smooth gasket surfaces. Always use new gaskets.
Valve guides worn excessively thus passing oil into combustion chamber.	Ream valve guide oversize and install 1/32″ oversize valve.
Cylinder wall worn or glazed, allowing oil to bypass rings into combustion chamber. Piston rings and grooves worn excessively	Bore hole or deglaze cylinder as necessary. Reinstall new rings and check land clearance and correct as necessary.
Piston fit undersized.	Measure and replace as necessary.
Piston oil control ring return holes clogged.	Remove oil control ring and clean return holes.
Oil passages obstructed.	Clean out all oil passages.

Figure 12-16. *Continued*

Tolerances and clearances

Engine *tolerances* and *clearances* are given in chart form in service manuals and/or bulletins. See **Figure 12-17**. When checking and adjusting spark plug gap, breaker point gap, ignition timing, etc., this chart gives the correct dimensions and piston locations.

Note that the spark plug gap setting is .030″ and breaker point gap (magneto point gap) is .020″.

Table 1. Tolerances and Clearances for the J-321 Engine			
Cylinder bore	2.1265 2.1260	Spark plug gap	.030
Piston skirt diameter	2.1227 2.1220	Magneto point gap	.020
Piston ring width	.0925 .0935	Ignition timing in degrees B.T.D.C.	22° (fully retarded)
Piston pin diameter	.5001 .4999	Piston skirt to cylinder clearance	.0033 .0045

Figure 12-17. *Typical tolerance and clearance chart. (Jacobsen Mfg. Co.)*

Ignition timing shows the crankshaft throw at 22° before top dead center (BTDC), which is the fully retarded position. Some engine charts show piston height in thousandths of an inch BTDC rather than in degrees.

When tolerance specifications show two values, the actual dimension must be within that range. In **Figure 12-17,** for example, the cylinder bore (diameter) must measure somewhere between 2.1260″ and 2.1265″, which is a range of .0005″.

Torque specifications

Before attempting troubleshooting or maintenance of any kind, you must be familiar with *torque specifications* for fasteners used in assembling different parts of the engine. Torque refers to the effort expended to turn something. Bolts and nuts must be *torqued* to a given tightness to hold mating parts together under specified tension.

Obviously, if a bolt or nut is too loose, vibration may loosen it more. If the same bolt is too tight, the threads may be stripped or the bolt may be broken off in the hole.

If the installed bolt does not break or strip the threads, expansion from heat will weaken the metal in the bolt. This expansion may go beyond the elastic limit of the bolt (a point where the bolt stretches but does not return to its original length upon cooling). When this happens, the bolt or part may fail.

There is still another reason why bolts and nuts should not be overtightened. Excessive internal stresses within a part may cause warpage or failure of the part.

When a number of bolts are required to fasten mating parts together (such as cylinder heads to cylinder block), it is necessary to tighten them evenly and according to a certain sequence (order)

to prevent warping and leaks. When gaskets are used, overtightening of fasteners will crush the gasket under the bolt heads and distort the metal between the bolts.

Figure 12-18 shows the correct cylinder head bolt tightening sequence in a typical small gasoline engine. To illustrate the procedure, the bolts are numbered in the particular order or sequence in which they are to be tightened. After the first bolt is tightened to 100 in-lb, the one directly opposite is tightened to the same torque reading. Next, the bolts that lie perpendicular to a line between the first two are similarly torqued. Then, the bolts 90° to the left and right are tightened, etc.

Generally, bolts are tightened in a sequence from one side to another in a kind of extending rotation, first tightening each bolt to 100 in-lb, then going back to tighten each bolt an additional 20 in-lb until specified torque is reached.

Figure 12-18. *Head bolts should be tightened evenly and in the sequence recommended by the manufacturer.*

Summary

Preventive maintenance will prevent premature wear of engine parts. Cleaning a small engine regularly will help prevent overheating. Keep the cooling fins free from dirt, leaves, grass clippings, and other obstructions.

Check crankcase oil (four-cycle engines) each time you add gas to the engine. Use only the type and viscosity grade oil recommended by the manufacturer. Oil change intervals are based on hours of running time. Time may vary from 10 to 50 hours. If necessary, replace oil filter when changing oil.

Although proper proportions are used, two-cycle engines can be damaged if the fuel and the oil is not mixed correctly.

If engine is equipped with a crankcase breather, it should be removed and cleaned periodically. The carburetor air cleaner should be cleaned at regular intervals. If necessary, the muffler may have to be cleaned or replaced to prevent power loss and overheating.

Water cooling systems are prone to the formation of rust and scale. Rust inhibitor should be added to these systems. Engine blocks and radiators can be cleaned periodically by reverse flushing the systems.

Outboard engines operated in saltwater are exposed to extremely corrosive conditions. These engines should be rinsed and flushed with freshwater. All lower unit parts should be wiped with a clean, oily rag.

When storing water cooled engines, antifreeze should be added to protect against the lowest possible freezing temperatures. If engine will not be started when in storage, drain the cooling system. Remove all plugs from gearcase and drive shaft housing on outboard engines.

When troubleshooting a small gas engine, always take a systematic approach. Check the simplest, most probable possibilities first. Make sure the engine meets the five fundamental operating requirements including: proper carburetion, correct ignition system operation, adequate lubrication, sufficient cooling, and proper compression.

Engine rpm can be checked with a wire-type tester or a stroboscope. Moving parts appear to be stopped when using a stroboscope. Do not touch moving parts.

A compression test can help to determine the internal condition of an engine. This information is valuable when engine lacks power, runs poorly, or does not respond to ignition and fuel system adjustments.

A differential pressure test can identify and locate specific problems with either piston rings and cylinder, intake, or exhaust valve. It is a superior test to that of the standard compression test with a compression gauge tester. It requires a special testing device with a calibrated orifice, gauges, and compressed air source.

A manufacturer's service manual should be used when servicing small gas engines. These manuals contain assembly drawings, troubleshooting charts, and information on engine tolerances and clearances.

Fasteners should always be tightened to the proper torque specifications. Refer to service manuals for manufacturer's recommendations.

Know These Terms

preventive maintenance
orifice
cooling fins
dipstick
filler plug
loaded oil
condensation
air cleaner
crankcase breather
muffler
rust inhibitor
reverse flushing

antifreeze
systematic trouble-shooting
stroboscope
compression test
differential pressure test
service manual
service procedures
exploded view
troubleshooting chart
tolerances
clearances
torque specifications

Chapter 12 Review Questions

Answer the following questions on a separate sheet of paper.

1. Preventive maintenance helps protect against premature engine wear. True or False?
2. Keeping an engine clean can help prevent _____.

3. Never run engine if the oil level is above full mark. True or False?
4. Although oil does not wear out, excess _____ make the oil useless.
5. When changing oil, the engine should be _____.
   a. cold
   b. warm
   c. running
   d. None of the above.
6. Fuel tanks should be kept full to prevent the formation of _____.
7. Clogged mufflers can cause power loss and overheating. True or False?
8. All mufflers can be cleaned by soaking them in solvent. True or False?
9. To prevent the formation of rust and scale, a rust _____ should be added to the water (antifreeze) in the cooling system.
10. Outboard engines operated in saltwater are exposed to _____ conditions.
11. When storing outboard engines, all plugs should be removed from the _____ and _____ _____ housing.
12. Systematic troubleshooting involves looking for the _____ problem possibilities first.
13. In order to run properly, an engine must have proper carburetion, correct ignition system operation, adequate lubrication, sufficient cooling, and _____.
14. Manufacturer's service manuals contain _____.
   a. service procedures
   b. troubleshooting charts
   c. tolerance and clearance specifications
   d. All of the above.

15. Engine rpm can be checked with a(n) _____, which flashes a light on and off at a controlled rate.
16. A compression test can be used to determine the condition of the spark plug. True or False?
17. During a differential pressure test, air leakage heard at the crankcase breather is an indication of defective _____.
18. A differential pressure test can identify and locate defective _____, _____, and _____.

Suggested Activities

1. Perform preventive maintenance procedures on several engines in your shop. Check the oil level and condition. If necessary, change the oil and the filter. Make sure the air filter element, crankcase breather (if applicable), and cooling fins are clean. If engine is equipped with a water cooling system, check engine block and radiator for signs of rust and corrosion.
2. Locate several malfunctioning engines. Try to determine the cause of the problems using the systematic troubleshooting method. Remember the five fundamental operating requirements.
3. Review several manufacturers' service manuals. Study trouble code charts and exploded-view assembly drawings. Check maintenance procedures for engines in your shop.
4. Perform a differential pressure test on an engine and determine the condition of the rings or valves.

Fuel System Service

After studying this chapter, you will be able to:
▼ Test a fuel pump for proper operation.
▼ Summarize basic carburetor adjustments.
▼ Test two-cycle engine reeds for leakage.
▼ Explain basic procedures for inspecting, over-hauling, and adjusting diaphragm and float-type carburetors.
▼ Troubleshoot float-type and diaphragm-type carburetors.

Troubleshooting the Fuel System

If symptoms of an engine's malfunction point to the fuel system, the problem could be located in several areas. It could involve the fuel pump, carburetor, reed valves (in two-cycle engines), fuel lines, filters, or air cleaner. Troubleshooting will involve checking and/or testing one part after another until the trouble is located and corrected.

Fuel pump

If fuel is not being delivered to the carburetor in a *gravity-fed system,* check the following:
• Is there fuel in the gas tank?
• Is fuel flow blocked by a clogged filter or obstructions in the line?
If engine is equipped with a *fuel pump* and fuel is not being delivered to carburetor, check to see that:
• There is fuel in the gas tank.
• Fittings connecting the fuel line to the tank and the pump are tight; otherwise, the pump will draw air.

• Pump filter is clean and gasket on the filter bowl is in good condition. The method of inspecting the filter and gasket is shown in **Figure 13-1.** Water and contaminants in the fuel are easy to see. Water, being heavier than fuel, separates and collects at the bottom of the bowl, along with other contaminants such as dirt and rust. Generally, visual inspection can be made without dismantling the pump. If disassembly is required, consult an exploded view of the pump. See **Figure 13-2.**

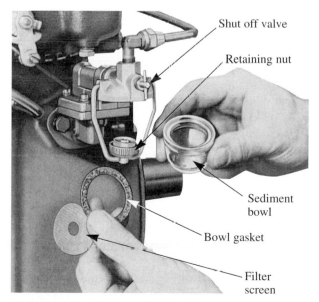

Shut off valve

Retaining nut

Sediment bowl

Bowl gasket

Filter screen

Figure 13-1. *Disassembled sediment bowl includes fuel bowl, strainer, and gasket. If water was present, it would settle to the bottom of the bowl.*

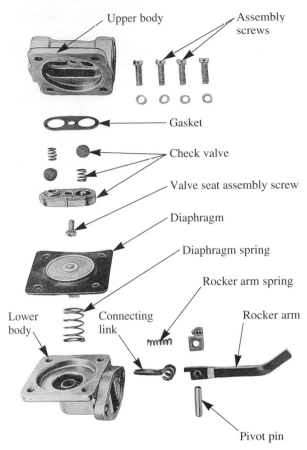

Figure 13-2. *Exploded view of fuel pump is typical of those found in engine service manuals. These are a useful aid to repair and reassembly. (Wisconsin Motors Corp.)*

Labels for Figure 13-2: Upper body, Assembly screws, Gasket, Check valve, Valve seat assembly screw, Diaphragm, Diaphragm spring, Rocker arm spring, Lower body, Connecting link, Rocker arm, Pivot pin

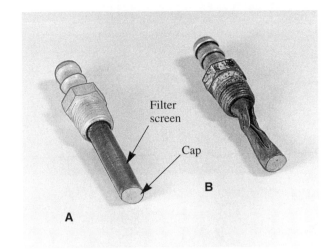

Figure 13-3. *Damaged gas tank filter can block fuel flow. A—New filter. B—Filter damaged when gasoline can nozzle was inserted too far into fuel tank.*

Labels for Figure 13-3: Filter screen, Cap, A, B

- Pump is actually working. Disconnect the fuel line between pump and carburetor. Turn the engine with the starter. There should be a well-defined spurt of fuel at every stroke of the pump (every two revolutions of the engine).

When fuel pump problems are suspected, make sure that the fuel flow from the supply tank to the pump is not interrupted. **Figure 13-3** shows a filter that allowed the engine to start and idle, but caused the engine to stall whenever the throttle was opened. A replacement filter (A) solved the problem. Varnish from stale gasoline can impede flow through filter screens.

Carburetor adjustments

If the fuel pump is working properly, but the engine still surges or lacks power, the cause may be poor carburetor adjustment or carburetor

defects. Most carburetors have two needle valve adjustments: the ***high speed adjustment*** and the ***idle mixture adjustment.*** A third adjustment that is found on carburetors is the ***idle speed stop screw.*** Needle valves are not always found in the same location on the carburetor body. It is good practice to refer to the manufacturer's manual for needle positions and instructions on proper settings.

Today, most carburetors are factory preset and sealed. They are not adjustable by the consumer. This ensures efficient performance. The purpose is to ensure a clean running engine and to minimize air pollution caused by exhaust emissions.

High speed and idle mixture adjustment

Each engine manufacturer will give *rough* settings for the high speed adjustment and idle mixture adjustment. This will permit the engine to be started. It should then be warmed up before further adjustments are made.

The general adjustment procedure is to open the throttle wide and turn the high speed adjusting needle forward and backward slowly until maximum speed is reached. See the carburetor in **Figure 13-4** for the general location of external parts. After reaching the maximum speed, turn the needle counterclockwise very slightly so the engine is running a little rich.

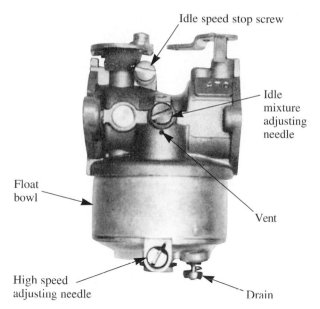

Figure 13-4. *External carburetor adjustments are the high speed adjusting needle, the idle mixture needle, and the idle speed stop screw. (Jacobsen Mfg. Co.)*

To adjust the idle mixture needle, move the throttle to the slow running position. Turn the idle mixture needle slowly, first in one direction and then in the other. Continue turning the needle until idle is smooth. If necessary, adjust the idle speed stop screw to obtain the idle speed recommended in the manual. Check the manufacturer's recommendations carefully. Each make and model may be adjusted differently.

Unlike carburetor A in **Figure 13-5,** not all carburetors have two needle settings. Carburetor B has only a high speed needle adjustment and carburetor C does not have fuel adjustment needles. Each carburetor has an idle speed adjustment screw. Some of the needle adjustments are omitted because the manufacturer has preset the adjustments and sealed them.

Testing two-cycle engine reeds

To determine if the *reeds* are leaking on a two-cycle engine, remove the air cleaner from the carburetor intake. Run the engine while holding a clean strip of paper about one inch from the carburetor throat. See **Figure 13-6.** If the paper becomes spotted with fuel, the reeds are not seating properly.

Operating an engine with leaking reeds will result in fuel starvation, poor lubrication, and overheating. The reeds should be replaced or repaired.

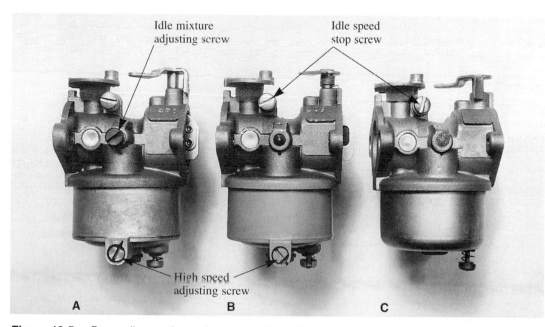

Figure 13-5. *Depending on the make and model, carburetors may have one or several adjustments.*
A—Carburetor with high speed adjustment, idle mixture adjustment, and idle speed stop screw.
B—Carburetor with two points of adjustment: high speed and idle speed.
C—Carburetor with idle speed stop screw only.

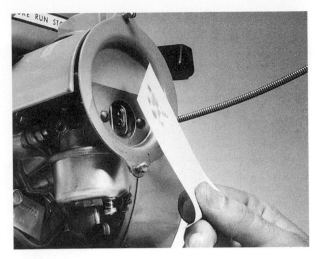

Figure 13-6. *Reeds can be tested for leakage by placing a clean strip of paper one inch from the carburetor intake. If spots of fuel show on the paper, the reeds are leaking.*

Carburetor Overhaul

A carburetor ***overhaul*** generally consists of disassembling, cleaning, and replacing parts as recommended by the manufacturer. ***Carburetor kits*** are generally available from small engine repair shops or manufacturer's distribution centers. These kits contain all the parts needed for a typical carburetor overhaul. A typical carburetor diagram showing part locations is illustrated in **Figure 13-7.**

⚠️ Do not allow flames, sparks, pilot lights, or arcing equipment near the fuel system. Ignition of fuel can result in severe personal injury or death.

Carburetor removal

The following removal procedures are typical to most carburetor configurations:
1. Remove air cleaner and set aside for later use.
2. Turn off fuel shut-off valve (located between fuel tank and carburetor). See **Figure 13-8.**
3. Drain carburetor bowl by removing drain screw.
4. Release fuel line hose clamp and disconnect hose.
5. Disconnect governor linkage and spring. (Do not deform spring or linkage.)

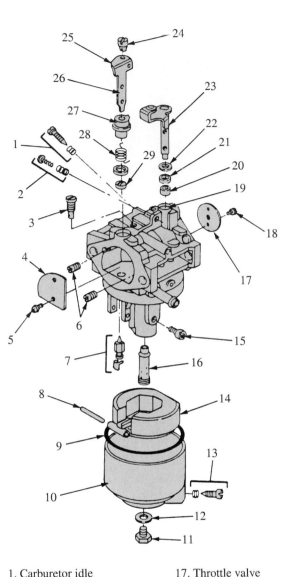

1. Carburetor idle valve assembly
2. Idle speed screw & spring assembly
3. Nozzle
4. Choke valve
5. (2) Choke valve screws
6. Carburetor nozzles
7. Fuel inlet needle valve
8. Float hinge pin
9. Bowl gasket
10. Bowl
11. Bowl mounting screw
12. Sealing gasket
13. Bowl drain
14. Float
15. Main jet
16. Nozzle
17. Throttle valve
18. (2) Throttle valve screws
19. Upper throttle shaft bushing
20. Lower throttle shaft bushing
21. Throttle shaft seal
22. Throttle shaft seal
23. Throttle shaft
24. Link retainer
25. Choke shaft
26. Upper choke shaft bushing
27. Choke shaft spring
28. Choke shaft seal
29. Lower choke shaft bushing

Figure 13-7. *Exploded view of a float-type carburetor, showing individual part locations. (Briggs and Stratton Corp.)*

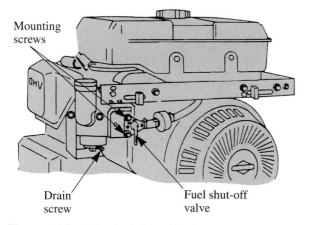

Mounting screws

Drain screw

Fuel shut-off valve

Figure 13-8. *The fuel shut-off valve is located between the fuel tank and the carburetor. (Briggs and Stratton Corp.)*

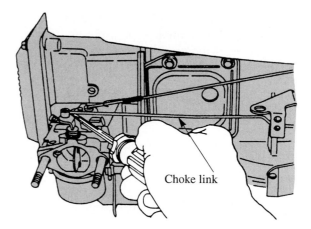

Choke link

Figure 13-9. *Choke linkage can be removed from the choke lever by prying it with a screwdriver. (Briggs and Stratton Corp.)*

6. Remove throttle linkage from throttle lever.
7. Remove carburetor mounting screws. (Hold body to prevent it from falling from engine.)
8. Remove choke linkage from choke lever. See **Figure 13-9.**
9. Lift carburetor from engine (should be free from engine and linkage).

Carburetor disassembly

 Carefully disassemble the carburetor, removing all nonmetallic parts (gaskets, O-rings, etc.). Clean all metallic parts with solvent or commercial carburetor cleaner. Never soak parts in solvent for longer than 30 minutes. Nonmetallic parts should never be exposed to solvents. Safety goggles and rubber gloves should always be worn when working with solvents and commercial carburetor cleaners. Many commercial carburetor cleaners are extremely caustic and can cause serious burns to skin and eyes.

On some carburetor models, *welch plugs* must be removed to expose drilled passages. To remove welch plugs, sharpen a small chisel to a wedge point and drive the chisel into the plug. See **Figure 13-10.** Push down on the chisel and pry the plug out of position. To install a new plug after cleaning the carburetor, place the plug into the receptacle with the raised portion up and flatten it

with a flat punch that is slightly larger in diameter than the plug itself. Do not dent the plug or drive the center of the plug below the top surface of the carburetor. See **Figure 13-11.**

Clean all carbon from the carburetor bore, especially where the throttle and choke plates seat.

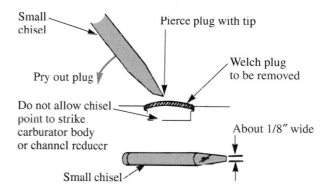

Small chisel

Pierce plug with tip

Pry out plug

Welch plug to be removed

Do not allow chisel point to strike carburator body or channel reducer

About 1/8" wide

Small chisel

Figure 13-10. *Before soaking carburetor in solvent, welch plug is removed from body to expose drilled passages. (Tecumseh Products Co.)*

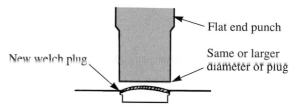

Flat end punch

New welch plug

Same or larger diameter of plug

Figure 13-11. *Use a flat punch to install new welch plugs. Only flatten plug. Do not dent it. (Tecumseh Products Co.)*

Be careful not to plug the idle or main fuel ports. Dry all passages with low-pressure air (35 psi). Avoid using wire or other objects for cleaning ports. These items frequently enlarge the diameter of important passages.

Inspection of parts

When all parts are clean and dry, organize them neatly on a clean cloth. Carefully inspect each part for damage and wear. Examine the idle and high-speed adjusting screw needles. The points should be straight and smooth. O-ring seals should be replaced if they are damaged. See **Figure 13-12.** Examine the choke and throttle plates for coded markings, which identify the way that they must be installed. See **Figure 13-13.** Study the choke shaft, throttle shaft, and bearing holes for wear.

The fuel bowl must be free of dirt and corrosion. Replace the bowl gasket or O-ring seal before installing the bowl on the carburetor body. See **Figure 13-14.** Some fuel bowls will interfere with the float if not positioned correctly. See **Figure 13-15.**

Examine the float for damage. Hollow, brass-type floats must be free from pinholes and dents.

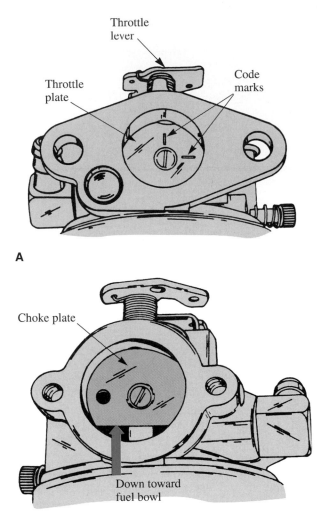

A

B

Figure 13-13. *A—This throttle plate must be installed with the code marks in the location shown. B—This choke plate must be installed with the flat side down toward the float bowl. (Tecumseh Products Co.)*

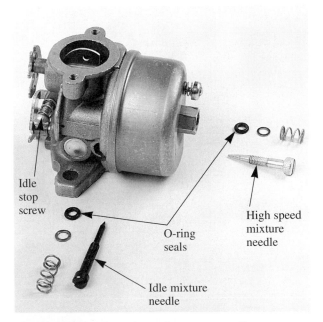

Figure 13-12. *Examine adjustment needles and replace O-rings if they are deformed. (Tecumseh Products Co.)*

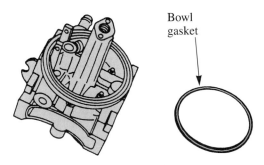

Figure 13-14. *When replacing float bowl, always install a new gasket or O-ring. (Briggs and Stratton Corp.)*

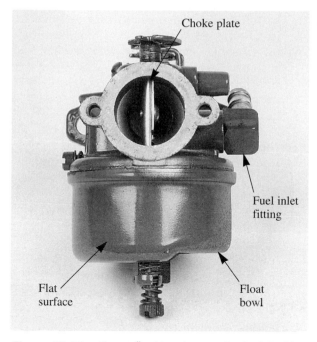

Figure 13-15. *Some float bowls must be installed in a particular way to avoid interference with the internal float. (Briggs and Stratton Corp.)*

Check the float hinge bearing surfaces for wear. The tab that contacts the inlet needle should also be inspected for wear. See **Figure 13-16.**

The inlet needle may seat on a synthetic rubber seat in the carburetor body. To remove this seat, pull it out with a short piece of hooked wire or force it out with a short blast of compressed air.

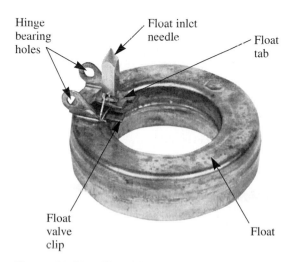

Figure 13-16. *Float hinge bearing holes should be examined for wear. Float tab is used to adjust float height per specifications. (Tecumseh Products Co.)*

When installing a new seat, moisten it with oil and insert it into the carburetor body (smooth side toward the inlet needle). Press the seat into the cavity using a flat punch that is the same diameter as the seat. Make sure it is firmly seated. See **Figure 13-17.** Check the condition of the spring clip that is used to attach the inlet needle to the float tab. See **Figure 13-18.**

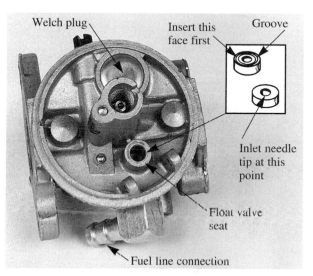

Figure 13-17. *Synthetic rubber needle seat must be installed with grooved face toward bottom of hole. Rubber seat is oiled before being pressed into hole with a flat punch. (Tecumseh Products Co.)*

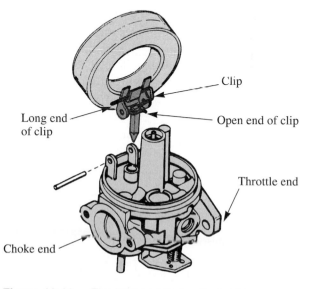

Figure 13-18. *The inlet needle is attached to the float with a specially shaped wire clip. (Tecumseh Products Co.)*

Float-type carburetor repair

Carburetion problems that cannot be corrected by adjusting mixture needles are usually the result of gummed-up fuel passages or worn internal parts. The most effective solution to these problems is to overhaul the carburetor.

When making float-type carburetor repairs, the following rules should be observed:
- Follow all instructions carefully.
- Never use drill bits or wires to clean passages.
- Never enlarge passages.
- Always remove all welch plugs before soaking carburetor in cleaner.
- Never soak carburetor in cleaner for more than 30 minutes.
- Never reuse gaskets.
- Always use new service replacement screws for choke and throttle valve plates (screws are treated with a dry adhesive to secure them).
- Always set carburetor float to manufacturer's specified height.

It should be noted that many carburetors are designed for specific engine models. Similar carburetors may contain hundreds of significant variations in detail. The procedures presented in this chapter are general and may not pertain directly to the carburetor that you are attempting to service. However, if care is taken during the disassembly phase, no difficulty will be encountered during reassembly.

Assembling float-type carburetors

After all carburetor parts have been cleaned, inspected, and replaced (if necessary), they should be reassembled in the following order:
1. Install new welch plugs (if applicable).
2. Install throttle shaft, bushings, seals, and washers.
3. Attach throttle valve plate with new screws.
4. Install idle-speed stop screw (if applicable).
5. Install choke shaft, spring, seal, washer, and bushing.
6. Attach choke valve plate with new screws.
7. Install metering nozzles and needle seats.
8. Install idle needle assembly.
9. Install high-speed needle valve assembly.
10. Attach inlet needle to float with wire clip.
11. Install float on carburetor with hinge pin.
12. Adjust float height per manufacturer's specifications. See **Figure 13-19.**
13. Install float-bowl over float assembly with mounting screw. Use a new gasket or O-ring to seal bowl. See **Figure 13-20.**
14. Install carburetor on engine. If a gasket is required, always use a new one.
15. Connect throttle, choke, and governor linkage. On some engines, linkage is connected before bolting carburetor to engine.

Figure 13-19. *Float must rest on needle at a specified height. If float is high, too much fuel will be used. If float level is low, lean mixture may cause overheating. (Tecumseh Products Co.)*

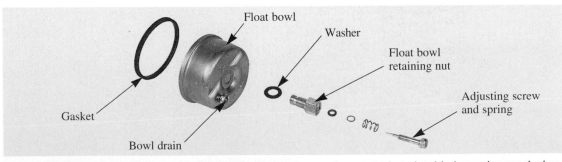

Figure 13-20. *Float bowl is assembled in the order shown. A new gasket should always be used when reinstalling the float bowl. (Tecumseh Products Co.)*

16. Connect fuel line.
17. Install air cleaner.
18. Adjust idle mixture and high-speed needles to initial starting settings per manufacturer's instructions.

Engine priming

A manually operated, plunger-type *primer* may be found on some carburetors. The main purpose of the primer pump is to force fuel through the main carburetor nozzle, producing a rich mixture for starting. **Figure 13-21** shows a bulb-type primer pump. When the rubber bulb is pressed, air pressure is increased in the float bowl, forcing fuel through the main nozzle and into the carburetor throat. See **Figure 13-22.**

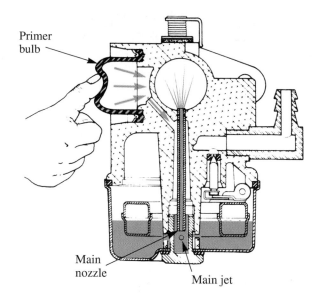

Figure 13-22. *Pressing the rubber primer bulb forces fuel through the main nozzle, providing a rich air/fuel mixture to help start a cold engine. (Tecumseh Products Co.)*

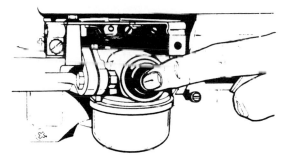

Figure 13-21. *Manually operated, bulb-type primer. (Tecumseh Products Co.)*

Troubleshooting float-type carburetors

The chart in **Figure 13-23** points out areas of possible float-type carburetor troubles and lists symptoms and suggested repairs. This figure is shown for illustrative purposes and is not representative of all float-type carburetors. Always consult

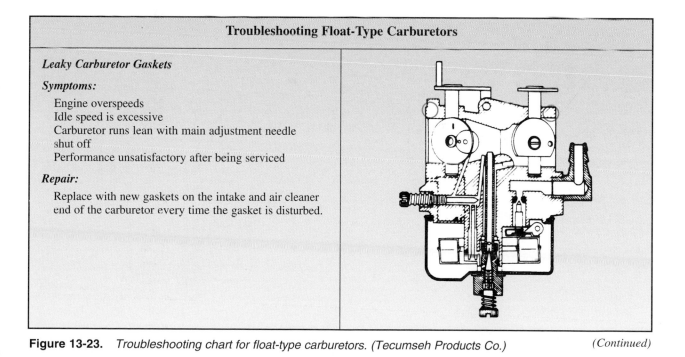

Troubleshooting Float-Type Carburetors

Leaky Carburetor Gaskets

Symptoms:

Engine overspeeds
Idle speed is excessive
Carburetor runs lean with main adjustment needle shut off
Performance unsatisfactory after being serviced

Repair:

Replace with new gaskets on the intake and air cleaner end of the carburetor every time the gasket is disturbed.

Figure 13-23. *Troubleshooting chart for float-type carburetors. (Tecumseh Products Co.)*

(Continued)

Troubleshooting Float-Type Carburetors

Throttle and/or Choke Shaft Worn, Throttle and/or Choke Springs not Functioning

Symptoms:

Engine will not start
Engine hunts (at idle or high speed)
Engine will not idle
Engine lacks power at high speed
Idle speed is excessive
Choke does not open fully
Performance unsatisfactory after being serviced

Repair:

Replace all worn parts, springs, dust seals(when so equipped). If carburetor body is worn out or round, causing the leak, a new service carburetor should be used.

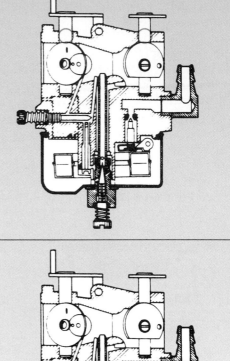

Fuel inlet is plugged or loose

Symptoms:

Engine will not start
Engine hunts (at idle or high speed)
Engine will not idle
Engine lacks power at high speed
Carburetor leaks
Engine starves for fuel at high speed (leans out)

Repair:

Clean fuel system completely. Refill with clean fresh fuel, as recommended. Replace loose clamps and fittings.

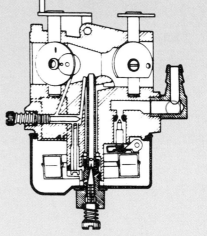

Dirty, stuck, or damaged needle and seat

Symptoms:

Engine will not start
Carburetor floods
Carburetor leaks
Poor engine performance

Repair:

Remove old needle and seat. Install a new needle and seat according to manual instructions.

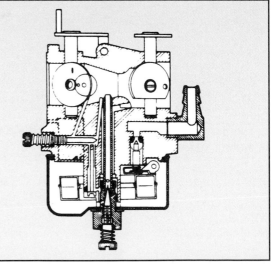

Figure 13-23. *Continued.*

(*Continued*)

Troubleshooting Float-Type Carburetors

Damaged or leaky O-rings

Symptoms:

Engine hunts (at idle or high speed)
Carburetor leaks
Engine overspeeds
Idle speed is excessive
Carburetor runs with main adjustment needle shut off

Repair:

All rubber O-rings should be removed before cleaning and should be replaced with new ones when rebuilding the carburetor.

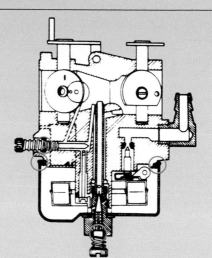

Main nozzle Restricted or plugged

Symptoms:

Engine will not start
Engine hunts at high speed
Engine starves for fuel at high speed (leans out)

Repair:

Soak carburetor in cleaner for no more than 30 minutes. Use compressed air to clean passages. CAUTION: Do not use compressed air with float on carburetor. The compressed air will crush the float.

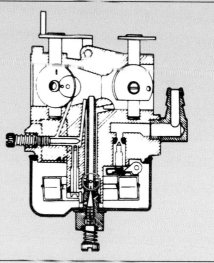

Damaged and/or Worn Hinge Pin or Float, Improper Float Height

Symptoms:

Engine hunts (at idle or high speed)
Engine will not idle
Engine lacks power at high speed
Carburetor floods
Engine starves for fuel at high speed (leans out)
Carburetor runs with main adjustment needle shut off
Poor starting

Repair:

Replace float and axle. If hinge pin area of casting is worn, the carburetor body must be replaced. The float height is set using specified tool.

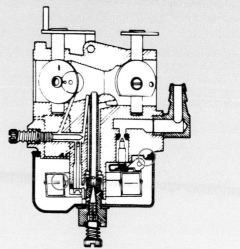

Figure 13-23. *Continued.*

(Continued)

Troubleshooting Float-Type Carburetors

Fuel Pick-Up Restricted or Plugged

Symptoms:

Engine will not start
Engine hunts (at idle or high speed)
Engine will not idle
Engine starves for fuel at high speed (leans out)

Repair:

After soaking carburetor in a commercial cleaner (no longer than 30 minutes), use compressed air to clean passages. CAUTION: Do not use compressed air with float on carburetor. The compressed air will crush the float.

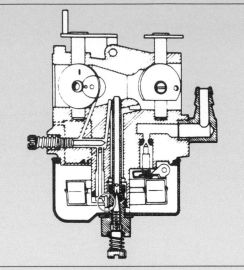

Damaged or Incorrect Fuel Adjustment Needles

Symptoms:

Engine will not start
Engine will not accelerate
Engine hunts (at idle or high speed)
Engine will not idle
Engine lacks power at high speed
Engine overspeeds
Engine starves for fuel at high speed (leans out)
Carburetor runs with main adjustment needle shut off
Performance unsatisfactory after being serviced

Repair:

Replace damaged needles with correct fuel adjustment needles. CAUTION: Do not over-seat needles.

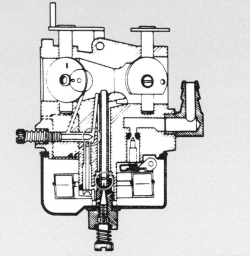

Restricted or Plugged Air Bleed or Idle System

Symptoms:

Engine runs rich
Engine hard to start
Engine will not accelerate
Engine hunts
Engine will not idle

Repair:

After soaking carburetor in a commercial cleaner (no longer than 30 minutes), use compressed air to clean passages. CAUTION: Do not use compressed air with float on carburetor. The compressed air will crush the float. The metering rod on Series I carburetors is not a serviceable part. If metering rod is not free, carburetor body must be replaced.

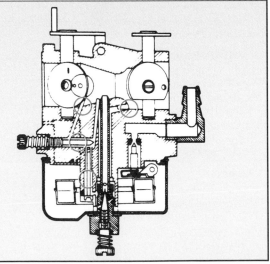

Figure 13-23. *Continued.*

(Continued)

Restricted Idle and/or Secondary Discharge Ports

Symptoms:

Engine will not start at idle
Engine will not accelerate
Engine hunts
Engine will not idle

Repair:

After soaking carburetor in a commercial cleaner (no longer than 30 minutes), use compressed air to clean passages. CAUTION: Do not use compressed air with float on carburetor. The compressed air will crush the float.

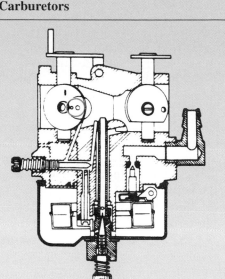

Restricted or Plugged Atmospheric Vent

Symptoms:

Engine will not start
Carburetor floods
Carburetor leaks

Repair:

After soaking carburetor in a commercial cleaner (no longer than 30 minutes), use compressed air to clean passages. CAUTION: Do not use compressed air with float on carburetor. The compressed air will crush the float.

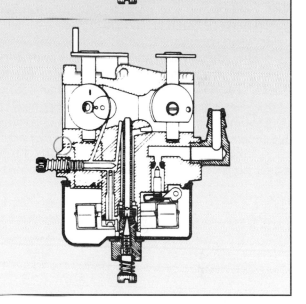

Figure 13-23. *Continued.*

an appropriate manufacturer's technical service manual when servicing a carburetor.

Diaphragm carburetor repair

Except for the float and the float chamber, the procedures for servicing ***diaphragm-type carburetors*** are similar to those used for float-type carburetors. After removing and disassembling the carburetor, wash each part in an appropriate solvent. Lay the clean parts out on a clean, white cloth so that they are not lost or damaged. If the

carburetor is very dirty, a commercial carburetor cleaner solution may be used.

Be careful not to get any solution on hands or clothing. Wear safety glasses.

Put only the metallic carburetor parts in the solution and let them soak. Nonmetallic parts can be damaged by harsh commercial cleaners.

After the parts have soaked for 30 minutes, rinse them with a milder cleaning solvent and dry with compressed air. Do not dry the parts with a rag or paper towel. Lint from the rags may get into the passages. Never clean holes or passages with wires or similar objects. These will distort the openings and may prevent the engine from running properly.

After cleaning, inspect the parts for wear, material failure, or other damage. Check the carburetor body and crankcase for cracks or worn mating surfaces. See **Figure 13-24.** If necessary, replace defective parts.

Check throttle and choke shafts. They must fit closely but turn easily in their bearing holes. See **Figure 13-25.** If loose, they will cause poor engine performance.

The diaphragm should be checked for defects that would cause leakage. The diaphragm needle valve must be straight and fit the seat so that it seals when closed.

The high speed and idle mixture needles should have straight, smooth tapers. See **Figure 13-26.** The O-rings or seals around the needles should be replaced if they are cut or deformed.

Some manufacturers supply diaphragm carburetor repair kits which include those items that would most likely need replacing. See **Figure 13-27.** Other parts can be purchased as they are needed.

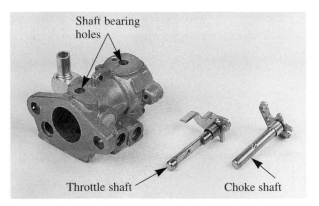

Figure 13-25. *Throttle and choke shafts must fit bores closely, but turn freely. (Tecumseh Products Co.)*

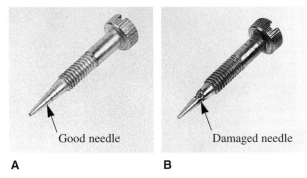

Figure 13-26. *A—Needle valve has straight, smooth taper. B—Needle valve is damaged and should be replaced.*

Assembling diaphragm carburetors

Because of the many different carburetor designs, it is highly recommended that the manual be followed when assembling a carburetor. After all carburetor parts have been cleaned, inspected,

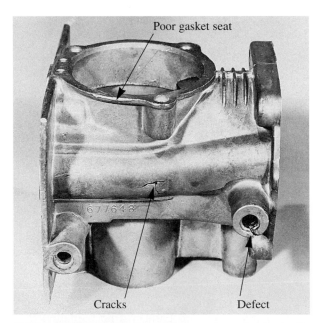

Figure 13-24. *This carburetor has several defects. If repair is not possible, carburetor should be replaced.*

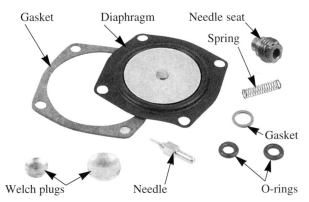

Figure 13-27. *Carburetor repair kit shown is for a diaphragm-type carburetor.*

and replaced (if necessary), the carburetor should be reassembled in the following order:

1. Place throttle shaft in the bearing hole with the return spring on the shaft as illustrated in **Figure 13-28.** The throttle plate must be placed properly in the carburetor bore. Usually, identifying marks of some sort are put on the plate to assist the assembler. On the carburetor shown in **Figure 13-28,** the marks have to be placed facing outward.
2. Assemble the choke plate in the same way as the throttle. It is *vented,* which means that it has openings to allow some air to enter even when it is closed. **Figure 13-29** shows the flat side facing the mixture needles and the indentations facing outward.

3. Install the spring and needle valve in the order shown in **Figure 13-30.** The spring and needle valve are inserted into the body. The gasket and seat are then screwed into the threaded hole in the carburetor body.
4. Assemble the diaphragm as shown in **Figure 13-31.** Four screws are placed through holes in diaphragm cover to tightly draw the assembly to the body. The rivet head in the center of the diaphragm is turned toward the needle valve.
5. Install high speed and idle mixture needles in the order shown in **Figure 13-32.**

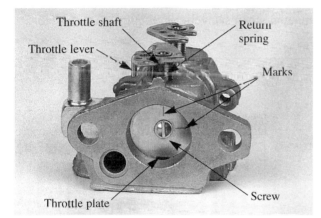

Figure 13-28. *Throttle shaft with return spring attached is placed in holes provided in the throttle body. Throttle valve must be fastened to shaft in the proper position. (Tecumseh Products Co.)*

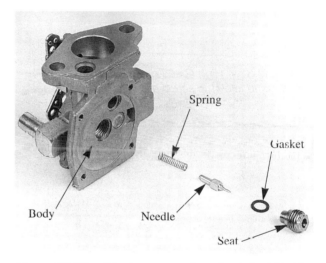

Figure 13-30. *Diaphragm needle valve is installed in carburetor body in sequence shown.*

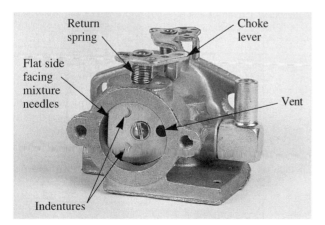

Figure 13-29. *Choke valve assembly is shown in place in carburetor air horn.*

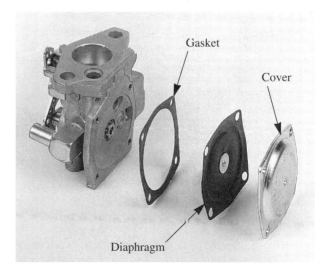

Figure 13-31. *In this application, gasket, diaphragm, and cover are fastened to the carburetor body with four screws. (Tecumseh Products Co.)*

Figure 14-1. *A normal spark plug will be clean and dry, showing a light tan to gray-tan color on the porcelain shell insulator.*

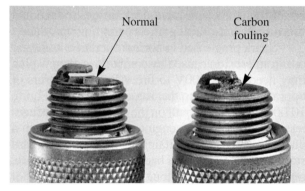

Normal

Carbon fouling

Figure 14-2. *Normal spark plug on left is compared to a carbon-fouled plug on right. (Jacobsen Mfg. Co.)*

Figure 14-3. *An oil-fouled spark plug can be an indication of a mechanical malfunction. (Champion Spark Plug Co.)*

Figure 14-4. *After many hours of use, spark plug electrodes tend to erode from normal combustion. (Jacobsen Mfg. Co.)*

Study the situation carefully before changing to a spark plug that is hotter than the one specified. The hot plug may stop carbon buildup, but other, more serious problems can develop. Spark plug deposits are usually caused by weak magneto voltage, incorrect carburetor adjustments, poor air cleaner maintenance, incorrect gasoline or oil, or incorrectly mixed gasoline and oil.

Spark plug removal

The following four steps for removing spark plugs are simple and should become a habit. They can prevent troublesome problems from occurring later.

1. Gently rotate and pull the spark plug boot from the spark plug. Do not grab or pull on the spark plug cable. Pull on the boot only.
2. Use a correct-size, deep spark plug socket with a rubber sleeve installed to protect the plug insulator from breaking.

> Before removing spark plugs from engines with aluminum heads, allow the engine to cool. The heat of the engine, in combination with a spark plug that has run for many hours, may cause the spark plug to seize.

3. Before removing the loosened spark plug, blast dirt away from the area around the plug with compressed air.
4. Remove spark plug and check its appearance. Refer to the *Spark Plug Analysis Table* in the *Appendix* section of this text.

Analysis of used spark plugs

Spark plugs that are cleaned and gapped at regular intervals can provide many hours of useful life in an engine. Spark plugs in high energy ignition systems (TCI or CDI) that burn unleaded fuel can provide more than twice as many hours of useful service.

Spark plugs in two different engines of the same make and model may show a wide variation in appearance. Engine condition, carburetor settings, and operating conditions, such as sustained high speeds or continual low-speed, stop-and-start operation, are all variables that affect spark plug life.

Spark plugs are sometimes incorrectly blamed for poor engine performance. Replacing an old plug may temporarily improve engine performance because of the lessened demand a new spark plug makes on the ignition system. However, it is not a cure-all for poor performance caused by worn rings or cylinders, improper carburetion, worn ignition system parts, or other engine problems.

Cleaning spark plugs

The following procedure should be observed when cleaning spark plugs:

1. Wipe all spark plug surfaces clean. Remove oil, water, dirt, and moist residues.
2. Check the firing end or tip of the spark plug for oily or wet deposits. If the spark plug has deposits, brush the plug with a nonflammable, nontoxic solvent. Then, dry the plug with compressed air to prevent caking of the cleaning compound deep within the spark plug shell.
3. Clean the spark plug in a spark plug cleaning machine.
4. File spark plug electrodes after cleaning to square them and to remove any oxide or scale from the surfaces. Open the gap so that a spark plug file can be put between the electrodes. See **Figure 14-5.**
5. Clean spark plug threads with a hand or powered wire brush. See **Figure 14-6.** Take care not to damage the electrodes or insulator. If threads are nicked or damaged, discard the plug.

When installing plugs, clean the spark plug seat on the cylinder head. Also, clean the threads and make sure the gasket is in good condition. Not all plugs require gaskets. **Figure 14-7** shows a spark plug with a tapered seat that does not require a gasket. The other one shown does require a gasket.

Figure 14-5. *Oxides should be removed from electrodes with a spark plug file.*

Figure 14-6. *Threads can be cleaned with a power wire brush.*

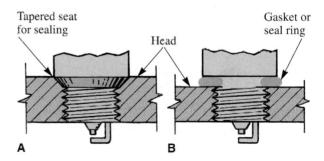

Figure 14-7. *Two types of plugs. A—Plug with tapered seat does not require a gasket. Clean the area around the plug hole for a good seal. B—Spark plug with gasket. Inspect plug gaskets carefully. A damaged gasket will not seal properly.*

Gapping spark plugs

When *gapping* a spark plug, bend the outer electrode toward or away from the center electrode. For best results, use a *gapping tool.* See **Figure 14-8.**

Standard *leaf-type feeler gauges* may be used if the plug is new or if the electrodes are in good condition. Otherwise, wire-type thickness gauges should be used. See **Figure 14-9.**

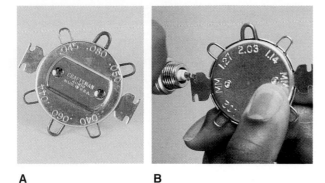

A　　　　　　**B**

Figure 14-8. *A—A common spark plug gapping tool. B—Use a spark plug gapping tool to bend the outer electrode toward or away from the center electrode.*

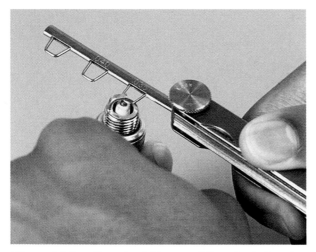

Figure 14-9. *Gap should be carefully adjusted and measured with wire gauges.*

The reason that ***wire-type thickness gauges*** are recommended for use on worn spark plugs is shown in **Figure 14-10.** Note that the flat, leaf-type thickness gauge would leave an additional gap between the worn electrodes.

Spark plug installation

Spark plugs must be installed properly. The heat dispersing properties of the spark plug depend on correct plug seating. If the spark plug is tightened excessively, the gasket will be crushed. Internal leakage may result. Attempting to remove an overtightened spark plug can strip cylinder head threads. Seating the spark plug too loosely can result in preignition and possible engine damage caused by spark plug overheating.

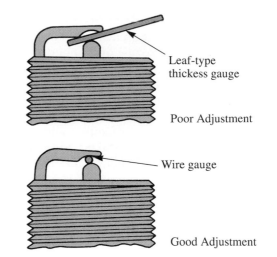

Figure 14-10. *Leaf-type thickness gauges may not give an accurate measurement of gap if electrodes are not flat and parallel.*

To install spark plugs:
1. Make sure cylinder head and spark plug threads are clean. If necessary, use a thread chaser and a seat cleaning tool.
2. Make sure that the spark plug gasket seat is clean. Thread the gasket to fit flush against the gasket seat on the spark plug.
3. Make sure the spark plug has the correct gap.
4. Screw the spark plug finger-tight into the cylinder head. Use a torque wrench to tighten the plug to 13-15 lb-ft. Do not overtighten.

Ignition testing procedure

Certain basic inspections and tests are necessary to determine whether an ignition system is working properly. The following procedure can be used to verify ignition system operation:
1. Make certain that the thin ignition ground wire is not grounding out the entire system. Examine the high tension spark plug lead for voltage leaks. Look for worn insulation or cuts where metallic contact is made.
2. Remove the spark plug and examine the electrodes and the porcelain insulator. If the plug is carbon grounded, replace it. Adjust the plug electrodes to the specified gap. Then, install the plug and tighten it to the specified torque. If torque specifications are not available, turn the plug finger-tight and tighten it a half turn with a wrench. This will allow for proper heat transfer from plug to cylinder head.

3. Hold the plug wire by the insulation (well away from metal connector) so that the connector is 3/16″ from the tip of the plug. See **Figure 14-11.**
4. Pull the starter cord. An orange-blue spark should jump the gap between the connector and the plug. If it does, the ignition system is good.
5. If no spark occurs, hold the wire 3/16″ from the base of the plug. See **Figure 14-12.** Pull the starter cord once more. If a spark occurs at base of the plug and not at the tip, the plug is failing under compression and should be replaced.
6. If a spark does not occur in either position, the problem is in the magneto system.

A spark tester can also be used to test the ignition circuit for proper operation. Most spark testers are simply attached to the secondary wire. See **Figure 14-13.** If current is flowing in the spark plug lead, it will induce a voltage in the tester. This voltage will energize a light in the window of the tester. The greater the voltage in the secondary wire, the brighter the light in the tester's window.

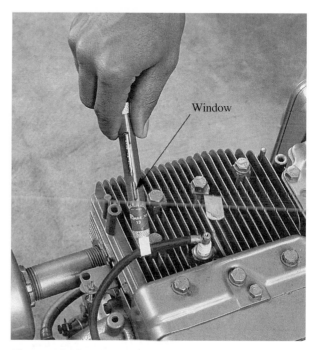

Figure 14-13. *A spark tester can be used to determine the intensity of secondary current. A light will flash in the window each time ignition spark occurs.*

Figure 14-11. *If spark docs not jump a 3/16″ gap at the plug tip but jumps the same gap at the base of the plug, the plug is defective.*

Figure 14-12. *If no spark occurs at the plug tip or the plug base, the problem is in the magneto system.*

Magneto service

In some *magneto systems,* the coil is mounted outside of the flywheel and the gap between the laminated core and the flywheel magnets is adjustable. This adjustment must be made carefully or the magnetic strength and voltage produced will be reduced. Loosen the adjustment screws and place a feeler gauge or nonmagnetic shim stock of the correct thickness in the gap. See **Figure 14-14.** After the correct gap has been obtained, tighten the adjustment screws.

Many magneto systems are completely contained under the flywheel. The flywheel must be removed from the crankshaft to service the ignition

Figure 14-14. *On some engines, magneto air gap is checked with a nonmagnetic thickness gauge. (Lawn Boy Power Equipment, Gale Products)*

components in these systems. The flywheel is mounted on the tapered end of the crankshaft. A keyway in the crankshaft is keyed for alignment. See **Figure 14-15.**

Removing the flywheel

Since the *flywheel* is fastened to the crankshaft with a nut, it may be hard to remove. One way is to remove the flywheel nut and other accessory parts and place a ***knock-off tool*** on the crankshaft. Thread the tool all the way onto the crankshaft and tap it with a hammer. See **Figure**

14-16. The sudden jolt will loosen the flywheel. Never use the flywheel nut as a knock-off tool or the crankshaft threads will be damaged.

Another way to remove the flywheel is to use a ***wheel puller,*** like the one shown in **Figure 14-17.** After installing the puller, tighten the center bolt snugly. If the flywheel does not break loose from the crankshaft when the center bolt is tightened, tap the bolt with a hammer. Use extreme care when removing aluminum flywheels.

Figure 14-16. *A knock-off tool is often used to remove a flywheel. The tool must be tight on the shaft before being hit. Pry bar provides a valuable assist.*

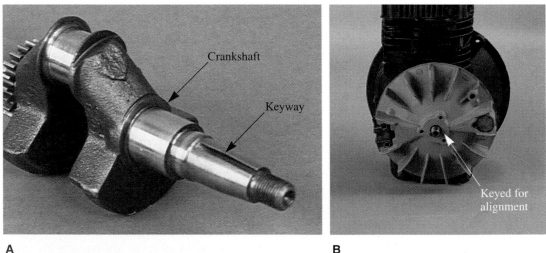

A

B

Figure 14-15. *A—The crankshaft end is tapered and has a keyway. B—The crankshaft is keyed to hold the flywheel in exact position. The flywheel often needs a special tool for removal from the taper.*

Figure 14-17. *A flywheel puller is often used to prevent the flywheel from distorting or cracking during removal. (Kubota Tractor Corp.)*

Figure 14-19. *Magnetism is the heart of the magneto system. Test magnets by placing a socket on them and shaking the flywheel. Socket should remain in place.*

Inspecting the flywheel

When the flywheel is removed, inspect it. Make sure that it is not cracked and that the mounting hole and keyway are not damaged. Inspect the key. If it has begun to shear, as in **Figure 14-18A,** or if the key is too narrow, as in **Figure 14-18B,** the engine will be out of time. In either case, the key must be replaced.

To check the strength of the flywheel magnet, place a 1/2″ socket on the magnet and shake the flywheel. The socket should remain in place on the magnet. See **Figure 14-19.** The magnet can lose its magnetism.

Testing ignition system components

After removing the flywheel, inspect the magneto system components. Check all parts for dents, cracks, and gouges. If system is equipped with breaker points, check for improper alignment and pitted contact surfaces. Make sure the insulation on all wire leads is in good condition.

A multimeter can be used to check the condition of coils and condensers. A ***multimeter*** is a meter that can measure resistance and amperage, as well as voltage. This type of meter is also referred to as a voltage/ohms/ampere meter (VOM). See **Figure 14-20.** The meter has a number of settings for reading voltage, resistance (ohms), or amperage.

To check a coil's primary windings for ***continuity,*** set the multimeter to its lowest resistance range setting. Connect one meter probe to the coil ground connection and the other to the coil primary lead. See **Figure 14-21.** The meter should indicate almost no resistance between these connections. To check the secondary coil for continuity, set the meter to the R × 100 or R × 1000 resistance range setting. Connect one meter probe to the ground connection and the other to the high tension output lead. See **Figure 14-22.** The resistance between these connections may range from several hundred ohms to several thousand ohms. Check manufacturer's specifications. See **Figure 14-23.** If the meter registers infinite resistance in either test, the coil is faulty and it must be replaced.

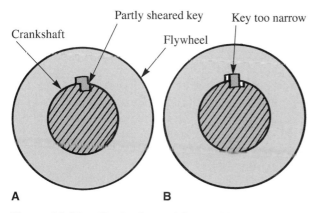

A   B

Figure 14-18. *Partly sheared flywheel keys or keys that do not fit well can put the ignition system out of time.*

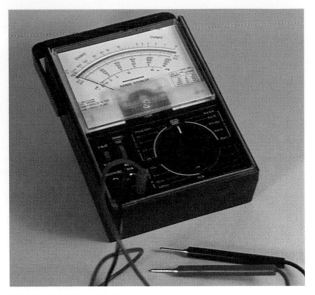

Figure 14-20. *This* multimeter *can test voltage, resistance, and amperage. This type of meter is also referred to as a voltage/ohms/ampere (VOM) meter.*

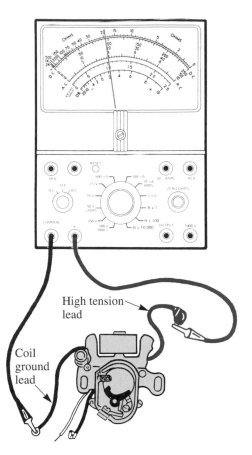

Figure 14-22. *Testing a secondary coil for continuity. Meter readings will vary, depending on coil type. Check manufacturer's specifications.*

To test a condenser for continuity, set the multimeter to the highest resistance scale. Connect the test leads as illustrated in **Figure 14-24.** The meter should initially indicate a low resistance, but the reading should rise rapidly as the condenser takes on a charge from the meter's battery. If the reading remains low, the condenser is faulty and it must be replaced.

Solid state ignition systems are offered in many small engines. They offer dependable service, with few moving parts to maintain or adjust. Certain solid state components are manufactured as assemblies and cannot be serviced.

If a solid state system fails to produce a spark between the plug cable terminal and the cylinder head, check the condition of the cable. If it has cracked insulation or shows the after effects of arcing (black powdery discoloration or deposits), replace the pulse transformer and the high tension lead assembly.

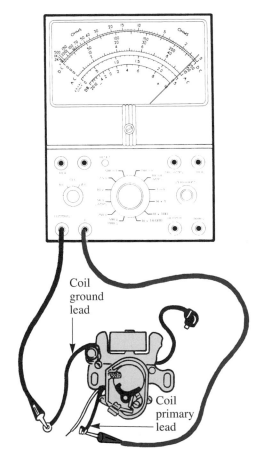

Figure 14-21. *Testing a primary coil for continuity with a multimeter.*

Small Engine Coil Specifications					
Coil No.	Mfg. Model No.	Mfg. No.	Operating Amperage	Primary Resistance Min.-Max.	Secondary Continuity Min.-Max.
26787	Fairbanks-Morse	HX-2477	1.90		50–60
28259	Wico	X-11654	2.1		40–55
29176	Phelon	FG-4081	2.8		40-60
29632	Lauson	5022 (Syncro)	2.3	.5–1.5	40–55
30560	Lauson VH60 John Deere (Replaces 30546)	5160 (Syncro)	2.9	.35–.45	40–55
32014	Lauson HH80	8	1.4	.37–.45	55–65

Figure 14-23. *Coil specifications are often provided by test equipment manufacturer. (Merc-O-Tronic Instrument Corp.)*

Examine the low-tension lead. If the insulation is cracked or in poor condition, replace it. The ignition charging coil, electronic triggering system, and mounting plate are available as an assembly only. If you must replace this assembly, carefully follow the manufacturer's instructions.

Before replacing the pulse transformer, attach the leads from a new transformer, ground the unit, and test for spark. If you have spark where none existed before, replace the old unit with the new one.

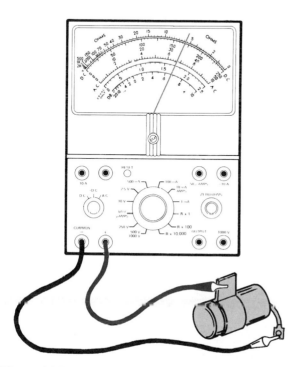

Figure 14-24. *A multimeter can be used to test a condenser for continuity.*

In some *solid state systems,* you can adjust the air gap between the trigger module and the flywheel trigger projection. Rotate the flywheel until the projection is next to the trigger module. Loosen the trigger retaining screw and move the trigger until the air gap is correct. Use a feeler gauge and make certain that the flat surfaces on the trigger and the projection are parallel before tightening the retaining screws.

 The trigger module may be located on the outside of the flywheel on some engines and on the inside of the flywheel on others.

In many cases, the condition of ignition components can be checked with a special ignition tester. The ignition tester checks components under conditions similar to operating conditions. These testers can be used to verify the operation of condensers, coils, and solid state ignition components. Recommended testing procedures vary among manufacturers. Always consult an appropriate manufacturer's and/or shop service manual before performing tests on ignition components. Follow the manufacturer's recommendations carefully.

Typical testing procedures are in *Typical Ignition Component Tests* in the *Appendix* section of this text. When testing ignition system components, perform tests on a wooden table or an insulated tabletop. This will prevent current leakage, which can cause *shorting* or shock hazards.

Adjusting breaker point ignition systems

The first step in adjusting a breaker point ignition system is to set the breaker points to the correct gap. Look in the engine manual for the proper setting. Remove the dust cover from the stator plate to expose the breaker points. Turn the crankshaft slowly until the high point of the cam lobe is directly under the wear block or cam follower. Then, slightly loosen the point adjustment screw, as demonstrated in **Figure 14-25.**

On the magneto system shown in these illustrations, a screwdriver can be used to move the stationary part of the breaker points. A fulcrum point of leverage is provided. See **Figure 14-26.**

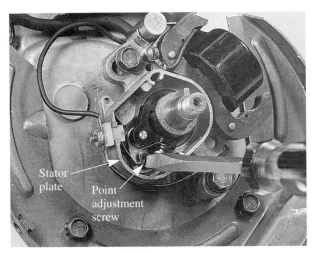

Figure 14-25. *To adjust breaker point gap, loosen stationary point set screw first.*

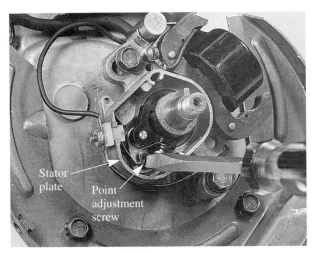

Figure 14-26. *Using a screwdriver as a lever resting on the fulcrum, move stationary breaker point to vary the size of the gap. Wear block should be on the high point of the cam.*

To adjust the breaker point gap, use a feeler gauge of the specified thickness. First, move the stationary point until it lightly touches the feeler gauge. See **Figure 14-27.** Then, tighten the adjusting screw that holds down the stationary breaker point. Finally, recheck the point gap. The stationary point may move when the adjusting screw is retightened. Repeat the adjusting procedure until the gap setting is right.

Adjusting piston height

Ignition spark occurs at the instant the breaker points open. When this happens, the piston must be at the proper position (near top dead center) in the cylinder. **Figure 14-28** shows piston heights for a number of Tecumseh engines. Model A, for example, requires the piston to be between .060″ and .070″ BTDC (before top dead center).

One method of setting piston height is to use a timing tool. See **Figure 14-29.** Install the cylinder head over the cylinder and fasten it with one or two head bolts. Do not use the gasket. Locate the head so that the spark plug hole is over the piston and both valves can move freely.

To use the timing tool, screw it into the spark plug hole with its graduated rod riding on the piston top. Note that each marking on the rod is 1/32″ (.031″). Place the nut on the crankshaft so it can be turned with a wrench. Then, turn the crankshaft in

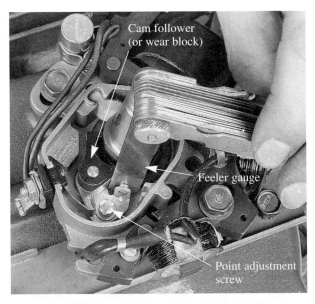

Figure 14-27. *A feeler gauge is used to set breaker point gap. When gap is correct, set screw is tightened to lock in adjustment.*

Small Engine Coil Specifications									
Model	A	B	C	D	E V & H40	F V & H60	G V & H50	H	I
Displacement	6.207	7.35	7.61	8.9	11.04	13.53	12.176	7.75	9.06
Stroke	1 3/4″	1 3/4″	1 13/16″	1 13/16″	2 1/4″	2 1/2″	2 1/4″	1 27/32″	1 27/32″
Bore	$\frac{2.125}{2.127}$	$\frac{2.3125}{2.3135}$	$\frac{2.3125}{2.3135}$	$\frac{2.5000}{2.5010}$	$\frac{2.5000}{2.5010}$	$\frac{2.6250}{2.6260}$	$\frac{2.6250}{2.6260}$	$\frac{2.3125}{2.3135}$	$\frac{2.5000}{2.5010}$
Timing Dimension Before Top Dead Center for Horizontal Engines	$H\frac{.060}{.070}$	$H\frac{.060}{.070}$	$H\frac{.060}{.070}$	$H\frac{.030}{.040}$	$H\frac{.090}{.100}$	$H\frac{.090}{.100}$	$H\frac{.090}{.100}$	$H\frac{.060}{.070}$	$H\frac{.030}{.040}$

Figure 14-28. *A table of specifications in the service manual will provide data concerning piston height (timing dimension before top dead center) for use in timing the opening of breaker points. (Tecumseh Products Co.)*

the direction the engine runs until the piston is at TDC. Since proper piston position is between .060″ and .070″ BTDC, two graduations or marks (.062″) will be required. However, to compensate for any backlash, back the piston down six graduations and bring it back up four marks. This places the piston exactly two graduations or .062″

BTDC. Finally, lock the thumbscrew and remove the wrench.

Another method of adjusting piston height involves using a ***depth micrometer***. With this technique, the micrometer is set for correct piston height BTDC and the piston is brought up to it. See **Figure 14-30.**

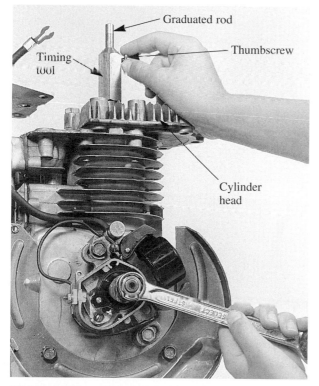

Figure 14-29. *A special timing tool can be used to adjust piston to specified location BTDC.*

Figure 14-30. *A depth micrometer can also be used to measure piston height BTDC. The piston can be raised or lowered by turning the wrench on the crankshaft nut.*

A third method of measuring and setting the piston height is shown in **Figure 14-31.** In this method, a special *dial indicator,* which is equipped with an extender leg, is used to reach over the piston top. A tip, which contacts the piston top, is installed on the end of the extender leg. To use this instrument, the correct tip must be installed. Use the small tip for engines with timing dimensions between Top Dead Center (TDC) and .050″ BTDC. Use the large tip for engines with timing dimensions of between .051″ BTDC and .150″ BTDC. If the engine is the four cycle type, make sure to secure the extender leg in position to locate the tip directly over the piston head. Loosen the sleeve screw on the side of the adapter to allow the sleeve to be turned into the threads of the spark plug hole. This will ensure the proper location of the tip. Once the adapter sleeve is secured in the hole, tighten sleeve screw to prevent the dial from moving up or down, which could give a false reading.

Find Top Dead Center (TDC) by rotating the crankshaft clockwise (when looking at the magneto end of the crank) until the needle on the dial stops and reverses direction. TDC is at the point where the needle stops. Loosen the dial screw and rotate the dial until the zero is lined up with the needle at TDC. Retighten the dial screw to secure it in place. See **Figure 14-32A.**

While watching the needle on the dial indicator, rotate the crankshaft counterclockwise (when looking at the magneto end of the crank) past the specified Before Top Dead Center (BTDC) dimension (Example: .090″). See **Figure 14-32B.** Then rotate the crankshaft back clockwise to the proper dimension (Example: .080″). See **Figure 14-32C.** This will compensate for any slack between the connecting rod and the crankshaft assembly.

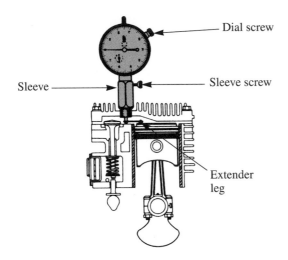

Figure 14-31. *A special dial indicator with an extender leg can be used to determine piston height. Use the correct type of extender leg for the engine being serviced. (Tecumseh Products Co.)*

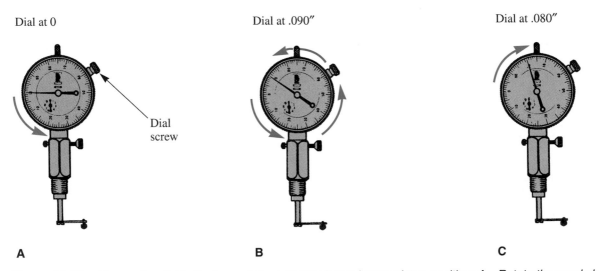

Figure 14-32. *To use the dial indicator and the extender leg to locate piston position: A—Rotate the crankshaft clockwise until needle stops. Loosen the dial screw and set zero under needle. Tighten dial screw. B—Rotate the crankshaft counterclockwise past the specified BTDC dimension. C—Rotate the crankshaft clockwise to the proper dimension to remove the slack between the connecting rod and the crankshaft.*

Timing ignition spark

With the correct breaker point gap set and the piston positioned at the point where spark should occur, the next step is to time the **breaking** of the points. You can do this with a simple, battery-operated continuity tester or a multimeter set on an ohm setting.

First, loosen the two stator adjustment bolts so the stator can be rotated. Disconnect the coil lead wire to the points and reinstall the securing nut. Connect one lead of a continuity tester or multimeter set on an ohm setting to the breaker point terminal and the other lead to a good ground location. Rotate the stator until the continuity tester light or the multimeter indicates a break in the circuit, showing that the breaker points have opened. See **Figure 14-33.** Carefully tighten the two stator adjustment bolts and reconnect the coil lead to the breaker point terminal.

Before replacing the dustcover over the stator plate, place one or two drops of oil on the felt cam oiler. See **Figure 14-34.** Lubrication reduces wear on the point wear block. Do not over oil. Excess oil will foul the breaker points. Clean the points by sliding a lint-free paper back and forth between the contacts. Manually open the points when removing the paper to eliminate paper fibers from remaining between the contact points. If you have carefully followed the recommended procedure, the engine ignition system will be *in time.*

Figure 14-34. *A felt oiler should have several drops of oil to lubricate the cam lobe. Some felt oilers are reversible. (Jacobsen Mfg. Co.)*

Reinstalling the flywheel

After performing the necessary ignition system testing and service, the flywheel can be replaced. To replace the flywheel, turn crankshaft to place the keyway in 12 o'clock position and insert the key. If the shaft uses a Woodruff key, position it as shown in **Figure 14-35.** Make sure that the key seats properly in the keyway before starting the flywheel on the shaft.

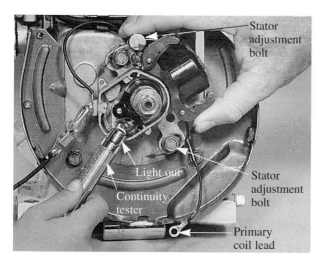

Figure 14-33. *A continuity tester is attached. The stator plate is loosened and turned until the continuity light goes out. This indicates that the points have opened. The stator plate is locked in place. The coil primary lead is disconnected during this test.*

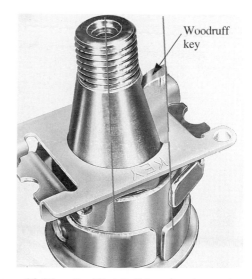

Figure 14-35. *If a Woodruff-type key is used on a crankshaft, it should be placed so that the top of the key is parallel to the centerline of the shaft. (Lawn Boy Power Equipment)*

Next, align the flywheel keyway with the crankshaft key and install the flywheel. Tighten the crankshaft nut to the correct torque.

Use a strap wrench or a spanner wrench to hold the flywheel when torquing the crankshaft nut. See **Figure 14-36.**

Reinstalling the cylinder head

If the cylinder head was removed to set piston position, it should be cleaned of all carbon deposits and checked for flatness by laying it on a surface plate. Use a .002″ feeler gauge between the mating surfaces to detect any warpage. See **Figure 14-37.**

Figure 14-37. *Cylinder head should be placed on a surface plate and tested for flatness with a thickness gauge.*

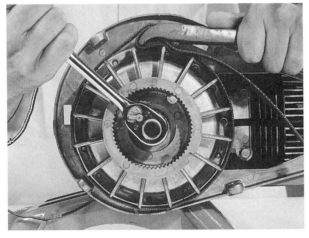

A

B

Figure 14-36. *A—The flywheel can be held with a strap wrench while tightening flywheel nut. B—A spanner wrench can also be used to hold flywheel when torquing flywheel nut.*

If cooling fins are broken, the head should be replaced. If there are cracks, nicks, or burrs on the machined surface of the head, it should be resurfaced. Always install a new head gasket and follow the bolt tightening sequence recommended by the manufacturer.

Servicing battery ignition systems

Battery operated ignition systems are much like magneto systems. Therefore, many of the components in the battery ignition systems are the same as those in the magneto systems. Refer to magneto system section of this chapter for service information on those similar components. However, battery ignition systems have additional components that require maintenance and service. When servicing a battery ignition system, check for the following problems before beginning an extensive system analysis:
1. Defective or undercharged battery.
2. Corroded or loose terminals and connections.
3. Wrong connections.
4. Cracked insulation or broken wires.
5. A wire *grounding out* in the system.
6. A defective switch.
7. Improperly functioning operator presence system.

The presence of a battery and a starter does not mean that the ignition system is battery operated. Some magneto ignition systems also use these parts. You can identify a battery ignition system by a can-shaped coil, in addition to a battery and a generator or alternator

 All lawn and garden tractors built after July of 1987 are required to have an operator presence system. Many implements were equipped with these systems prior to this date. If an engine will not start or is cutting out, check for a malfunction in the operator presence system.

Batteries

Storage batteries need regular maintenance to keep them in good operating condition. Remove the battery caps about twice a month and, if necessary, add distilled water to bring the electrolyte above the plates. This prevents sulfating. *Sulfating* is the forming of salt-like deposits when air combines with the electrolyte. Sulfating will ruin the battery. These deposits are the same as the green deposits that collect around the posts and cable clamps of an automobile battery.

The main storage or electrical power in *electric start* lawnmowers is the battery. See **Figure 14-38.** With proper setup and maintenance the battery will last for years. However, in some remote cases even with proper maintenance a battery can lose power.

Figure 14-38. *A battery pack for electric start lawnmower mounted on mower handle so key switch is convenient to operator.*

A chemical reaction between the battery's electrolyte and plates, or electrodes, will supply electrical energy to an external circuit. When the battery is discharging the positive plate (lead dioxide) and the negative plate (sponge lead) are both changed to lead sulfate. At the same time, part of the electrolyte (diluted sulfuric acid) is changed to water. This conversion of diluted sulfuric acid to water reduces the specific gravity (density) of the electrolyte. By measuring the specific gravity with a hydrometer, a direct measure of how far the discharge process has progressed can be made.

Battery starting circuits

Battery starting circuits consist of the following components:
1. A battery is the source of electrical energy.
2. A starter solenoid switch to transfer high starting current from battery to starter (starter relay).
3. A key start switch or other switch to energize the starter solenoid.
4. A starter which is a series wound, low resistance, high current draw direct current motor.

Make sure a circuit breaker has not disconnected the circuit. If the circuit breaker has disrupted the circuit it will not allow the starter to crank. Also, check diode wires to see if they are crossed. If crossed, reverse the diode wires.

Testing the battery

A visual inspection of the battery includes looking for the following:
1. A broken or leaking cover.
2. A broken case.
3. Damaged post(s).
4. Missing caps. See **Figure 14-39.**

Handle a battery with care. Never leave a battery setting in a discharged condition. If a battery must be carried, use an approved battery carrier, wear rubber gloves and goggles. Keep the battery away from the body and clothing. A battery may be quite heavy depending upon it's size. Never test a battery by striking a cable or metal strap across the output terminals. An internally shorted battery could explode.

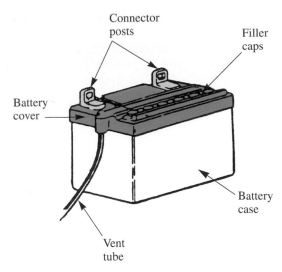

Figure 14-39. *Visual inspection of a lead-acid storage battery should include looking for a broken or leaking cover, a broken case, damaged posts, and missing caps.*

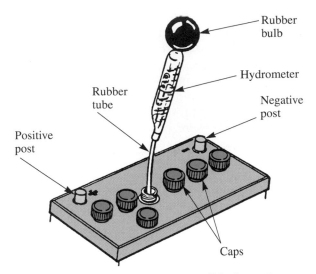

Figure 14-40. *A hydrometer test will indicate the condition of each cell and the battery in general. All cells should read equally if the battery is good. A fully charged battery should read 1.265 or more.*

Hydrometer test

If battery condition is doubtful, test it. Use a *hydrometer* to determine if the battery is fully charged. The hydrometer determines charge by measuring the specific gravity (density) of the electrolyte. The hydrometer test will also indicate if a cell is shorted which, in some cases, cannot be charged or will not hold a charge. See **Figure 14-40.**

1. Specific gravity tests must be performed before adding water to the battery.
2. In the event the electrolyte level is too low to test with the hydrometer, add water and charge before testing.
3. A correct specific gravity reading can be measured only when the electrolyte temperature is 80°F. If the electrolyte temperature varies from this temperature, compensation must be made in the reading as follows:
   a. Add four gravity points (.004) for each 10° electrolyte temperature is above 80°F.
   b. Subtract four gravity points (.004) for each 10° below 80°F.
4. During the reading the float bulb must be floating freely and the eye must be even with the liquid level to obtain an accurate reading.
5. When all cells are tested, if the readings between the highest and lowest cells vary 50 points (.050) or more, the battery should be removed from service and properly discarded.

6. If there is less than a 50 point variation between the highest and lowest cell, and the specific gravity in one or more cells is below 1.235, recharge the battery.
7. Inability to bring the specific gravity of any one cell up to 1.235 after charging indicates an unusable battery and it should be removed from service.
8. After the battery is recharged, let it stand for at least 24 hours, and repeat the hydrometer test on all cells. If there is a 50 point variation or more between the highest and lowest cell, remove the battery from service.

The hydrometer test procedure is quite easy to perform. Remove the vent caps and squeeze the rubber bulb. Place the rubber tube from the tester into the first opening and release the bulb. Electrolyte is drawn into the glass until the float rises. Read the specific gravity at the level of the fluid on the float markings. See **Figure 14-41.** Squeeze the rubber bulb to replace the electrolyte in the cell. Record the cell number and it's reading. Repeat this step for each cell.

Maintenance free or sealed batteries. *Maintenance free* or *sealed batteries* cannot be tested with a hydrometer because there is no access to the electrolyte. These batteries are liquid filled or have a gel electrolyte and must be tested under load with special equipment.

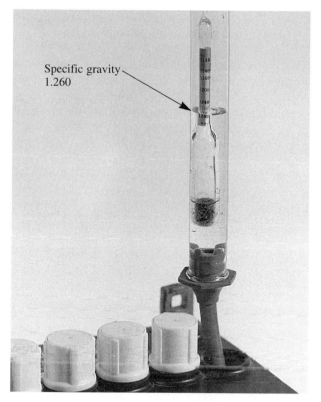

Figure 14-41. *This hydrometer reading shows a fully charged condition of the battery. Specific gravity reading is in the 1.260 to 1.280 range.*

Keep the battery case and terminals clean. Remove the battery cables occasionally and clean all parts thoroughly. Reinstall the cables and coat them with petroleum jelly to retard further corrosion. Keep these connections clean and tight to avoid short circuits and/or voltage loss.

Checking voltage

Another way of testing a battery is to measure cell voltage with an accurate multimeter set on a voltage setting. Cell voltage can also be tested with a high-rate discharge tester. The tester puts a heavy load on the battery for three seconds.

Check each cell in turn. If voltage checks out from 1.95V to 2.08V, the battery is good. If there is a difference of .05V between any cells, replace the battery.

Sealed lead acid batteries can also be checked with a multimeter set on a voltage setting. When checking the voltage, the male terminal of the connector plug is the negative terminal, and the female terminal of the connector plug is the positive terminal. See **Figure 14-42.**

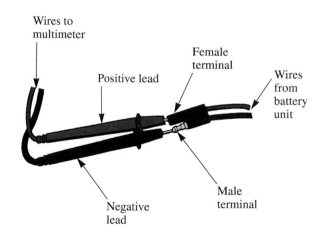

Figure 14-42. *Checking sealed lead-acid battery in the system with a multimeter. Connect negative lead to male terminal of connector and connect the positive lead to female terminal of the connector.*

When checking battery in the system check in line fuse for continuity. If good, then:

1. Connect the negative lead of multimeter into the male end of the plug on the wire harness.
2. Connect the positive lead of the multimeter into the female pin of the wire harness.
3. A fully charged battery will indicate 13.0V on the meter.
4. A battery that reads between 11.0V and 13.0V needs charging.
5. If the battery indicates less than 10.0V on the meter, the battery probably will not accept a charge and should be replaced.

Battery installation

The following is the procedure for installing a battery:

1. Remove old battery. Mark which cable is connected to positive (+), and negative (−) terminals. Positive cable is usually coded red, negative black.
2. Clean cable connectors with wire brush to remove oxidation.
3. After filling with acid and charging (follow manufacturers instructions), install new battery. Connect cables to the proper terminals. Positive cable to positive terminal (+) and negative cable to negative terminal (−).

    Connect negative cable last.

> Reversing polarity by connecting positive cable to negative terminal and negative cable to positive terminal can seriously damage the electrical system. Diodes in the system may be destroyed.

4. Check vent tube to avoid crimping or obstruction.
5. Securely fasten the battery to the unit with the battery holddown clamp. A shaking or vibrating battery can be damaged.

Causes for battery failure

There are numerous causes for battery failure. The following is a list of these causes:

Overcharging. Charging a battery in excess of what is necessary is harmful. The following is a list of the results:

1. Severely corrodes the positive plate grids with weakening and loss of electrical conduction.
2. Decomposes water of electrolyte into hydrogen and oxygen gas. Gas bubbles tend to wash active material from plates and carry moisture and acid from the cells as a fine mist.
3. Decomposition of water leaves acid more concentrated and is harmful to cell components, particularly at high temperatures over a prolonged time period.
4. High internal heat is created, which accelerates the corrosion of positive plate grids and damages separators and negative plates.
5. Overcharging alone, or in combination with a previous condition of undercharging, may cause severe buckling and warping of positive plates with accompanying perforation of separators.
6. May cause corrosion damage to battery box, cables and other critical electrical and engine parts by forcing liquid from the cells.

Undercharging. Undercharging a battery is harmful. The following is a list of the results:

1. A battery with insufficient charge over a prolonged period may develop a dense, hard, coarsely crystalline sulfate which cannot be electrochemically converted to normal active material again. This condition can cause distortion and buckling of the positive plates.
2. An undercharged battery is unable to provide full power and is subject to freezing during

severe winter weather. This may cause cracking of the case and leakage of acid.

Lack of water. Water is essential to a lead-acid storage battery and normally is the only component that is lost as a result of charging. Water should be added as soon as it falls to the top level of the separators. If water is not replaced the plates become exposed and the acid reaches a very high concentration. This condition may char and disintegrate the separators and can permanently sulfate the plates and impair performance of the battery. Plates must always be completely covered by electrolyte.

> Sulfuric acid *must never* be added to a cell unless it is known that acid has been spilled out.

Loose holddowns. A battery that is not held down securely in an implement may be shaken around. The bridges upon which the elements rest can notch the bottom of the separators and the lead plates can notch the bridges causing a severe disarrangement of the plates and separators. A battery that is bouncing may crack or wear the container causing acid to leak. Acid leaks corrode terminals and cables causing high resistance at battery connections weakening power and shortening the life of the battery. Holddowns that are too tight can distort or crack the battery case.

Frozen electrolyte. Electrolyte will begin to freeze at various temperatures depending upon the condition of charge. A 3/4 charged lead-acid battery is not in any danger of freezing. During winter weather a battery should be kept at least 3/4 charged. The freezing points of electrolytes is shown in **Figure 14-43.**

> Never attempt to charge a frozen battery. It may explode.

Battery recharging

Batteries generate *hydrogen* and *oxygen* during charging. Always observe the following rules:

1. Never check a battery fluid level with a flame.
2. Do not use jumper cables on a battery unless the ignition wiring is disconnected from the battery.

Freezing Points of Electrolytes	
Specific Gravity	*Freezing Point*
1.265	–75°F (–59.5°C)
1.225	–35°F (–37°C)
1.200	–17°F (–27°C)
1.150	5°F (–15°C)
1.100	18°F (–7.8°C)
1.050	27°F (–3°C)

Figure 14-43. *As specific gravity of electrolyte decreases, the temperature at which freezing begins increases.*

3. Do not use electric welding equipment on an implement with a battery without first disconnecting the wires to the battery terminals.
4. Connect the negative cable last and disconnect it first (in negative ground systems).

 Caution must be exercised when charging a battery. Hydrogen and oxygen gases combine to form a highly explosive mixture.

Recharging is necessary when you find lights get dim, and/or when a battery is out of use for more than one month. Charging a 12V battery should be done with a 12V, 1 amp (A) automotive type charger. Charging 6V, or 12V batteries should not exceed 1A. Charging should continue until the battery gases freely and specific gravity of electrolyte reaches 1.265 or more. Exceeding the recommended charging rate can cause warping of the plates and will affect the life of the battery.

 To avoid sparks, do not disturb connections to battery while charging: first, throw switch *Off* at charger.

If batteries are used in temperatures below 32°F (0°C), it is important to keep them fully charged. A full charge can prevent the electrolyte from freezing and cracking the battery case. In cold temperatures, battery power decreases while the need for engine cranking power increases. Sub-zero temperatures can reduce a fully charged battery's capacity to 30% of its normal power, while increasing the cranking load beyond the warm weather cranking requirements. See **Figure 14-44.**

Only direct current can be used to recharge a battery. The battery charger automatically rectifies

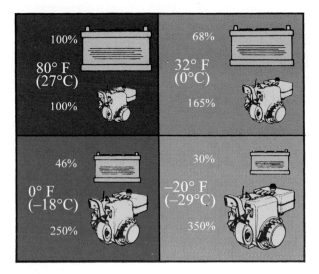

Figure 14-44. *Cold temperatures can reduce cranking power of battery, while increasing the cranking load.*

(converts) the alternating current (ac) available from electrical receptacles to direct current (dc). Before charging a battery, add water to bring the electrolyte in the battery cells up to the right level. Make sure the outside of the battery is clean. Connect the positive charger lead to the positive battery terminal and negative lead to negative terminal. See **Figure 14-45.** Charging times will

Figure 14-45. *The positive charger cable must be attached to the positive battery post and the negative cable must be connected to the negative post. Otherwise, the battery could be ruined. If battery markings are unreadable, dip leads from the battery terminals in a weak solution of sulfuric acid. More gas bubbles will collect around negative lead than the positive lead.*

vary, depending upon battery condition and charging rate. Follow manufacturer's instructions.

Badly sulfated batteries can sometimes be reclaimed by recharging them very slowly. This reconverts the sulfate to electrolyte.

Most new batteries are dry-charged. To place this type of battery in service, add the electrolyte solution according to manufacturer's instructions. Some newly activated batteries require a short period of charging; others can be placed in service immediately.

Maintenance and service

The life of a battery can be greatly extended by proper maintenance and service. The following is proper procedure:
1. Clean battery case top with a stiff bristle brush. Do not use a wire bristle brush. Wear safety goggles, rubber gloves, and be careful not to scatter corrosion particles. Wipe off with a cloth wetted with ammonia or baking soda in water. Finish by wiping with a cloth wet with clear water.
2. Inspect cables. Replace cables if unserviceable. Inspect the terminal posts for damage.
3. Clean the battery posts and cable clamps or connectors to bright metal whenever they are disconnected. A wire bristle brush can be used for this purpose. Coat the contact surfaces with mineral grease or petroleum jelly before reconnecting the cables.
4. Examine the battery box and adjust the hold-down bracket, clamps, or straps. Replace if damaged or severely corroded.
5. Check level of electrolyte once a month. Add clean distilled water if below plates, or below *Upper* and *Lower* level indicators.

 Never add acid to refill a battery.

6. Make a hydrometer and/or voltage test. Remember, a voltage test alone will not give an accurate indication of battery condition. Even a partially discharged battery will display correct voltage under a no-load condition.
7. Keep vent tube free of kinks and obstructions.
8. Store a battery only with a full charge. A discharged battery can become sulfated and/or freeze.
9. Carefully inspect and recharge the battery at the beginning of each working season.

Wiring

Electrical system *wiring* must have good insulation between all points of connection. Wires should be securely fastened and connecting points should be free of corrosion, rust, and oil. Loose and corroded connections can severely diminish battery potential. A pinhole in a wire's insulation can cause electricity to leak and *ground out* on the engine frame. This condition can be amplified if water or oil is present on the insulation. A wire that is grounding out can make starting impossible. It can also cause an engine to run erratically. The thickness of copper wire is expressed in gage numbers. The larger the gage number, the smaller the diameter of the wire. The smaller the number, the larger the diameter of the wire. See **Figure 14-46.** Six gage wire (minimum) must be used for the starter circuit. Sixteen gage wire (minimum) is generally used for the charging circuit. (20A system requires fourteen gage wire.) Eighteen gage wire (minimum) is required for the magneto circuit (ground circuit).

Ammeters

Some battery ignition systems are equipped with an *ammeter,* which is used to measure the rate of current flow from the alternator to the battery. If no current flow is indicated by the ammeter, remove the meter from the circuit and check all components in the system. If the system is operating properly, use a multimeter set on an ohms setting to check continuity across the ammeter terminals. If continuity does not exist, the ammeter is faulty and should be replaced.

Switches

Switches are used to control many functions in a battery ignition system. The switch is a common point where most of the wiring comes together on an implement such as a small riding tractor. Many varieties of switches are available. When a

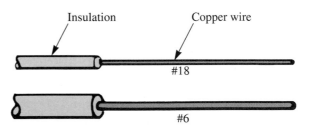
Figure 14-46. *The larger a wire's gage number, the small the wire.*

switch fails, replace it according to the manufacturer's specifications. Never substitute an automotive switch for small engine applications.

In some switching applications, current is very high and normal switches are too small to continuously *make* and break the circuits without burning their electrical contacts. In these situations, a solenoid should be installed in the circuit. A **solenoid** is a heavy-duty, electromagnetic switching mechanism that is used to handle large amounts of current. It consists of a heavy metal strip or disk, which is activated by an electromagnet. The metal strip or disk connects two contact points and *makes* or *breaks* the electrical circuit. Because the metal strip (disk) is heavier than most electrical switch contacts, it will not pit or burn away due to arcing. See **Figure 14-47.**

Switches and solenoids that are used for small engine applications should be waterproof to prevent shorting, which could occur if the unit is left out in the rain or washed with a garden hose. Several common switches are shown in **Figure 14-48.** Most switches and solenoids can be tested with a standard multimeter set on an ohm setting or another type of continuity tester. See **Figure 14-49.**

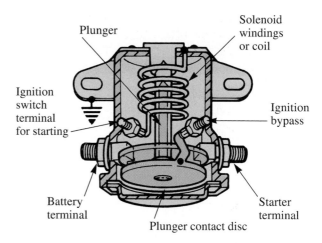

Figure 14-47. *Study the construction of this starter solenoid. As the plunger moves, it pulls the disk in to contact the battery terminals. When contact is made, the starter is energized.*

Checking the dc starter-generator circuit

There are certain steps to follow when checking output or troubleshooting the direct current (dc) starter-generator circuit found on some small engines.

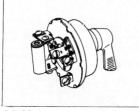

Multiple contact switches can be turned to different positions to open or close complex circuits (combination lighting and variable speed control circuits)

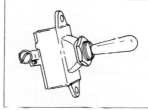

Toggle switches are simple on-off switches used to control auxiliary circuits.

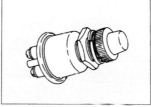

Push button switches are momentary switches that are pushed in one direction to open or close a circuit.

Pressure switches sense high or low pressure conditions and close a circuit to provide audible and/or visual warning signals.

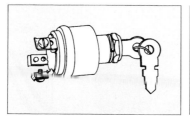

Ignition key switches must have a key inserted to turn them on and off.

Push-pull switches are used for light switches and emergency switches.

Cutout switches are used to break electrical circuits during emergencies (if operator involuntarily leaves implement).

Figure 14-48. *There are many switches used in a battery ignition system. (Deere & Co.)*

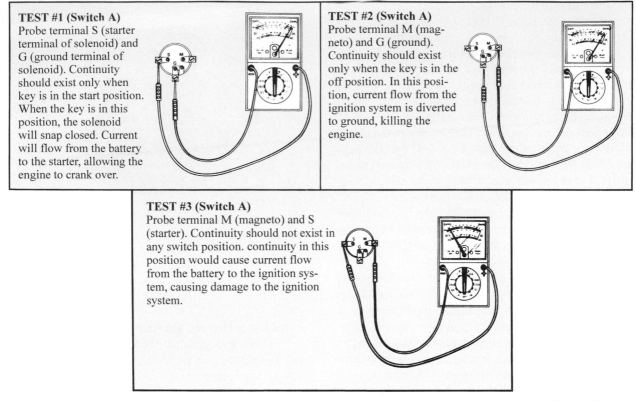

TEST #1 (Switch A)
Probe terminal S (starter terminal of solenoid) and G (ground terminal of solenoid). Continuity should exist only when key is in the start position. When the key is in this position, the solenoid will snap closed. Current will flow from the battery to the starter, allowing the engine to crank over.

TEST #2 (Switch A)
Probe terminal M (magneto) and G (ground). Continuity should exist only when the key is in the off position. In this position, current flow from the ignition system is diverted to ground, killing the engine.

TEST #3 (Switch A)
Probe terminal M (magneto) and S (starter). Continuity should not exist in any switch position. continuity in this position would cause current flow from the battery to the ignition system, causing damage to the ignition system.

Figure 14-49. *This solenoid is being checked for continuity with a multimeter. Follow manufacturer's instructions when testing switches and solenoids. (Tecumseh Products Co.)*

1. Disconnect all equipment, place the transmission in neutral (if applicable), and turn on the ignition switch.
2. If the generator warning light comes on or the ammeter shows a discharge, these units are working and the battery is supplying current.
3. Start the engine. If the generator warning light stays on or the ammeter continues to show a discharge, look for trouble in the generator circuit.
4. If the system is charging, check out the starting circuit. There are four separate starting circuit checks:
   a. Test the ground connection at the starter-generator or at the battery with a multimeter set to a voltage setting. If the negative battery cable goes to ground, attach the negative meter lead to the starter-generator mounting frame, and clip the positive meter lead to the negative post of battery. See **Figure 14-50.** Reverse this procedure if the system is a positive ground. If the meter shows about 10V, move on to the

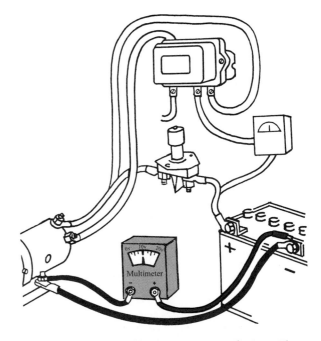

Figure 14-50. *A multimeter set on a voltage setting can be used to check battery and starter-generator system ground connections.*

next step. A lower reading indicates a poor ground. Clean and tighten all ground connections. Replace the ground cable if it is worn or corroded. Operate the starter and check the voltage.

b. If the voltage is still low, check the circuit between the battery and the starter switch. To do this, leave the negative cable grounded to the mounting frame. Move the positive lead to the switch terminal nearest battery. Then, turn on the starter switch. A low reading on the meter indicates a poor connection between the battery and the starter switch. If the engine will not crank, clean and tighten the cable connections. Then turn on starter switch and recheck voltage.

c. Check the starter switch. Leave the negative lead of the meter attached to the starter-generator frame. Move the positive lead to the switch terminal nearest the starter-generator. Turn on the switch. If there is little or no voltage, the starter switch is not closing the circuit. Repair or replace. If the voltage is still low, go to the next step.

d. Move the positive lead to the armature post of the starter-generator. Leave the negative lead grounded. The starter should operate and the meter should read about 11V. If the engine does not start and the voltage is normal, the starter is faulty. A no-start condition and little or no voltage indicates a loose or broken connection between the starter switch and the starter-generator connection. Clean, tighten, and inspect or replace wiring. Recheck voltage.

Commutators and brushes

All dc starters and generators have *commutators* and *brushes,* which occasionally need service. Start by cleaning the unit's metal housing. Avoid getting cleaning solvent on insulated wiring.

Check for worn bearings at both ends of the armature shaft. (You can feel play with your hands, and worn bearings are usually noisy when operating.)

If there is a cover band, remove it. A ring of solder along the inside of the band indicates that the unit has overheated. Further repair must be done by an experienced technician.

If the generator has no cover band, remove the long bolts running through the housing and pull off the end plate nearest the commutator. The commutator consists of a group of bars arranged in a cylinder-like fashion inside the generator or starter. Spring-loaded brushes rub on the commutator. See **Figure 14-51.** Check the brushes for wear. They should move freely and press firmly against the commutator.

A generator will have either two or three brushes. A starter-generator will have two. If the brushes are worn to half of their original length or the clips are resting on the brush holders, replace the brushes. If brushes are binding in the holders, wipe the holders with a clean, dry rag.

Check and tighten all electrical connections. Inspect the commutator for damage and/or wear. If the bars are rough and out of round, the armature will have to be chucked in a lathe and turned to a smooth finish. After turning, the mica (insulating material) between the commutator bars should be undercut. Follow all of the manufacturer's specifications.

If the commutator is only dirty and glazed, clean it with fine sandpaper. Do this while the engine is running slowly.

> ⚠ Do not leave any wires disconnected during this operation. This will burn out the generator.

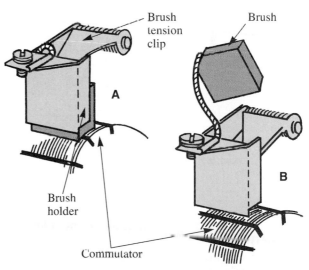

Figure 14-51. *Brushes are held against the commutator by a brush clip. A—Brush is worn down so clip rides on holder. B—Brush is removed for replacement. Never pull on brush leads.*

Clean the starter commutator while the starter is turning over the engine. Disconnect the spark plug wire so the engine does not accidentally start. Wrap No. 00 sandpaper over the squared end of a stick and move the sandpaper back and forth on the commutator until all dirt and glaze is removed. Never use emery cloth or solvent. Emery will cause arcing. Solvents will soften the insulation between the bars.

If necessary, install new brushes. If they do not seat squarely, pull sandpaper back and forth between brush and the commutator. Let the sandpaper work the brush down to the shape of the commutator. Blow out the dust and replace the band.

Polarize the generator at the regulator by placing one end of a jumper wire on the battery terminal. Momentarily, touch the other end of the wire to the generator terminal. This is only necessary if you have disconnected any of the wire leads to the generator.

If you do not do this and the polarity has reversed, you may burn out the generator, damage the cutout relay points in the regulator, or run down the battery.

Maintaining an alternator

Small engine *alternators* come in many different sizes. Maintenance generally involves lubrication (on some units, but not on others) and inspection. Periodically, check that brushes, slip rings, and bearings are in good condition. If the battery is run down, test alternator output and, if necessary, test separate parts of the alternator. Check battery polarity first. Is the proper terminal going to ground? If not, reverse them.

 Never reverse polarity when servicing the battery. It could burn out alternator diodes and damage wiring by overheating. Do not try to polarize an alternator. It is *not* necessary. Polarity of an alternator cannot be lost or reversed.

Most small engine alternator systems use permanent magnets that are attached to the inner rim of the flywheel. These magnets are similar to the magneto magnets. The *stator assembly* consists of a series of coils that are mounted on a circular plate and attached to the engine inside the flywheel. The alternator produces alternating current (ac), which is converted to direct current (dc) to charge the battery. This is accomplished through the use of a rectifier or diode arrangement, which is located in the circuit between the alternator and the battery. The size of the magnets and stator coils determines the output current of the alternator. **Figure 14-52** illustrates an alternator ring and a regulator-rectifier.

Dual, unregulated alternator systems charge the battery with dc current and provide ac current for operating the lights and other accessories. **Figure 14-53** shows a wiring diagram for a dual, unregulated alternator system using a diode in the direct current circuit. The current output is limited by the construction of the coils in the stator. Therefore, no regulator is used. Current to the electrical accessories is available only when the engine is running, and the brightness of the lights varies with the speed of the engine. The ac and dc units are separate, so the load on one does not affect the other.

Another common type of alternator system is the unregulated ac-only type. In these systems, the alternator is only used to operate the lighting system. See **Figure 14-54.** Current is available only when the engine is running and the brightness of the lights varies with engine speed. A battery is not used in these systems. The troubleshooting chart in **Figure 14-55** points out common alternator system problems.

Alternator output tests

A few common tests can be performed to determine the output of most small engine alternators. A multimeter is recommended for carrying

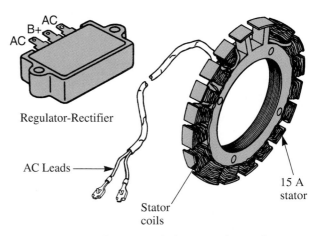

Figure 14-52. *One type of alternator ring and a regulator-rectifier.*

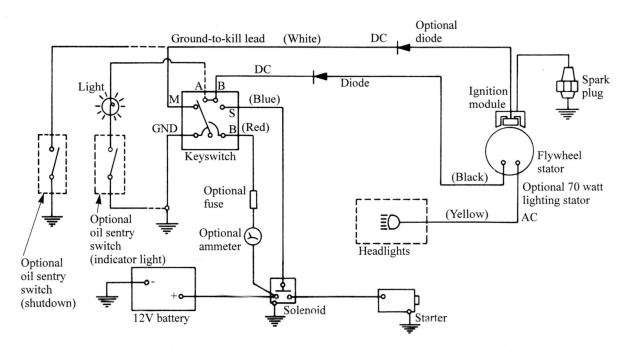

Figure 14-53. *This wiring diagram Is for a dual-purpose alternator. It provides ac for headlights and dc for battery charging.*

out these tests. See **Figure 14-56.** When checking alternators, make the tests in the following sequence:

1. Test alternator output.
2. Test diode(s) or regulator-rectifier (if equipped).

Most meter receptacles and test lead connector ends are color coded to ensure correct test lead attachment. See **Figure 14-57.** Before testing the alternator's output (volts, amps), use an accurate tachometer to temporarily adjust the engine speed to the rpm specified in the test instructions.

Upon completion of the alternator output test, always readjust the engine rpm to its correct no-load governed speed as specified in the engine service manual.

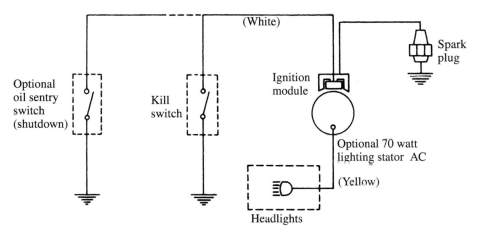

Figure 14-54. *Wiring diagram for a manual-start engine with an ac-only lighting stator for the headlights. A battery is not used in this system. (Kohler Co.)*

<table>
<tr><th colspan="1">Alternator Systems Troubleshooting</th></tr>
</table>

Alternator Systems Troubleshooting

Battery not Charging

Engine RPM too low.
Inline fuse "blown" (if equipped).
Defective battery.
Loose, pinched, or corroded battery leads.
Open, shorted, or grounded wires between output
    connector and battery.
Defective diode (open or shorted).
Defective or improperly grounded regulator-rectifier.
Diode installed incorrectly (reversed).
Damaged battery (shorted battery cells).
Excessive current draw from accessories.
Low magnetic flux or damaged alternator magnets.

Battery in State of Over Charge

Severe battery vibration (missing or broken tie-down
    straps).
Battery rate of charge not matched to alternator output.
Damaged battery (shorted battery cells).
Defective regulator.

Headlamps not Working

Inline fuse "blown" (if equipped).
Defective headlamps.
Loose or corroded wires.
Open, shorted, or grounded wires between output
    connector and headlamps.
Low magnetic flux or damaged alternator magnets.

Figure 14-55. *Alternator system troubleshooting chart.*

Figure 14-57 illustrates a typical output test for a 9A regulated alternator. This system provides alternating current to a regulator-rectifier.

The regulator-rectifier converts the ac to dc and regulates the current to the battery. The charging rate will vary with engine RPM and temperature. When testing the regulator-rectifier for output, a 12V battery with a minimum charge of

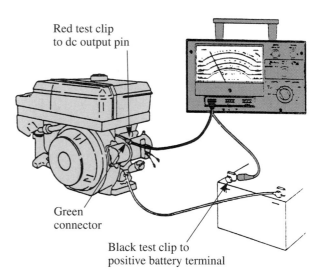

Figure 14-57. *Typical test connections for checking the output of a 9A, regulated alternator. Output for this particular model should not be less than 40V. (Briggs and Stratton Corp.)*

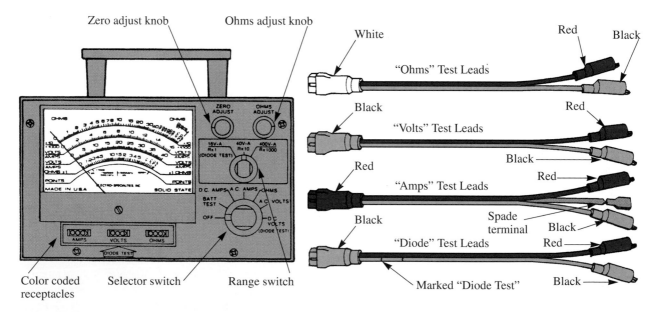

Figure 14-56. *A multimeter can be used when performing alternator tests. Note color-coded test leads. (Kohler Co.)*

5V is required. There will be no charging output if battery voltage is below 5V. See **Figure 14-58.**

> When testing a regulator-rectifier, connect test leads before starting engine. Be sure connections are secure. If a test lead vibrates loose while the engine is running, the regulator-rectifier may be damaged.

A dc-only and dual-circuit alternator system uses a *diode* to convert ac to dc. The diode is located in the output wire on most alternators. A typical test hookup for one type of diode is shown in **Figure 14-59.** After the test is made in one direction, the leads are reversed and the diode is tested again. The meter should only show a reading in one direction. If meter shows a reading in both directions, the diode is defective (closed). If the meter does not show a reading (needle movement) in either direction, the diode is defective (open).

> Replacement diode harnesses are available. When installing a new harness, use rosin-core solder. Use shrink tubing or electrician's tape on connections. Do not use crimp connectors.

Voltage regulator service

Do not attempt to service voltage regulators until the engine has run for about 20 minutes. This

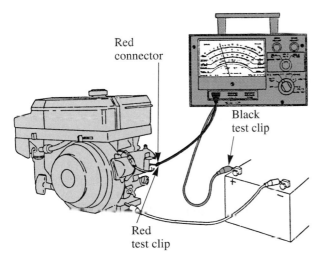

Figure 14-58. *Typical test connection for checking the output of one type of regulator-rectifier. Typical output should range from 3A to 9A, depending on battery voltage. (Briggs and Stratton Corp.)*

Note: It may be necessary to pierce the wire with a pin if the connections are not in the open.

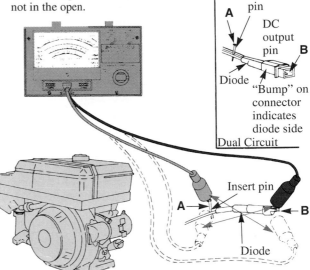

Figure 14-59. *Testing a diode for proper operation. Meter should show a reading in one direction only. (Briggs and Stratton Corp.)*

will allow the temperature of the regulator to stabilize. Then check:

1. Condition of the ground strap running to the engine. Connections must be clean and tight.
2. Appearance of the breaker points. They tend to wear in a concave pattern and should be filed as necessary. Wipe filed points with a clean, lint-free cloth soaked in methyl ethyl ketone (MEK). Never use emery cloth or sandpaper on regulator points.

Adjusting current voltage

There are two points of adjustment in the current voltage regulator. See **Figure 14-60.** Disconnect battery, then:

1. Check the armature air gap. Push down on the armature until the contact points just touch. Measure the air gap and check it against the manufacturer's specifications (usually .075"). Adjust by loosening the contact mounting screws and raising or lowering the contact bracket. Align and adjust the points and retighten the screws. This is always done before adjusting the voltage setting.
2. Adjust voltage setting by turning the adjusting screw. Clockwise adjustment will increase the voltage setting. Counterclockwise adjustment will decrease the setting. Replace the cover.

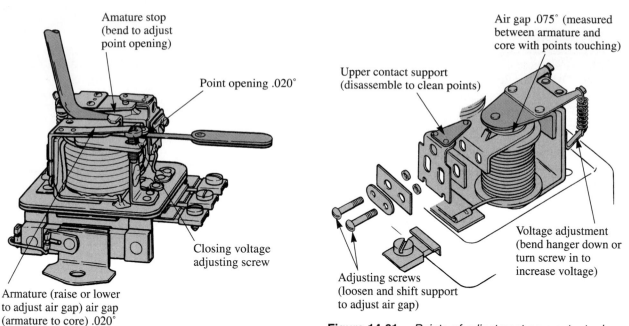

Figure 14-60. *Current-voltage adjustments on typical voltage regulator. (Kohler Co.)*

Figure 14-61. *Points of adjustment on a cutout relay are shown. To clean the points, loosen the upper contact support.*

Run the engine to stabilize the voltage and recheck.

Cutout relay service

The generator regulator usually has a cutout relay unit with three adjustments. See **Figure 14-61.** Disconnect battery, then:

1. Measure and adjust the air gap. Place finger pressure on the armature, directly above core until points close. Measure gap and adjust to manufacturer's specifications. Generally, this is about .020″ Retighten screws.
2. To adjust the point opening, bend the armature stop. See the service manual for the correct gap.
3. Adjust the closing voltage by turning the screw clockwise to increase tension and voltage; counterclockwise to decrease tension and voltage. Closing voltage must be at least .5V less than current voltage regulator unit setting.

Distributors

Small engines with more than one cylinder and a battery ignition system use a ***distributor*** to bring spark to the right cylinder at the right time. During tune-ups, the distributor cap should be removed and inspected for cracks, carbon tracking, and pitted contacts.

The rotor also requires inspection. This is the small plastic *arm* that is mounted on top of the distributor shaft and revolves during engine operation. The metal tab on top of the rotor must make good contact with the metal inset in the center tower of the distributor cap. Therefore, the firing end of the rotor should not be worn or irregular.

The breaker points should be filed flat or replaced as described for magneto points. Spark timing can be set by rotating the distributor. Usually, the distributor is clamped in place to lock in the ignition timing adjustment.

Ignition timing

To set the ignition ***timing*** on the engine shown in **Figure 14-62,** the number one spark plug must be removed. Next, while turning the engine over slowly by hand, hold your finger over the spark plug hole to determine when the piston is on the compression stroke. Then, with the piston at TDC, the rotor should be approximately in line with a timing notch in the distributor housing. Loosen the clamp and rotate the distributor until the points just begin to open. Hold the distributor in place and retighten the clamp. This procedure is accurate enough to start and operate the engine.

More accurate timing can be done with a neon ***timing light.*** In this method, the timing light

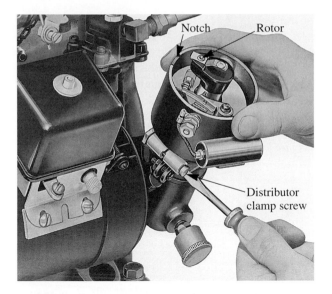

Figure 14-62. *With the number one piston at TDC, the rotor should point to the notch in the distributor housing. To time ignition, loosen clamp screw and rotate distributor housing until breaker points close. Then, turn distributor in the opposite direction until the points begin to open and retighten clamp screw. (Wisconsin Motors Corp.)*

is aimed at a fixed mark on the engine block or the engine shroud. When the correct timing is reached, a mark on the flywheel will appear to be aligned with the fixed mark when the engine is at idle. When using a timing light, follow manufacturer's recommendations.

Lubrication

Distributors should be lubricated at several points. Put a small amount of oil or grease in the reservoir or cup provided for the shaft, a film of grease on the breaker cam, and a drop or two of oil on the pivot for the breaker points. Use care; too much lubricant can cause the points to burn.

Summary

The ignition systems in small engines require periodic inspection and maintenance. Each ignition system component should be checked for proper operation and replaced if defective.

When an engine is hard to start, a new spark plug may seem to solve the problem. A new plug may require 5000V to fire. After many hours of use, the same plug may require 10,000V to fire. Changing the plug means that less voltage is need-

ed to fire the plug and, therefore, the engine is easier to start. Often, however, a less obvious problem has caused the plug to fail.

You can analyze the quality of combustion by examining the carbon deposits on a spark plug. Normal combustion produces a beige or gray-tan deposit. Plugs can be cleaned in a spark plug cleaning machine. When gapping spark plugs, only use leaf-type feeler gauges on new plugs. Wire-type thickness gauges are recommended for use on worn plugs. When installing plugs, tighten only to the torque specified by the manufacturer.

A basic ignition test can help determine the condition of the ignition system. Remove the plug wire from the plug. Hold the wire 3/16″ from top of the plug and pull the starter cord. If a spark is produced between the plug and the wire, the ignition is good. If a spark does not appear between top of the plug and the wire, hold the wire near the base of the spark plug and pull the cord. If a spark occurs, then the plug is failing under compression. If a spark does not occur, then the problem is in the magneto system.

In some magneto systems, the gap between the coil or trigger module and the flywheel is adjustable. In some systems, the ignition components are completely contained under the flywheel. A knock-off tool or a wheel puller can be used to remove the flywheel from the crankshaft.

After the flywheel is removed, all magneto parts should be checked for defects. Look for dents, cracks, and gouges. Make sure all wire insulation is in good condition. A multimeter set on an ohm setting can be used to check the condition of coils and condensers. In many cases, a special ignition tester must be used to check the condition of ignition system components.

In breaker point ignition systems, the points must be adjusted, the piston height must be set, and the ignition spark must be timed. Follow manufacturer's recommendations. An engine can be timed using a continuity tester or an multimeter set on an ohm setting. After performing necessary ignition system service, the flywheel can be replaced.

If the cylinder head was removed to adjust piston height, it should be cleaned of carbon deposits and checked for flatness. A new head gasket should be used and the head bolts should be tightened as specified by the manufacturer.

Battery operated ignition systems operate much like magneto systems. However, they contain many additional components that require service. Before beginning extensive tests, check for a defective battery, corroded or loose terminals, incorrect connections, faulty wires, defective switches, or a malfunctioning operator presence system.

The dc generators and starters occasionally require service. Check for worn bearings and overheating. Inspect the commutator and brushes for wear. Check and tighten all electrical connections. Polarize dc generator if wire leads are disconnected.

Alternators require occasional lubrication and inspection. Make sure brushes, slip rings, and bearings are in good condition. A multimeter can be used to determine the output of most small engine alternators. Check service manual for proper meter connections.

If the charging system is equipped with a voltage regulator, adjust the unit per manufacturer's instructions. An engine should run for approximately 20 minutes before conducting regulator tests. The will allow the temperature of the regulator to stabilize.

Multi-cylinder small engines use a distributor to send spark to the appropriate cylinder at the correct time. During service, the distributor cap should be removed and checked for cracks, carbon tracking, and pitted contacts. Distributors may require lubrication in several locations. In addition to the distributor cap, the rotor should also be checked for wear.

Spark timing is set in distributor ignition systems by rotating the distributor. When ignition timing meets specifications, the distributor clamp must be tightened

Know These Terms

ignition system tune-up	knock-off tool
spark plug	wheel puller
magneto system	multimeter
spark plug gap	continuity
gapping	solid state system
gapping tool	depth micrometer
leaf-type feeler gauge	dial indicator
wire-type thickness gauge	breaking
	battery operated
flywheel	ignition system

sulfating	switches
electric start	solenoid
hydrometer	commutator
maintenance free battery	brush
sealed battery	alternator
hydrogen	stator assembly
oxygen	diode
multimeter	distributor
wiring	timing
ammeter	timing light

Chapter 14
Review Questions

Answer the following questions on a separate sheet of paper.

1. Which of the following determines the amount of voltage induced by the magnets when the engine is running?
   a. breaker points
   b. spark plug
   c. condenser
   d. engine speed
2. What color should the internal porcelain insulator be on a normal, used spark plug?
3. What five conditions cause spark plug deposits?
4. If a torque specification is not given, what is the procedure for tightening a new spark plug?
5. When performing an ignition system test, spark does not occur between the high tension lead and the plug tip when the engine is cranked. However, spark does occur when the lead is held near base of the plug. The problem lies with the _____.
   a. magneto
   b. high tension lead
   c. coil
   d. plug
6. Name two tools used to remove flywheels from small engine crankshafts.
7. If the key that positions the flywheel is deformed or partly sheared, the engine will most likely _____.
   a. lose the flywheel
   b. be out of time
   c. run exceptionally fast
   d. burn fuel excessively

8. The three major steps in timing a small engine are as follows: adjust the piston height BTDC, rotate the stator plate until the points open, set the breaker point gap. List these steps in the correct order.

9. What are the seven basic things to check for before starting an extensive battery ignition system analysis?

10. In a lead-acid battery the electrolyte is _____ acid and water.

11. What are the four components of a battery starting circuit?

12. What is included in a visual inspection of a battery?

13. Which test will determine the condition of each cell of a battery?

14. If one or more battery cells shows a specific gravity of less than 1.235 the battery should be _____.
    a. discarded
    b. recharged
    c. discharged, then recharged

15. Which cable should be connected last when installing a battery?
    a. negative cable
    b. positive cable
    c. Either A or B.

16. Charging 6V and 12V batteries should not exceed _____ amp(s).

17. If the fluid level in a battery is below the separators, _____ should be added, but never add _____.

18. What two gases are generated during battery charging?

19. Why is it important to keep batteries fully charged in below-freezing temperatures?

20. Why is engine cranking load increased significantly in cold weather?

21. Name the single component that indicates for certain that an engine has a battery ignition system.
    a. battery
    b. distributor
    c. can-shaped coil
    d. starter
    e. generator

22. What kind of testing unit can be used to test switches?

23. What is the type of current generated by an alternator?

24. Explain why emery cloth should not be used to clean a commutator of a generator or a starter.

25. Under what conditions would you polarize an alternator?

26. Two electronic components that convert alternating current to direct current are _____ and _____.

27. Why should you put only one or two drops of oil on a felt cam oiler?

28. Why should a wooden table or an insulated bench top be used when performing electrical tests?

Suggested Activities

1. Perform the complete timing procedure on a small engine.

2. Demonstrate the proper technique for removing, inspecting, and replacing a flywheel.

3. Examine some old spark plugs. Attempt to analyze engine condition by their appearance. Clean, gap, and test them.

4. Make a cutaway from a defective coil. Open a condenser and examine its construction.

5. Place a bad plug in an engine. Test for spark at the plug tip and, then, at the base. Explain the difference.

6. Demonstrate a hydrometer test.

7. Demonstrate proper maintenance procedure for a battery.

8. Demonstrate proper installation procedure of a battery.

9. Demonstrate the proper procedure for charging a battery. Be sure to follow all safety rules when working with batteries and chargers.

10. Perform continuity tests on several types of switches.

11. Test several electrical system components with the test equipment described in this chapter and in the *Appendix* section of this text. If other equipment is available, follow manufacturer's instructions for the equipment at hand.

To keep this utility vehicle in proper working condition, the ignition system must be maintained and serviced properly. (Deere & Co.)

Engine Inspection, Disassembly, and Cylinder Reconditioning

After studying this chapter, you will be able to:

▼ Inspect engines for problems.
▼ Describe the procedure for removing an engine from an implement.
▼ List the steps involved in disassembling an engine.
▼ Measure cylinder conditions such as wear and out-of roundness.
▼ Explain the procedures involved in reboring a cylinder.
▼ Summarize the reasons for honing a cylinder.

Introduction

When repairing an engine, it is best to work in a clean, well-lighted area. Tools should be clean and close at hand.

 When using tools, safety rules should be observed. For example, safety glasses should be worn to protect the eyes. Oily and gasoline-soaked rags should be placed in flameproof containers. Heat or flames should never be allowed near solvents or other materials that will burn. A workplace is not safe unless good judgment is used while working in it.

Some engines are easier to work on if they are mounted on an engine stand. Other engines in need of major repairs can be removed from the implement and torn down right on the workbench. See **Figure 15-1.**

You will need a *service manual* for the engine you are working on. Every engine make and model is built to certain dimensions and specifications that are different from other engines. The kind of job you do depends a great deal upon the care and accuracy used in following the engine service manual.

Figure 15-1. *Small gasoline engines are easier to work on when removed from implement.*

Engine Inspection

It is good practice to look for causes of engine problems even before removing the engine from the implement. Loose or broken engine mounts, misaligned pulleys, or unevenly worn drive belts can cause excessive vibration. Wet oil on the outside of the engine may mean that there are loose parts, leaking gaskets, leaking oil seals, or a cracked casting.

Wires may need to be disconnected before the engine can be removed. If so, make flags of masking tape for the wire ends. Identify them with matching numbers. See **Figure 15-2.** This will prevent damage from wrong connections and will save time during reassembly.

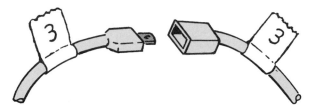

Figure 15-2. *If wires are coded with masking tape, they can be properly identified and reconnected.*

Engine disassembly

Before removing the engine, take out the spark plug. It could fire accidentally or break during disassembly of other parts. When working on lawn mowers, snow throwers, and other implements with exposed moving parts, never touch the blades or driven parts until the spark plug has been removed from the engine.

After the engine has been removed, the starter unit can be unbolted. It may be a simple pulley-rope starter, a retractable rope starter, or an electronic starter. Usually, only a few bolts fasten the starter unit to the engine. See **Figure 15-3.**

The exhaust manifold pipe and muffler can be taken off next. Set them aside and out of the way. The carburetor and intake manifold pipe can also be removed. See **Figure 15-4.** It often helps to sketch the carburetor and linkage locations. This can save time in reassembly. Check all gasket surfaces for defects. On some engines, the fuel tank is fastened to the carburetor. In any case, the tank can be removed, and the fuel lines can be disconnected.

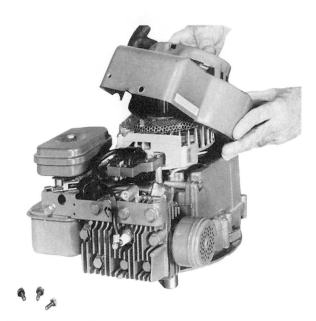

Figure 15-3. *With the engine out of the implement, the starter unit can be removed.*

The air shroud, blower housing, and baffles can be removed to uncover the flywheel. Flywheel removal is explained in Chapter 14. Always use the proper puller to prevent damage to the crankshaft or the flywheel. After the flywheel is removed, the magneto components can be disassembled and set aside.

Figure 15-4. *When removing carburetor and fuel tank, sketch carburetor linkage hookup for later reference.*

Organizing the job

Organizing the job saves time and effort. At this point, the outer parts have been taken off the engine. They should be set aside in their own groups. Magneto, flywheel shroud, and starter parts should be in one group; carburetor, fuel tank, and exhaust manifold in another. See **Figure 15-5.**

Keeping the work clean is part of good organization. Outside surfaces can be cleaned at this point of tear-down. See **Figure 15-6.** Inside surfaces will be cleaned later. Grass clippings and other debris should be removed by scraping and brushing the fins and housings before using cleaning fluid.

A safe engine cleaning solvent should be used to remove grease, oil, and grit. Some parts, such as the coil and condenser, may be cleaned by wiping with a clean cloth moistened with solvent. This is better than total immersion. Never use solvents that burn easily or those that may be harmful to humans.

As each part is washed, wipe it with a clean cloth and set it aside to dry. If the workbench is oily and greasy, clean it before doing disassembly work on it. Hands should be washed often to keep dirt off the cleaned parts.

Figure 15-6. *Grass and other debris should be removed from cooling fins before using solvent.*

Cylinder Reconditioning

Cylinder reconditioning is an important part of keeping an engine running efficiently. Reconditioning consists of inspection, measurement, reboring, and honing of the cylinder.

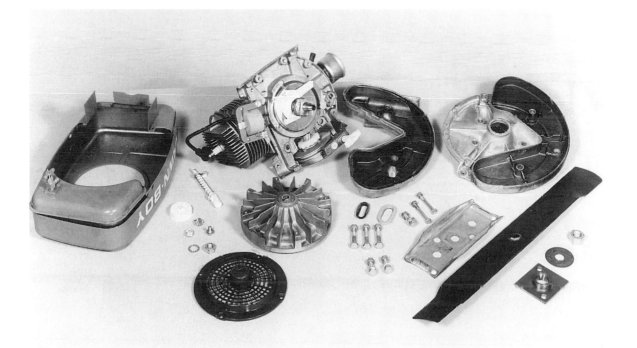

Figure 15-5. *A clean workbench and an organized work procedure can speed the job.*

Cylinder inspection

Several types of cylinders may be found on small engines. One type is the separate cylinder block shown in **Figure 15-7.** Another variation is the integral (one-piece) cylinder block and crankcase. See **Figure 15-8.**

If the engine is of the separate cylinder type, remove the bolts and lightly tap the cylinder block with a soft leather hammer. See **Figure 15-7.** A slow, smooth pull will remove the cylinder from the piston assembly. An alternate procedure is to remove the connecting rod cap and pull the cylinder and piston from the crankcase as a unit. Then, pull the piston from the cylinder.

The piston must be removed from the top of the cylinder in the integral cylinder block and crankcase type engine. First, remove the cylinder head. Next, take the crankcase cover from the side of the block and remove the connecting rod cap. Then, push the piston out of the top of the cylinder. If there is a heavy ridge around the inside top of the cylinder, the piston rings will not pass the ridge without damage. Special *ridge reaming tools* are available for cutting the ridge to make this part of the cylinder the same diameter as the worn portion below it. See **Figure 15-9.**

Figure 15-8. *On a one piece cylinder block, the piston may be removed from cylinder if there is no noticeable ridge at top of cylinder above the ring wear area.*

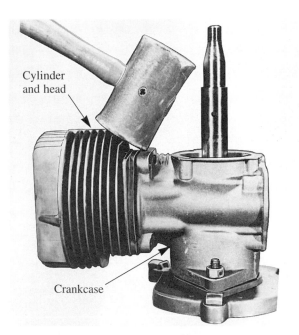

Figure 15-7. *After bolts are removed, a light tap with a soft hammer will loosen cylinder block from crankcase. (Lawn-Boy Power Equipment, Gale Products)*

Figure 15-9. *Maximum cylinder wear takes place near the top of the cylinder. The unworn portion at the top of the cylinder (ring ridge) must be removed. Cut it flush to the cylinder wall with a ridge reamer.*

With the piston out, inspect the cylinder block. Look for areas of scuffing or scoring on the walls. Check for nicks or grooves in gasket surfaces. Inspect it for chipped or broken fins. Examine head bolt holes and spark plug holes for damaged or stripped threads.

A worn cylinder has a narrow, unworn ridge at the top. The bottom of this ridge indicates the extent of the top piston ring travel. Right below the ridge is the area of most cylinder wear. The wear will be the greatest on two opposite sides of the cylinder, 90° from the crankshaft centerline. The cylinder, therefore, wears into an oval shape. This increased wear is due to several things:

- Less lubrication at this portion of the cylinder wall.
- The diluting effect of raw gas on the engine oil.
- The pressure that builds up behind the rings at their highest position.

Below the point where cylinder diameter is the greatest, cylinder wear lessens rapidly. Because of this wear pattern, there is a gradual taper toward the bottom of the ring travel. Below ring travel there is almost no wear. This lack of wear results because this area is well lubricated and only receives light wall pressure from the piston skirt.

Cylinder measurement

If the cylinder wall looks smooth and free of scuff and score marks, you are ready to measure the cylinder for wear and out-of-roundness.

Finding the amount of *cylinder taper* is the first important measurement in determining cylinder condition. First, measure the cylinder diameter below the ring travel. Then, measure it just below the ring ridge. The difference between these measurements is an accurate indication of the amount of cylinder taper. The taper measurements should be taken both parallel and at right angles to the crankshaft to determine the greatest amount of wear and out-of-roundness.

The amount of taper allowed before reboring is required depends on engine design, its general condition, and the type of service in which it is used. There are no rules that apply to all engines regarding cylinder taper. The manufacturer, however, may set a limit. Beyond a certain point, the engine manufacturer will advise reboring or cylinder replacement.

Out-of-roundness for small engines is generally limited to .005″ or .006″. Beyond this limit, engine performance is greatly reduced.

Several methods can be used to measure cylinders. **Figure 15-10** shows an *inside micrometer* equipped with an extension handle. This precision instrument must be carefully adjusted to cylinder size. It must give the exact diameter of the cylinder.

Figure 15-11 shows a *telescoping gauge* being used to measure cylinder size. The gauge head is spring loaded to expand when the thumbscrew is released. Once located in the cylinder, the gauge is locked in place by tightening the thumbscrew. Then, it is removed from the cylinder, and an outside micrometer is used to measure the length of the telescope head. This measurement represents the cylinder size (diameter).

There is another convenient tool designed for measuring small engine cylinders. **Figure 15-12** shows the various sections of the *set gauge* part of this tool.

To use the tool, look up the standard cylinder size in the engine manual. Then place the right spacers on the shaft. Place the stop bar on the shaft with the remaining spacers and the nut.

The *cylinder gauge* attaches to the set gauge as shown in **Figure 15-13.** The needle can be set

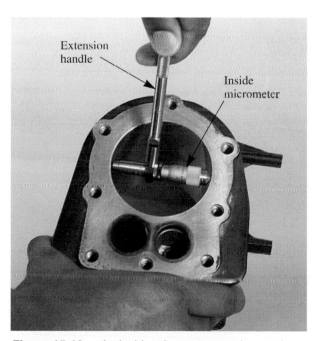

Figure 15-10. *An inside micrometer can be used to measure cylinder diameter.*

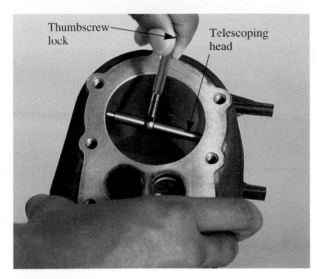

Figure 15-11. *A telescoping gauge can also be used to measure cylinder diameter. Gauge is adjusted to cylinder size and is measured with an outside micrometer.*

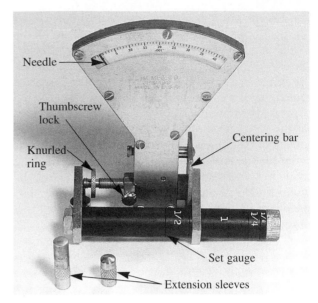

Figure 15-13. *When located in set gauge, cylinder gauge is adjusted to read zero by turning the knurled ring.*

to zero by turning the knurled ring to the left. Lock the ring with the thumbscrew and remove the cylinder gauge from the set gauge. The extension sleeves are used only when necessary.

When the cylinder gauge is adjusted properly (set zero at specified diameter), put it into the cylinder. See **Figure 15-14.** The centering bar automatically centers the gauge in the cylinder. Since the gauge is spring loaded, it is free of the measuring spindle. The amount indicated by the needle is cylinder diameter oversize in thousandths of an inch. Take readings in three or four directions across the diameter. Check these measurements against manufacturer's specifications.

Cylinder service needs are determined by cylinder condition. If the bore is not damaged, and if taper and out-of-round readings are within specified limits, only a light deglazing with a fine emery cloth may be needed. The engine block must be thoroughly washed afterward.

Figure 15-14. *Cylinder gauge is self centering and will provide a direct reading. Needle deflection from zero indicates the amount of wear in thousandths of an inch.*

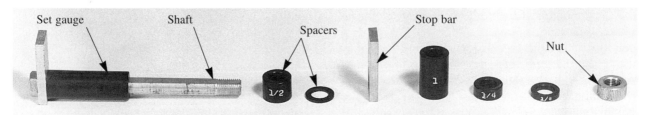

Figure 15-12. *Parts of a disassembled set gauge are shown. Gauge is preset to a specified size.*

Reboring the cylinder

There are several ways to repair a cylinder that shows too much wear. Repair procedures depend on engine type. Some engines have chrome-plated, aluminum cylinders. See **Figure 15-15.** Worn or damaged cylinders of this type should be thrown away and replaced with a whole new cylinder. Other engines have a pressed-in, flanged *cylinder sleeve* that can be removed and replaced with a new sleeve. See **Figure 15-16.** The cast-in sleeve or the solid cast iron cylinder can be rebored to a larger size. See **Figure 15-17.**

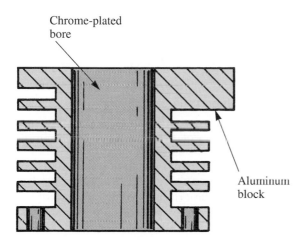

Figure 15-15. *Special cylinder construction is shown with a chrome-plated surface.*

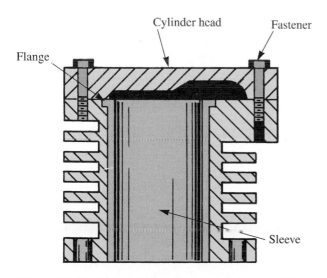

Figure 15-16. *Pressed-in sleeve can be secured by forming a flange on the upper end. Cylinder head will hold sleeve in place.*

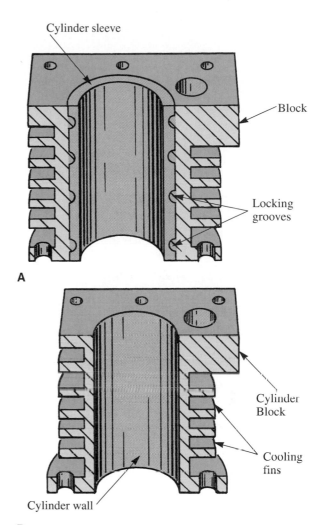

Figure 15-17. *A—Some cast iron cylinders are cast in place in the block. B—Most cast iron cylinders are part of the engine block.*

Two problems must be solved when *reboring* a cylinder. The first is to resize and maintain original alignment while producing a round, straight bore. The second is to produce the correct cylinder wall finish.

Cylinders are usually rebored in .010″ steps. If the cylinder is being rebored for the first time, its diameter will be increased .010″ over the standard size. If this does not *clean up* the cylinder (remove imperfections), the next step is .020″ over the standard size. When replacing pistons and rings, order .010″ or .020″ over standard size to match the new cylinder bore.

The *boring machine* shown in **Figure 15-18** is designed for use on small engines. In setting up

Figure 15-18. *A small engine cylinder boring machine is prepared for operation. Engine block is clamped in place. (Cedar Rapids Engineering Co.)*

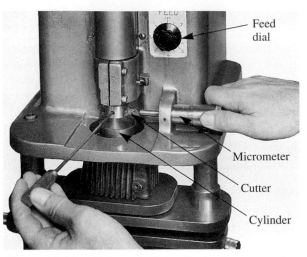

Figure 15-19. *Using a built-in micrometer gauge, the cutter on this boring machine is set to give a precise bore diameter. (Cedar Rapids Engineering Co.)*

Honing the cylinder

Honing is an abrasive (sandpaper-like) finishing process that removes the boring tool marks and surface fractures. Honing also produces the desired cylinder wall finish. The smoothness of the finish depends upon the grit size of the stones used.

Ideally, a newly honed cylinder will wear smooth at just about the same rate the new piston rings wear in. When this *break-in* process is complete, both the cylinder walls and the piston rings will be smooth and should last for hundreds of hours of engine operation.

Figure 15-20 illustrates a typical cylinder honing tool for small engine work. It has two abrasive stones and two guides to keep the tool aligned with the cylinder. The stones can be removed and replaced with new ones as necessary.

An electric drill is used to rotate the cylinder hone. In operation, the assembly is slowly and steadily moved in and out of the cylinder. See **Figure 15-21.** Stones should not be permitted to extend out of the cylinder end as uneven wearing of the stones may occur. For best results, the honing process should produce a fine surface pattern, like that shown in **Figure 15-22.**

After reboring and honing the cylinder, use a piece of fine emery cloth to remove any burrs that may have developed around the ports. Clean the cylinder walls and the block thoroughly with kerosene and a brush (do not use a rag). To prevent cylinder rusting, apply a light coat of SAE 10 oil.

the machine, the engine block or cylinder block can be clamped in place below the boring head. **Figure 15-19** shows the cutter being adjusted to the correct diameter for the new bore. A built-in micrometer provides accurate setup to .0001" (one ten-thousandth of an inch).

An electric motor drives the spindle to rotate the cutter. The feed rate controls the distance the cutter advances into the bore during each cutter revolution and can be changed by moving the feed dial. See **Figure 15-19.**

Boring the cylinder will produce a straight, round bore. However, the boring operation does not produce a satisfactory surface finish. The boring tool leaves microscopic furrows and surface fractures, so it is recommended that an overall minimum of .0025" of stock should be left for finish honing.

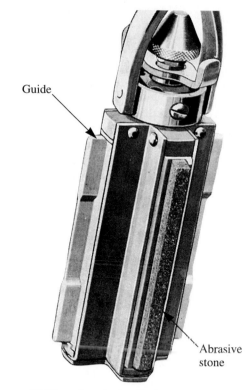

Figure 15-20. *A typical cylinder honing tool has two abrasive stones and two guides. (Sunnen Products Co.)*

Figure 15-22. *A fine cross-hatched surface pattern is created by the in-and-out motion of the revolving hone.*

Summary

Engine work should be performed in a clean, well-lighted area. An engine stand can be used if necessary. Always refer to a manufacturer's service manual for exact dimensions and specifications.

Before removing the engine from an implement, look for causes of engine problems. If wires need to be disconnected before removing engine, flag ends with masking tape. Remove the spark plug to avoid accidental firing.

After the engine has been removed, unbolt the starter unit, exhaust manifold pipe, muffler, carburetor, intake manifold pipe, fuel tank, and fuel lines. Check all gasket surfaces for defects. Remove the air shroud, blower housing, and baffles to expose the flywheel. Use a proper puller to remove the flywheel. After the flywheel is detached, remove the magneto components and set them aside.

Clean the outer engine parts by scraping off loose debris and soaking the components in solvent. Some parts, such as the coil and the condenser, should not be immersed in solvent.

If an engine is the separate cylinder type, remove bolts and tap the block with a soft leather hammer. Slowly pull the cylinder from the piston assembly. In an integral cylinder block and

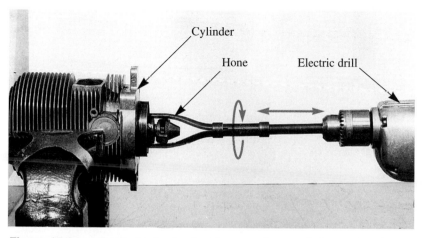

Figure 15-21. *A cylinder boring tool leaves microscopic furrows and surface fractures. These are removed and the cylinder is brought down to the correct size by honing. (Kickhaefer Mercury)*

crankcase, the piston must be removed from the top of the cylinder.

When the piston is removed, inspect the cylinder walls for scuffing and scoring. Check gasket surfaces for nicks and grooves. Examine the head bolt holes and spark plug holes for damaged threads.

If the cylinder wall looks smooth, measure it for wear and out-of-roundness. Check for excessive cylinder taper.

Cylinder service is determined by the cylinder condition. If the bore is within specifications, only a light deglazing is necessary.

If the cylinder is worn excessively, repair will depend on engine type. Chrome-plated aluminum cylinders must be replaced. Cast-in sleeve and solid cast iron cylinders can be rebored. Cylinders are usually rebored in steps of .010″. The cylinder must be honed after boring. Honing is an abrasive finishing process that removes boring tool marks and produces a desirable cylinder wall finish.

Know These Terms

service manual	set gauge
ridge reaming tool	cylinder gauge
cylinder taper	cylinder sleeve
out-of-roundness	reboring
inside micrometer	boring machine
telescoping gauge	honing

Chapter 15 Review Questions

Answer the following questions on a separate sheet of paper.

1. When should the small engine mechanic begin looking for engine defects or problems?
2. Excessive vibration could be caused by _____.
   a. loose engine mounts
   b. pulleys out of line
   c. worn drive belts
   d. All of the above.
   e. Only a and c.
3. Before removing an engine from an implement, it is safe practice to _____.

   a. drain the oil
   b. turn off the ignition switch
   c. remove the spark plug
   d. crank the engine slowly to remove all the fuel from the cylinder(s)

4. When you remove a carburetor and tear it down for repair, what will help you remember how it goes back together?
5. In what ways can you save time and effort when reconditioning an engine?
   a. Consult the service manual.
   b. Keep workbench and engine parts clean.
   c. Organize your work by grouping certain engine parts.
   d. All of the above.
6. Explain how the piston is removed if there is a heavy ridge at the top of the cylinder in an integral cylinder block and crankcase type engine.
7. The location of cylinder diameter measurements should be _____.
   a. at the very top and bottom of the cylinder
   b. taken 90° to each other at the center of the cylinder
   c. below the ridge and at the bottom of the cylinder to determine taper
   d. only below the ridge, but at 90° to each other
8. What do you call the special tool designed to remove cylinder ridges?
9. After a cylinder is rebored, it must be _____.
   a. lapped   c. deglazed
   b. reamed   d. honed

Suggested Activities

1. Design a well-organized workbench and tool panel for repairing small engines.
2. Completely disassemble an engine that has been in service for many hours and is in need of reconditioning.
3. Inspect disassembled parts for wear or damage. Discuss possible causes.
4. Measure a worn cylinder with an inside micrometer or other measuring tool. Record readings for taper and out-of-round.
5. Rebore and hone or deglaze a cylinder.
6. Ream a cylinder ridge.

Pistons and Piston Rings

After studying this chapter, you will be able to:
▼ Describe piston and piston ring construction.
▼ Differentiate between compression rings and oil control rings.
▼ Explain the purpose of ring end gap.
▼ Identify common types of piston damage and list possible causes.
▼ Summarize what happens during piston ring wear-in.
▼ Explain the purpose of a piston pin.

Piston and Piston Ring Service

In reconditioning small gasoline engines, pistons and piston rings are critical service items. Generally, reboring and/or honing of the cylinders is necessary, followed by the thorough inspection and repair of the parts closely fitted to them.

To do the job well, the small engine technician must understand the stresses to which a piston and its rings are subjected. A technician also must know the kinds of materials they are made from. Lastly, a technician must know what to do to put these parts back into top shape and to reassemble them for efficient, long-lasting engine operation.

The condition of the rings and pistons can be learned by observing and inspecting the parts during disassembly of the engine. The need for service is evidenced by low compression, blow-by, oil pumping, and fouled plugs.

The piston slides up and down in the cylinder. It sucks in the air-fuel mixture, compresses it, and then carries the force of the burning fuel to the crankshaft through the connecting rod. On the final stroke of the engine cycle, the piston pushes burned gases from the cylinder.

In normal operation, the piston travels up and down in the cylinder more than a thousand times per minute. Subjected to heat, pressure, and friction, the piston must be lightweight, strong, and properly fitted. It has a lot of work to do.

Piston Construction

Pistons can be made of aluminum or steel. Aluminum is by far the most popular metal for this application. The surface may be coated with a special break-in finish (tin or other coating). Sometimes pistons are chrome plated for installations where they operate directly on aluminum-alloy cylinder walls.

The type of piston often used in a four-cycle engine is shown in **Figure 16-1.** The head is quite thick, giving this hardworking part strength and resistance to overheating. The area below the head has grooves for the piston rings. The full-diameter ridges between the grooves are called *lands.* The wall or bottom of the oil ring groove is either slotted or pierced with holes. Oil wiped from the cylinder wall by the oil ring flows through these holes and back into the oil sump. The *sump* is the low area of the engine block where the oil collects.

Pistons may have grooves for one to four rings. Generally, the two-cycle engine piston has one or two grooves. Both are compression ring

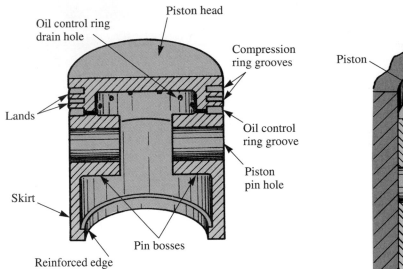

Figure 16-1. *A typical four-cycle engine piston is cut away to show construction details.*

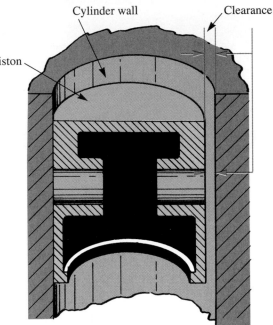

Figure 16-2. *Sufficient clearance must be allowed between piston skirt and cylinder wall to permit adequate lubrication and to allow for expansion of parts due to high temperatures.*

grooves. Four-cycle engine pistons will generally have three grooves, two for compression rings and one for an oil control ring.

The section of the piston surrounding the piston pin hole is called the *pin boss.* It is thick and often reinforced with cast-in webs. The *piston skirt* is the part of the piston below the bottom of the lower ring groove. The skirt is designed to be as light as possible to hold down the weight of the assembly.

The skirt actually guides the piston and keeps it from tipping from side to side. Portions of it may be cut away for lightness. Also, in some two-cycle engines, portions may be cut away to allow the air-fuel mixture to pass through the piston skirt into other parts of the cylinder.

Piston fit

The piston is subjected to high temperatures, causing it to expand during operation. To allow for this increase in size, there must be a specific amount of clearance between the piston skirt and cylinder wall.

The cylinder also expands, but not as much as the piston. Normal clearance must be great enough to allow for lubrication and piston expansion. Different engines have different clearances. The amount depends upon engine design and use. Most small engines call for .003″ to .005″ piston-to-cylinder wall clearance. See **Figure 16-2.**

Cam-ground pistons

When the designer wants the smallest possible clearance between the piston skirt and the cylinder wall, skirts are often cam ground to an elliptical (oval) shape. See **Figure 16-3A.** The oval shape of a *cam-ground piston* allows the

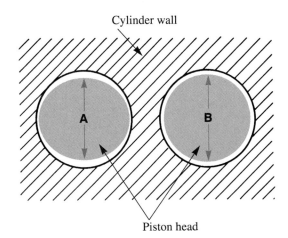

Figure 16-3. *Exaggerated top views of a cam ground piston as it would fit in a cylinder. A—Cold. B—Hot. Arrows indicate piston pin position.*

thrust surfaces (sides of skirt forced against cylinder during compression and firing) to fit more closely, even when cold. As the piston heats up, the diameter across the thrust surfaces remains constant and the piston enlarges parallel to (in the same direction as) the piston pin. See **Figure 16-3B.** These exaggerated views illustrate how a cam-ground piston expands to a round shape as it becomes hot.

Piston thrust surfaces

During the compression stroke, the pressure of the confined air-fuel mixture forces the piston toward one side of the cylinder. See **Figure 16-4A.** When the crankshaft throw passes TDC, burning and rapidly expanding gases push hard on the piston, forcing it against the opposite side of the cylinder. See **Figure 16-4B.**

In each instance, the sides of the piston forced against the cylinder wall are called thrust surfaces. These surfaces are at right angles (90°) to the centerline of the crankshaft and piston pin.

If the piston has too much clearance in the cylinder, side thrust during the compression and firing strokes will make it move, or *slap,* from one side of the cylinder to the other. As it moves sideways, the piston will tend to tip or cock in the cylinder. This *loose fit* can be very harmful to the piston and rings. The piston must fit the cylinder properly to avoid slapping.

Piston head size

The piston head receives the brunt of combustion heat, so it runs hotter than the skirt and expands more. Because of this, the head of the piston often is made with a smaller diameter than the skirt. **Figure 16-5** shows an exaggerated view of a piston with the head smaller than the skirt. The actual difference is only a few thousandths of an inch.

Piston head shape

Piston heads are manufactured in many different shapes, depending on the type of small engine and its use. On four-cycle engines, the piston head can be flat, domed, or wedge-shaped. Pistons used in two-cycle engines generally are flat when used with a loop-scavenging design. Cross-scavenging designs use a raised baffle or deflector head piston. See **Figure 16-6.**

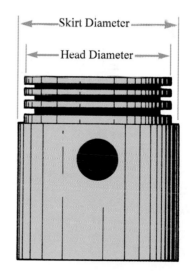

Figure 16-5. *Piston head receives greatest heat and is sometimes made smaller to compensate for expansion.*

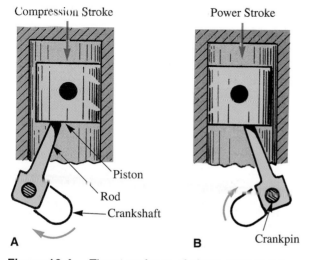

Figure 16-4. *Thrust surfaces of piston must resist heavy side pressure against cylinder walls. A—Upstroke. B—Downstroke.*

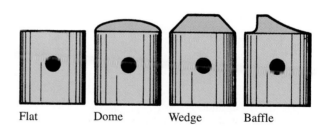

Figure 16-6. *Small gasoline engine piston heads are manufactured in a wide variety of shapes.*

Piston Ring Design

All small engine pistons must have clearance for lubrication and expansion. At the same time, they must have rings to help do the job of sealing the cylinder(s). Without *piston rings,* the piston could not compress the fuel charge properly. Also, burning gases would leak out between the sides of the piston and the cylinder wall.

In performing their job, the piston rings ride against the cylinder wall, separated from it only by a thin film of oil. The rings rub freely against the sides of the ring grooves, which hold the rings squarely to the bore and force them to slide up and down the cylinder with the piston. See **Figure 16-7**. Since the ring face is in steady contact with the cylinder walls an effective seal is formed. See **Figure 16-7.**

Figure 16-8 shows a compression ring in its groove. The sides of the ring groove are flat, parallel, and smooth. The ring has proper side clearance. In operation, expanding gases force the ring down against the lower side of the groove. At the same time, gases behind the ring force it against the cylinder wall. These forces help to form a good seal.

Piston ring construction

Piston rings are made of cast iron or steel. Both may be plated with chrome or other long-wearing materials. See **Figure 16-9.** Most pistons use cast iron compression rings. Steel, when used, general-

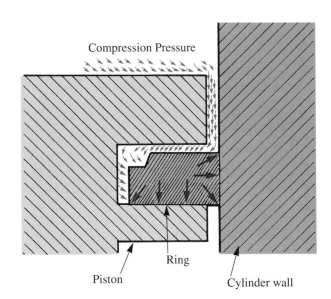

Figure 16-8. *Combustion chamber pressure forces ring against cylinder wall and bottom side of groove.*

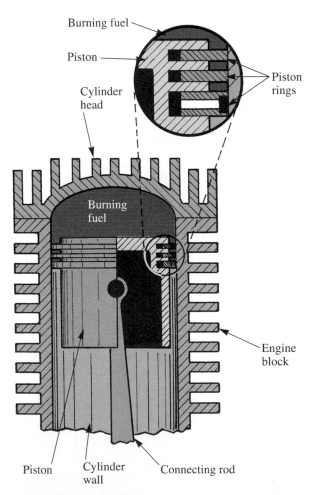

Figure 16-7. *Piston rings form a seal between the piston and cylinder wall.*

Figure 16-9. *Piston rings are often plated with chrome or other materials to reduce wear.*

ly goes into the construction of the oil control ring. In some installations, a cast iron center spacer-scraper may be combined with steel side rails.

Piston ring types

Most pistons use three rings. These three rings consist of two ring types. Generally, the two upper rings are compression rings and the lower ring is an oil control ring. See **Figure 16-10.**

Compression rings

The first and second rings from the top of the piston are compression rings. *Compression rings* are designed to provide a strong seal, keeping the compressed fuel mixture and the burning gases above the piston by preventing passage between the piston and the cylinder wall. Compression ring shapes vary somewhat with the scraper grooves, beveled faces, and grooves or bevels on the inner side of the ring. See **Figure 16-11.**

The various bevels and grooves are designed to create an internal stress in each compression ring. The stress causes the ring to twist slightly in its groove during the intake stroke of the piston. The twisting action places the lower edge of the ring, rather than the face, in contact with the cylinder wall. This allows compression rings to act as a mild scraper to aid in oil control. See **Figure 16-12A.**

On the compression and exhaust strokes (four-cycle engine), the rings are in a tipped position and

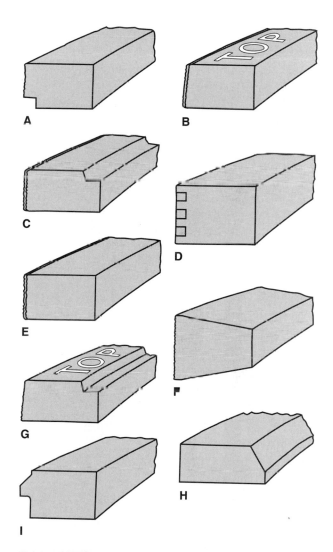

Figure 16-11. *Compression ring shapes.*
A—Outer groove. B—Chrome-plated, tapered face. C—Inner groove, chrome face. D—Ferrox-filled, grooved face. E—Plain chrome face. F—Keystone. G—Inner groove, tapered face. H—Inner chamfer, molybdenum-filled, grooved face. I—Scraper face. (Perfect Circle Products)

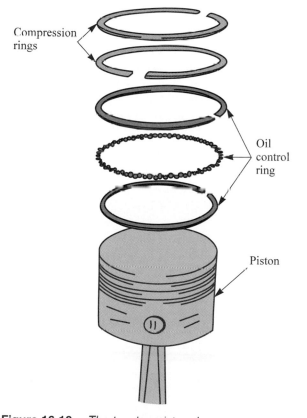

Figure 16-10. *The two top piston rings are compression rings and the bottom ring is the oil control ring. They fit into the grooves cut into the piston.*

tend to slip lightly over the oil film on the cylinder. See **Figure 16-12B.** On the power stroke, the pressure of the gases forces the ring flat so that the entire edge bears firmly against the cylinder wall. Maximum sealing is provided during this critical time. See **Figure 16-12C.**

Oil control rings

The oil control rings are designed to remove surplus oil from the cylinder walls. They do this through a light scraping action against the walls. The ring and groove are both slotted and perforated (having holes). Oil trapped by the ring passes through the slots or holes of the ring and the groove. See **Figure 16-13.** It then flows down inside the piston where it drops into the oil pan or the crankcase. A three-piece oil control ring with a hump-type, spring-steel expander is shown in **Figure 16-14.**

Ring tension

To permit the piston rings to expand and contract under varied temperatures and operating conditions, the rings are cut through at one place at the time of manufacture. See **Figure 16-15.** The size of this opening between the ends of the ring (with piston and rings in the cylinder) is called the ring end gap. Although a great number of gap joint designs have been used in an effort to seal against gas leakage, the plain butt joint is the most common.

In another design feature, the outside diameter of a piston ring is made slightly larger than cylinder bore diameter. This causes the ring to exert force on the cylinder wall when installed. This force is called *ring tension.* Because the outside diameter of the ring is larger than the cylinder bore diameter, the ends of each ring in a single-piston set must be squeezed together with a compressor tool to get the piston assembly into the cylinder.

Ring movement

Piston rings are free to move inward and outward in their respective grooves in the piston. See **Figure 16-16B.** In addition, the rings will gradually work their way around (float) in the grooves, unless each ring is pinned in place, as shown in **Figure 16-16A.** In most four-cycle engines, the rings float. Some two-cycle engines have the rings pinned in position. This is to prevent the ring ends from catching on the edge of the intake or exhaust ports and cutting into the cylinder wall.

Floating rings are installed with the ring end gaps staggered to avoid gap alignment and possible oil flow through the series of gaps to the combustion chamber. *Pinned rings* are held in position by a short pin manufactured into the ring groove in the piston. The ring ends are cut out to straddle the pin. See **Figure 16-16.** Obviously, the pin prevents rotation of the ring around the groove.

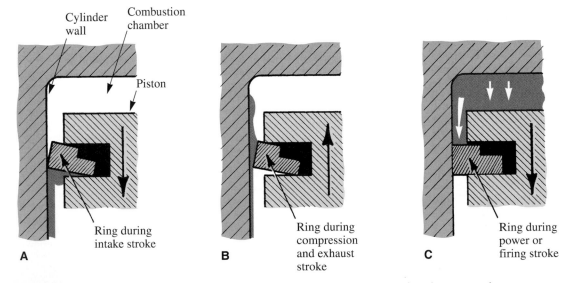

Figure 16-12. *Inner groove causes ring to twist slightly, aiding oil control and compression. It also reduces wear.*

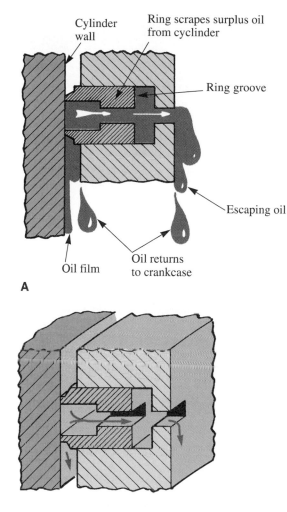

A

B

Figure 16-13. *An oil control ring removes surplus oil from cylinder walls. A—Oil being removed. B—Path of oil during removal.*

Figure 16-14. *Three-piece oil control ring with flat steel (hump-type) expander in place. (Perfect Circle Products)*

Figure 16-15. *End gap is cut through the piston ring to permit the ring to enter the cylinder and still exert tension on the cylinder wall.*

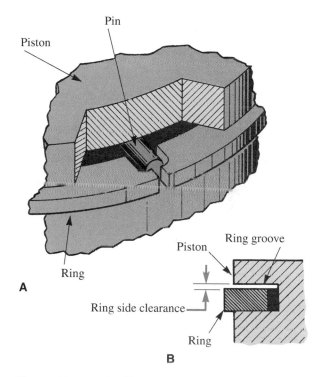

A

B

Figure 16-16. *A—Many two cycle engines have pinned rings to prevent ring rotation. B—Ring side clearance allows movement, admits lubricating oil, and permits expansion of parts due to heat.*

Ring side clearance

Piston rings must have the right amount of *side clearance,* which permits them to move in and out in the groove while exerting tension on the cylinder wall. Side clearance also provides for adequate lubrication and heat expansion. **Figure 16-17A** shows how the groove is cleaned with a special tool having different size scrapers that are pulled around the grooves. **Figure 16-17B** illustrates how side clearance is checked with a feeler gauge.

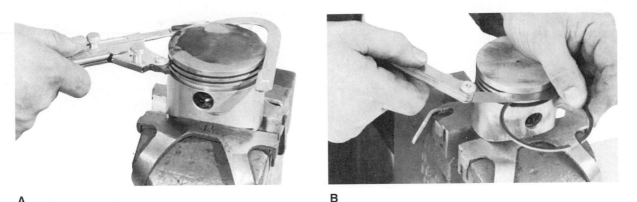

A   **B**

Figure 16-17. *A—A special tool is being used to clean carbon from ring grooves of piston. Do not widen or deepen any part of grooves. B—Side clearance of groove is checked with a piston ring and a thickness gauge.*

Ring end gap

The inside diameter of a piston ring is always made smaller than the piston's diameter. This being the case, each ring must be expanded to get it over the piston head and into the ring groove. The amount of **end gap** is critical (vital to success of rebuilding operation), so manufacturer's specifications should be followed. As a rule of thumb, however, allow .004″ of end gap for every inch of cylinder diameter. For example, the minimum end gap for a 2.5″ cylinder is .010″.

Too much end gap will allow the gases to leak between the ring ends. Too little gap is even more serious. When the rings heat up in service, they will expand and close up. If the rings continue to heat and expand, they will break and score the cylinder wall.

To measure ring end gap, place the ring in the cylinder. Then, turn a piston upside down and push the ring to the lower end of the cylinder. When the ring reaches the proper depth, remove the piston and try various size feeler gauges in the gap until one fits. See **Figure 16-18.** If the ring gap is too small, dress the butt ends of the ring with a file. See **Figure 16-19.**

When piston ring-to-groove side clearances and end gaps are satisfactory, install the rings on the piston. See **Figure 16-20.** Next, use a piston ring compressor to compress the rings flush with the grooves. See **Figure 16-21.** Hold the compressor firmly against the top of the block and use a wooden dowel or hammer handle to tap the piston out of the compressor and into the cylinder. Once free of the compressor, the rings will maintain firm contact with the cylinder wall.

Figure 16-18. *Ring end gap is measured by pushing ring into cylinder with an inverted piston. Then, piston is removed and a thickness gauge is used to measure gap.*

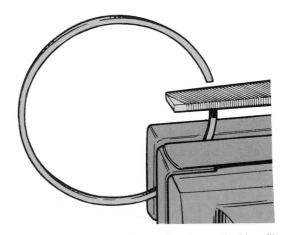

Figure 16-19. *Ring ends can be dressed with a file if end gap is too small. Use copper vise jaws to protect ring.*

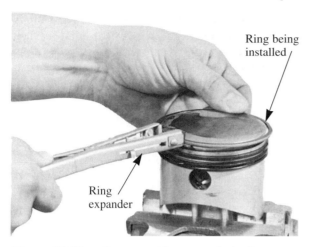

Figure 16-20. *Rings must be expanded with a special tool. Otherwise, danger of ring breakage is increased. (Tecumseh Products Co.)*

Figure 16-21. *A ring compressor is used to squeeze ring ends together while the piston is pushed into cylinder.*

Rings and cylinder walls

When cylinders wear, they become larger at the top. This is due to the wearing action of burning gases and to the dust and grit brought in with the air-fuel charge. Lack of lubrication in the upper cylinder also increases wear in the upper area of ring travel.

Even though the cylinder walls have a tapered wear pattern, the rings must stay tight against them. When taper becomes greater than .008″ or .010″, ring tension on the cylinder wall is lost and the end gaps open wide to allow oil to pass through to the combustion chamber. See **Figure 16-22A.**

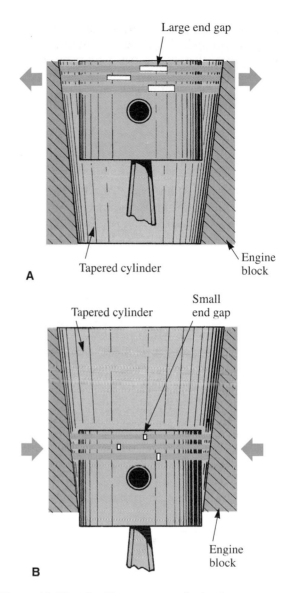

Figure 16-22. *A—Piston at top of a badly tapered cylinder. Rings must expand a great distance to stay in touch with cylinder walls. B—Piston at bottom of a badly tapered (exaggerated) cylinder. Rings are forced into grooves.*

When the piston is at the bottom of its stroke, the rings are forced inward by the walls. See **Figure 16-22B.** When the piston travels to the top of the cylinder, the rings must constantly expand to maintain contact with the ever-widening cylinder.

When the engine is running at high speed, the piston can move to the top of a badly tapered cylinder and back down again so quickly that the rings do not have a chance to expand. This leaves the rings completely free of the walls at the top of the stroke. An engine with cylinder walls in this condition would burn large amounts of oil. It is possible, too, that the piston would tip and slap, causing ring wear and damage.

Once the tapered condition exceeds .010″, it can only be corrected by boring the cylinder over-size or by replacing the unit.

Piston damage

Pistons can be damaged in many ways. They should be cleaned, examined, and measured after removal. The accompanying illustrations are good examples of conditions to look for.

Most piston and cylinder damage can be traced to one or more of the following causes: lack of oil, use of the wrong oil or oil-fuel mixture, use of the incorrect type of gasoline, foreign particles in the cylinder, overheating caused by clogged cooling fins, excess carbon buildup in the cylinder exhaust ports, and improperly fitted rings and pistons.

Figure 16-23 shows scoring and light scuffing of both a piston and rings due to overheating. This occurs when high friction and combustion cause temperatures to approach the melting temperature of the piston materials.

Check and correct:
- Dirty cooling shroud and cylinder head fins.
- Lack of cylinder lubrication.
- Poor combustion.
- Improper bearing or piston clearance.
- Overfilled crankcase causing fluid friction.

Figure 16-24 illustrates a piston with rings stuck and broken from lacquer, varnish, and carbon buildup, caused by abnormally high operating temperatures.

Check and correct:
- Overloading the engine.
- Ignition timing.
- Lean fuel mixture.
- Clogged cooling fins.
- Wrong oil.
- Low oil.
- Stale fuel.

Figure 16-25 reveals vertical scratches on the ring faces and the piston, caused by the presence of abrasive particles.

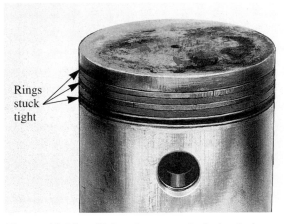

Figure 16-24. *Piston rings are stuck due to lacquer, varnish, and carbon buildup, resulting from high temperatures.*

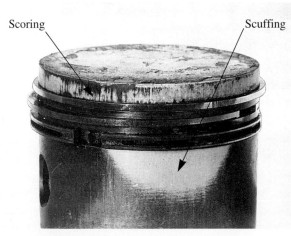

Figure 16-23. *Piston scored and scuffed due to overheating. (Tecumseh Products Co.)*

Figure 16-25. *Vertical scratches on ring faces and piston are caused by abrasive materials entering the engine.*

Check and correct:
- Damaged or improperly installed air cleaner.
- Air leaks between air filter and carburetor.
- Leak in gasket between carburetor and block.
- Air leak around throttle shaft.
- Improperly cleaned cylinder bore after reconditioning engine.

Compare the oil ring (bottom ring) in **Figure 16-26** with the one in **Figure 16-25.** The rails of the oil ring in **Figure 16-26** are worn down to the drain holes, and the ring surface is flat. This type of wear results from extended use and, possibly, from abrasives in the engine.

Check and correct:
- Rails worn down.
- Low ring tension.

The piston in **Figure 16-27** has a burned top land, resulting from detonation. Detonation is abnormal combustion which causes too much pressure and excessively high temperatures in the combustion chamber. Detonation is sometimes referred to as carbon knock, spark knock, or timing knock. It occurs when the air-fuel mixture ignites spontaneously and interferes with the normal combustion flame front.

Check and correct:
- Lean fuel mixture.
- Low octane fuel.
- Over-advanced ignition.
- Engine lugging.
- Excessive carbon deposits on piston and cylinder head (increasing compression).
- Milled cylinder head (increasing compression ratio).

Figure 16-27. *Burned top land results from detonation.*

Preignition is the burning of the air-fuel mixture before normal ignition occurs. Preignition creates a pinging sound, resulting from a severe internal shock. It is accompanied by vibration, detonation, and power loss. If allowed to continue, it could cause severe damage to the piston, rings, and valves. See **Figure 16-28.**

Check and correct:
- Internal carbon deposits remaining incandescent (red hot).
- Spark plug heat range too high.
- Spark plug ceramic shell broken.
- Thin edges on valves or elsewhere in combustion chamber too hot.

If the connecting rod and piston are not aligned, a diagonal wear pattern will show on the piston skirts, **Figure 16-29.** This condition can occur, along with poor ring contact, if the cylinder is bored at an angle to the crankshaft.

Figure 16-26. *Piston has extremely worn rings because of long use and possible abrasives. (Tecumseh Products Co.)*

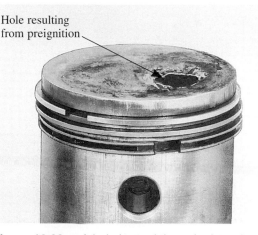

Figure 16-28. *A hole burned through piston head was caused by preignition.*

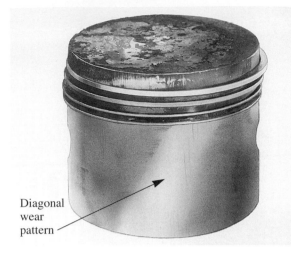

Figure 16-29. *A diagonal wear pattern indicates improper alignment of connecting rod and piston. (Tecumseh Products Co.)*

Check and correct:
- Rapid piston wear.
- Uneven piston wear.
- Excessive oil consumption.

Piston damage can be caused by foreign objects carelessly left inside an engine during reconditioning. **Figure 16-30** shows the results of a needle bearing that became embedded in a piston. Do not be careless. Do the job with painstaking care from the start to final assembly.

Figure 16-30. *Objects left inside engine can cause serious damage. A needle bearing is shown embedded across ring. (Jacobsen Mfg. Co.)*

Piston ring wear-in

After the small engine is reconditioned, a short wear-in period occurs. **Wear-in** is the process in which the face of each ring wears off until it fits perfectly against the cylinder wall. To help the rings seat quickly, the face is covered with microscopic grooves. During the first few hours of operation, these grooves rub against the cylinder wall and all high spots are worn off.

As the grooves wear away, the faces of rings and the cylinder wall become very smooth. Under normal operating conditions, very little wear occurs beyond this point.

Piston Pins

A **piston pin** is used to secure the connecting rod to the piston. These pins are made of case-hardened steel and are ground to exact size. They may be hollow or solid.

A typical solid piston pin is shown in **Figure 16-31.** Note the finer grain size of the metal at the surface. This is due to a heat treatment that provides a hard, durable case with a soft, tough center.

Many different piston pin assemblies have been used. The full-floating pin arrangement shown in **Figure 16-32** is free to turn in the rod as well as in the piston bosses. When both the connecting rod and piston are of aluminum alloy, the pin can operate directly against this material. If the rod is steel, either a bronze bushing or a needle roller bearing is used in the rod. The piston bosses

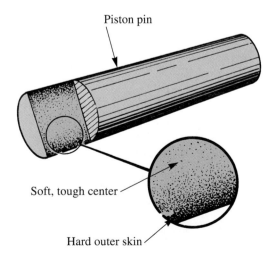

Figure 16-31. *Solid piston pin has hardened and ground surfaces.*

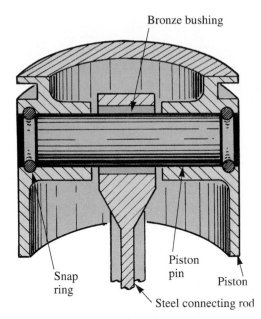

Figure 16-32. *A full-floating piston pin used in a steel connecting rod requires use of a bushing.*

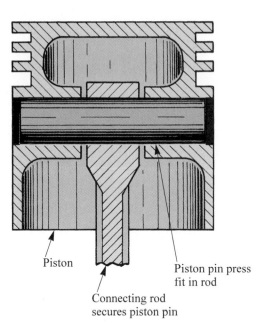

Figure 16-34. *This piston pin is pressed into connecting rod, but is allowed to turn in pin bosses of piston.*

also may have bronze bushing inserts for the pin. The snap rings in each boss prevent the pin from rubbing on the cylinder surface. **Figure 16-33** shows a full-floating pin with a steel connecting rod. This setup requires the use of a bushing or bearing.

Retaining *snap rings* are compressed and placed in grooves in the piston pin bosses. See **Figure 16-33.** Some piston pins are a tight, press fit in the connecting rod. See **Figure 16-34.** The pin may turn in the piston bosses, bushings, or needle bearings, depending on the type of construction used.

Removing piston pins

If snap rings are used, they must be removed first. Some are formed, spring-steel wire; others are stamped flat spring steel. The two common methods of removing the wire-type snap rings are with a screwdriver or with needle nose pliers. See **Figure 16-35.** Removal of stamped snap rings usually requires specially designed snap ring pliers.

⚠️ Be careful when removing or replacing snap rings. They can slip out of the pin boss or the jaws of the pliers and cause severe eye injuries. Wear safety glasses when working with snap rings.

If the piston pin is full floating and somewhat worn, it may slide out easily. On the other hand, it may be quite snug. A soft-faced mallet and a dowel rod can be used to tap the pin out. Be careful not to hit the piston. Let the pin fall into a soft cloth.

Press-fit pins must be removed with a mechanical or hydraulic press. Refer to the engine manual for the proper support of the piston. If needle bearings are used, be careful not to lose them or misplace them.

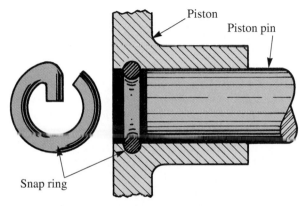

Figure 16-33. *Snap rings keep full-floating pin in place in piston.*

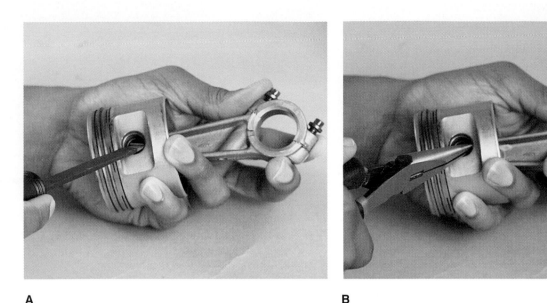

A   B

Figure 16-35. *A—Some rings can be removed from their grooves using a screwdriver. B—Some rings can be removed using needle nose pliers. Always wear safety glasses for this operation.*

Measuring piston pins and bosses

When the piston pin has been removed, measure its outside diameter with a micrometer. Make a note of the measurement. Next, measure the inside diameter of the bosses using a small hole gauge. Expand the gauge in the boss until it gently contacts the inner surfaces. Withdraw the gauge and measure it with the micrometer. Then, subtract the pin diameter from the boss diameter. The difference must be within the limits specified for the engine being serviced.

If the pin-to-boss clearance is excessive, a new pin and bushing or an oversize pin and reaming of the bosses are needed. Most piston pins fit snugly in the piston. If the pin turns by hand, inspect it for wear.

Summary

The condition of the piston and the rings can be observed by inspecting the parts during disassembly of the engine. Low compression, blow-by, oil pumping, and fouled plugs all signal the need for piston and ring service.

Pistons can be made of aluminum or steel. Pistons have grooves for one to four rings. Two-cycle engine pistons have one or two compression ring grooves. Four-cycle engines have two compression ring grooves and one oil control ring groove.

Pistons expand during operation. Therefore, a specific amount of clearance between the piston skirt and the cylinder wall must be present.

A piston's thrust surfaces are forced against the sides of the cylinder wall. The thrust surfaces are at right angles to the centerline of the crankshaft and the piston pin.

The piston head is often smaller than the skirt because it receives the brunt of combustion heat and expands more than the rest of the piston. Pistons are manufactured in many different shapes, depending on their application.

Piston rings seal the cylinder. Without rings, the piston could not compress the fuel charge properly. Rings also keep burning gases from leaking between the sides of the piston and the cylinder wall. Piston rings ride against the cylinder wall, separated only by a thin film of oil.

Generally, the first and second rings from the top of the piston are compression rings. These rings are designed to keep the compressed fuel mixture and burning gases from escaping between the piston and the cylinder wall. Oil control the rings remove surplus oil from the cylinder walls.

Pistons can be damaged in many ways. They should be cleaned, examined, and measured after removal. A piston pin is used to secure the piston to the connecting rod. The pin may be held in place by a snap ring or can be press fit in the connecting rod.

Know These Terms

pistons   compression rings
lands   ring tension
sump   floating rings
pin boss   pinned rings
piston skirt   side clearance
cam-ground piston   end gap
thrust surfaces   wear-in
slap   piston pin
piston rings   snap rings

Chapter 16 Review Questions

Answer the following questions on a separate sheet of paper.

1. Explain why the piston head diameter may be less than the skirt diameter.
2. If a piston has too much clearance in the cylinder, it will produce a motion and sound called _____.
   a. knocking
   b. slapping
   c. pinging
   d. hammering
3. Piston rings are made of _____.
   a. cast iron
   b. steel
   c. aluminum alloy
   d. cast iron, aluminum alloy
   e. cast iron, steel
4. Name the tool used to squeeze the piston rings together so they can be installed in the cylinder.
5. Briefly list the results of excessive ring end gap and a lack of ring end gap.
6. Pinned piston rings would most likely be found in _____.
   a. four-cycle engines
   b. two-cycle engines
   c. None of the above.
7. Piston rings can be one of two basic types. Can you name them?
8. Piston pins are prevented from contacting the cylinder wall by use of _____.
   a. cotter pins
   b. setscrews
   c. tapered pins
   d. snap rings

Suggested Activities

1. Remove the rings from a piston, clean the piston and measure it.
2. Measure ring end gap and side clearance.
3. Using an old ring, demonstrate the method of dressing ring ends with a file to increase ring end gap.
4. Recondition piston bosses by reaming, and replace the worn piston pin with an oversize pin.
5. Replace piston rings with a ring expander.
6. Using ring compressor, replace reconditioned piston assembly in the cylinder.

This cutaway shows the relationship between the piston and the other internal components of an engine.
(Tecumseh Products Co.)

Bearing, Crankshaft, Valve, and Camshaft Service

After studying this chapter, you will be able to:
▼ Describe the function of the connecting rod and the bearings.
▼ Define bearing spread and bearing crush.
▼ Differentiate between friction bearings and antifriction bearings.
▼ Summarize the function of the crankshaft.
▼ Service conventional and overhead valve assemblies.
▼ Explain the operation of ports, reeds, and rotary valves.
▼ Describe the purpose of the camshaft.
▼ Explain the purpose of an automatic compression release.

Introduction

Like pistons and piston rings, connecting rods, bearings, and valves are used in areas of the engine that demand *close fits*. Due to high temperatures, however, some clearance must be allowed for part expansion. While there are differences between makes of engines, maintenance of the rods, bearings, and valves is much the same for all.

Special attention must be given to four-cycle engines because they contain more parts that require service. Rod and bearing service is the same for both two-cycle and four-cycle types. The valve system of four-cycle engines (major area of difference) will be covered in detail near the end of the chapter.

Connecting Rods and Bearings

The *connecting rod* attaches the piston to the crankshaft. The upper end of the connecting rod

has a hole through which the piston pin is passed. The lower end contains a large bearing that fits around the crankshaft journal. See **Figure 17-1.**

The lower end of the connecting rod is usually split when friction bearings are used. *Friction bearings* use smooth, sliding surfaces to reduce friction between moving parts. The place at which the halves separate is called the parting line.

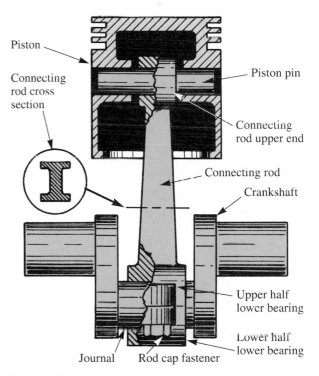

Figure 17-1. *Connecting rod attaches piston to crankshaft. Bearings are used at both ends of rod to reduce friction.*

297

When needle or roller bearings are used, the rod end can be split or solid. See **Figure 17-2.**

A variety of connecting rods, both split and solid, are pictured in **Figure 17-3.** Roller bearings, needle bearings, and precision inserts are also shown.

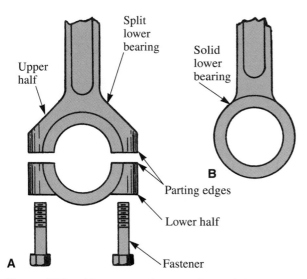

Figure 17-2. *Two types of connecting rod designs (crankshaft end). A—Split construction. B—Solid construction.*

Friction-type rod bearings

There are three types of friction bearings commonly used in the big end of connecting rods. The three types of friction bearings are:

- Rod metal (used when rod is made of aluminum alloy). See **Figure 17-4.**
- Bearing bronze (cast into rod end, bored, and finished).
- Removable precision insert bearings (steel shells lined with various materials).

The thin lining material on removable bearing inserts can be lead-tin babbitt, aluminum, or copper-lead-tin. **Figure 17-5** shows a steel-backed insert (1) that is coated with cast babbitt (2). This type of bearing is called a *precision* insert because it is made to an exact size for proper fit.

Bearing inserts are kept from turning in the rod end by a locating tab on the parting line edge of each insert. The tab fits into a slot in the rod itself. **Figure 17-6** illustrates this tab and slot arrangement.

Bearing spread

The diameter across the parting surfaces of insert bearing halves is slightly larger than the

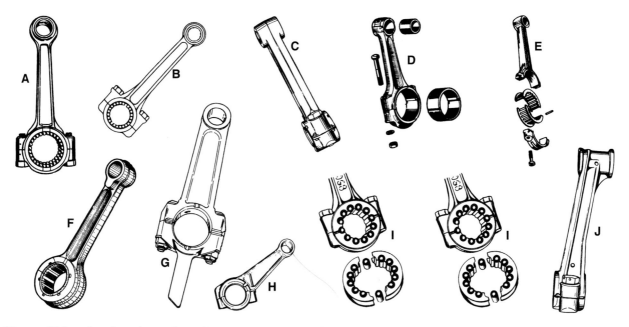

Figure 17-3. *Small engine rods and bearings. A and B—Split rod, roller bearings. C—Split rod, cast-in bearing. D—Split rod, precision inserts. E—Split rod, needle bearings. F—Solid rod, paired roller bearings. G—Split rod with dipper, precision inserts. H—Rod split at an angle, precision inserts. I—Split rod, roller bearings. J—Split rod with offset cap, cast-in bearing.*

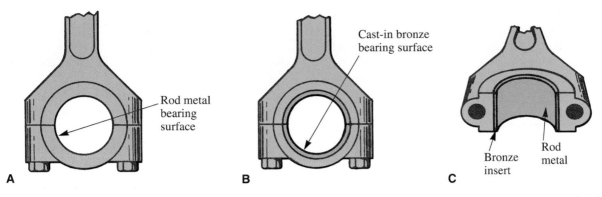

Figure 17-4. *Friction-type connecting rod bearings. A—Rod metal forms bearing surface. B—Bronze bearing is cast into rod metal. C—Replaceable precision insert bearing.*

Figure 17-5. *Construction of a typical precision insert bearing. 1—Steel back. 2—Cast babbitt (about .004″ thick). (Clevite Corp.)*

diameter across the curve machined into the rod and rod cap. This condition is called ***bearing spread.*** To seat the insert, the ends must be forced down and snapped into place. Never press down in the center of the insert to seat it in the rod bore. The correct amount of bearing spread gives tight insert-to-bore contact around the entire bearing and provides support and alignment. It also helps to carry heat away through the rod and bearing cap and holds the bearing in place during assembly.

Bearing crush

When precision inserts are snapped into the rod bore, the ends will protrude slightly above the parting surface. See **Figure 17-7A.** This built-in design feature is called ***bearing crush.*** Generally, bearing crush varies from .001″ to .002″.

When the rod cap is installed and drawn into place, the insert ends meet first and force the insert

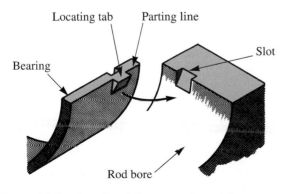

Figure 17-6. *Locating tabs prevent precision inserts from turning.*

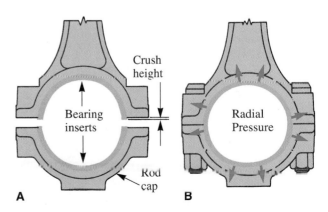

Figure 17-7. *Effect of bearing crush. A—Rod and cap separated. B—Rod and cap drawn together, creating radial pressure on inserts. (Sunnen Products Co.)*

halves tightly against the rod bore. This provides firm support for the insert. The forced fit makes the insert round and, through close metal-to-metal contact, allows heat to be carried away through the rod. **Figure 17-7B** shows how radial pressure is exerted against the rod bore.

Inserts are matched

Precision inserts must be kept in matched pairs. Never mismatch bearing inserts. Always use the exact size needed. Bearings cannot be made larger or smaller in the shop. Standard and various undersizes are available.

Antifriction bearings

Many small gasoline engines use an antifriction bearing in the big end of the connecting rod. *Antifriction bearings* use rollers or balls to reduce friction between moving parts. Often, needle rollers are used. These roller elements can be held together by a roller cage or separator. See **Figure 17-8A.** The rollers can also be left free as in **Figure 17-8B.** Antifriction bearing assemblies are hardened and ground to an exact size. They must fit accurately, but still have some clearance for expansion.

During manufacture, the *rod cap* is bolted into position on the rod. Then, the assembly is bored to an exact size. It is important, therefore, that the rod cap is always put back in its original position. If the cap is turned 180°, the upper and lower halves can be offset. This error in assembly will cause bearing and shaft failure. See **Figure 17-9A.**

Connecting rods are usually marked with either a line, punch mark, number, special boss, or chamfered edge to show correct cap alignment during assembly. If a *mark* is not apparent, punch

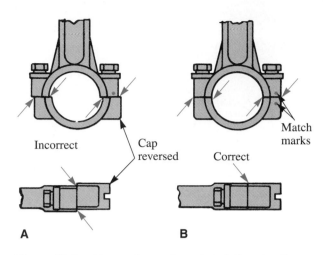

Figure 17-9. *Connecting rod cap installation. A—If cap is turned 180°, rod bore will be offset. B—Match marks on rod and cap signal correct assembly.*

mark the rod end and cap for later reference, **Figure 17-9B.**

> Caps must never be switched from one rod to another.

Rod bolt locking devices

To stop connecting rod bolts or cap screws from loosening in service, locking devices are used. One common device is a thin sheet metal strip with locking tabs. See **Figure 17-10.** The cap screw is inserted through holes in the locking strip, holding it in place against the rod. After the cap screw is tight, the metal tabs are bent up against the flat sides of the screw head. See **Figure 17-11**.

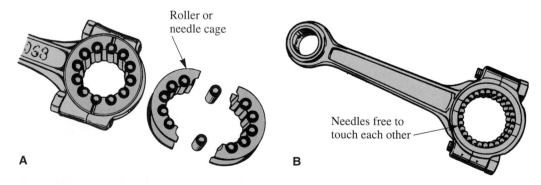

Figure 17-8. *Two types of connecting rod needle bearings. A—Caged needles. B—Free needles. (Evinrude Motors)*

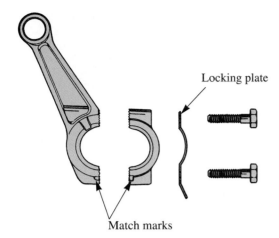

Locking plate

Match marks

Figure 17-10. *A locking plate is often used between connecting rod cap and cap screws. (Tecumseh Products Co.)*

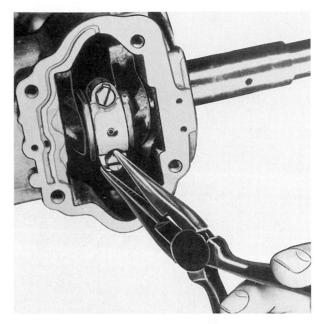

Figure 17-11. *After cap screw has been tightened, tab on locking plate is bent up against head of screw to lock it in place. (Lawn Boy Power Equipment, Gale Products)*

Self-locking nuts, lock washers, and specially shaped cap screws are also used to prevent loosening. The final tightening of the cap screws is especially important. Always use a torque wrench to tighten rod fasteners to the exact torque specified by the manufacturer.

Crankshaft

The ***crankshaft*** converts the reciprocating (back and forth) motion of the piston into rotary (circular) motion. It transmits engine torque to a pulley or gear, so that some object may be driven by the engine. The crankshaft also drives the camshaft (on four-cycle engines), supports the flywheel, and, in many engines, operates the ignition system.

Crankshafts can be made of cast or drop-forged steel. One-piece and multi-piece crankshafts are used. **Figure 17-12** shows a typical one-piece small engine crankshaft. A multi-piece crankshaft is shown in **Figure 17-13**.

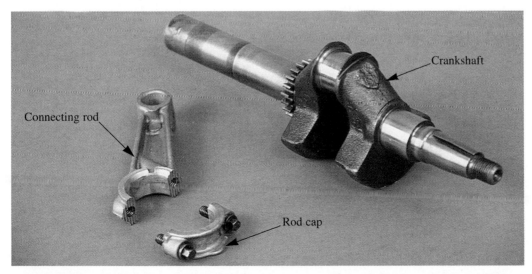

Crankshaft

Connecting rod

Rod cap

Figure 17-12. *Single-piece crankshafts are most popular in small gasoline engine applications.*

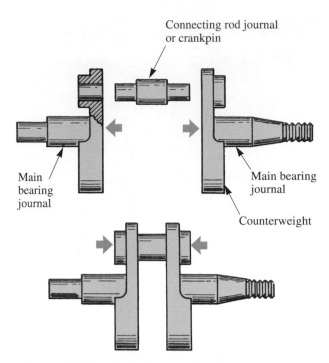

Figure 17-13. *Multi-piece crankshafts have various parts pressed together under heavy pressure.*

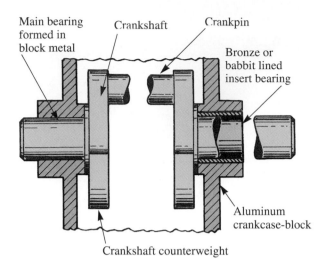

Figure 17-14. *Friction-type crankshaft main bearings. Shaft at left uses bore in aluminum crankcase as a bearing surface. Shaft at right uses a precision insert bearing.*

The **crankshaft throw** is the offset portion of the shaft measured from the centerline of the main bearing bore to the centerline of the connecting rod journal. The connecting rod journal is commonly referred to as the *crank throw* or *crankpin*.

Crankshaft balance

To help offset the unbalance created by the force of the reciprocating mass (connecting rod, piston, and crankpin), *counterweights* are added to the crankshaft. By placing these weights opposite the crankpin, engine vibration is greatly reduced. As shown in **Figures 17-12** and **17-13,** the counterweights are usually forged as an integral part of the crankshaft.

Crankshaft main bearings

The crankshaft is supported by one or more **main bearings.** Often, the main bearing journal surfaces are hardened by an induction hardening process to provide long service life. The three types of main bearings used are:
- Sleeve or bushing. See **Figure 17-14.**
- Roller bearing. See **Figure 17-15.**
- Ball bearing. See **Figure 17-16.**

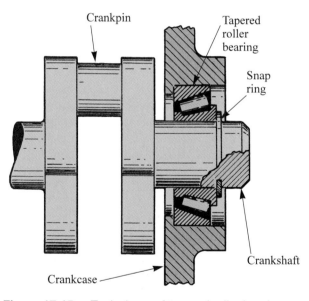

Figure 17-15. *Typical use of tapered roller bearing as a crankshaft main bearing.*

Crankshaft clearances

To allow space for lubricant between the moving parts, as well as to provide room for expansion when heated, crankshaft bearings must have a slight end clearance. See **Figure 17-17.** Shaft movement from end to end is controlled by the bearing adjustment when tapered roller bearings or ball bearings are used.

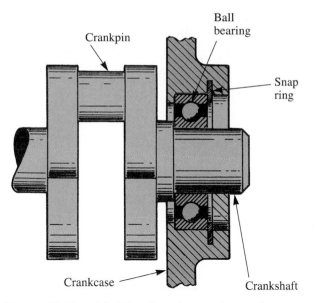

Figure 17-16. *A ball bearing also can be used as a crankshaft main bearing.*

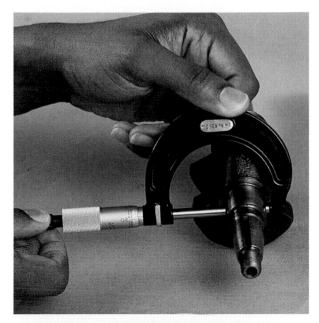

Figure 17-18. *A micrometer is required to accurately measure bearing surface diameter on a crankshaft.*

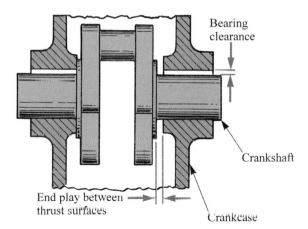

Figure 17-17. *Crankshaft bearings and thrust surfaces must have some clearance (end play) to provide space for lubricant and for heat expansion.*

With friction bearings, a thrust surface on the shaft rubs against a similar surface on the crankcase. A precision insert main bearing may have a thrust flange for the crank to rub against. In some applications, a bronze thrust washer is used.

Clearances will vary with engine type, design, and use. Bearing and thrust surface clearances are critical. They must be held to exact tolerances as recommended by the manufacturer.

Figure 17-18 illustrates the method of measuring the bearing surfaces on a crankshaft with a micrometer. A measurement must be taken in at least two positions 90° to each other. If any of the dimensions are smaller than specified, or if there are any score marks, the bearing surfaces should be reground. Basically, wear and taper should not exceed .001″.

Measuring bearing clearance

Bearing clearance is the space between the inner bearing surface and the crankshaft main or rod journal. When checking bearing clearance, use a special compressible plastic material called *Plastigage*. This material is color coded and selected according to the recommended clearance range. It comes in a thin, round strand, which is stored in a paper package.

To use Plastigage, select the correct color for the specified clearance. Cut a piece of plastic equal to the width of the bearing and lay it across the bearing surface. Torque the bearing cap in place. Then, remove the cap and compare the compressed width of the plastic with the comparison chart printed on the Plastigage package. Clearance is given alongside the matching marks on the chart. In effect, the wider the plastic, the less clearance there is.

If bearing clearance is too great, undersize inserts will have to be used. If the crank journal is worn, it will require grinding to clean it up. After grinding the journal, recheck the clearance with a Plastigage.

On many small engines, the main bearings are simply machined bores in the crankcase halves or pressed inserts. Plastigage will not measure wear in these bearings. To check clearance, first measure the crankshaft diameter with a micrometer. Next, measure the inside diameter of the main bearing with a telescoping gauge. See **Figure 17-19.** Lock the gauge, remove it from the bearing, and measure the setting with a micrometer. The difference between the shaft reading and the bearing reading is the amount of bearing clearance.

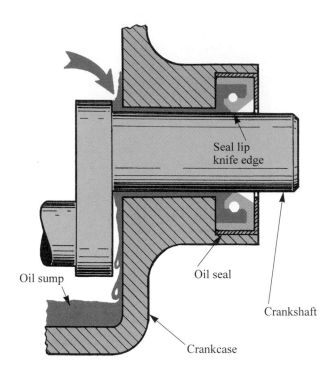

Figure 17-20. *Typical neoprene oil seal has sealing lip with sharp edge, providing increased pressure and reduced friction.*

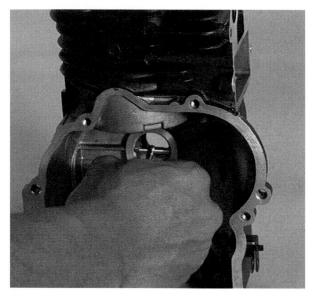

Figure 17-19. *Measuring diameter of a pressed main bearing with a telescoping gauge.*

Note in **Figure 17-20** that the sealing lip must face the fluid being sealed in. In this application, it faces the crankcase. In this way, the pressure of the oil will tend to force the lip against the shaft. If the seal is installed backwards, oil pressure will force the sealing lip away from the shaft and oil leakage will occur.

When removing the crankcase cover from the crankcase and crankshaft, as in **Figure 17-22,**

Crankcase seals

Crankcase seals prevent leakage of oil from the areas where the crankshaft and crankcase come together. The shell of the seal makes fixed contact with the crankcase, while the knife edge of the sealing lip rubs lightly against the crankshaft. See **Figure 17-20.**

Seals are made of neoprene, leather, graphite, or other materials, depending on how they are used. A typical crankcase seal has a steel outer shell with a neoprene center. A small coil spring keeps the sealing lip in constant contact with the shaft it seals. See **Figure 17-21.**

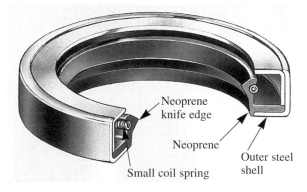

Figure 17-21. *This crankshaft oil seal has an outer steel shell with a neoprene center. A small coil spring produces contact pressure. (Chicago Rawhide Mfg. Co.)*

place tape over the keyway. This will keep the sharp keyway edges from cutting the neoprene oil seal.

Figure 17-23 shows how a press is used to push the old seal out of the backplate. In **Figure 17-24,** the seal is being readied for installation. A liquid sealant is applied to the outside of the shell

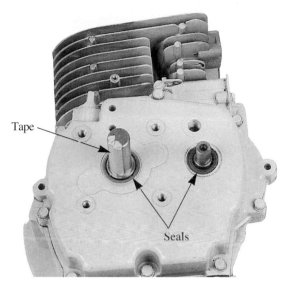

Figure 17-22. *Taped keyway edges will protect oil seal when cover is removed. (Deere & Co.)*

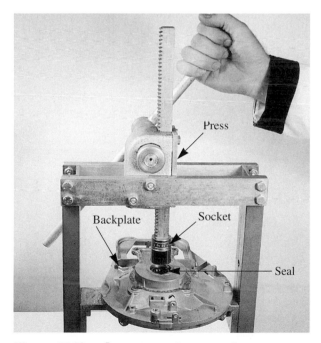

Figure 17-23. *On some engines, an arbor press can be used to push out old oil seals. Press with care to avoid damaging housing. (Jacobsen Mfg. Co.)*

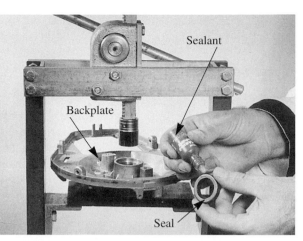

Figure 17-24. *Apply sealing compound to shell of seal before pressing seal into bore of housing.*

of the seal before pressing it in place in the backplate. Often, seals can be replaced by tapping them into the bore of the backplate with a special driving tool. See **Figure 17-25.**

Valve Service

Four-cycle engines contain *poppet valves,* which are subjected to tremendous heat. The normal operating temperature of the exhaust valve exceeds 1000°F. To withstand this heat, high-

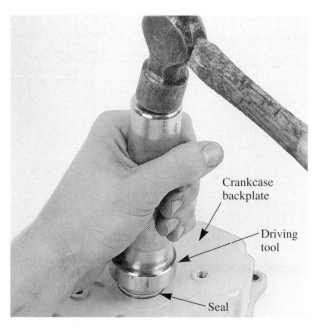

Figure 17-25. *Use a special driving tool to tap seal in place in housing. (Deere & Co.)*

quality, heat-resistant steel must be used and the correct operating clearances must be maintained.

Removing valve assembly

The engine valve assembly includes the valve, valve spring, and one or more retainers, **Figure 17-26.** The locking-type retainers are called *valve keepers.* Once the cylinder head has been removed, remove the valve by compressing the valve spring with a compressor. See **Figure 17-27.** Remove the valve spring retainer from the groove in the valve

Figure 17-28. *When spring is fully compressed, use a pliers to remove valve keepers. Then, release valve compressor and remove valve, spring, and retainers.*

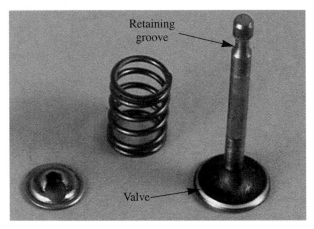

Figure 17-26. *A poppet valve assembly is used in four-cycle engines.*

stem using a pair of pliers. See **Figure 17-28.** Then, slowly release pressure on the spring and remove the compressor. The valve can be pulled out the top and the spring taken from the side.

Inspecting valves and seats

When the valves have been removed, clean them with a power-operated wire brush and inspect them for the following defects:
1. Eroded, cracked, or pitted valve faces, heads, or stems.
2. Warped head. See **Figure 17-29.**
3. Worn or improperly ground valve stems. See **Figure 17-29.**

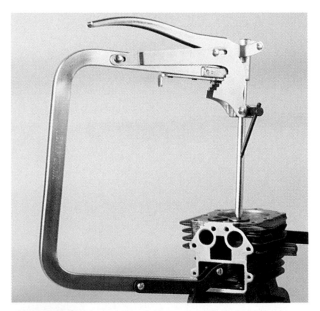

Figure 17-27. *A valve spring compressor squeezes the spring to uncover keepers in the valve stem.*

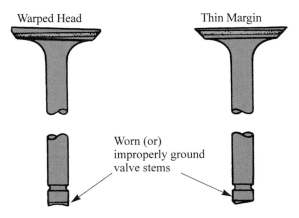

Figure 17-29. *Warped valve head, thin margin, or worn stem may mean refacing or replacement is needed.*

4. Bent valve stems.
5. Margin less than 1/64″.
6. Partial seating.

Heavy carbon deposits on intake valves sometimes cause faulty valve operation by restricting the flow of fuel into the cylinder. If any serious defects are observed, the valve should be replaced. In any case, valve faces should be machined to a smooth, true finish.

Inspecting valve springs

Through overheating and extensive use, *valve springs* can lose their elasticity and become distorted (warped or bent). Check each spring for squareness and proper length with a square and a surface plate. See **Figure 17-30.** Replace all springs that are badly distorted or reduced in length. Test spring tension and compare with the specifications in the engine manual. See **Figure 17-31.** Lack of spring tension can cause *valve flutter* or incomplete closing and sealing of the valves.

Figure 17-31. *A special tester is available to test a valve spring for adequate tension.*

Valve guides

Valve guides align and *steer* the valves so that they can open fully and close completely. Guide-to-valve stem clearance must not exceed tolerances, since this would permit the valve to tip. Tipping causes the valve face to strike the seat at an angle, allowing hot combustion gases to escape. Some clearance is required, however, to allow for heat expansion and lubrication. Generally, guide-to-valve stem clearance should run about .002″ to .003″.

Valve guides can be a replaceable insert or an integral part of the block. See **Figure 17-32.** Replaceable guides are cast iron. The old guide must be driven out and the new guide pressed in place. Integral guides can be reamed to fit a valve with an oversize stem.

Inspecting valve guides

Valve guides must be cleaned before inspection. A special cylindrical wire brush, driven by a power drill, is made for this job. After cleaning the guide, measure the bore with a small hole gauge. See **Figure 17-33.** Expand the gauge until it

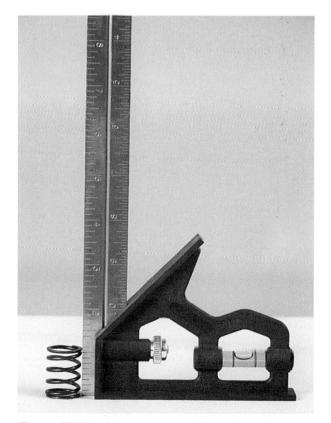

Figure 17-30. *Use a square and a surface plate to check a valve spring for proper length and squareness.*

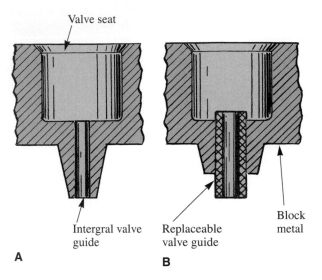

Figure 17-32. *There are two types of valve guides. A—Bored in block. B—Pressed in block. Guides that are pressed in the block can be replaced.*

lightly touches the sides of the bore. Remove the gauge and measure it with a micrometer.

Next, measure the valve stem diameter with a micrometer. See **Figure 17-34.** Subtract the stem diameter from the guide diameter to find the precise amount of clearance. Compare this with the clearance specified.

Valve guide reaming

If the clearance between the stem and the guide exceeds the allowable limit, enlarge the guide to the next oversize dimension with an adjustable reamer. See **Figure 17-35.** Select and install a valve with the correct oversize stem. Do not enlarge the tappet guides, because oversize tappet stems are seldom available.

Alternate valve guide replacement

There are many engine makes and models available that may need valve reconditioning. Some

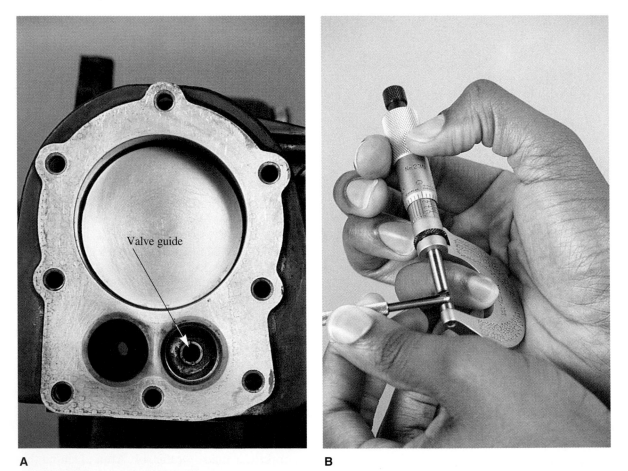

Figure 17-33. *A—Valve guide diameter can be measured with a small hole gauge. B—The small hole gauge measurement is transferred to a micrometer.*

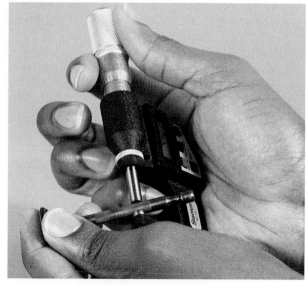

Figure 17-34. *Measure valve stem diameter to determine valve stem-to-guide clearance.*

engines have valve guides that are replaceable by pulling out the worn guide bushings and pressing in new bushings. For each make and model, the manufacturer's service manuals should be examined to determine the proper method, bushings, and tools to use.

Figure 17-35. *Use an adjustable reamer to recondition and resize integral valve guides. Replacement valves must have oversize stems.*

Figure 17-36 is one example of a method used for several selected engine models and is meant to be illustrative only.

Rebushing worn aluminum guides

1. If this special plug gauge tool can be inserted into the valve guide a specified distance, the valve guide is worn and should be replaced. See **Figure 17-36A.**
2. Place pilot of counterbore reamer in valve guide. Slide pilot bushing down over counterbore reamer until bushing rests on valve seat. See **Figure 17-36B.**
3. Place proper replacement bushing on top of pilot bushing and mark reamer 1/16″ above top of bushing.
4. Ream worn valve guide until mark on counterbore reamer is even with top of pilot bushing. Use kerosene or aluminum cutting fluid to lubricate reamer.
5. After guide is counterbored, withdraw reamer while turning it in the same direction used to ream guide. See **Figure 17-36B.**

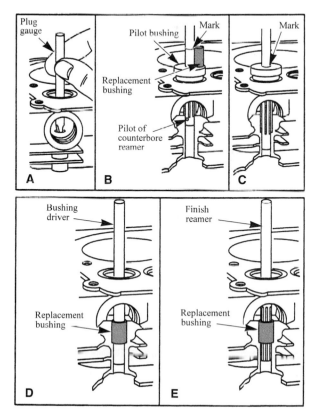

Figure 17-36. *Replacing worn aluminum valve guides with new valve guide bushings. (Briggs & Stratton Corp.)*

6. Position new bushing in counterbored guide. Press bushing with valve guide bushing driver until bushing is flush with top of guide. See **Figure 17-36D.**
7. Finish ream bushing with proper size finishing reamer through to breather chamber. Use kerosene or similar lubricant. See **Figure 17-36E.**
8. Flush all chips away before removing reamer.
9. Withdraw reamer by turning it in the same direction used for reaming while pulling up on reamer.

Replacing worn brass or sintered iron guides

1. Use modified reamer guide to align 7mm tap. See **Figure 17-37.** Measure shank of 7mm tap and drill reamer guide to assure tap will be square to the bushing to be pulled.

Figure 17-37. *A self-aligning reamer guide like this one can be easily made on a metal cutting lathe. (Briggs & Stratton Corp.)*

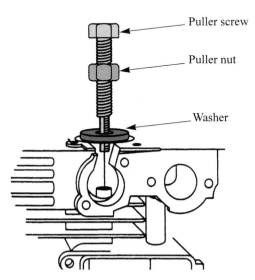

Figure 17-38. *After old valve guide is tapped with threads this tool can pull out the old guide. A tool like this can be easily made from a bolt turned and threaded on a metal lathe. (Briggs & Stratton Corp.)*

2. Using a tap wrench and pilot guide bushing, turn tap into bushing clockwise until tap is 1/2″ deep.
3. Remove tap and wash chips out of bushing.
4. Rotate puller nut up to head of puller screw. Insert puller screw down through puller washer. See **Figure 17-38.**
5. Thread puller screw into threaded bushing until screw bottoms in tapped hole.
6. Back off screw 1/8 to 1/4 turn and place a drop of engine oil on threads of puller screw.
7. Hold puller screw stationary and turn puller nut down on washer until guide bushing is removed.
8. Press correct new bushing into cylinder with bushing driver tool until bushing bottoms in hole. See **Figure 17-39.**
9. Finish ream bushing with finish reamer and reamer guide bushing.
10. Wash chips away and remove reamer.

Valve seat angle and width

The correct *valve seat angle* is necessary for proper valve seating. Valve seats are generally cut to a 45° angle, although 30° seat angles are used in a few engines. Follow all of the manufacturer's recommendations.

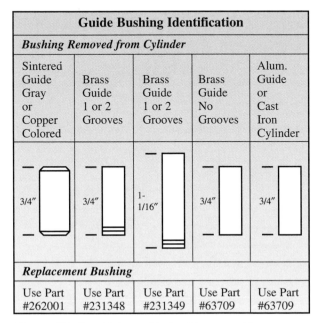

Figure 17-39. *Valve guide bushings used in some Briggs & Stratton engines. Note code system. (Briggs & Stratton Corp.)*

The **valve seat width** is important for effective valve system operation. The seat must be wide enough to prevent cutting into the valve face. It also must provide enough contact area to provide for adequate heat dissipation. On the other hand, the seat must not be too wide. If it is, carbon will pack between the seat and the valve face, holding the valve off the seat. A valve that fails to seat produces a rough-running engine and will quickly warp and burn. Specified seat widths range from .030″ to .060″ (1/32″ to 1/16″). See **Figure 17-40.**

Some valve seats are ground to an angle of 44° and the valve face is ground to an angle of 45°, or vice versa. The 1° variation produces a hairline contact that gives fast initial seating. Some manufacturers believe that, upon heating, the valve will form a perfect seal. The difference in the angle between the valve face and the valve seat is called an **interference angle.** See **Figure 17-41.** Valve seat contact must be near the center of the valve face. See **Figure 17-42.**

Valve seats should be cut with a special valve seat cutting tool that has sharp carbide blades, such as the one in **Figure 17-43.** These tools can be purchased separately or in a kit like the one in **Figure 17-44.** If the carbide cutter blades become dull, they can be easily replaced. The cutting blades have angular teeth to give a smooth shearing cut as they are turned. See **Figure 17-45.** The pilot rod is placed in the valve guide first and the valve seat cutter is placed over the pilot. The

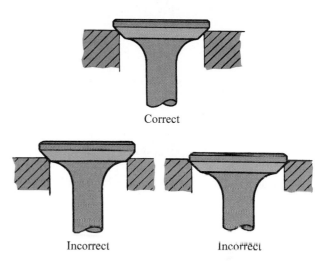

Correct

Incorrect   Incorrect

Figure 17-42. *Comparison of correct and incorrect location of seating area on valve face. (Deere & Co.)*

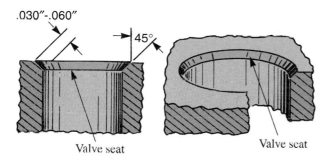

.030″-.060″

45°

Valve seat   Valve seat

Figure 17-40. *A typical integral valve seat. Hole for seat is bored in the block metal. Note valve seat angle and valve seat width.*

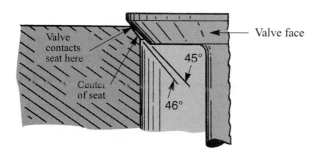

Valve contacts seat here

Valve face

Center of seat

45°

46°

Figure 17-41. *A 1° difference between valve face and valve seat provides better seating.*

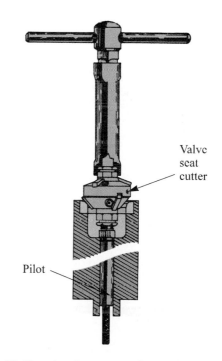

Valve seat cutter

Pilot

Figure 17-43. *A valve seat cutting tool with carbide cutting blades. This tool is used to recondition valve seats by hand. (Neway Mfg. Co.)*

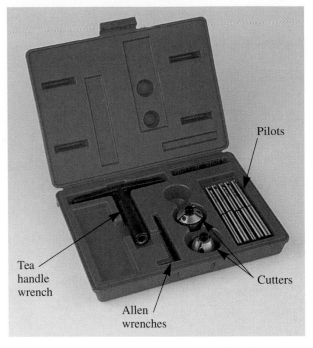

Figure 17-44. *A valve seat cutting tool kit. Cutter heads have carbide cutter blades that are very hard and will cut smooth and precise angles. Pilot rods accommodate various valve guide sizes. (Neway Mfg. Co.)*

Pilots

Cutters

Tea handle wrench

Allen wrenches

tee handle wrench is used to turn the cutter. Apply moderate downward pressure and turn the cutter clockwise only. Cutting action is controlled by steady pressure and even, smooth turning of the handle. When doing any cutting operation such as reaming valve guides, or cutting valve seats, it is important to thoroughly wash away any chips or grit afterwards. **Figure 17-46** illustrates the correct procedure to follow when using a valve cutter and pilot. **Figure 17-47** shows how to correct poor valve seat geometry.

Figure 17-45. *Valve cutter heads have angled cutting edges that provide a chatterless shearing cut for smooth valve seat surfaces. (Neway Mfg. Co.)*

Lapping valves

Good used valves may be reseated by *lapping*. Some manufacturers do not recommend lapping. When a valve expands, it does not seat in the same place that it does when it is cold. Therefore, any benefits from the lapping process will cancel out. Nevertheless, the procedure can be followed if you wish to seat valves this way.

Some engine manufacturers recommend hand lapping of the valve seats. Lapping compound is available from engine parts distributors. Some suppliers package it in two-compartment canisters. One compartment contains a coarse silicon carbide abrasive combined with a special grease. The second compartment contains a finer compound. The condition of the valve will dictate which grade to use.

If the coarse lapping compound is used, follow up with the finer compound. **Figure 17-48** shows how to apply the compound to the valve face only. The compound should not be allowed to contact the valve stem or guide. Next, a lapping tool is attached to the valve head by means of a suction cup. See **Figure 17-49.** The tool shown has a spring-loaded piston in the handle to help create suction. With the tool attached, the valve is placed in the guide and twirled back and forth. See **Figure 17-50.**

The lapping process is complete when a dark gray, narrow band, which is equal to the seat width, can be seen all the way around the valve face. Do not lap more than is necessary to show a complete seat.

After lapping, thoroughly clean the valve and valve seat chamber so that none of the abrasive finds its way into the engine. The best way to get the cleaning job done is to turn engine upside down and wash the chamber with solvent, from the bottom.

Valve seat inserts

Valve seat inserts can be removed and new inserts installed to replace them. The following illustrations show the procedure for removing and installing inserts for certain models of *Briggs & Stratton* engines. They are examples only, and it is imperative that the manual pertaining to a specific engine make and model be studied first.

If the engine is a cast iron block, the exhaust valve may already have an insert. The intake valve may, or may not, have an insert. If not, one may

Valve Seat Cutter Kit

GENERAL INSTRUCTIONS

SELECTION AND USE OF PROPER PILOT
A. SOLID PILOTS.

1. Select a pilot same diameter (fractional or metric) as valve guide.
2. Insert pilot in valve guide, twisting slightly, until very snug. Pilot shoulder should not touch valve guide. (Fig. 1)
   - If small, try next size larger.
   - If too large, try next smaller size.

DETAILED CUTTING INSTRUCTIONS

A. Slowly lower cutter to valve seat. DO NOT DROP CUTTER.

B. Turn clockwise and apply very light pressure. Release the down pressure at end of each cut. Make one or two turns with no pressure.

C. BOTTOM NARROWING OUT.

1. Cut lighly with narrowing cutter (usually 60°).

2. Cut until a fine continuous line is formed with valve seat. (Fig. 2).

D. TOP NARROWING CUT.

1. Cut lightly with narrowing cutter (usually 15°. For engines with hemispheric combustion chambers, use 30°/31°.)

2. Cut until seat width is slightly less than required. (This operation LOWERS THE SEAT.)

E. FINAL SEAT CUT.

1. Cut lightly, with seat cutter (usually 31° or 46°).

2. Cut seat to proper width. This should take only a few turns (Fig. 3)

F. INSPECT SEAT.

1. Remove pilot, using pilot puller. (Fig. 4)

2. Insert valve in valve guide.

3. Tap valve slightly up and down in the guide (holding it with fingers top and bottom —above and below the cylinder head). Do this until seat contact ring shows on the **valve face.**

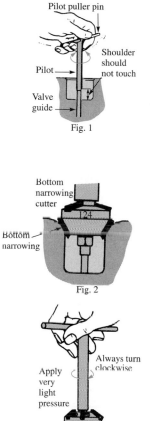

Fig. 1

Fig. 2

Fig. 3

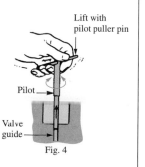

Fig. 4

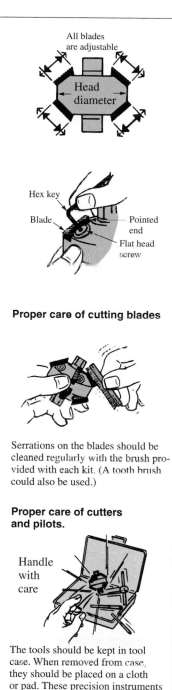

All blades are adjustable

Head diameter

Hex key
Blade
Pointed end
Flat head screw

Proper care of cutting blades

Serrations on the blades should be cleaned regularly with the brush provided with each kit. (A tooth brush could also be used.)

Proper care of cutters and pilots.

Handle with care

The tools should be kept in tool case. When removed from case, they should be placed on a cloth or pad. These precision instruments will last a long time if reasonable care is used.

Figure 17-46. *Proper procedures for using a valve seat cutting tool. (Neway Mfg. Co.)*

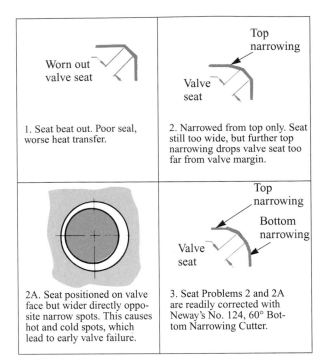

	Top narrowing
Worn out valve seat	**Valve seat**
1. Seat beat out. Poor seal, worse heat transfer.	2. Narrowed from top only. Seat still too wide, but further top narrowing drops valve seat too far from valve margin.
2A. Seat positioned on valve face but wider directly opposite narrow spots. This causes hot and cold spots, which lead to early valve failure.	**Top narrowing** / **Bottom narrowing** **Valve seat** 3. Seat Problems 2 and 2A are readily corrected with Neway's No. 124, 60° Bottom Narrowing Cutter.

Figure 17-47. *Poor valve seat geometry can be corrected with proper reconditioning procedures. (Neway Mfg. Co.)*

Figure 17-49. *To use a lapping stick, attach stick to valve head with suction cup. (Power-Grip Co.)*

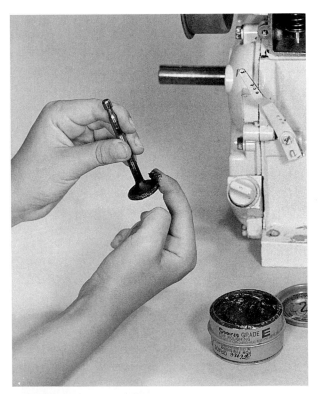

Figure 17-48. *Apply lapping compound to valve face before lapping face to seat.*

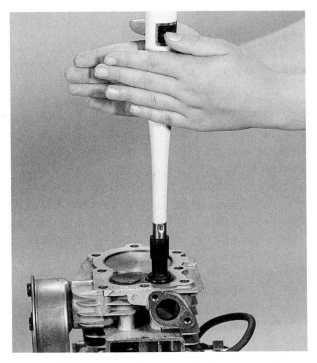

Figure 17-50. *Lap valve to seat by twirling lapping stick between palms of hands. Lift lapping stick and valve occasionally to increase cutting action of compound.*

have to be installed. To do so will require a pilot to be inserted into the valve guide as shown in **Figure 17-51.** A counterbore tool of correct diameter will be needed to cut a recess for the valve seat insert. See **Figure 17-52.** The counterbore cutter in **Figure 17-53** is designed so it will cut only to the correct depth. If another tool is used, a depth micrometer will be necessary to measure depth. Do not cut too deep.

To counterbore insert cylinder
1. Select the correct seat insert, cutter shank, counterbore cutter, pilot, and driver. Refer to engine manual, model number, and tool list.
2. Insert pilot in valve guide. See **Figure 17-51.**
3. Assemble the counterbore cutter shank. See **Figure 17-52.**
4. Counterbore cylinder by hand until cutter stop touches top of cylinder. See **Figure 17-53.**

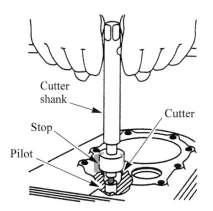

Figure 17-53. *When the stop of the counterbore tool reaches the surface, the correct depth has been obtained for a specific insert. The tool is turned clockwise with the tee handle cutter shank and is guided by the pilot. (Briggs & Stratton Corp.)*

Do not apply side pressure to the cutter to avoid cutting cylinder oversize.

5. Blow out all chips with compressed air.
6. Remove cutter from cutter shank with a knockout pin.
7. Remove pilot from valve guide.

Removing a valve seat insert
1. **Figure 17-54** shows a special tool used for pulling an insert out of an engine block. The puller nut must be of correct size so it does not score or damage the insert seat. It should pull only on the lower surface of the insert. It must also pass through the insert.
2. Hold the valve seat bolt and nut in place with one finger inserted through the port. Slide and center the puller body over the bolt. Tighten the bolt by hand.
3. The puller body must not rest on any part of the valve insert.
4. Turn the bolt with a wrench until the insert is pulled out of the cylinder. See **Figure 17-55.**

Installing new valve insert (cast iron block)
1. Select proper valve seat insert, pilot, and driver according to the engine manual. One side of the insert will be chamfered at the outer edge. This side should go down into the cylinder. See **Figure 17-56.**

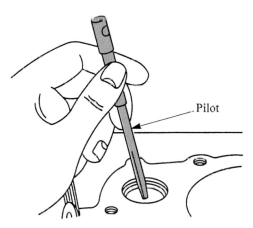

Figure 17-51. *Pilot rod being inserted in valve guide. (Briggs & Stratton Corp.)*

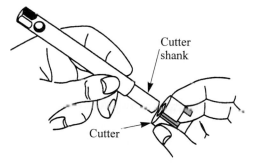

Figure 17-52. *A counterbore tool being installed on the cutter shank. The cutter cuts a cylindrical recess to accept a new valve seat insert. (Briggs & Stratton Corp.)*

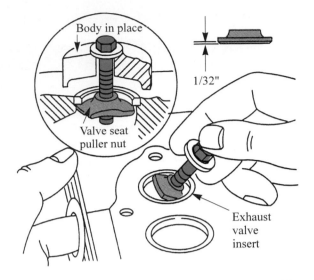

Figure 17-54. *A valve seat puller tool removes old valve seat inserts. The puller nut is inserted through the hole and held in place with a finger in the port. The puller body is placed on the bolt and over the insert. (Briggs & Stratton Corp.)*

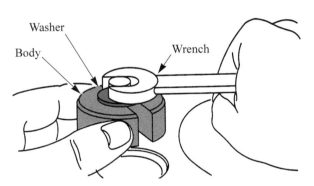

Figure 17-55. *When the nut on the puller is tightened with a wrench the insert is pulled up and out. (Briggs & Stratton Corp.)*

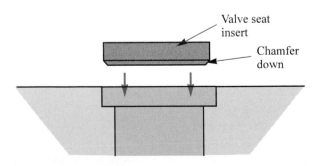

Figure 17-56. *When installing the insert the chamfered edge should be placed down in the cylinder recess.*

2. Insert the pilot in the valve guide.
3. Place the insert over the cylinder and the driver over the pilot. All should be aligned.
4. Drive the valve insert into place with the driver. See **Figure 17-57.**
5. Reface the seat as previously described in this chapter with a carbide valve seat cutter.

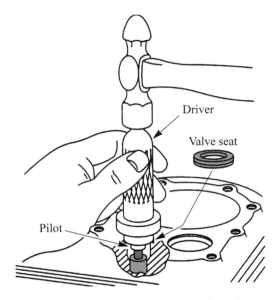

Figure 17-57. *Driving the valve insert into place with a driver tool and ball peen hammer. A pilot guides the driver tool. (Briggs & Stratton Corp.)*

Installing new valve insert (aluminum alloy block)

1. Use the old insert as a spacer between the driver and the new insert (aluminum alloy engine only). Drive new insert until it bottoms. Top of insert will be slightly below cylinder head gasket surface. See **Figure 17-58.**
2. Use a 1/8″ diameter flat ended drift or pin punch to peen around the insert to secure it as shown in **Figure 17-59.**

Valve lifter-to-stem clearance

Valve clearance refers to the space between the end of the valve stem and the top of the valve lifter when the valve is closed. The amount of clearance needed depends upon engine design and use. Due to hotter operation, the exhaust valve often requires more clearance than the intake valve. Clearances of around .008″ for the intake valve and .012″ for the exhaust valve are fairly common. Follow manufacturer's specifications.

When there is too little valve clearance, the valve may be held open when the valve stem heats up and lengthens (expands). As a result, engine performance is poor and both the valve face and valve seat will burn. See **Figure 17-60.** Insufficient clearance can also alter valve timing, making it too far advanced.

Too much valve clearance, on the other hand, will make valve timing late and reduce valve lift. This results in sluggish engine performance. It can also cause rapid lifter wear because of the pounding action involved. Under these conditions, the engine will be noisy and the valve could break. **Figure 17-61** shows a complete small engine valve train.

Refacing valves

Valve refacing can be done on a specially designed grinder. See **Figure 17-62.** The valve is revolved while being fed over an abrasive wheel. The collet that holds the valve is adjusted to achieve the desired face angle. Coolant flows over the valve head during grinding to reduce heat and produce a

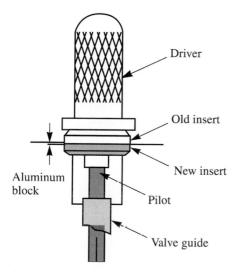

Figure 17-58 The old insert is used to drive in the new insert and seat it so that it will be slightly below the surface. (Briggs & Stratton Corp.)

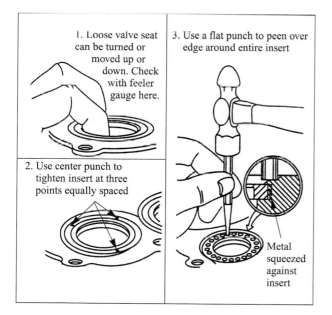

1. Loose valve seat can be turned or moved up or down. Check with feeler gauge here.

2. Use center punch to tighten insert at three points equally spaced

3. Use a flat punch to peen over edge around entire insert

Metal squeezed against insert

Figure 17 59. 1 If the space between the insert and cylinder are more than .005″ a new insert must be installed. 2—A center punch is used in three equally spaced locations to hold insert for peening operation. 3—A flat end punch about 1/8″ diameter is used to force metal against insert and hold it tightly in place. (Briggs & Stratton Corp.)

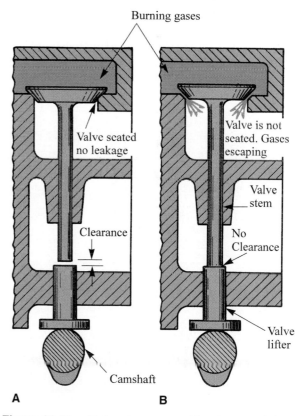

Burning gases

Valve seated no leakage

Clearance

Valve is not seated. Gases escaping

Valve stem

No Clearance

Valve lifter

Camshaft

A   B

Figure 17-60. Valve clearance setting is essential to good engine performance. A—Correct clearance permits valve to seat. B—Lack of clearance keeps valve open.

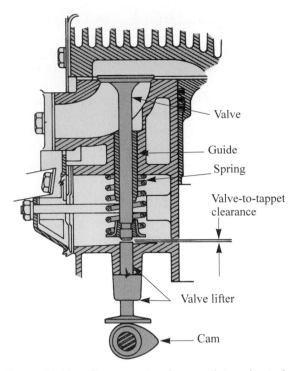

Figure 17-61. *Components of a complete valve train. (Kohler Co.)*

Labels in figure: Valve, Guide, Spring, Valve-to-tappet clearance, Valve lifter, Cam

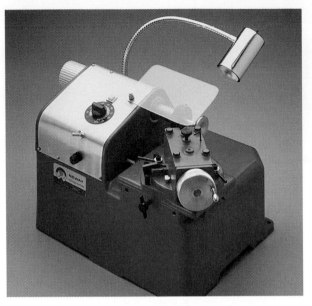

A

B

Figure 17-63. *A—A valve lathe is specially designed to cut and true the valve face with a carbide cutting tool. B—Closeup of a valve face being turned and trued on a valve face lathe.(Neway Mfg. Co.)*

good surface finish. The infeed wheel is used to precisely control the amount of material being ground from the valve face. In some cases, a lathe can be used to reface valves. See **Figure 16-63.**

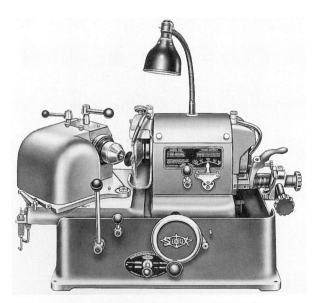

Figure 17-62. *During refacing operation, valve rotates slowly while being fed back and forth across the abrasive wheel. (Sioux Tools, Inc.)*

Valve refacing can also be done using a manual valve refacer. See **Figure 16-64.** The valve is placed tightly into a machinist's vise. Then, the valve refacer is placed on the valve. The crank lever turns the 45° (or 30°) cone against the stationary valve face. The cone has carbide blades that cut the valve face to the desired angle. The crank lever provides enough leverage to make the

Figure 17-64. *The handle of the manual valve refacer is turned by hand until the desired face angle is achieved. (Neway Mfg. Co.)*

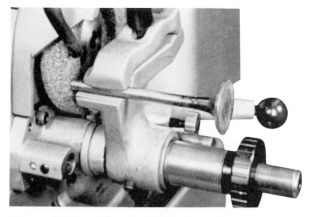

Figure 17-66. *Grinding the end of the valve stem trues the end surface and removes stock for clearance adjustment.*

job effortless. The second lever is used to control the feed rate of the blades.

When valve lifters become concave or deformed, they can be ground flat as in **Figure 17-65.** Likewise, the end of the valve stem can be dressed and shortened to produce correct valve clearance. See **Figure 17-66.**

Adjusting valve clearance

When a valve has been refaced, it rides lower in the guide, and, therefore, valve-to-tappet clearance is reduced. If the engine does not have adjustable tappets, the end of the valve stem must be ground to obtain correct clearance. To check clearance, turn the camshaft until the lobe is away from the tappet. Hold the valve against its seat while testing clearance with a thickness gauge, **Figure 17-67.** If there is too little clearance, remove the valve and grind .001″ or .002″ off of the stem. Repeat the clearance check and grinding operation until the clearance is correct.

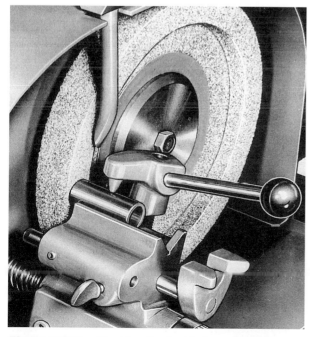

Figure 17-65. *Machine the head of a valve lifter as shown.*

Figure 17-67. *Valve stem-to-lifter clearance is checked with a thickness gauge.*

After the valves and seats have been properly reconditioned, place each valve in its respective guide. Use a valve spring compressor to compress the spring and, then, install the keepers. Make sure that the oil drainback holes are open before reinstalling the valve breather cover. See **Figure 17-68.**

Figure 17-68. *Make sure that all drainback holes are open before installing breather cover.*

Overhead Valve Systems

Many four-cycle engines use an ***overhead valve system,*** which transmits motion through ***pushrods*** and ***rocker arms*** to open and close valves. See **Figure 17-69.** The overhead valve

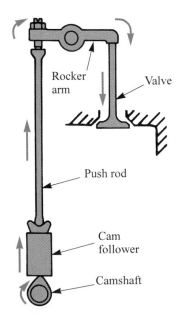

Figure 17-69. *Simplified overhead valve system.* (Deere & Co.)

design improves volumetric efficiency and eliminates combustion chamber hot-spots, which can cause cylinder distortion. Overhead valve systems increase fuel efficiency by as much as 25% and improve engine service life.

Overhead valve system disassembly

The following procedure should be observed when disassembling an overhead valve system:
1. Remove engine accessories that interfere with the removal of the valve cover.
2. Remove the valve cover bolts and the cover. See **Figure 17-70.** The rocker arm assembly should now be exposed.
3. Remove the rocker arm locking screws and nuts as shown in **Figure 17-71.** Remove the rocker arms. Some engines have rocker arms

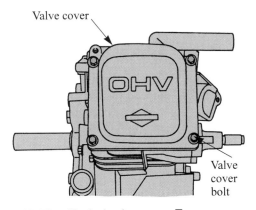

Figure 17-70. *Typical valve cover. To remove cover, loosen bolts on each corner. (Briggs & Stratton Corp.)*

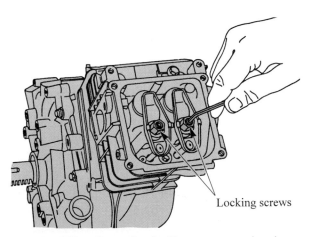

Figure 17-71. *Removing locking screws and rocker arms. (Briggs & Stratton Corp.)*

that pivot on a rocker arm shaft. See **Figure 17-72.** To remove, loosen the adjusting screws, remove the retaining rings, and slide the rocker arms off the shaft.

4. When the rocker arms are removed, lift out the pushrods. See **Figure 17-73.** (Note the location of each pushrod and do not interchange them when reassembling the engine.) After cleaning the pushrods, check them for straightness by rolling them on a flat, machined surface. If they are bent, replace them. Do not try to straighten pushrods.

5. Remove the cylinder head bolts and the cylinder head. See **Figure 17-74.**

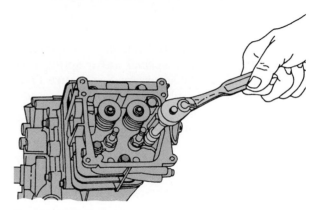

Figure 17-74. *Remove cylinder head by removing head bolts evenly and in slight increments. (Briggs & Stratton Corp.)*

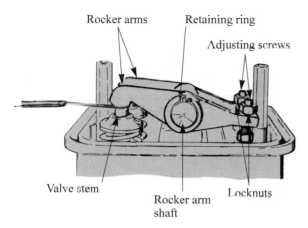

Figure 17-72. *Rocker arms on a rocker arm shaft. Loosen adjusting screws, remove retaining rings, and slide rocker arms off shaft. (Tecumseh Products Co.)*

Removing valves

To remove the valves in an overhead valve system, locate the cylinder head on a workbench and place small wooden blocks under the valve faces to hold them in place. If there are wear buttons or caps on the valve stems, remove them.

> When removing valves and valve springs, identify the parts to prevent interchanging them during reassembly.

On some engines, you can compress the valve springs by pressing on the spring retainers with your thumbs. Push the spring retainer toward the large end of its slot and release pressure. See **Figure 17-75.** Remove the retainer, spring, and valve stem seals. Discard the seals.

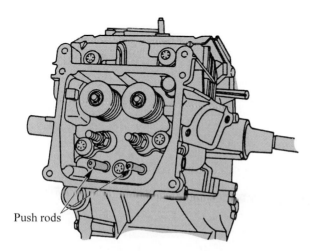

Figure 17-73. *Remove pushrods by lifting them out of holes. Check for straightness by rolling the rods on a flat surface. (Briggs & Stratton Corp.)*

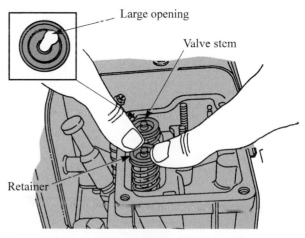

Figure 17-75. *Compress valve spring and move retainer to large opening to release. A wooden block placed under the valve head prevents valve movement. (Briggs & Stratton Corp.)*

If the valve springs cannot be compressed with your thumbs, a special valve spring compressing tool may be necessary. See **Figure 17-76**. In this illustration, split-type retainers are used to secure the valve springs to the valve stems.

When removing the valve springs, note that the coils are closer together on one end of the spring than on the other. These are called *dampening coils* and they should be located opposite the valve cap and retainers. See **Figure 17-77**.

Servicing overhead valves, seats, and guides

The valves, seats, and guides used in overhead valve systems are serviced in the same way

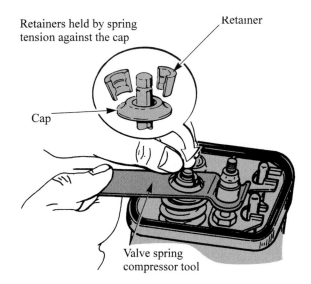

Figure 17-76. *Compressing valve spring with a valve spring compressor to remove split-type retainers. (Tecumseh Products Co.)*

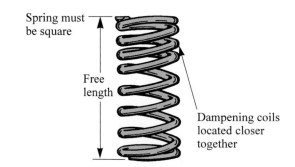

Figure 17-77. *Valve spring with dampening coils. Measure for correct length and tension. (Tecumseh Products Co.)*

as those in conventional systems. Valves should be cleaned and resurfaced to a 45° angle (or a 30° angle) on a valve grinding machine.

Valve seats can be reconditioned with a valve seat cutting tool. Valves should be lapped if recommended by the manufacturer. Thoroughly clean lapping compound from valve seats and faces. Inspect, measure, and test valve springs. Replace any parts that do not meet specifications.

Measure intake and exhaust valve guides. See **Figure 17-33**. If dimensions are not within specifications, the guides must be replaced. To remove worn guides, use a bushing driver or flat-ended pin punch. Support the cylinder head and press the guides out. See **Figure 17-78**. When pressing new guides into a cylinder head, press only to the specified depth. See **Figure 17-79**. This dimension will vary from one engine model to another.

Installing overhead valves

Before starting assembly, inspect valve stems for foreign material and burrs, which can cause sticking and damage the new stem seals. Coat the valve stems with valve guide lubricant. Do not allow the lubricant to contact the valve face, valve seat, or end of the valve stem.

Place the cylinder head on a workbench and support the valve faces with wooden blocks. Place the valve springs over the valve stems and set the retainers on the springs. Compress the springs and install the retainers. See **Figure 17-80**. If stem seals are used, place them over the stems as

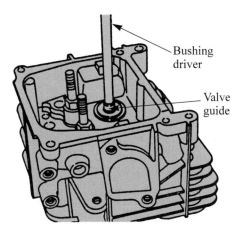

Figure 17-78. *Pressing valve guide out of cylinder head with a bushing driver or a flat punch. (Briggs & Stratton Corp.)*

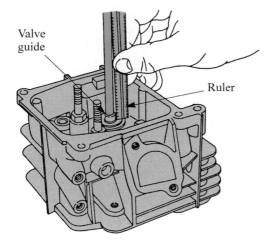

Figure 17-79. *Pressing new valve guides into cylinder head. Press to specified height above hole as shown. (Briggs & Stratton Corp.)*

required. Do not attempt to install the rocker arms until after the cylinder head is installed on the engine.

Installing cylinder head

The mating surfaces of the cylinder and the cylinder head should be completely clean. Use a new head gasket and place the cylinder head on the cylinder. See **Figure 17-81.** Never use gasket cement or sealer on a head gasket. Lubricate the cylinder head bolt threads with oil. Install the bolts

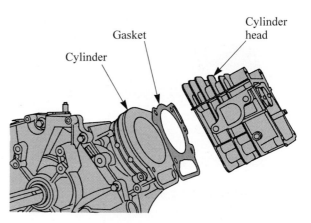

Figure 17-81. *Use a new gasket when installing the cylinder head on the block. Head and block surfaces must be clean. (Briggs & Stratton Corp.)*

through the head and into the cylinder block holes. Tighten the bolts evenly by hand and, then, use a torque wrench to tighten them to the proper torque specifications. See **Figure 17-82.** Torque the head bolts in sequential increments to avoid causing the cylinder head to warp. Finally, place the pushrods into their respective holes.

Installing rocker arms

Place the rocker arms on the studs and install the rocker arm nuts. Turn the nuts until they just touch the rocker arms. Carefully rotate the crankshaft to verify proper pushrod operation.

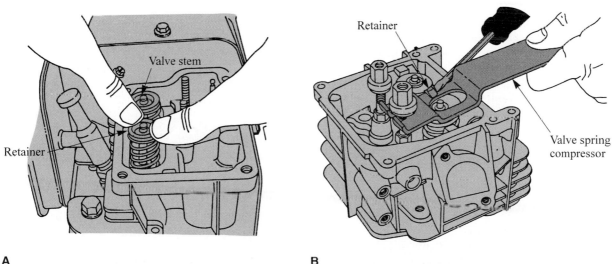

A   **B**

Figure 17-80. *A—Installing valve spring retainers. B—Replacing split-type retainers. A magnetized screwdriver or a bit of grease helps to place retainer onto the valve stem recess. (Briggs & Stratton Corp.)*

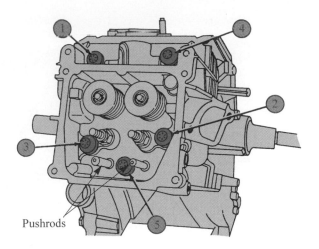

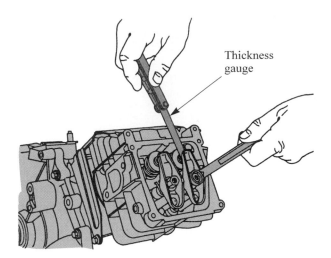

Figure 17-82. *Tighten cylinder head bolts with a torque wrench in proper sequence and in gradual increments to avoid head warpage. Place pushrods in proper holes.*

Figure 17-83. *Adjusting rocker arm-to-valve stem clearance. Very slight drag should be felt on the thickness gauge. (Briggs & Stratton Corp.)*

Adjusting valve clearance

Proper clearance between the rocker arm and the valve stem is essential. Too much clearance will reduce volumetric efficiency. Too little clearance can cause valve burning or warpage.

Before checking valve clearance, position the piston as recommended by the manufacturer. To accomplish this, simply rotate the crankshaft until the piston reaches the position specified. Top Dead Center may be the correct piston position for some engines; others may require the piston to be a certain distance beyond Top Dead Center. Always check specifications. If necessary, the distance past Top Dead Center can be measured through the spark plug hole with a ruler, dial indicator, or similar tool. Once the piston position is attained, place the proper feeler gauge leaf between the rocker arm and the valve stem. See **Figure 17-83.** Check engine specifications for the required clearance. Some engines require equal clearance for both intake and exhaust valves. However, some engine manufacturers use a different metal for exhaust valves than intake valves, and, therefore, the coefficient of thermal expansion is not the same for each valve. Therefore, clearances must be different for each valve.

Turn the locking/adjusting nut clockwise to reduce clearance or counterclockwise to increase clearance. The feeler gauge should drag slightly when pulled out. Hold the adjusting nut with a

wrench and tighten the locking screw slightly. Recheck clearance with the feeler gauge. If necessary, readjust until correct clearance is obtained. Tighten the locking screw.

Replace the valve cover and gaskets. See **Figure 17-84.** Tighten the valve cover bolts to the recommended torque setting.

> ⚠ Overtightening the valve cover bolts can warp the cover flanges, causing oil to leak. Be careful! See **Figure 17-85.**

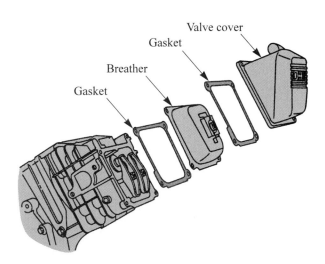

Figure 17-84. *Replacing valve cover. All mating surfaces should be clean and new gaskets should be used. (Briggs & Stratton Corp.)*

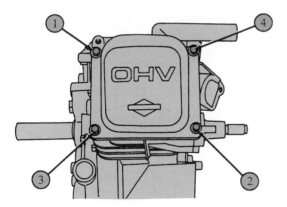

Figure 17-85. *When installing the valve cover bolts, tighten them in the proper sequence. Tighten to specified torque to avoid warping the cover flanges. (Briggs & Stratton Corp.)*

Ports, Reeds, and Rotary Valves

Two-cycle engines generally use *porting* of the cylinder wall (instead of poppet valves) in the fuel-feed and exhaust systems. Porting basically consists of two holes (ports) in the cylinder wall. One port admits the air-fuel mixture and the other port allows exhaust gases to escape.

Ports have no service requirements other than keeping them free of carbon. Unlike poppet valves, they cannot be replaced since they are part of the cylinder. *Reed valves* are used in two-cycle engines to control fuel flow from the carburetor to the crankcase, which serves as a second combustion chamber. A reed valve, operating on vacuum, opens during the compression stroke of the piston and closes before the start of the power stroke.

A reed valve is usually mounted on a plate to separate it from the carburetor. **Figure 17-86** shows a reed valve assembly used in an outboard engine.

The openings through the reed plate and the reeds must be kept clean. The surface of the plate must be smooth so that there is a good seal when the reed closes against it. If the reed valve has lost its springiness or tension, install a new one. High-speed, two-cycle engines use reed stops to prevent distortion and damage to the reeds. See **Figure 17-87.** If the reeds are bent, they must be replaced.

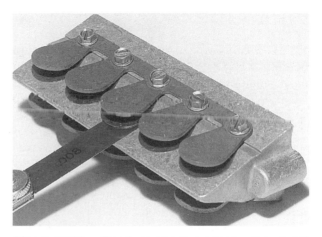

Figure 17-86. *Reed valves are not apt to stick shut if they are adjusted to stay open from .005″ to .010″ when at rest.*

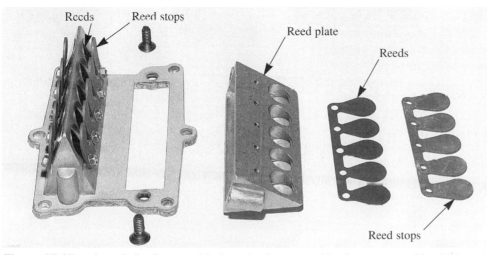

Figure 17-87. *A partially disassembled reed valve assembly shows ease with which parts can be replaced.*

Some two-cycle engines use rotary valves (instead of reeds) to control air-fuel intake. These valves are generally attached to the end of the crankshaft, although some run on a separate shaft geared off of the crankshaft. In operation, a valve rotates against a wear plate. Both the valve and plate have holes in them. When the holes align, an air-fuel charge enters the crankcase.

Inspect rotary valve ports for wear or damage. Replace defective parts. Keep openings clean and see that the surface of the wearplate is smooth and flat. If the wearplate or the mating surface of the valve is pitted or bent, replace the assembly. If a spring holds the parts in contact, check it against specifications for proper length and tension.

Camshafts and Gears

The *camshaft* of an engine is designed to operate the valves. A single camshaft is used in most small engines, with a cam (lobe) for each valve. When the camshaft rotates, the lobe of the cam lifts the valve from its seat.

Camshafts are made of steel or cast iron. The surface of the shaft is hardened to improve wearability. The ends of the camshaft may turn in bearings or in the block metal. See **Figure 17-88A.** Some small engine camshafts are hollow and have a second shaft running through them, **Figure 17-88B.** With this setup, the inner shaft is fixed and the hollow camshaft revolves on it.

Most small gasoline engines use gears to turn the camshaft. A gear on the crankshaft meshes with and drives a gear on the camshaft. Since the camshaft gear is exactly twice the size of the crankshaft gear, it runs at half crankshaft speed. See **Figure 17-89.** Also note that the timing mark on the cam gear is aligned with the keyway on the crankshaft. This is the correct procedure for timing valve operation to the crankshaft on this particular engine.

Automatic Compression Release

To make hand cranking easier, some small engines have an *automatic compression release* mechanism on the camshaft. This device lifts the exhaust valve slightly during cranking and releases part of the compression pressure.

One manufacturer's compression release mechanism is pictured in **Figure 17-90.** In view A, the camshaft is at rest and springs are holding the

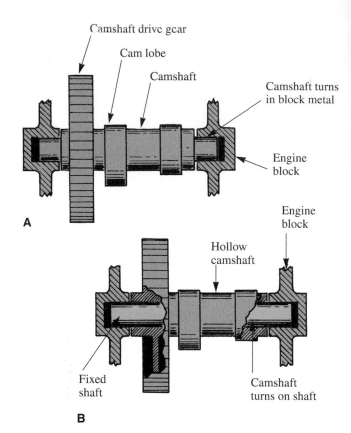

Figure 17-88. *Typical small engine camshafts. A—Solid camshaft. B—Hollow camshaft turning on a fixed shaft.*

Figure 17-89. *Camshaft gear is meshed with crankshaft gear so that timing marks are aligned. Camshaft turns at half crankshaft speed. (Deere & Co.)*

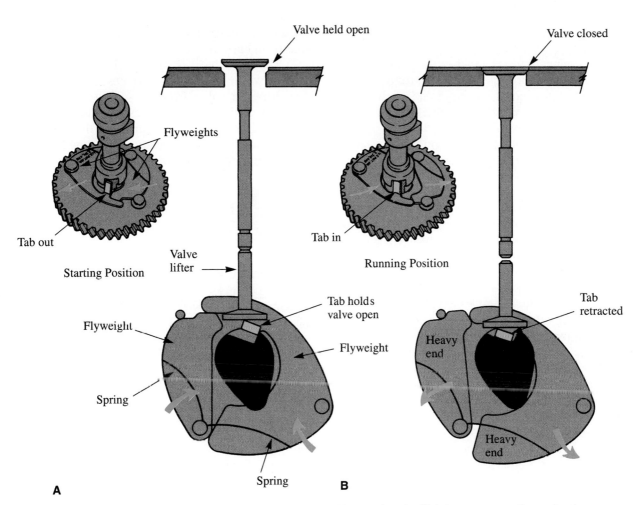

Figure 17-90. *Automatic compression release makes cranking easier. A—Tab is out, preventing valve from closing completely. B—When engine starts and reaches 600 rpm, the flyweights move out, the tab retracts, and the valve functions normally. (Kohler Co.)*

flyweights in. In this position, the tab on the larger flyweight protrudes above the base circle of the exhaust cam, holding the exhaust valve partially open. In view B, the tab prevents the exhaust lifter from resting on the cam.

After the engine starts and its speed reaches about 600 rpm, centrifugal force overcomes spring pressure and the flyweights move outward. Movement of the flyweights causes the tab to be retracted, and the exhaust valve seats fully. See views C and D in **Figure 17-90.** The flyweights remain in this position until the engine is stopped.

The compression release mechanism in the starting position is shown in **Figure 17-91A** and in the running position in **Figure 17-91B.** Automatic compression release is one of the many advances in small engines that ease the chore of engine start-up.

Figure 17-91. *Automatic compression release. A—Starting position. B—Running position.*

Summary

The connecting rod attaches the piston to the crankshaft. There are three types of friction bearings commonly used in the large end of the connecting rod: rod metal bearings, cast bronze bearings, and precision insert bearings. Many small engines use antifriction bearings in the large end of the connecting rod.

The crankshaft converts the reciprocating motion of the piston into rotary motion. To help offset the unbalanced condition created by the force of reciprocating mass, counterweights are added to the crankshaft. The crankshaft is supported by one or more main bearings. Crankcase seals prevent leakage of oil from areas where the crankshaft and crankcase come together.

A four-cycle engine's valve assembly includes the valve, the valve spring, and one or more retainers. After valves are removed, clean and inspect them for defects. Valves with serious defects must be replaced. Valve springs should be checked for squareness, length, and tension. Replace all springs that are not within specifications.

Check valve guides with a small hole gauge. If clearance between guide and stem exceeds the allowable limit, enlarge the guide with an adjustable reamer. A new valve with an oversize stem must be installed.

Valve seats are generally cut to a 45° angle. Seat contact must be near the center of the valve face. A valve seat cutter is used to recondition seats. Good used valves can be reseated by a hand-lapping process.

Valve clearance refers to the space between the end of the valve stem and the top of the lifter. Valves must be closed when measuring clearance. If there is too little clearance, a valve may be held open when the stem expands.

Valve refacing can be done on a specially designed grinder or with a manual valve refacer. Valve clearance is reduced when a valve is refaced. Therefore, the tappets must be adjusted or the valve stem end must be ground to obtain the correct clearance.

Overhead valve systems transmit motion through pushrods and rocker arms to open and close the valves. These systems improve volumetric efficiency and eliminate hot spots.

Two-cycle engines generally use intake and exhaust ports instead of poppet valves. Ports have no service requirements other than keeping them free from carbon.

The camshaft is designed to operate the valves. When the camshaft rotates, the lobe of the cam lifts the valve from its seat. To make hand cranking easier, some engines have an automatic compression release mechanism on the camshaft. This device lifts the exhaust valve during cranking to release compression pressure.

Know These Terms

connecting rod	valve guides
friction bearings	valve seat angle
bearing spread	valve seat width
bearing crush	interference angle
antifriction bearings	lapping
rod cap	valve clearance
crankshaft	valve refacing
crankshaft throw	overhead valve system
counterweights	pushrods
main bearings	rocker arms
bearing clearance	dampening coils
Plastigage	porting
crankcase seals	reed valves
poppet valves	camshaft
valve keepers	automatic compression
valve springs	release

Chapter 17 Review Questions

Answer the following questions on a separate sheet of paper.

1. Properly fitted friction bearing ends protrude slightly above the parting surface of the connecting rod cap. This characteristic produces what is commonly called _____
   a. bearing crush c. bearing seat
   b. bearing spread d. bearing swell
2. Bearing caps must never be _____ when being replaced on the rods.
3. What tool must always be used to tighten rod caps?
4. Name three types of crankshaft main bearings.
5. What do you call the special plastic substance used to measure bearing clearance?

6. When removing the crankcase cover from the engine _____.
   a. always pry it loose with screwdrivers
   b. pry out the oil seal first
   c. place tape over the keyway to protect the oil seal
   d. hold the cover while hammering on the crankshaft end
7. When replacing oil seals, the knife edge of the seal lip should face fluid being sealed. True or False?
8. After reaming a valve guide, bushing, or cutting a valve seat it is important to immediately _____.
9. Why is it important to select the correct bushing before reaming the block?
10. A good cutting fluid for reaming and seat cutting is _____.
11. Valve margins should not be allowed to be less than _____".
    a. 1/64   c. 1/16
    b. 1/32   d. 3/32
12. Valve seats that are too wide will _____. *Select correct answer(s).*
    a. cause valves to stick closed
    b. cause valves to stick open
    c. transfer too much heat to the block
    d. warp and burn
13. What is the 1° difference between the valve face angle and the valve seat angle called?
14. Name the process of placing abrasive compound on the valve face and twirling it back and forth in the valve seat.
15. Too little valve clearance will cause the valve to _____.
    a. break   c. burn
    b. be noisy   d. open late
16. The camshaft revolves at _____.
    a. twice crankshaft speed in all engines
    b. four times crankshaft speed in four-cycle engines
    c. one-half crankshaft speed in four-cycle engines
    d. one-half crankshaft speed in two-cycle engines
17. An automatic compression release is used on some engines to _____.
    a. make cranking easier
    b. prevent knocking due to excessively high compression
    c. control speed
    d. prevent overheating

18. What are the advantages of overhead valves?
19. Why should parts be identified as they are disassembled?
20. How should pushrods be checked for straightness?
21. When the coils are closer together on one end of a valve spring than on the other, they are called _____.
22. When new or reconditioned valves are being installed in guides, what should be placed on the valve stems first?
23. What tool is used to check valve clearance?
24. Why should valve cover screws never be tightened excessively?

Suggested Activities

1. Measure crankshaft bearing clearances with Plastigage and telescoping gauges.
2. Install new main and rod bearing inserts. Observe rules of cleanliness and torque tighten rod bolts to specified value.
3. Replace oil seals in the crankcase.
4. Grind valves on a grinding machine or turn and true valve faces on a valve lathe.
5. Recondition old valve seats with a valve seat cutter.
6. Lap valves into seats after the valve faces and seats are reconditioned by grinding.
7. Ream new valve guides in an aluminum block engine having aluminum guides.
8. Remove old valve guide bushings and install new guide bushings.
9. Demonstrate removing old valve seat insert.
10. Demonstrate counterboring cylinder for a new valve seat insert.
11. Demonstrate peening a valve seat on aluminum engine block.
12. Test valve springs for length, straightness, and tension.
13. Ream valve guides to fit an oversize valve stem with proper clearance.
14. Adjust valve lifter to valve clearance by grinding valve stems or adjusting tappets.
15. Adjust valve clearances in an overhead valve assembly.
16. Time the camshaft to the crankshaft.

Shown is a valve seat cutter in use. (Neway Mfg. Co.)

Lawn Equipment

After studying this chapter, you will be able to:
- ▼ List and conform to safe work practices.
- ▼ List features to consider when purchasing a lawn mower.
- ▼ Summarize basic lawn mower maintenance procedures and safety precautions.
- ▼ Describe proper method for storing a lawn mower for long periods of time.
- ▼ List the features to consider when purchasing a chain saw, a leaf blower, a string trimmer, or an edger/trimmer.
- ▼ Summarize the maintenance, safety, and storage procedures for chain saws, string trimmers, brushcutters, and edger/trimmers.
- ▼ Identify a variety of cutting blades for trimmers and brushcutters.

Working Safely

Safety is of primary importance when working on small gas engines and the implements they power. The following rules of safety are provided by a reputable manufacturer of lawn and garden implements. The safety rules pertain to a wide variety of implements that may occasionally need adjustment, or repair services. The list is only partial because it is impossible to predict every possibility that might be hazardous, cause accident or injury. The rules listed are broad. It is expected that persons servicing their own equipment, or the equipment of others, should at all times use good judgment and be familiar with safety precautions provided in manufacturer's technical service manuals relating to the specific piece of equipment being serviced.

1. To prevent accidental starting, always pull the high tension wire(s) off the spark plug(s) and remove plug(s) before servicing and/or adjusting the machine. If possible ground the high tension wire and tie it out of the way. See **Figure 18-1.**
2. To prevent injury, do not allow children or bystanders around the machine while it is being adjusted and/or serviced.

Figure 18-1. *Always pull the high tension wire off the spark plug and remove plug before servicing.*

3. Do not wear rings, wrist watches or loose fitting clothing when working on machinery; they could catch on moving parts causing serious injury. Wear sturdy, rough-soled work shoes. Never adjust and/or service a machine in bare feet, sandals, or sneakers.

4. Always wear safety glasses when using a hammer, chisel, or other tools that may cause chips or parts to fly.

5. Be sure to reinstall safety devices, guards, or shields after adjusting and/or servicing the machine.

6. When operating a power washer to clean a machine before servicing, be careful at all times to avoid injury. Maintain proper footing and balance at all times. Never direct the spray at people or animals, as high pressure spray can cause serious injury.

7. If a portable heater is used to heat the service area, the following precautions must be observed:

    a. Do not use portable heaters in presence of volatile materials such as gasoline or paint, as fire or explosion may result.

    b. To avoid being burned, do not touch the heater during operation.

    c. Portable heaters consume oxygen and combustion fumes can be hazardous. Heaters should be used only in a well ventilated area. Keep a window or door partially open to provide ventilation.

    d. Keep the heater at least four feet from combustible materials.

    e. Never use gasoline as a heater fuel.

8. Handle gasoline with care—it is highly flammable.

    a. Use approved gasoline container. See **Figure 18-2.**

    b. Never remove the fuel tank cap or fill the fuel tank when the engine is running, hot, or indoors. Also, do not permit smoking when working around flammable fuel.

    c. Avoid fires. Be sure fuel container or funnel, if metal, does not touch the battery. Do not overfill the fuel tank. Wipe up spilled gasoline and dispose of rags safely.

    d. Replace fuel tank cap securely.

9. Never use trouble lights or electric powered tools that have cut and/or damaged cords or plugs. Be sure all electric tools are properly grounded.

10. Never run an engine in a enclosed area such

Figure 18-2. *Always use an approved gasoline container for storing fuel.*

as a garage or storage building any longer than is necessary for immediate moving of the machine out of, or into, the area.

 Exhaust gases are toxic; carbon monoxide (CO). Opening doors and windows may not provide adequate ventilation.

11. After servicing be sure all tools, parts, or servicing equipment are removed from the machine.

12. Electrical storage batteries give off highly inflammable and explosive hydrogen gas when charging and continue to do so for some time after receiving a steady charge. Do not under any circumstances allow an electric spark or an open flame near the battery. Always disconnect a battery cable before working on the electrical system.

13. Hydraulic fluid escaping under pressure can have enough force to penetrate the skin. Hydraulic fluid may also infect a minor cut or opening in the skin. If injured by escaping fluid, see a doctor at once. Serious infection or reaction can result if medical treatment is not given immediately.

   Do not attempt to repair or tighten hoses that are under pressure, when a boom is raised, or with a tractor engine running. Cycle all hydraulic control valves to relieve all pressure before disconnecting lines or performing other work on a hydraulic system. Make sure all connections are tight and hoses and lines are in good condition before applying pressure to the system. To locate a leak under pressure, use a small piece of cardboard or wood. Never use hands.

14. When using an oxyfuel gas torch, always wear welding goggles and gloves. Keep a charged fire extinguisher within reach. Do not weld or heat areas near fuel tanks or fuel lines, and utilize proper shielding around hydraulic lines.

15. Always use safety stands in conjunction with hydraulic jacks or hoists. Do not rely on the jack or hoist to carry the load; it could fail. Always use a safety bar to block hydraulic cylinders.

16. When splitting tractors or disassembling machines, be sure to use safety stands and adequate supports to prevent tipping over.

17. Use a safety catch on all hoist hooks. Do not take a chance; the load could slip off the hook.

18. Use pullers to remove bearings, bushings, gears, cylinder sleeves, etc. when applicable. Use hammers, punches, and chisels only when absolutely necessary. Then, be sure to wear safety glasses.

19. Be careful when using compressed air to dry parts. Use an OSHA approved air blow guns, do not exceed 30 psi, wear safety glasses or goggles and use shielding to protect everyone in the work area.

20. Petroleum-base solvents, often used for cleaning parts, are flammable. Use care to avoid fire or explosion when using petroleum-base solvents.

Lawn Mowers

This chapter will examine a few of the more common small gas engine powered implements. It will discuss purchasing considerations, safety features, service warnings, and maintenance methods. Since there are hundreds of specialized powered implements, you should always study the owner's manual before operating and servicing these devices.

 Lawn mowers have caused a great number of injuries. The more severe injuries were lacerations to hands and feet and injuries from objects thrown from under the mower housing or out of the discharge opening.

Recently, laws have been enacted forcing manufacturers to provide certain safety devices on lawn mowers to help prevent accidents.

Like many other implements, there is not an age, skill, or intelligence requirement for using a lawn mower, However, young children should never be allowed to operate a power mower. The human body is no match for a sharp steel blade rotating at hundreds of miles per hour. Objects have reportedly been thrown a distance of one quarter of a mile by a power mower. A *grass discharge chute guard* is an important safety device that can prevent lawn mower injuries.

Knowing Mowing program

More than 60,000 injuries are treated by hospital emergency rooms each year due to unsafe operation of power lawn mowers according to the United States Consumer Product Safety Commission. Children under 15 years of age are the most frequent victims. To help youngsters learn safe mowing practices, Briggs & Stratton Corporation and the American Red Cross have teamed-up to create the national *Knowing Mowing* program. This program is designed to teach youngsters 12 years and older to safely operate and maintain power lawn mowers. The program, funded by Briggs & Stratton Corporation, consists of a 90-minute course offered by Red Cross chapters across the United States.

Course information includes:
- Parts of a power mower.
- Basic lawn mower maintenance.
- Safe mowing practices.
- Basic first aid.
- Yard waste recycling.
- A mowing obstacle course.

Participants receive:
- Certificate of accomplishment.
- Wallet card.
- Laminated garage poster of mowing *Do's* and *Don'ts*.
- *Knowing Mowing* baseball cap.

For information:
National Knowing Mowing Office
606 East Wisconsin Avenue
Milwaukee, WI 53202

Purchasing considerations

Because of the large variety available, certain considerations should be made when purchasing a power mower. The operator's strength may determine whether the mower should be the push type or the self propelled type. **Rotary mowers** (those having horizontally rotating blades) for large areas have large diameter blades with large, heavy engines and housings. Also, safety and accessory items may add to mower weight. Remember, a mower may roll easily on a smooth floor. However, it may be difficult to push on a deep, rough lawn when the grass catcher is loaded with grass.

If you have a super-quality, *showpiece* lawn that is very level, you may want to consider a **reel-type mower** (helical blades rotate around horizontal shaft). This is the type used on golf course greens. See **Figure 18-3.** They produce a high-quality job but may be more difficult and expensive to maintain. Special equipment is needed to sharpen the reel. Adjustments of the reel blades to the cutter bar are critical, and occasional readjustments are necessary as the blades wear. Cutter height is controlled by raising or lowering the rollers.

For the average size yard, a rotary-type mower with blade diameter of 22″ is usually satisfactory. For small yards, a 20″ mower is more maneuverable and takes less storage space. The length that the grass is cut to is controlled by raising or lowering the entire mower through adjustments on the wheels. See **Figure 18-4.**

Push-type and self-propelled mowers

Selecting a push-type or self-propelled mower is a matter of personal preference. Engines for self-propelled mowers have to rotate the blade and drive mechanism to the wheels. This requires additional engine horsepower and a drive mechanism to the wheels. In some mowers, power is transmitted to the wheels through shafts and gears. Today, most self-propelled mowers use a belt and pulley system. See **Figure 18-5.** Depending on which type of drive mechanism is used, greasing bearings and gears, or replacing belts are additional maintenance considerations.

Two-cycle or four-cycle engines

As discussed earlier in this text, two-cycle engines have fewer moving parts than four-cycle engines and require a proper mixture of oil and gasoline. Refer to Chapter 12 of this text for information on mixing oil and fuel. Four-cycle engines retain the oil in the crankcase and only gasoline is placed in the fuel tank. Crankcase oil must be drained and replaced at recommended intervals.

Both types of engines are quite reliable but be aware of the type you are purchasing. If it is a two-cycle engine, purchase two-cycle oil. You should also have an approved type of gasoline can that is labeled so you do not confuse it with another can of plain gasoline. Without the oil in the fuel, the two-cycle engine will be ruined in only a few moments of running.

Figure 18-3. *This walk-behind greens mower is a reel-type mower powered by a Kawasaki 3.7 horsepower engine. (Deere & Co.)*

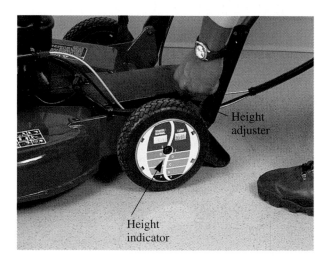

Figure 18-4. *Each wheel on this mower has an adjuster to raise or lower housing and blade for desired length of grass. Always stop engine before making these adjustments.*

A

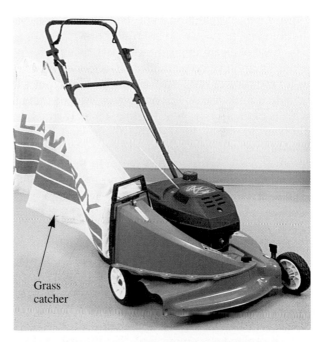

install. If mowing in narrow quarters, a rear bag may be slightly more maneuverable, particularly if the mower must be backed out of the space.

Using either a side or rear bag reduces thatch (compacted dead grass cuttings) in the lawn. Thatch, if allowed to accumulate, tends to smother the lawn and create growth problems.

> ⚠ To prevent serious injury, always stop the engine or blade when emptying the grass catcher. Never start the engine or blade without the grass catcher in place. See **Figure 18-6.**

Grass cuttings are excellent additions for the compost pile or garden. Some prefer to use a bagless mower and rake the grass cuttings after mowing. Again, the cuttings must be removed. Some mowers have blades and housings designed to cut and mulch (cut into fine particles) grass and leaves for composting.

Another optional accessory is a ***dethatcher blade.*** See **Figure 18-7.** This can be purchased to fit any mower and is installed in place of the cutting blade. There are spring-like fingers attached to the blade that reach down into the lawn. As the

B

Figure 18-5. *A—Self-propelled mowers are commonly driven by belts and a pulley system. B—The belts drives the rear axle of the mower.*

Optional features

One optional feature for mowers is side-bagging, rear-bagging, or no bagging at all. Grass bags may be fabric or molded plastic containers. Side-bagging mowers tend to extend the width of the mower but may be a little easier to remove and

Grass catcher

Figure 18-6. *Never have engine or blade running when grass catcher is removed. On this mower, blade brake and clutch stop blade while allowing engine to continue running.*

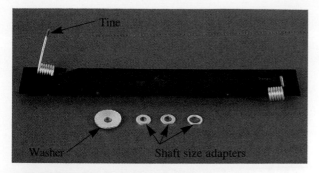

Figure 18-7. *A dethatcher blade uses spring tines (pointed prongs) to rake out dead grass. Adapters with various hole sizes permit bar to fit on any mower shaft diameter.*

Figure 18-8. *Bottom of mower should be kept clean for efficient cutting. Steel housings should be cleaned and painted occasionally to prevent rust.*

blade turns, the fingers rake the thatch up to the surface. The thatch is then collected in the grass bag or hand raked later.

Conveniences

Most manufacturers try to make bag removal and replacement as easy as possible. The bag usually must be removed a number of times before mowing is completed. This means stopping the mower, removing the bag, emptying the bag, replacing the bag, and restarting the mower. This is enough work without having to *wrestle* with a clumsy bag connection. The bag should also be durable and not wear on chaffing points.

Ease of cleaning the mower, particularly underneath, is another convenience. It is good practice to let the mower cool after mowing; then tip it on its side (crankcase down) and wash out the residue with a garden hose. **See Figure 18-8.** If grass is allowed to build up on the blade and in the housing, it will dry and become very difficult to remove. Also, cutting efficiency will be lessened. The upper parts of the mower and engine should also be kept clean to maintain engine cooling and proper functioning of the carburetor and governor parts.

Air filter maintenance should be easily done. See **Figure 18-9.** Also, the spark plug should be easily reached for cleaning or changing as needed. See **Figure 18-10.** Mufflers, like those on any internal combustion engine, eventually deteriorate to the point where they must be replaced. Ease and cost of replacement may vary. See **Figure 18-11.**

The *business end* of the mower is the cutting blade. After prolonged use, it will become nicked and dull. There are many styles of blades. To remove the blade for sharpening, the mower is tipped on its side (crankcase down) and several bolts are removed. See **Figure 18-12.** The mower should be set on its side so that it cannot fall upside down.

> ⚠ Always remove the spark plug wire and tie it back before tipping the mower. Wear gloves or wrap a cloth around the blade to protect hands from the sharp edges.

Storage and portability may be important. Today, many mowers are designed with handles that fold down so that little space is needed for

Figure 18-9. *Air filter must allow large amounts of air to enter the engine easily, while trapping dirt and debris. It should be easily removed for service.*

A

B

Figure 18-10. *A and B show two easy-to-access locations of spark plugs. Spark plugs should be serviced or replaced at least once each season. A spark plug socket should be used to prevent damage to porcelain insulator.*

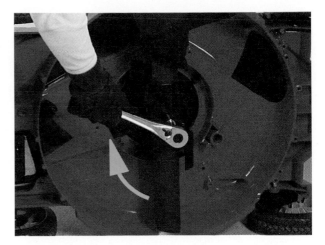

Figure 18-12. *Before removing a blade, always remove the spark plug wire and tie it away from the plug. Hands and knuckles should be protected with gloves while loosening or tightening bolts.*

storage. If transporting the mower in a car, this may also be convenient so that you can shut and latch the trunk lid.

Engine starting

Various mechanical means have been devised for starting small engines. The ***recoil starter*** is still used today. It utilizes a rope, a ratchet mechanism, and a rewind spring. When the rope is pulled, the ratchet engages the flywheel and rotates the crankshaft. When the engine starts, the ratchet disengages from the flywheel. The rewind spring

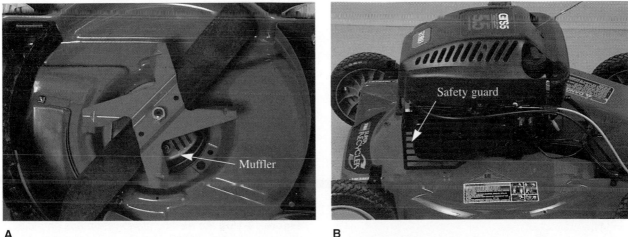

A   B

Figure 18-11. *A—Muffler on this mower requires removal of blade for replacement. It is tuned to the engine for quietness and efficiency. B—On this mower, the safety guard must be removed so the muffler can be replaced.*

retracts and recoils the rope for the next starting. The proper technique for starting is shown in **Figure 18-13.**

Because of convenience and safety concerns, the *extended rope starter* has become very common today. The advantage to this system is that the operator does not have to bend over as far and feet are clear of the blade housing. See **Figure 18-14.** Also, the blade brake must be released with one hand while the opposite hand pulls the rope.

Electric starts are becoming more common. The electric start mower has the additional components of a starter motor, switch, battery, and wiring. A key is required to turn on the switch. See **Figure 18-15.**

Inertia starters is a method that once was common. The inertia starter utilizes a coil spring attached to a ratchet mechanism and a crank or lever. The coil spring is wound tightly with the crank and held with a locking pin. When the control knob or lever is released, the coil spring engages the ratchet with the flywheel to start the engine. See **Figure 18-16.**

The simplest method has been the *independent rope starter.* A rope with a knot in one end is wrapped around the flywheel pulley and given a quick pull. This is no longer in use due to its many problems.

Blade brakes

One of the major safety features of every mower manufactured today is that the blade must automatically stop within three seconds after the operator's hands leave the handle. This requires a highly efficient braking system. There are basically two kinds of blade brakes.

In one system, a brake band wraps around and grips the flywheel to stop the engine and the blade. See **Figure 18-17.** The second system, as shown in **Figure 18-18,** has a clutch release that allows the engine to remain running while a brake stops the blade. Both systems utilize a *bail* (hand lever) hinged to the handle. In the position shown in **Figure 18-19,** the brake is engaged and the

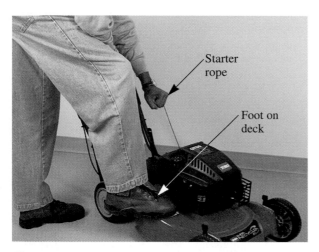

Figure 18-13. *When hand starting a vertical-pull engine, place one foot on the deck and the other foot away from the mower. Pull the rope briskly.*

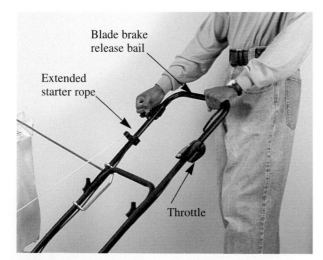

Figure 18-14. *Extended rope starting system helps keep hands and feet away from mower while starting.*

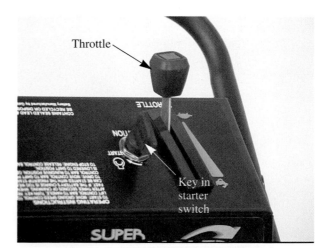

Figure 18-15. *An electric start system has an electric motor, battery, and switch, like an automobile. Most of these mowers have a hand start backup system in case the battery is discharged.*

x

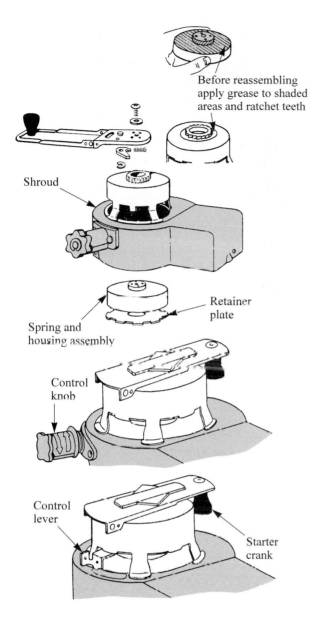

Before reassembling apply grease to shaded areas and ratchet teeth

Shroud

Spring and housing assembly

Retainer plate

Control knob

Control lever

Starter crank

Figure 18-16. *The inertia starter is used by winding a coil spring with the starter crank. When ready to start the engine, turn the control knob or move the control lever. Spring will then rotate the engine flywheel for starting.*

blade will not turn. When the bail is pulled into the handle, the blade begins to spin.

 In all cases, operators should thoroughly study the owner's manual and be familiar with all safety precautions and operating procedures before using a power mower.

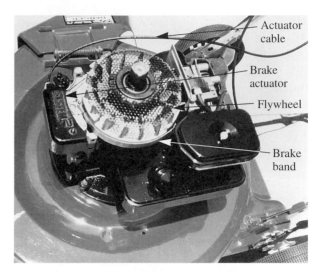

Actuator cable

Brake actuator

Flywheel

Brake band

Figure 18-17. *To stop engine and blade within three seconds of release of handle, this engine has a flywheel brake. Cover has been removed for clarity to expose brake band around flywheel.*

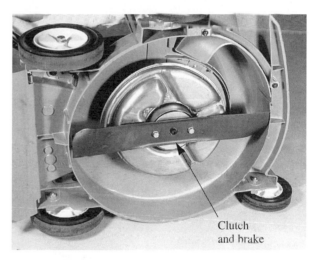

Clutch and brake

Figure 18-18. *This mower has a clutch that disengages the engine from the blade while the brake stops the blade. The engine can safely remain running when the operator releases the handle and bail.*

Procedure for starting an engine

The following is a basic procedure for starting a cold engine:

1. Fill fuel tank with proper fuel for engine type. Refer to **Figure 18-20.**
2. Check oil level and condition of oil. Add or change if necessary. See **Figure 18-21.**
3. Close choke or prime engine with fuel. See **Figure 18-22.**

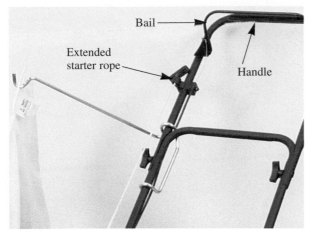

Figure 18-19. *Bail must be held against handle to release engine or blade brake.*

Figure 18-22. *Depressing primer forces fuel into cylinder for quickly starting a cold engine. Engines without a primer utilize a choke.*

Figure 18-20. *Check fuel level in tank before starting. (Vemco, Inc.)*

4. Turn key on or advance throttle to *Start* position. See **Figure 18-23.**
5. a. Vertical or Horizontal Rope Start: Place foot on mower deck and pull rope. See **Figure 18-13.**
   b. Extended Rope Start: Hold bail to handle and pull rope. See **Figure 18-14.**
   c. Electric Start: Turn key in switch to *Start*. Release key as soon as engine starts. See **Figure 18-15.**
6. When engine starts, open choke and adjust throttle to operating speed (about 1/2 to 2/3 maximum speed).

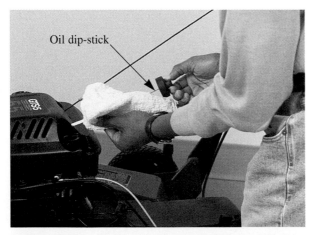

Figure 18-21. *This engine has a dip-stick for checking oil level. Oil is added through dip-stick opening.*

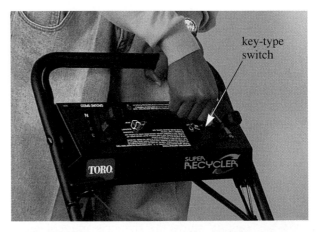

Figure 18-23. *Electric start mowers have a key-type switch. To start, insert key and turn to start position until engine begins to run. Release key immediately upon starting.*

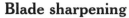

For restarting a warm engine, priming or choking should not be necessary. A cold two-cycle engine may need several prime pump strokes after the engine starts to keep it running.

Minor checks

If engine idles too fast or too slow, refer to manufacturer's service manual. In most cases, the idle speed can be adjusted with the idle speed screw on the carburetor with a screwdriver. The engine should idle slowly and smoothly.

If the engine will not start, remove the spark plug as shown in **Figure 18-24.** Test for spark as described in Chapter 14 of this text.

If a spark jumps the gap (3/16″) from the spark plug wire to the engine when the engine is cranked, replace the wire on the spark plug and ground the base of the plug against the engine (not close to gasoline tank). Have someone hold the bail and crank the engine. Watch for spark between the electrodes. If there is no spark, service or replace the spark plug.

If the engine still does not start, refer to the troubleshooting chart near the end of this chapter.

General maintenance

As mentioned earlier, cleaning the mower after each use is an important maintenance procedure. Additional maintenance procedures will be described here.

Figure 18-24. *If engine will not start, remove spark plug. With spark plug wire about 3/16″ from engine or mower, a blue spark should jump gap when engine is cranked. No spark indicates ignition system problem.*

Blade sharpening

When mower blades become dull and nicked, it makes the engine work harder and the lawn is cut poorly. Blades can be sharpened by clamping them to a table or in a vise and filing. Always retain the same edge angle. Blades can also be sharpened on a grinding wheel. See **Figure 18-25.**

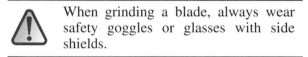

When grinding a blade, always wear safety goggles or glasses with side shields.

When sharpening the blade, balance it at the same time. Grind or file a small amount off of the heavy end until the blade balances horizontally. Test the blade by using a blade balancer, **Figure 18-26** or balance on a sharp edge, as in **Figure 18-27.** An unbalanced lawn mower blade will create damaging engine vibration.

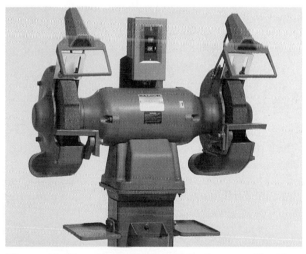

Figure 18-25. *Always wear safety glasses with side shields when grinding cutting edges of blade. (Baldor)*

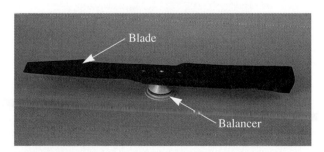

Figure 18-26. *This blade balancer is being used to test blade balance. An unbalanced blade can cause vibration and shorten engine life.*

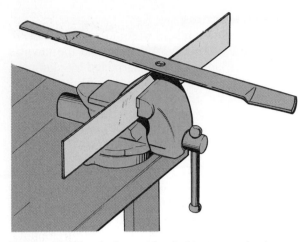

Figure 18-27. *A sharp-edged object can also be used for blade balancing. Center the blade hole over sharp edge.*

Hitting solid objects with the blade, such as large rocks, cement edges, pipes, etc., can be very damaging to a mower. See **Figure 18-28.** Besides damage to the blade, the engine crankshaft can be bent, crankcase housings distorted, connecting rods bent, and flywheel keys sheared.

Spark plugs

A spark plug is normally easy to change. The spark plug should be carefully removed with a spark plug socket and ratchet wrench so the porcelain insulator is not damaged.

Spark plugs can be cleaned and gapped as described in Chapter 14 of this text. When replacing

the spark plug, clean the seat and do not over-tighten the plug. Spark plugs should be serviced about every one-hundred hours of operation.

When replacing the spark plug, use only the kind specified for the engine with the proper reach (thread length) and heat range (insulator tip length).

Air Cleaners

Engines consume a tremendous amount of air, which passes through the air cleaner. When mowing, dust and dry grass materials are trapped by the filter. When too much debris collects in the air filter, it can restrict airflow. Therefore, the air cleaner element should be cleaned frequently; about every twenty-five hours of operation. Many lawn mowers use the oil-wetted type element, which is a spongy plastic saturated with oil.

To clean the oil-wetted filter:
1. Remove the filter cover.
2. Remove the foam element from the base.
3. Wash the element thoroughly in liquid detergent and water. Follow the steps in **Figure 18-29.**
4. Wrap the foam in cloth and squeeze dry.
5. Saturate the foam with engine oil.
6. Squeeze out the excess oil.
7. Replace the foam and filter cover.

Others use a pleated paper air filter cartridge. For descriptions of the other filter types, refer to Chapter 8 of this text.

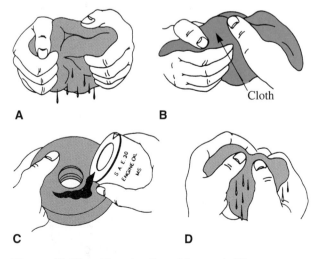

Figure 18-29. *Cleaning the oil-foam air filter. A—Wash foam element thoroughly in liquid detergent and water. B—Wrap foam in cloth and squeeze dry. C—Saturate foam with clean oil. D—Squeeze out excess oil.*

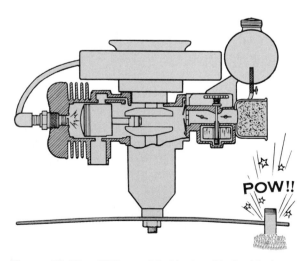

Figure 18-28. *Hitting solid objects with the blade can severely damage an engine.*

Changing oil

Four-cycle engines should have the oil changed about every twenty-five hours of operation. To drain oil from lawn mower engines:

1. Run engine (warm oil).
2. Stop engine; remove and ground spark plug wire.
3. Place a pan under the mower housing.
4. Locate and remove the oil drain plug.
5. Drain all of the oil and replace the drain plug.

To replace oil:

1. Obtain the kind of oil recommended by the manufacturer of the engine.
2. Pour oil into oil-fill neck that is provided and labeled. Fill to the level indicated by engine manufacturer. Some engines have dip-sticks. See **Figure 18-21.** Fill to *Full* line.
3. Replace the filler cap.

Muffler replacement

Engine mufflers get extremely hot and are subjected to the acidic products of combustion. They eventually need replacement. **Figure 18-30** shows a new muffler and one that needed to be replaced. The mufflers shown are removed by loosening the fasteners that hold them in place.

There are many styles and shapes of mufflers. The muffler shown in **Figure 18-11A** requires the removal of the blade before it can be replaced. The muffler shown in **Figure 18-11B** requires removal of safety guard before it can be replaced.

The muffler shown in **Figure 18-31** simply requires unscrewing tapered pipe threads on one end of muffler. When replacing these, coat threads with antiseize compound.

Threaded end of muffler

Figure 18-31. *Most mufflers are inexpensive and easy to replace. This type of muffler threads directly into the engine.*

V-belts

V-belts are a widely used means of power for self-propelled mowers. V-belts may look very much alike, but there are many unseen differences that may affect their quality and life span.

Some belts have a fabric cover; others do not. The cords embedded inside the belt may be rayon, polyester, or Kevlar. Rayon will not last as long as the tremendously strong Kevlar. Polyester tends to shrink as it gets hot. The shrinking of the polyester could present a danger, causing unintentional engagement due to shrinkage. Always, use only the type and size belt specified by the manufacturer. See *Appendix* for V-belt failures and corrections.

> ⚠ Always keep hands and objects away from any exposed belts when engine is running. Remove spark plug wire when servicing belts. Always keep belt guards in place when operating implements.

Storing a power lawn mower

Several things can be done when preparing to store a lawn mower for extended periods. Proper preparation will help ensure long mower and engine

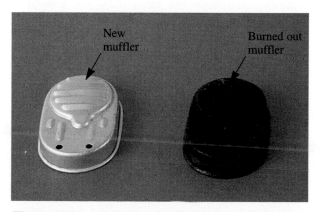

New muffler

Burned out muffler

Figure 18-30. *Compare the new muffler and the one that was in need of replacement. Combustion products and heat cause mufflers to corrode.*

life. It will also ensure easy starting the following season.

1. Clean the grass bag and hang it in a dry location.
2. Clean the mower of grass cuttings, mud, etc.
3. Avoid storing gasoline for long periods of time. Store only in approved *safety* containers. Never store fuel or mower in an enclosure where there is an open flame. If fuel must be stored, add a stabilizer to it. A fuel stabilizer is available from implement dealers.
4. Try to plan ahead and run the engine dry of fuel at the last use.
5. Drain the oil from the crankcase. Do not refill it now. Refer to Chapter 12 of this text for additional information on draining oil. Place a tag on the engine that says *No Oil.*
6. Rotate the engine so the piston is at bottom of the cylinder. Remove the spark plug. Spray storage oil through the spark plug hole, or squirt about one tablespoon of clean motor oil through the spark plug hole with an oil can. Rotate the engine slowly several times to distribute the oil on the cylinder walls. Replace the spark plug.
7. Leave the spark plug lead disconnected. Using the pull rope, rotate the engine slowly until compression resistance is felt. Then rotate the engine an additional one-quarter turn to close off its ports. This seals the cylinder and prevents moisture entry.
8. Leave the throttle in the off position (closed) and close the choke.
9. Lubricate the mower, drive system, etc., as described by the manufacturer.
10. Coat the cutting blade with chassis grease to prevent rusting.
11. Store the mower in a dry, clean area.

Removing a mower from storage
1. Replace the grass bag.
2. Fill crankcase with new oil, or make new mixture for two-cycle engine. See Chapter 12 for additional information on mixing oil and fuel.
3. Remove the spark plug. Using the pull rope or starter, spin the engine rapidly to remove excess oil from the cylinder. Clean or replace the spark plug.
4. Clean and oil the air filter or replace cartridge if necessary.
5. Fill fuel tank.

6. Start the engine and idle until warm. Adjust the idle speed if necessary.
7. Increase the engine speed in normal manner.
8. Make a brief test run while listening to the engine and watching the condition of all parts.
9. If the engine does not start, review the troubleshooting chart in **Figure 18-32** and the other chapters of this text.

Chain Saws

Small gasoline engine powered *chain saws* have become very popular for cutting firewood and trimming trees. See **Figure 18-33.** Like any cutting tool, they work best when properly maintained. To avoid hazards, all safety devices should be in place and safe operating procedures must be carefully followed.

Purchasing considerations

When investing in a chain saw, there are a number of things to consider so that you will be satisfied with its performance and operation. Chain saws are manufactured in a number of sizes from about 10″ to over 40″ of blade length. The type of work to be done is the best indicator for size selection. A smaller saw works well for cutting branches, small trees, and fireplace logs. A large professional model would be for big trees and continuous rugged work.

The lighter a chain saw is, the easier it is to handle, control, and carry. Chain saws range in weight from less than 10 pounds (lb) to over 25 lb. Weight is a consideration that should be planned according to the type of work to be done. When trimming tree branches from a ladder, a lighter saw will be less tiring.

Chain saws also come with different size engines. Rugged work and long continuous cutting require powerful engines.

Some states have laws that require certain safety devices to be used on every chain saw. Check which devices are required and be sure the saw is properly equipped.

Safety features

Safe operation of a chain saw comes from a good knowledge of correct operating procedure. However, there are a few safety features on many chain saws that are very important.

Troubleshooting Chart

Problem	Cause	Remedy
1. Engine fails to start.	A. Blade control handle disengaged. B. Check fuel tank for gas. C. Spark plug lead wire disconnected. D. Throttle control lever not in the starting position. E. Faulty spark plug. F. Carburetor improperly adjusted, engine flooded. G. Old, stale gasoline. H. Engine brake engaged.	A. Engage blade control handle. B. Fill tank if empty. C. Connect lead wire. D. Move throttle to start position. E. Spark should jump gap between center electrode and side electrode. If spark does not jump, replace the spark plug. F. Remove spark plug. Dry the plug. Crank engine with plug removed, and throttle in off position. Replace spark plug and lead wire and resume starting procedures. G. Drain and refill with fresh gasoline. H. Follow starting procedure.
2. Hard starting or loss of power.	A. Spark plug wire loose. B. Carburetor improperly adjusted. C. Dirty air cleaner.	A. Connect and tighten spark plug wire. B. Adjust carburetor. See separate engine manual. C. Clean air cleaner as described in separate engine manual.
3. Operation erratic.	A. Dirt in gas tank. B. Dirty air cleaner. C. Water in fuel supply. D. Vent in gas cap plugged. E. Carburetor improperly adjusted.	A. Remove the dirt and fill tank with fresh gas. B. Clean air cleaner as described in separate engine manual. C. Drain contaminated fuel and fill tank with fresh gas. D. Clear vent or replace gas cap. E. Adjust carburetor. See separate engine manual.
4. Occasional skip (hesitates) at high speed.	A. Carburetor idle speed too slow. B. Spark plug gap too close. C. Carburetor idle mixture adjustment improperly set.	A. Adjust carburetor. See separate engine manual. B. Adjust to .030". C. Adjust carburetor. See separate engine manual.
5. Idles poorly.	A. Spark plug fouled, faulty, or gap too wide. B. Carburetor improperly adjusted. C. Dirty air cleaner.	A. Reset gap to .030" or replace spark plug. B. Adjust carburetor. See separate engine manual. C. Clean air cleaner as described in separate engine manual.
6. Engine overheats.	A. Carburetor not adjusted properly. B. Air flow restricted. C. Engine oil level low.	A. Adjust carburetor. See separate engine manual. B. Remove blower housing and clean as described in separate engine manual. C. Fill crankcase with the proper oil.
7. Excessive vibration.	A. Cutting blade loose or unbalanced. B. Bent cutting blade.	A. Tighten blade and adapter. Balance blade. B. Replace blade.

Figure 18-32. *Study this engine troubleshooting chart.*

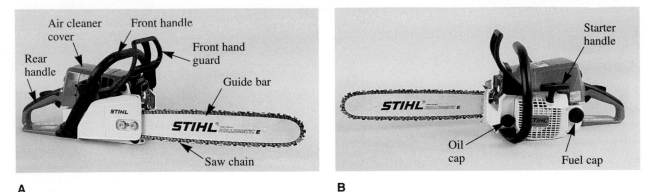

Air cleaner cover

Front handle

Rear handle

Front hand guard

Guide bar

STIHL

STIHL ROLLOMATIC E

Saw chain

A

Starter handle

STIHL ROLLOMATIC E

STIHL

Oil cap

Fuel cap

B

Figure 18-33. *Study the parts of a gasoline engine-powered chain saw.*

When the unshielded nose of a chain saw hits a solid surface, the spinning chain may cause the saw to fly back toward the operator. This is known as kickback and may be very dangerous. Some chain saws have a tip guard device that is attached to the end of the blade that helps prevent kickback. It can be installed or removed very quickly.

Another safety feature on chain saws is the quick-stop device. Illustrated in **Figure 18-34,** this device stops the chain to reduce the possibility of injury. With a sudden kickback, the operator's left hand moves forward to make contact with the front hand guard and the chain is stopped. The front hand guard is the quick-stop activating lever.

When carrying the chain saw by hand, the engine must be stopped and the saw must be in the proper position. Grip the front handle and place the muffler to the side away from the body. The chain guide bar should be behind you.

Some chain saws come with a case that protects the saw during transportation and storage. See **Figure 18-35.** Another safety feature is the *chain guard (scabbard)* shown in **Figure 18-36.**

The chain guard protects the operator from the sharp blade when not in use and protects the blade from moisture and rust.

Hold the chain saw with both hands when cutting. See **Figure 18-37.** You should also wear heavy gloves to protect your hands and dampen vibration.

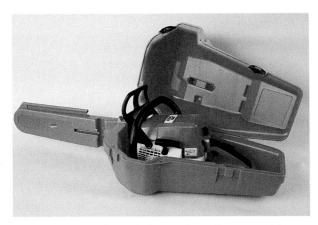

Figure 18-35. *This chain saw is well protected in a tough, plastic carrying case.*

Front hand guard

STOP

Figure 18-34. *When operator's left hand makes contact with front hand guard, it will activate the quick-stop to stop chain and reduce risk of injury.*

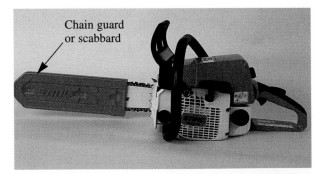

Chain guard or scabbard

Figure 18-36. *A chain guard (scabbard) protects saw blade during transportation and storage.*

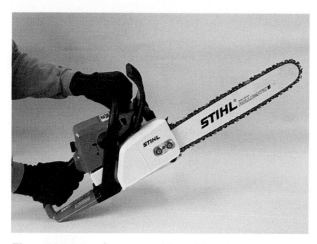

Figure 18-37. *Correct position of hands when operating chain saw. Note use of heavy gloves for protection.*

Another safety feature required by some states is the **spark arrestor.** The spark arrestor is built into the exhaust system to prevent sparks from the system from catching dry grass or wood chips on fire. These devices are sometimes known as fire arrestor screens.

Rules for safe operation

The following rules for the operation of a chain saw should be followed carefully.

1. Never operate a chain saw when you are tired.
2. Use safety footwear; snug-fitting clothing; and eye, hearing, and head protection devices.
3. Always use caution when handling fuel. Move the chain saw at least 10′ (3m) from the fueling point before starting the engine.
4. Do not allow other persons near the chain saw when starting or cutting. Keep bystanders and animals out of the work area.
5. Never start cutting until you have a clear work area, secure footing, and a planned retreat path from the falling tree.
6. Always hold the chain saw firmly with both hands when the engine is running. Use a firm grip with thumb and fingers encircling the chain saw handles.
7. Keep all parts of your body away from the saw chain when the engine is running.
8. Before you start the engine, make sure the saw chain is not contacting anything.
9. Always carry the chain saw with the engine stopped, with the guide bar and saw chain to the rear, and the muffler away from your body.

10. Never operate a chain saw that is damaged, improperly adjusted, or is not securely assembled. Be sure that the saw chain stops moving when the throttle control trigger is released.
11. Always shut off the engine before setting the chain saw down.
12. Use extreme caution when cutting small brush and saplings because slender material may catch the saw chain. This could fling the saw toward you or pull you off balance.
13. When cutting a limb that is under tension, be alert for springback so that you will not be struck when the tension in the wood fibers is released.
14. Keep the handles dry, clean, and free of oil or fuel mixture.
15. Do not operate the chain saw with a deteriorated or removed muffler system. Fire-preventing mufflers (fire arrestor screen types) should be used in dry areas.
16. Operate the chain saw only in well ventilated areas.
17. Do not operate a chain saw in a tree unless you are specially trained to do so.
18. Guard against kickback. Kickback can lead to severe injuries.

Avoiding kickback

1. Hold the chain saw firmly with both hands. Do not reach too far.
2. Do not let the nose of the guide bar contact a log, branch, ground, or any other obstruction.
3. Cut at high engine speeds.
4. Do not cut above shoulder height.
5. Follow manufacturer's sharpening and maintenance instructions for the saw chain.

Chain saw maintenance

Careful maintenance of the chain saw engine and chain provide for long-lasting service life and safe use. Never operate a chain saw that is damaged, improperly adjusted, or not completely assembled.

 Always stop the engine and be sure that the chain is stopped before doing any maintenance or repair work on the saw.

Fuel and carburetor

Always use the correct gasoline and oil mixture for a two-cycle engine as recommended by the manufacturer. Before refueling, carefully clean the filler cap and the area around it to ensure that no dirt falls into the tank. See **Figure 18-38.**

Do not adjust the carburetor unless it is necessary. The high speed and low speed carburetor adjustments on a chain saw engine are very critical. Incorrect settings of these speeds can cause serious damage to the engine. If adjustments become necessary, follow the suggested manufacturer's recommendations.

If the engine stops while idling, the exhaust smokes, or engine does not run smoothly, try adjusting the carburetor. Normally, you must turn the adjusting screw clockwise when the idle setting is too rich and counterclockwise when the setting is too lean. See **Figure 18-39.**

Figure 18-38. *Clean area around filler cap before refueling engine.*

Figure 18-39. *Turn idle speed adjusting screw with a screwdriver to obtain correct idle.*

Apart from minor adjustments, carburetor repairs should be made by a trained technician who has all the necessary service tools and equipment.

Cylinder fins

Check the cylinder fins periodically since clogged fins result in poor engine cooling. Remove the dirt and saw dust from between the fins to allow cooling air to pass freely. This can be done with a small stick, brush, or compressed air. See **Figure 18-40.**

Figure 18-40. *Cylinder fins can be cleaned with a compressed air.*

Air cleaner (filter)

The function of the air filter is to catch dust and dirt in the inlet air and reduce wear on engine components. Clogged air cleaners cut down on engine power, increase fuel consumption, and make starting more difficult. The air cleaner should be cleaned every day in very dusty operating conditions.

Before removing the air cleaner, close the choke valve so that no dirt can get into the carburetor. See **Figure 18-41.** Remove the air cleaner cover and remove the element. See **Figure 18-42.** Lightly brush off dust or wash cleaner element in cleaning solvent if extremely dirty. It should be dried completely before replacing it on the engine.

Fuel filter

Check the fuel filter periodically. A clogged fuel filter will cause engine trouble, such as hard starting. Remove the fuel filler cap and fish for the

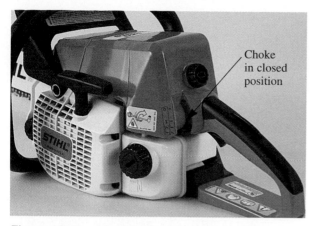

Figure 18-41. *Move choke into closed position so dirt cannot enter carburetor.*

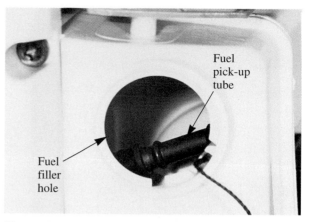

Figure 18-43. *Flexible fuel pick-up tube is removed through fuel filler hole.*

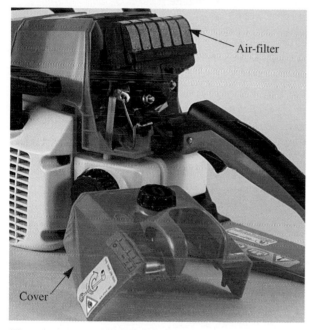

Figure 18-42. *This air filter cover was removed and the air filter is ready for cleaning.*

flexible pick-up tube with a hook. Pull the fuel pick-up out. See **Figure 18-43.** Remove the old filter sleeve and slide a new sleeve in place. Return the flexible pick-up tube to the gas tank.

Lubrication

The saw chain and guide bar must be continuously lubricated during operation to protect them from excessive wear. This is provided for by the automatic chain oiling system. Clean the lubricating oil supply hole daily. It is located at the base

of the blade. Always fill the oil tank with chain oil each time the engine is refueled. See **Figure 18-44.** *Never* use waste oil for this purpose.

Muffler

The muffler should be kept clean and open. Do not run a chain saw without the muffler. See **Figure 18-45.** If regulations require the use of a spark arrestor, check its condition periodically. Carbon deposits in the muffler and cylinder exhaust port will cause lower engine power output and sparking from the muffler. If needed, remove the muffler and clean the cylinder exhaust port. Use a wood stick or dowel to protect the metal surfaces.

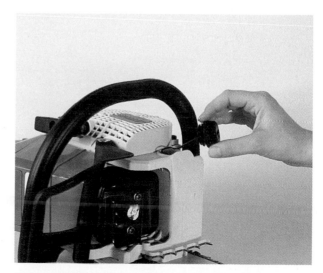

Figure 18-44. *Removing oil tank cap to fill tank with chain oil.*

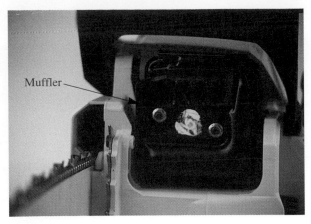

Figure 18-45. *Do not run a chain saw without the muffler.*

Spark plug

If the engine does not start, it may be due to a wet, fouled, or faulty spark plug. Check the spark plug periodically and clean or install a new one as necessary. See **Figure 18-46.** Adjust the spark gap if it is wider or narrower than the standard gap. Be sure the stop-start switch is in the *Off* position when checking the spark plug.

Guide bar

Clean the *guide bar* daily or before each use of the chain saw. Remove any burrs that may be found along the bar rails. On most chain saws, the roller nose bearing must be lubricated. Place the chain saw on its side so that the bar nose is firmly supported. Clean the grease hole and pump grease in as shown in **Figure 18-47.**

Figure 18-46. *The spark plug should be frequently checked and cleaned. If necessary, install a new one.*

Figure 18-47. *Lubricate roller nose bearing with a grease pump as shown.*

Storage

Inspect and make adjustments of every part of the chain saw before storage. Clean all parts and apply a thin coat of oil to all metal surfaces to prevent rust. Remove the chain and the guide bar. Apply a sufficient oil coat and wrap them in a plastic bag. Drain the fuel tank and pull the starter a few times to drain the carburetor.

Pour a small amount of oil in the spark plug hole and replace the spark plug. Slowly pull the starter to crank the engine a couple of revolutions. Place the saw in its case and store it in a dry, dust-free area.

String Trimmers and Brushcutters

String trimmers and *brushcutters* are very popular items for homeowners, landscaping contractors, and foresters. See **Figure 18-48.** As with any type of small gas driven machines, safety and proper use of string trimmers and brushcutters are major concerns. These machines also require proper maintenance for assurance of efficient performance.

Purchasing considerations

When selecting a string trimmer or brushcutter, the kind and amount of use will determine the features that are most important. Listed are some of the features available and other items to consider:

Figure 18-48. Shown is a person wearing proper safety gear and operating a string trimmer properly. (Stihl)

- Engine power should be adequate for the kind of work. Brushcutting requires more power than grass cutting for example. The manufacturer determines proper power for each model.
- Construction should be durable, yet lightweight. If use will be for long periods, weight becomes very important and a shoulder harness may be needed.
- Interchangeable cutting heads. Changing from nylon string to circular saw blades should be easily accomplished.
- Air-cooled, two-cycle engine with electronic ignition for easy, dependable starting in all weather conditions.
- Antivibration system.
- Smooth running engine with a minimum of vibration.
- Rear discharge, shielded muffler.
- Easy to service air filter.
- Straight drive shaft with gear drive at cutter head for straight shaft machines.
- Straight drive shaft with a flexible shaft for machines with curved end at cutter head.
- Centrifugal clutch that stops blade when engine idles.

- Fuel tank that is easy to fill.
- Engine controls and throttle should be conveniently built into the grip.
- Handles adjustable to suit operator's reach and height with good balance.

Handles and cutters

The string trimmers and brushcutters may have a straight shaft or one that is curved at the end. String trimmers most commonly have a loop-type handle. See **Figure 18-49.** Bike- and J-type handles are available for brushcutters with interchangeable cutting blades.

> ⚠ The loop style handle should never be used for cutting blades.

The versatility of string trimmers and brushcutters is due to the great variety of cutters that can be installed on the cutting head of the machine. See **Figure 18-50.** Cutters are available from nylon string types for grass, small weeds, and large weed to multiple toothed steel blades with from three teeth to circular saw blades for small trees.

Grass cutting

The most basic cutting end is the rotating string cutting head. The cutting head contains nylon string, which is available in various strengths and rigidity. The nylon string cutting heads are intended to supplement a lawnmower. They are used primarily for trimming grass around obstacles where a lawnmower will not cut, such as around flower beds, shrubbery, and fences. When replacing the string, always use the manufacturer's recommended type for the specific make and model of trimmer.

A deflector mounted to protect the operators feet and legs is fastened at the cutting end. The deflector has a cutting blade, or chopper blade, to cut the string to the correct length. See **Figure 18-49.**

Brush cutting

A brushcutter may be fitted with a brush knife. The brush knife is suitable for applications ranging from cutting matted grass to clearing weeds, wild growth, and scrub. Brush knives are used for cutting young saplings or other woody materials up to 3/4″ (2cm) in diameter.

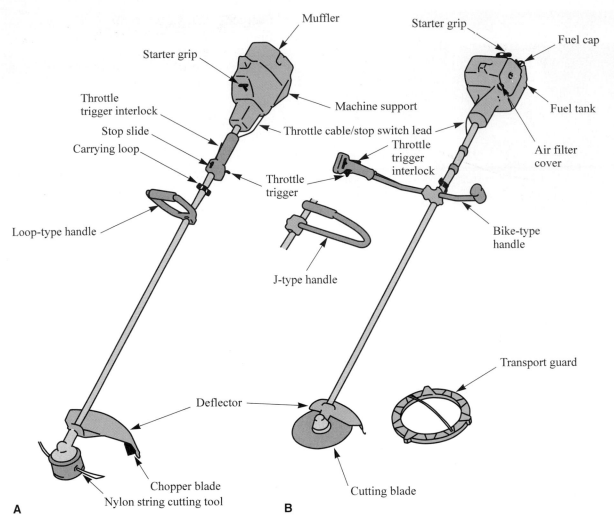

Figure 18-49. *Major parts of a string trimmer. A—This string trimmer is equipped with a loop-type handle. B—This brushcutter is equipped with a bike-type handle. The J-type handle is shown between the A and B views. (Stihl)*

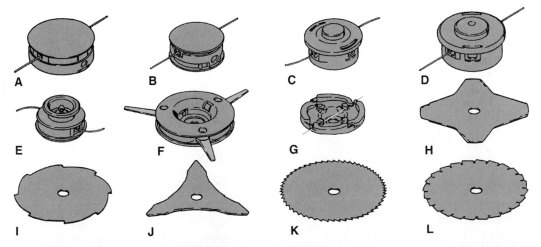

Figure 18-50. *Samples of the many kinds of cutting heads for string trimmers and brushcutters. (Stihl)*

A brushcutter may also be fitted with a circular saw blade. Circular saw blades are suitable for thinning brush and cutting small trees up to a diameter of 2 3/4" (7cm). Do not attempt to cut trees with larger diameters, since the blade may catch or jerk the brushcutter forward.

Operator safety

String trimmers and brushcutters are high-speed cutting tools. Special precautions must be practiced to reduce the risk of personal injury. Careless use can cause serious or even fatal injury.

Basic safe clothing consists of leather work shoes, heavy fabric work pants, leather work gloves, hearing protection, safety goggles or glasses, and shoulder harness. See **Figure 18-48**. In some instances additional safety clothing such as high-top boots, chaps or leggings, face shield, and hard hat are required. See **Figure 18-51**.

All hand-held machines have two-cycle engines, and because they are close to the body, noise from the engine can definitely cause hearing loss over a period of time if hearing protection is not used. The soft, padded, earphone-type ear protector should be worn. See **Figure 18-52.** The ear protector should have a headband that maintains consistent pressure and is comfortable to wear. The ear protector should provide a high noise reduction of 22 decibels (dB) and meet *ANSI (American National Standards Institute) S3.19* standards.

Operators of hand held machines should be alert and in good health. If fatigue occurs during operation, a break should be taken before resuming. Prolonged use of a trimmer or brushcutter exposing the operators hands to vibrations may produce white finger disease (Raynaud's phenomenon) or carpal tunnel syndrome. These conditions reduce the hands ability to feel and regulate temperature, produce numbness and burning sensations, cause nerve and circulation damage, and tissue necrosis (deadening).

Starting and safe operation

All trimmer and brushcutter engines are two-cycle. They will need oil mixed with the gasoline in proper proportions. Refer to the manufacturer's technical service manuals for proper oil/fuel mixing amounts. Once mixed, the fuel should be poured into the fuel tank well away from any

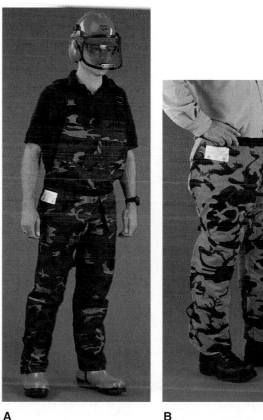

A   **B**

Figure 18-51. *A—Basic safe wearing apparel: heavy work clothes, work boots, safety goggles or glasses, work gloves. B—Shown are of wrap-around chaps. (Stihl)*

Figure 18-52. *The earphone-type protector should have a headband that maintains consistent pressure and is comfortable to wear. The ear protector should provide a high noise reduction of 22 decibels (dB) and meet* ANSI (American National Standards Institute) S3.19 *standards.*

flame or sparks. Once fueled, the following is the procedure for starting, operation, and shut down:

1. Place the machine on the ground in the starting position. See **Figure 18-53.**
2. Press the primer 5 to 6 times.
3. Move the choke lever to the *Choke* position.
4. While holding the unit down with your left hand, also hold the throttle in the open position.
5. With your right hand, pull the starter until the engine tries to run. If the engine starts proceed to Step 6. If the engine does not start in 7 pulls, or starts and then stops, proceed as follows:
   a. Move the choke lever to half *Choke* position.
   b. With throttle in open position, pull the starter until engine runs. Once it is started proceed to Step 6.
6. After warming up for 5 to 45 seconds, move the choke lever slowly to the *Run* position.
7. When using a string trimmer, there are a few *Do's* and *Don'ts* that should be known. See **Figure 18-54.**
8. When using a brushcutter to cut grass, sweep it left and right in an arcing motion. See **Figure 18-55A.** When using a brushcutter equipped with a brush knife to cut wild growth and scrub, lower the rotating brush

Figure 18-53. *To start engine, lay the trimmer on a flat bare surface. For right-handed people, hold the trimmer down with the left hand and pull the starter cord with the right hand. For left-handed people, hold the trimmer down with the right hand and pull the starter cord with the left hand. (Homelite)*

knife down onto the growth to achieve a chopping effect. Make sure to keep the cutting attachment below waist level. See **Figure 18-55B.**

When using a brush knife to cut young saplings or other woody materials up to 3/4″ (2cm), or when using a circular saw blade for thinning brush and cutting small trees up to a

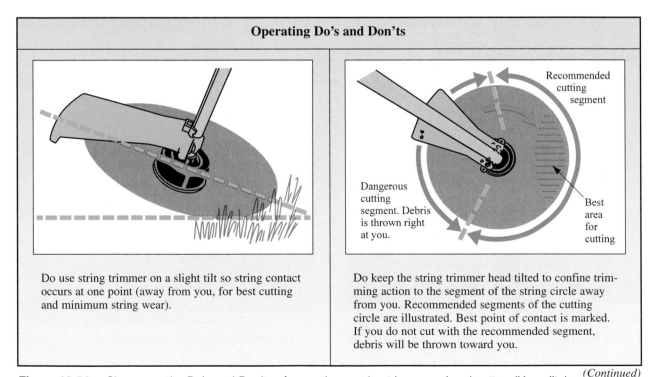

Operating Do's and Don'ts

Do use string trimmer on a slight tilt so string contact occurs at one point (away from you, for best cutting and minimum string wear).

Recommended cutting segment

Dangerous cutting segment. Debris is thrown right at you.

Best area for cutting

Do keep the string trimmer head tilted to confine trimming action to the segment of the string circle away from you. Recommended segments of the cutting circle are illustrated. Best point of contact is marked. If you do not cut with the recommended segment, debris will be thrown toward you.

Figure 18-54. *Shown are the Do's and Don'ts of operating a string trimmer or brushcutter. (Homelite)*

(Continued)

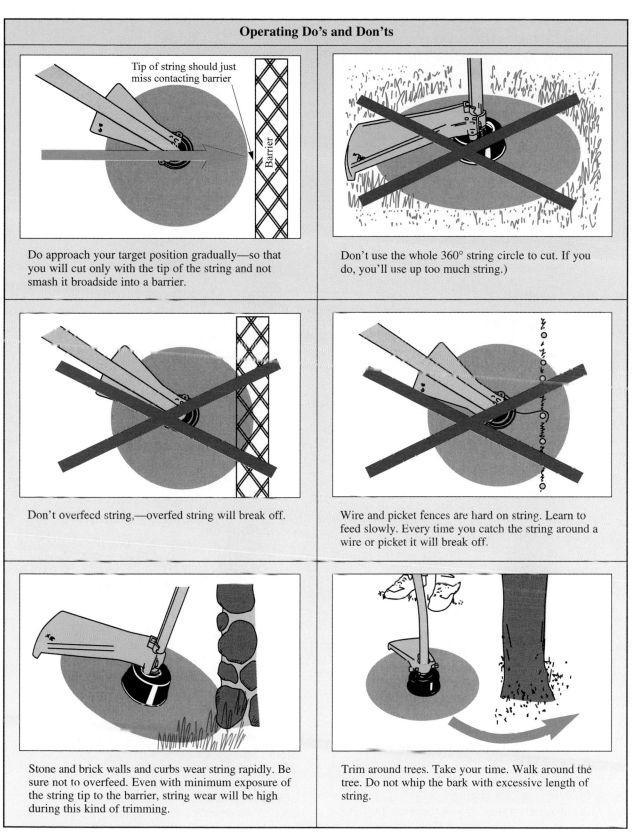

Operating Do's and Don'ts

Tip of string should just miss contacting barrier

Barrier

Do approach your target position gradually—so that you will cut only with the tip of the string and not smash it broadside into a barrier.

Don't use the whole 360° string circle to cut. If you do, you'll use up too much string.)

Don't overfeed string,—overfed string will break off.

Wire and picket fences are hard on string. Learn to feed slowly. Every time you catch the string around a wire or picket it will break off.

Stone and brick walls and curbs wear string rapidly. Be sure not to overfeed. Even with minimum exposure of the string tip to the barrier, string wear will be high during this kind of trimming.

Trim around trees. Take your time. Walk around the tree. Do not whip the bark with excessive length of string.

Figure 18-54. *Continued.*

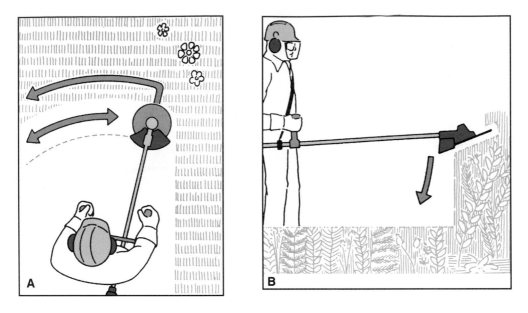

Figure 18-55. *A—A brushcutter can be used for cutting grass by sweeping it left and right in an arcing motion. B—To cut wild growth with a brush knife, lower the blade down onto the growth to achieve a chopping effect. Always keep the attachment below waist level. (Stihl)*

diameter of 2 3/4″ (7cm) in diameter, use the left side of the blade to avoid a *kickout* situation. See **Figure 18-56.** Also, do not attempt to cut material that has larger diameters than recommended for the cutter. The blade may catch or jerk the brushcutter forward. This may cause damage to the blade or brushcutter, or loss of control of the brushcutter that can result in personal injury.

9. When using a string trimmer it will become necessary to adjust the string to the correct length. The following is the procedure for extending and trimming the string to the correct length.
   a. The engine should be running at high speed.
   b. Hold the rotating cutting head horizontal above the ground and tap it on the ground.
   c. The blade on the deflector will trim the string to the correct length. Avoid tapping the head more than once.
   d. If the string does not extend, stop the engine and examine the spool to see if it may be out of string. A new nylon string can be installed. See *Spool replacement* section in this chapter.
10. To stop the engine, press the stop button or turn off ignition switch. Allow engine to cool before returning it to storage.

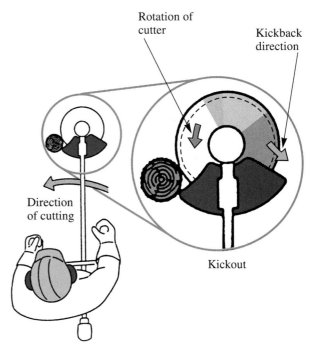

Figure 18-56. *When cutting young saplings up to 3/4″ (2cm) in diameter, use the left side of blade to avoid* kickout. Kickout *is the sudden and uncontrolled motion towards the operators right or rear that can occur if the shaded area of the rotating blade comes n contact with a solid object like a tree, rock, bush, or wall; or if it snags, stalls, or binds.*

String trimmer and brushcutter maintenance

Careful maintenance of a string trimmer or brushcutter will provide for long-lasting service life and safe use. There are a number of parts that need periodic inspection and maintenance. Never operate a string trimmer or brushcutter that is damaged, improperly adjusted, or not completely assembled.

 Always stop the engine and be sure that the cutter is stopped before doing any maintenance or repair work on a string trimmer or brushcutter.

Because of the shape of string trimmers and brushcutters, they can be difficult to hold while servicing them. The apparatus shown in **Figure 18-57** is designed to securely hold a string trimmer or brushcuttter while servicing it.

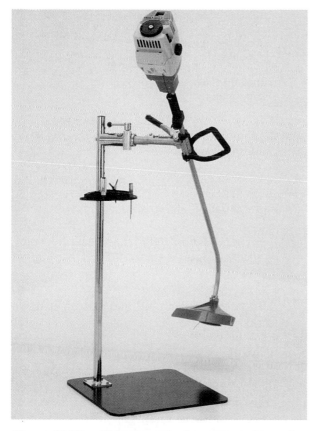

Figure 18-57. *This apparatus is used to hold a string trimmer or brushcutter, so it can be serviced. (Park Tool Co.)*

Spool replacement

Before starting a string trimmer, always visually check the spool and string. When using a string trimmer, lengthening the string is common. If the string does not extend when lengthening it, stop the engine and examine the spool. It may be out of string. A new nylon string can be installed. Refer to the manufacturer's technical service manuals for installation procedures for the specific make and model. The following is the method for installing new string or new spool for one particular type of cutting head:

1. Remove retainer cap by turning it clockwise. Lift out the spool and string. See **Figure 18-58A.**
2. The spool will hold a maximum of 20′ (6.1m) of string. Insert one end of the string into one of the small holes in the spool flange and center the spool on the length of the string. Thread the string back through the closest hole. See **Figure 18-58B.**
3. Pull the loop tight and wind the string counterclockwise on the spool. Wind tightly and evenly. After winding, 1/4″ should be between the wound string and the outer edge of the spool. See **Figure 18-58C.**
4. Place strings into slots in the spool lugs. Leave 5″ to 6″ of string protruding from the lugs. See **Figure 18-58D.**
5. Reinstall the retainer. Once a year, lightly grease the threads on the retainer.
6. To test the operation, pull on the string while alternately pressing down on and releasing the retainer.

Carburetor adjustment

On older and some current models, the high speed and idle speed carburetor needles may need adjustment. Newer models have factory adjusted and sealed carburetors. No adjustment can be made on these carburetors. The carburetor is replaced if it is a suspected cause of poor performance.

Lubrication

Other than the engine, which is lubricated by the fuel mixture, the flexible drive shaft and gear box on some models may need greasing. The shaft must be removed from the shaft housing, cleaned, coated with molybdenum disulfide grease. Reinstall the shaft in reverse direction to extend service life of the shaft.

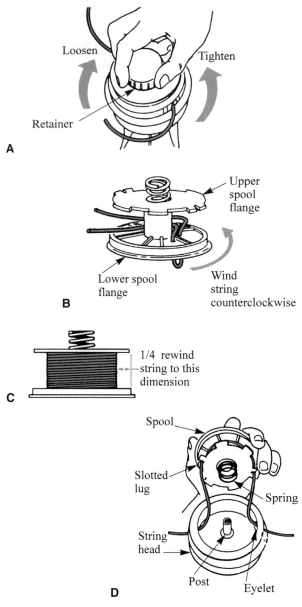

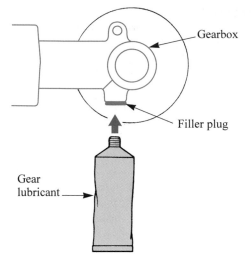

- Check grease level after about every 50 hours of operation.
- Unscrew the filler plug.
- If no grease can be seen on the inside of the filler plug, screw the tube of gear lubricant into the filler hole.
- Squeeze grease into the gear housing—about 5–10g (1/4 oz).

  Do not completely fill the gear housing with grease.

- Refit the filler plug and tighten it down firmly.

Figure 18-59. *Trimmers and brushcutters with straight shafts have a gearbox. The gearbox should be greased at specified intervals. Do not completely fill the gearbox.*

3. Squeeze grease into the gear housing.

⚠ Do not completely fill the gear housing with grease.

4. Reinstall the filler plug and tighten it down firmly.

Figure 18-58. *Procedure for rewinding new nylon string on spool for string trimmer. (Homelite)*

Straight shaft machines have a gear box at the cutter end with a filler plug for grease. See **Figure 18-59.** Check the grease level after every 50 hours of operation. To check and lubricate the gearbox, proceed as follows:
1. Unscrew the filler plug.
2. If no grease can be seen on thc inside of the filler plug, screw the tube of gear lubricant into the filler hole. Refer to manufacturer's technical service manuals for specific lubricant type.

Spark plug

If the engine does not start, it may be due to a wet, fouled, or faulty spark plug. Check the spark plug periodically and clean or install a new one as necessary. Adjust the spark gap if it is wider or narrower than the standard gap. Be sure the stop-start switch is in the *Off* position when checking the spark plug.

Air filter

The function of the air filter is to catch dust and dirt in the inlet air and reduce wear on engine components. Clogged air cleaners cut down on engine power, increase fuel consumption, and make starting more difficult. The air cleaner should be cleaned after every operation in very dusty operating conditions.

Muffler

The muffler on most models is located to the rear of the engine. It is usually shielded. The muffler should be kept clean and open. Check the condition of the muffler periodically and change it as needed. Do not run a string trimmer or brushcutter without the muffler.

Starter rope/spring breakage

After extensive use, it is not uncommon for a starter rope or the flat coil spring to break. Either of these items can be repaired. Because there are variations in recoil starters, refer to the manufacturer's technical service manuals for the procedure for a specific make or model.

Edger/Trimmers

The combination *edger/trimmers* powered by small gasoline engines are versatile units for lawn care maintenance. See **Figure 18-60.** Since the edger has an exposed, fast-spinning blade, it is very important that safety devices designed for the unit are always in place. Use proper operating procedures to assure safe, dependable service.

Purchasing considerations

When purchasing a combination edger/trimmer, a number of factors should be considered. Edgers come in a variety of sizes with respect to the construction and power of the unit. Small engines in the range of two to five horsepower are generally used to drive a belt to the blade.

The most popular edgers are the small units designed for home lawn care and weed cutting. Commercial edgers for heavy-duty work are built to be more rugged. They incorporate a larger, more powerful engine and are considerably more expensive.

Some edgers are single units designed for edging along sidewalks and driveways. Others are designed so the blade can be tilted from vertical to

Figure 18-60. *This is a combination edger/trimmer set for edging operation.*

horizontal for trimming long grass or weeds under fences and close to buildings. Consider the type of work the edger will be expected to perform when planning a purchase.

Safety features and adjustments

The blade of a gasoline engine-driven edger is belt driven from the engine. The blade clutch should always be disengaged when starting the engine or when doing maintenance work on the unit. See **Figure 18-61.** This loosens the belt drive to the blade and allows the engine shaft to turn freely.

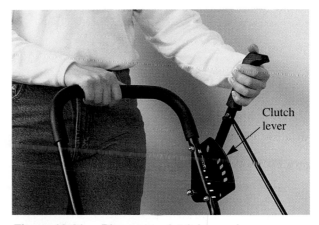

Figure 18-61. *Disengage clutch lever when you are not trimming or edging.*

Never attempt to make adjustments on an edger/trimmer while the engine is running. Serious injury could result.

The guide wheel can normally be adjusted horizontally by loosening the knob, or lever, on the front of the frame and moving the wheel to either side of the frame as needed. See **Figure 18-62.** The lever should then be tightened to hold the wheel firmly in place.

With some designs, the depth of the edger blade can be controlled by raising and lowering the guide wheel. The height of the wheel is adjusted by placing the lever on the right of the frame in the desired notch. See **Figure 18-63.**

One lever, either on the handle or on the engine, may operate the choke, regulate engine speed from slow to fast, and stops the engine. See **Figure 18-64.** When starting the engine, place this lever in the choke position, or run position, and pull the starter handle rapidly. See **Figure 18-65.** Grip the handle firmly and place your foot behind the rear wheel so that the unit will not move during starting. Before starting the carburetor may need to be primed. Depending on the type of priming system, you may have to turn a knob or press a plunger to prime the carburetor. See **Figure 18-66.**

The *blade guard* should always be in place during use. The adjusting arm releases the blade guard and blade, so the guard can be rotated to cover the blade in any edging or trimming position.

For edging along a sidewalk or driveway, the blade is in the vertical position. See **Figure 18-67.** When the unit is to be used for trimming, the blade is set horizontally. It can also be set at an angle for

Figure 18-63. *Raising front guide wheel by placing adjusting lever in correct notch.*

Figure 18-64. *The speed adjustment and stop lever is located on the engine on this model.*

Figure 18-62. *Locking front guide wheel in a newly adjusted position.*

Figure 18-65. *Pulling starter handle to start engine. The belt should be in loose position.*

Figure 18-66. *The carburetor may need to be primed before starting the engine.*

Figure 18-68. *Blade is set at an angle for special trimming jobs.*

special edging jobs. See **Figure 18-68.** The notches in the bracket will hold the blade firmly in any position.

Rules for safe operation

1. Thoroughly inspect the area where the equipment is to be used and remove all stones, sticks, wire, bones, and other foreign objects.
2. Do not operate equipment when barefoot or when wearing open sandals. Always wear substantial footwear.
3. Check the fuel before starting the engine. Do not fill the gasoline tank indoors, when the engine is running, or while the engine is still hot. Wipe off any spilled gasoline before starting the engine.

4. Disengage the blade clutch before starting the engine.
5. Never attempt to make a wheel adjustment while the engine is running.
6. Never operate the equipment in wet grass. Always be sure of your footing; keep a firm hold on the handle.
7. Do not change the engine governor settings or overspeed the engine.
8. Do not put hands or feet near or under rotating parts. Keep clear of the discharge opening at all times.
9. Stop the blade when crossing a gravel drive, walk, or road.
10. After striking a foreign object, stop the engine. Remove the wire from the spark plug. Thoroughly inspect the edger for any damage

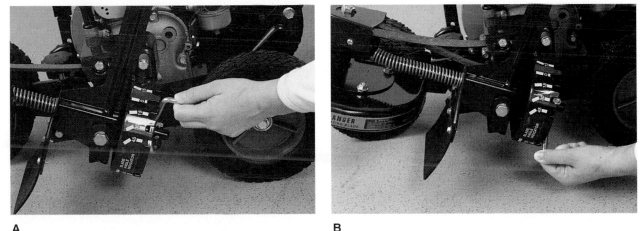

A　　　　　　　　　　　　　　　　　　　**B**

Figure 18-67. *A—Blade is set vertical for edging jobs. B—Blade is set horizontal for trimming jobs.*

and repair the damage before restarting and operating the edger.

11. If the equipment should start to vibrate abnormally, stop the engine and check for the cause. Vibration is generally a warning of trouble.

12. Stop the engine whenever you leave the equipment, before cleaning the guard assembly, and when making any repairs or inspections.

13. When cleaning, repairing, or inspecting, make certain the blade and all moving parts have stopped. Disconnect the spark plug wire, and keep the wire away from the plug to prevent accidental starting.

14. Do not run the engine indoors.

15. Shut the engine off and wait until the blade comes to a complete stop before unclogging guard assembly.

16. Safety glasses or other eye protection should always be used when operating an edger.

Edger/trimmer maintenance

The standard edger blade is 10″ long and is notched on the ends. Since the blade scrapes the edges of driveways or sidewalks during operation, it wears down quickly and should be replaced as needed.

To change the blade, raise the front wheel and loosen the nut on the drive shaft. Remove the old blade, and replace it with a new one. Be sure the blade nut is tightened properly. Always wear a glove to hold the blade to prevent injury to your hand. See **Figure 18-69.**

Edger blades are never sharpened. They are just replaced when they become too short to make good contact with a surface being edged.

To replace a worn belt, remove the belt guard on the engine pulley and the belt guard on the spindle housing. Remove the belt. Replace with a proper size V-belt and secure the belt guards.

Lubrication

Check the engine oil level before starting the engine and after every six hours of use. Add oil as necessary to keep level on full. Before removing the filler plug, clean the area around the plug to prevent dirt from entering the oil fill opening.

Check oil before operating. See **Figure 18-70.** Change the oil after 30 hours of operation by draining oil through the lower oil drain plug. Refill with the correct amount and weight of fresh oil.

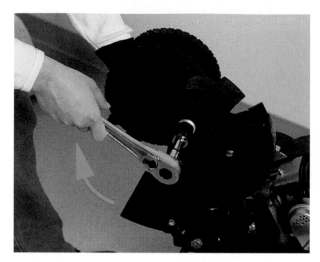

Figure 18-69. *Loosening blade nut on drive shaft. Hold blade with a heavy glove to protect hand.*

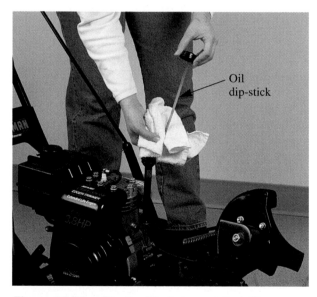

Oil
dip-stick

Figure 18-70. *Check oil before operating. Change the oil after 30 hours of operation by draining oil through the lower oil drain plug.*

Lubricate all moving parts of the edger with engine oil periodically. Check and clean or replace the spark plug each operating season. Remove and clean the air filter at recommended intervals.

Edger/trimmer storage

The following steps should be taken to prepare edgers and trimmers for storage:

1. Clean and lubricate the unit thoroughly.

2. Loosen the belt so it will not stretch during extended storage.
3. Coat the cutting blade with oil to prevent rusting.
4. Remove the spark plug and pour a tablespoon of clean engine oil into the spark plug hole. Rotate the crankshaft a few times and replace the spark plug.
5. Check the blade and engine mounting bolts for proper tightness.
6. Never store the edger with gasoline in the tank when inside of a building where fumes may reach a spark or open flame.
7. Store the edger in a dry, clean area.

Summary

There are many considerations when purchasing a power lawn mower. For the average yard, a rotary-type mower with a 22″ diameter blade is satisfactory. Selecting a push-type or self-propelled mower is a matter of personal preference. Both two-cycle and four-cycle engines are reliable.

There are several mechanical methods for starting small engines including the recoil starter, inertia starter, and electric starter. Every mower manufactured today must be equipped with a blade brake.

After each use, the mower should be cleaned. Blades can be sharpened when they become dull or nicked. Spark plugs should be cleaned and gapped as recommended by the manufacturer. Air filters should be cleaned and oil should be changed after every twenty-five hours of operation.

Always follow safe operating procedures when using a gasoline powered chain saw. When purchasing a saw, make sure it is suited for the type of work to be done. Safe operation of a chain saw comes from a thorough knowledge of correct operating procedures.

Always stop the engine and make sure the chain is stopped before doing maintenance work on a chain saw. Follow all manufacturer's maintenance instructions.

Always keep safety devices in place when using a gas-powered string trimmer, brushcutter, or edger/trimmer. Never attempt to make adjustments on the this type of equipment when the engine is running.

Common maintenance procedures on string trimmers, brushcutters, or edger/trimmers include replacing broken strings, worn blades and belts, changing oil, and lubricating moving parts.

Special precautions should be taken when storing any small engine powered implement.

Know These Terms

grass discharge chute guard	chain saws
rotary mowers	chain guard
reel-type mowers	scabbard
dethatcher blade	spark arrestor
recoil starter	guide bar
extended rope starter	string trimmer
electric starts	brushcutter
inertia starts	edger/trimmer
independent rope starter	blade guard

Chapter 18 Review Questions

Answer the following questions on a separate sheet of paper.

1. Name five safety features used on power lawn mowers.
2. How should accidental engine starting be prevented when working on an engine driven implement?
3. List at least eight safety precautions discussed in this chapter.
4. What invisible, odorless, toxic gas designated CO, is generated from running gasoline engines?
5. What explosive gas is generated when charging lead-acid storage batteries?
6. When using an acetylene torch what protection should be worn?
7. If compressed air is used to clean and dry solvent from parts, a blow gun should be used that is approved by _____ .
8. The majority of lawn mower accident victims each year are children under _____ years of age.
9. The *Knowing Mowing* program is a lawn-mower safety course for youngsters over 12

years of age and is sponsored by Briggs & Stratton Corporation and _____.

10. List eight considerations for purchasing a power lawn mower.
11. List five safe operating practices when using a power lawn mower.
12. Name seven lawn mower conveniences.
13. List the procedures for preparing an engine before starting.
14. List the cold engine starting procedure for an extended rope type lawn mower with blade brake.
15. If the engine idles too fast, you can adjust the idle speed _____ on the _____.
16. If the engine will not start, though proper procedures have been used, the first trouble check should be _____ _____.
17. List the five maintenance steps that will keep the power lawn mower in proper working condition.
18. When the unshielded nose of the chain saw hits a solid surface, it may jump toward the operator. This is called _____.
19. The device often built into the muffler of a chain saw to prevent sparks from causing a fire is called a(n) _____ _____.
20. A small chain saw works well for cutting _____, small _____, and _____ _____.
21. List eight articles of safe wearing apparel when using brushcutters or trimmers.
22. What does the abbreviation ANSI stand for?
23. Explain the procedure discussed in this chapter for adjusting nylon string length on a trimmer.
24. When cutting a small tree with a saw blade on a brushcutter, which edge of the blade would do the cutting in a safe manner?
25. List five maintenance services to string trimmers and brushcutters.
26. The _____ _____ should always be in place when operating an edger/trimmer.
27. A standard edger/trimmer blade is _____ long.

Suggested Activities

1. Demonstrate proper safety precautions for preparing to work on an engine or implement.
2. Show safe use of oxyfuel gas welding equipment.
3. Demonstrate safe use of compressed air blow gun for cleaning parts.
4. Change the oil in a power lawn mower.
5. Service an air filter.
6. Clean and gap a used spark plug.
7. Sharpen and balance a lawn mower blade.
8. Demonstrate engine starting procedures.
9. Adjust engine idle setting.
10. Replace a burned out muffler.
11. Go to a local implement store and compare lawn mowers as though you were planning to purchase one.
12. From literature obtained from dealers, make a list of chain saws produced by different manufacturers. Describe the advantages and disadvantages of each in terms of safety.
13. Make a list of maintenance features to be done before placing a chain saw in storage.
14. Check and change, if necessary, the gas line filter in a chain saw fuel tank.
15. Check the muffler of a chain saw to see if it has a spark arrestor device.
16. Demonstrate proper safe wearing apparel and use of a string trimmer or brushcutter.
17. Remove a string trimmer spool and replace it with a new nylon string.
18. Disassemble a trimmer drive shaft and demonstrate how it should be greased and reassembled. Demonstrate how to lubricate the gear box on a string trimmer.
19. Demonstrate how to replace a string cutting head with a metal cutter such as a saw blade.
20. Replace a worn out blade on an edger/trimmer.

Lawn and Garden Tractors

After studying this chapter, you will be able to:

▼ Describe tractor operating safely.
▼ List features to look for when purchasing a lawn and garden tractor.
▼ List the various kinds of work done with lawn and garden tractors.
▼ Identify principles of good design for lawn and garden tractors.
▼ Describe the kinds of accessories that can be used with lawn and garden tractors.
▼ Identify several transmission systems used for lawn and garden tractors.
▼ Describe electrical systems and components used on lawn and garden tractors.

Operating Safely

Lawn and garden tractors, compared to heavy farm type tractors, are generally smaller in physical size and weight. They are designed to do the kinds of jobs associated with residential homes or light commercial work. See **Figure 19-1.** Before attempting to operate a lawn or garden tractor, all safety instructions provided by the manufacturer should be read and understood. Safety signs and labels placed on the machine and implements should be read and understood before use.

Learn how to properly operate the tractor and controls. Keep the machine in proper operating condition. Do not let anyone operate the tractor who has not had proper instruction about its use.

Figure 19-1. *Garden tractors are generally smaller in size and weight than farm tractors. They can do many jobs required with residential homes and light commercial work. (The Toro Co.)*

Protect children

When operating a tractor and approaching blind corners, shrubs, trees, or other objects that may block vision, be alert for children. *Never* carry children or let children ride on the tractor or any attachment.

While a tractor is being operated, children should be kept in the house under supervision. If children enter a work area turn the tractor off. Take the children out of the area to a safe location where they will be supervised.

Protective clothing

Wear loose fitting clothing and safety equipment appropriate to the job. Leather steel-toed boots are recommended. Leather work gloves can provide added protection for the hands and improve gripping some items. See **Figure 19-2.**

Wear protective safety glasses or goggles. Hearing protective earphone-type protectors or earplugs can lessen noise and make work more comfortable. Prolonged exposure to excessive noise can cause serious hearing impairment.

Depending on the type of work being done, protective masks or respirators may be required. Also depending on the location of the work area, hard hats and safety vests may be required.

 Safety requires full attention of the operator. Do not wear radio or music headphones while operating a machine.

Figure 19-2.   *Wear safe work clothing. (Deere & Co.)*

Considerations while mowing

Rotating mower blades can cut off arms or legs. The following are a number of safety considerations while mowing

1. Clear mowing area of objects that might be thrown by mower blades.
2. Keep people and pets out of mowing area.
3. Study the area to be mowed. Determine a safe mowing pattern. Avoid conditions where traction or stability is doubtful.
4. First, test drive area with ***power-take-off (PTO)*** in the *Off* position and mower lowered.

5. Slow down when traveling over rough ground.
6. *Do not* mow in reverse unless it is absolutely necessary.
7. Back very carefully:
    a. First: disengage the mower.
    b. Then look carefully over the entire area behind the tractor.
    c. Look especially for children.
8. Drive forward very carefully:
    a. People, especially children, can move quickly into the mowing area.
    b. Be alert at all times when driving forward.

Avoiding injury while adjusting mower blades

1. Before dismounting to disconnect or adjust mower:
    a. Shift to neutral.
    b. *Stop* the engine.
    c. *Lock* the parking brake.
    d. Put PTO lever in *Off* position.
    e. Remove key.
    f. Wait for mower blades to *Stop*.
2. Keep hands, feet, and clothing away from mower when engine is running. *Never* place fingers or feet under mower deck.
3. *Stop* mower blades when not mowing.

Avoid tipping the mower

If not handled properly, mowers may tip. To avoid tipping a mower, follow these rules:

1. Do not drive where machine could slip or tip.
2. Be alert for holes, rocks, roots, and other terrain hazards. Stay clear of drop-offs.
3. Slow down before making sharp turns.
4. Do not stop or start suddenly when going uphill or downhill.
5. If the machine stops going forward uphill, back down slowly. Do not attempt to turn around to descend the hill.

Parking a tractor safely

The following is the procedure for parking a tractor safely:

1. Shift to neutral.
2. Disengage PTO.
3. Lower equipment to ground.
4. Engage parking brake.
5. *Stop* the engine.
6. Remove the key.

Transporting a tractor

Transporting a tractor may become necessary. Use either a heavy-duty trailer or a truck to transport the tractor. Use the following guidelines:

1. Do not tow a lawn or garden tractor without it trailered.
2. When transporting with a truck, or with a trailer that does not have a tilting bed, use sturdy ramps secured to the truck or trailer bed. See **Figure 19-3.** Ramps should be adequately wide to accommodate tractor wheels.
3. Lower deck completely and block wheels.
4. *Lock* the parking brake.
5. When using a truck, close the tailgate and secure the tractor to the bed of the truck. When using a trailer, fasten the tractor to trailer with straps, chains, or cables.
6. Check trailer for all necessary lights and signs required by law.
7. Be sure trailer is properly secured to tow vehicle with safety chains.

Purchasing Considerations

When selecting a lawn and garden tractor a number of factors should be considered. It is important to know the kind of work that the tractor will be used for. Some of the most common uses of lawn and garden tractors are:

- Mowing and mulching lawn.
- Leaf collection.
- Plowing, throwing, and blowing snow.
- Preparing soil for planting.
- Pulling a trailer for hauling.
- Rotary broom sweeping walks, and driveways.
- Front end loading of sand, gravel, soil, etc.

Lawn and garden tractors are used for other applications such as the following:

- Specialized grass care (golf course maintenance).
- Pulling heavy items, such as stumps and rocks.
- Augering (drilling) holes for posts.
- Backhoeing (digging holes and trenches).

Mowing and mulching

If mowing a lawn is the only job to be done and is not more than about one-half acre, a riding mower may be more appropriate and economical than a larger tractor type mower. However, for larger areas and a variety of jobs a tractor should be considered. A mower deck should be easy to install and remove from the tractor. When purchasing, ask for a demonstration of this procedure. The area to be mowed is again an important factor.

A

B

Figure 19-3. *A—Loading a tractor safely on a truck. B—Loading a tractor safely on a trailer. (Easy-Up Industries, Inc.)*

The width of the mower deck should be wider for larger areas so that fewer trips around, or back and forth, will be necessary. Fewer trips means less time on the tractor and engine, less fuel consumption, and less time required to get the work completed.

Options for mowing include cutting and letting the cuttings lie, cutting and collecting the cuttings, or cutting and mulching the cuttings to refertilize the earth. Cutting and *mulching* requires specially designed blades that will cut and chop the cuttings into very fine particles and deposit them evenly over the path being cut. **Figure 19-4** shows the mulching blades. The small grass particles filter into the lawn and decompose organically into natural nitrogen fertilizer to enrich the soil and stimulate thicker growth and a healthier lawn. Depositing long cuttings without mulching may cause a smothering effect as it blankets over the shorter blades of grass preventing sunlight and moisture from penetrating very deeply.

If mulching is not accomplished, the cuttings should then be collected in some sort of container connected to the tractor. See **Figure 19-5.** To move the cuttings from the mower deck to the container, which is usually located at the rear of the tractor, there must be a strong blower and a pathway tube to the container. This pathway tube should be resistant to clogging. The container should have a large capacity, but should be easy to empty when it is full.

Figure 19-6 shows grass cuttings traveling through the pathway to the containers at the rear of the tractor. The grass cuttings can be discharged and utilized as *compost* for future fertilizer.

Figure 19-5. *Cuttings can be collected for composting. A blower forces cuttings into containers which can be emptied. (Deere & Co.)*

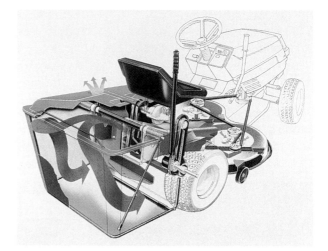
Figure 19-6. *An unrestricted pathway is provided for grass cuttings to travel easily to containers. (MTD Products, Inc.)*

Proper mowing

Each parcel of land is different and will require different procedures when mowing. When mowing a parcel of land for the first time, use the following as a guide:
- Travel *slow* and with cut height on *high* so the terrain is learned, the best mowing pattern is learned, and hitting hidden objects in the grass is prevented.
- Mow grass only when it is dry.
- Mow grass often. Short clippings decay quickly.
- Mow at full throttle for best performance.

Figure 19-4. *Special mulching blades cut grass into fine particles and return it to the soil to become fertilizer. (MTD Products, Inc.)*

- Travel at a speed that fits the conditions: *Slower* in thick, tall grass. *Fast* may leave stripes or uneven cut.
- Cutting grass too short may kill grass and let weeds grow easily.
- Before making a turn, slow down. Short fast turns may skim the ground and pull grass out by roots.
- When trimming, turn left around trees, bushes, and other obstacles. Drive slowly and avoid all obstacles.
- To avoid scalping ground that slopes, approach trees or bushes head on.
- Keep blades sharp. Dull blades tear grass and tips of grass turn brown.
- If cut is uneven check mower level and make adjustments if necessary, slow down before making turns, sharpen blades often, and check gauge wheels for adjustment.

Mower deck and blade adjustment

The mower deck must be level from side to side to obtain an even cut. From front to rear it should be *almost* level. A better cut is obtained if the blade slants *forward* an 1/8″ to 1/4″. See **Figure 19-7.**

As mowing progresses the sharpened tip (A) cuts a forward arc through the grass. Tip (A) then swings to the rear cutting little grass. Tip (B) follows tip (A) and cuts a forward arc of grass also. The grass is therefore cut only once.

> The blades should never be level or slant toward the rear.

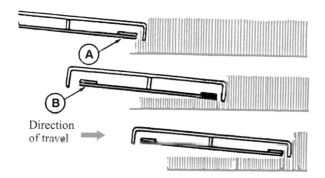

Figure 19-7. *Blades should slant slightly toward direction of travel for best cutting action. The drawing is exaggerated for illustration purposes. (Deere & Co.)*

Proper mulching

When mulching, there are a number of rules to follow. The following is a list of these rules:
- Cut only about 1/3 of the blade of grass when using a mulcher.
- Mulch only when the grass and leaves are dry.
- Mow a different direction each time you mow.
- Overlap mowing paths 2″ to 4″.
- Slow the mower travel speed. Mulching requires more power.
- Keep the underside of the mower deck clean. Clean the deck after each use with a garden hose.

Leaf collecting

Leaf collecting is similar to collecting grass. Leaves are shredded and blown into the container at the rear of the tractor mower. The leaves can be mixed with grass cuttings and other organic materials. This mixture then can be composted into fertilizer.

Plowing, throwing, and blowing snow

Plowing snow with a tractor requires a concave snowplow blade attached to the front of the tractor. See **Figure 19-8.** The blade should be easily raised and lowered, and angled to either side hydraulically or mechanically. The tractor should have enough horsepower to move heavy snow

Figure 19-8. *Snowplowing can be accomplished with a curved plow blade. The blade can be raised and lowered mechanically. (The Toro Co.)*

without lugging or stalling. Chains and wheel weights may be needed on the rear drive wheels to attain more traction on slippery, icy surfaces. See **Figure 19-9.**

The snowplow blade may also be used to spread and level gravel for driveways. See **Figure 19-10.** Blade edges wear and preferably should be replaced with a new metal edge.

Single-stage snow throwers of varying widths can be attached to the front of tractors. These throwers have a horizontal auger that gathers and throws the snow through an adjustable deflector. The deflector should be easily directed to the right or left from the seat of the tractor.

Two-stage snow blowers are snow blowers that incorporate a high speed blower in conjunction with the auger of the single stage type. The auger draws the snow in where the blower then discharges the snow through the deflector chute. A two-stage snow blower is a more efficient system that can handle larger quantities and heavier snow than the single-stage thrower. A two-stage snow blower may require more horsepower. See **Figure 19-11.**

Figure 19-11. *Snow blowers can be attached to the front of some tractors. Note the tire chains, wheel weights, and enclosure to protect the driver. (The Toro Co.)*

Figure 19-9. *Chains and wheel weights may be needed to increase traction for some jobs such as plowing snow. (The Toro Co.)*

Figure 19-10. *A plow blade can be used for moving and grading operations. (Deere & Co.)*

Preparing soil for planting

Tractors are excellent for tilling, or preparing, soil for gardens. *Rotary garden tillers* are available that connect to the rear of the tractor and the tines, or blades, are driven mechanically from a power-take-off (PTO) from the rear drive unit of the tractor. See **Figure 19-12.** Tines should be easily replaced when they become worn too short to be effective. A *three-point hitch* allows the tiller to be raised and lowered mechanically or hydraulically and controlled with a lever located so the driver can easily reach it. Cultivators, harrows, and discs are available but are generally reserved for tractors with at least 18 horsepower (hp) or 20 hp.

Figure 19-12. *Soil can be prepared for planting with a garden tiller. (The Toro Co.)*

Pulling a trailer for hauling

A trailer for hauling can be a very economical and useful accessory. See **Figure 19-13.** Trailers can be easily attached to the tractor and will serve to carry all types of equipment and materials. Logs, dirt, gravel, mulch, and garden tools are just a few of the items that can be hauled. There are trailers available to accomplish specific jobs such as collecting grass clippings. Most trailers will tilt and dump the load by releasing a lever located on the tongue. Some trailers have hinged tailgates that make the trailer easy to load or unload. Trailers may be constructed from steel or plastic.

Figure 19-13. *A trailer is a very useful accessory for hauling behind a tractor. They may be made of metal or plastic. (Deere & Co.)*

Rotary broom sweeping

Rotary bristled brooms are available for some tractors. These brooms are useful for sweeping sidewalks and driveways. See **Figure 19-14.** The rotary broom is fitted to the front of the tractor. The broom can be angled to the right or left so debris can be cleared to either side. The broom rotates so that debris is pushed forward and to the side of the tractor's path. Rotary brooms can also be used for clearing light, dry snow.

Figure 19-14. *A rotary broom can be used to sweep debris and light dry snow. The broom can be directed to the right, straight, or to the left. (Deere & Co.)*

Front end loading

Generally, front end loading devices are often used under adverse conditions. For this reason, the devices are used with the light commercial/construction size tractors that have greater horsepower engines and are designed to handle heavy loads for long periods of time. Front end loading utilizes a horizontal bucket located on the front of the tractor. It is raised and lowered with hydraulic cylinders attached to each side of the bucket and tractor. The bucket can be tilted with two additional cylinders. The tractor should have wheel weights on the rear wheels and/or weights attached to the back of the tractor to help offset the front load when it is raised. Additional hydraulic control levers are needed to manipulate the bucket.

Engine Components

Nearly all yard and garden tractor engines are four-cycle, electric start engines with **overhead valves** and electronic ignition systems. Electronic ignition systems are essentially trouble free. Electric starter motors require a 12V battery and alternator battery charging system. Engines may range from 12 hp to 22 hp. Engine horsepower for a tractor depends on the size and weight of the tractor, any accessories it must drive, and the kind of work the tractor must do. For example, less horsepower is required to operate a dual-bladed mower deck for lawn care than is needed for tilling land, or back-hoeing a trench.

Twelve, and some 14 hp engines, are single-cylinder engines used for light-duty tractors. These engines, like larger ones, may have cast-iron cylinder sleeves or a solid cast-iron block. If the engine has cast-iron sleeves, they can be replaced with standard size sleeves when overhauling. Standard size pistons and rings are used. Cast-iron engine blocks require boring and honing to the next larger standard size and corresponding oversize pistons and rings must be installed. Refer to Chapter 16 of this text.

Tractor engines from 14 hp to 22 hp generally have two cylinders. They are either the vee, or opposed cylinder configuration. They may have a vertical or horizontal output shaft. The engine type depends on the manufacturer and their decision on the best way to drive the tractor and its accessories. Overhead valves and hydraulic valve lifters are used on all tractor engines, because of their greater efficiency and service life.

Most tractor engines are easy to maintain. Some tractors have a hood that covers the entire engine. The hood protects the engine from external elements. Other tractors have partial hoods over the top of the engine compartment leaving the sides open for better cooling and ease of access. When tilling dry ground, a considerable amount of dirt may accumulate on the engine. This may hamper engine cooling and clog air filters. Make sure to clean the engine compartment, and check and service the air filter regularly.

Tractor hoods are hinged so they can be opened forward to expose the engine compartment. Locate the oil dipstick. Become familiar with oil level markings on the dipstick. Check the oil regularly. Locate the oil filler tube and cap, and the oil filter. Oil should be poured without spillage, and the filter should be changed every time the oil is changed.

Locate the in-line fuel filter. Observe and change the filter if needed. Most all tractors have fuel tanks that can be reached easily with the spout of the fuel container. The majority of these tanks are located at the rear so that if there is an accidental spill, the fuel will not run onto a hot portion of the engine.

Locate the air filter housing. Become familiar with the air filter replacement procedures. Refer to the manufacturer's technical service manual for this information.

Make sure the muffler and exhaust system are in good condition. The engine should run comfortably quiet. The system should direct the exhaust away from the operator. The exhaust system normally gets very hot. It should be located so that the operator is protected from accidental burns. Spark arresting mufflers are available for most tractors. These type of mufflers are required by law in some areas to prevent accidental fires.

Some tractor engines are liquid cooled. The coolant level needs to be checked periodically. It is good practice to flush and add new coolant seasonally. Locate the coolant drain point. It must be located at a low point on the engine. Check the coolant and the thermostat regularly. Replace thermostat and coolant as needed. Check the manufacturer's service manual for recommendations. Occasionally clean the radiator cooling fins to prevent dirt or debris from clogging them.

Become acquainted with the engine controls, such as the throttle, choke, and ignition switch. On most tractors these controls are within easy and comfortable reach of the operator. Gauges and system indicators should also be visible to the operator. See **Figure 19-15.**

Chassis and Steering

The **chassis** is the main framework of the tractor around which the entire tractor is assembled. The chassis must be strong and rigid enough to withstand bending and torsional forces imposed upon it during operation. The engine, transmission, and work accessories must all be held in exact proper alignment regardless of terrain. Heavy steel components secured with high quality welds and locking type threaded fasteners are indicators of a well designed chassis. Sheetmetal components

Figure 19-15. *Throttle, choke, ignition switch and other controls and instruments should be easy to reach and see. (The Toro Co.)*

that are secured to the chassis should be smooth and of a heavy gage thickness. The sheetmetal should protect the operator from contact with any moving parts.

The steering components must function smoothly and easily in all driving configurations. Strength in the front axle is essential, because it must be able to pivot up and down laterally to conform to changing terrain while allowing controlled steering of the front wheels. In many cases, cast-iron is used for front axles. Heavy cast-iron is an excellent material for this application, because of its inherent strength and rigidity. ***Grease (zerk) fittings*** are used where lubrication is required for king pins and tie rods. The prescribed type of grease can be applied regularly with a hand operated grease gun. Excess grease should be wiped away to prevent mixing with dirt.

The steering wheel should be of such size, angle, and location that the operator can easily reach it without straining arm or back muscles. Some steering wheels can be tilted for comfort. See **Figure 19-16.** Steering linkage between the steering column and front wheels should be free of excess looseness and provide positive control of direction at all times. Easy steering on a smooth display room floor may not be representative of how it will react on an irregular terrain. Customers should request, and be allowed, to thoroughly test drive a tractor outdoors on irregular terrain before making a purchase.

Four wheel steering is available on some tractors to provide greater maneuverability around obstacles. See **Figure 19-17.** When the front wheels

are turned less than 45°, the rear wheels remain straight ahead. For tight turning, when front wheels are turned more than 45°, the rear wheels will turn the opposite direction with about a 15° delay from the front wheels. See Figure **19-18.** The delay is intentional and helps prevent accidental

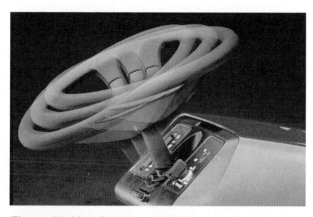

Figure 19-16. *An adjustable, tilting steering wheel lifts to provide comfortable steering for the driver. (The Toro Co.)*

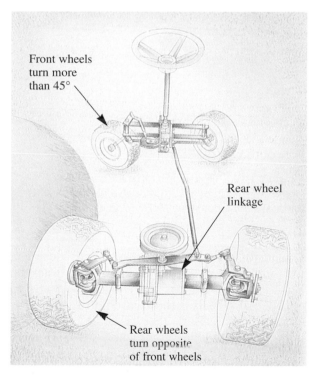

Figure 19-17. *Four wheel steering can increase maneuverability around trees and shrubbery. Notice additional linkage and joints at rear wheels. (MTD Products, Inc.)*

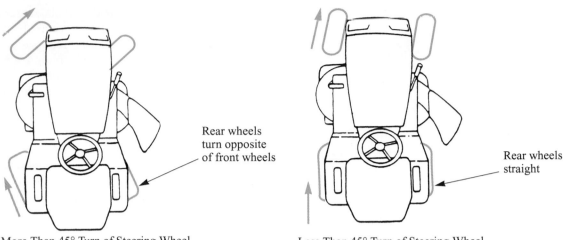

More Than 45° Turn of Steering Wheel   Less Than 45° Turn of Steering Wheel

Figure 19-18. *Front wheels turn 45° before rear wheels begin to turn opposite direction. (MTD Products, Inc.)*

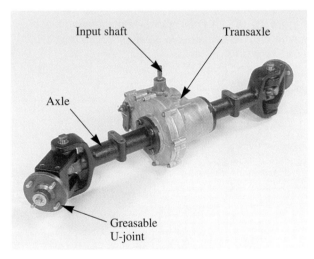

Figure 19-19. *Rear wheel drive for four wheel steering consists of axles, greasable universal-joints, transmission, input shaft. (MTD Products, Inc.)*

striking of obstacles. The rear axle is designed with universal drive joints and steering brackets. See **Figure 19-19.** The entire steering system to typical rear wheels is shown in **Figure 19-20.**

> ⚠️ If operating a four wheel steering lawn tractor near a drop-off in the ground, *do not* turn sharply. The rear wheels turn outward and could lose ground contact. This could result in the tipping over of the tractor.

Transmissions

The transmission and rear axles are very important, hard working components of any tractor. They are called ***transaxles.*** The transaxle unit may be driven by the engine from a V-belt, a series of V-belts and pulleys, or a direct drive shaft. See **Figures 19-21** and **19-22.** The major function is to allow the operator to control motion and power from the engine to the rear wheels. Mechanical, geared transaxles produce greater pulling power and slower tractor speed by gearing down (low-speed gear). They produce less pulling power with greater tractor speed by gearing up (high-speed gear). Mechanical transmissions contain a variety of gear sizes and ratios. The operator can shift gears with a lever. This selection provides the necessary driving force without slowing the engine excessively. Heavy pulling requires a lower speed gear ratio. A clutch mechanism is utilized to disconnect the engine from the transmission while the shifting of gears is being done. The clutch is operated by pressing a foot pedal. A reverse gear and brake are also a part of the transmission. They allow the operator to back up or stop the tractor.

Variable-speed transmission

The ***variable-speed pulley*** is a torque converter used on some tractors, riding mowers, and rotary tillers. The variable-speed pulley allows the operator to vary the tractor speed while main-

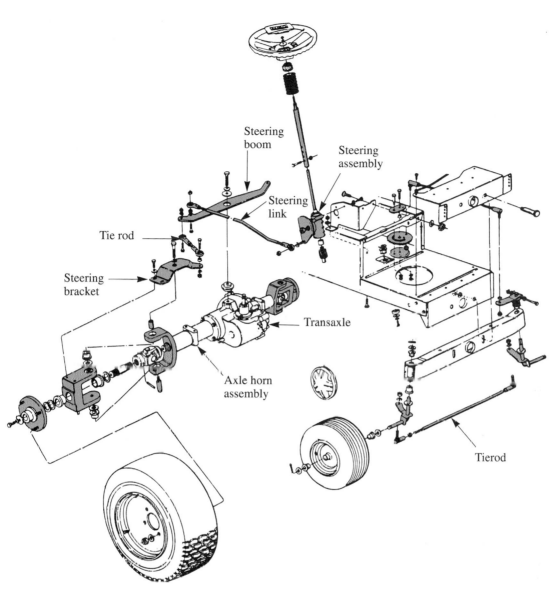

Four Wheel Steer Rear Axle Assembly

Figure 19-20. *Four wheel steering assembly. (MTD Products, Inc.)*

taining a constant engine speed. The variable drive pulley system coupled with the transaxle permits setting the speed selector for a given speed, then allows slowing down for a turn by depressing the clutch/brake pedal. See **Figure 19-23.** The tractor automatically resumes speed when the pedal is released. This type of drive system permits no-clutch-on-the-go speed control. This enables the operator to match the tractor speed to various mowing conditions without slowing the RPM of the engine or mowing blades.

The variable-speed pulley illustrates how the pulley ratios are changed and thereby alter the RPM of the driven pulley. See **Figure 19-24.** The center part of the pulley, called a *moveable sheave,* is free to move up and down. Remember, V-belts normally do not stretch or change length. See **Figure 19-23.** When the speed control pedal is pressed it mechanically pulls the variable-speed pulley toward the engine pulley and the V-belt runs outward on the variable-speed pulley. At the same time, the rear belt runs closer to the center of the

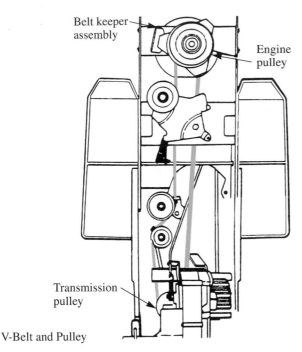

Belt keeper assembly

Engine pulley

Transmission pulley

V-Belt and Pulley

Figure 19-21. *Single V-belt drive system from engine to transmission pulley. Idler pulley maintains proper belt tension. (MTD Products, Inc.)*

variable-speed pulley. This is accomplished by a tension spring and an idler pulley. Because the large driving pulley is now turning slower, the belt on the smaller pulley is running slower. In this case, it is running 1/2 speed.

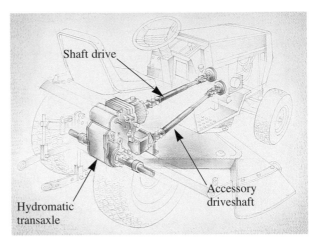

Shaft drive

Accessory driveshaft

Hydromatic transaxle

Figure 19-22. *Drive shaft from engine to hydrostatic transaxle. Power-take-off (PTO) unit also driven by second drive shaft. (MTD Products Inc.)*

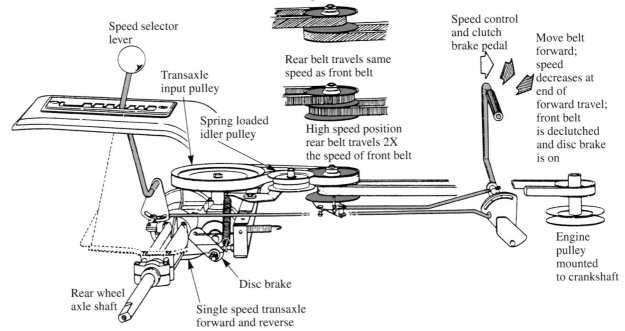

Slow speed position rear belt travels 1/2 speed of front belt

Speed selector lever

Transaxle input pulley

Spring loaded idler pulley

Rear belt travels same speed as front belt

High speed position rear belt travels 2X the speed of front belt

Speed control and clutch brake pedal

Move belt forward; speed decreases at end of forward travel; front belt is declutched and disc brake is on

Engine pulley mounted to crankshaft

Rear wheel axle shaft

Disc brake

Single speed transaxle forward and reverse

Variable-Speed Drive System

Figure 19-23. *A variable-speed drive system. Speed selector lever selects speed range. Tractor must be stopped when shifting speed selector lever. Speed control pedal varies speed through variable-speed pulleys. (MTD Products, Inc.)*

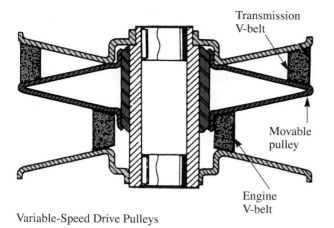

Variable-Speed Drive Pulleys

Figure 19-24. *Variable-speed drive pulleys. If movable pulley moves down, the lower belt will run closer to the periphery while the upper belt will move inward thereby changing the pulley diameter ratios and changing the speed of the driven belt. This is accomplished by linkage from the tractor pedal. (MTD Products, Inc.)*

To speed up, the pedal is released. The speed selector lever provides a variety of *speed ranges*. See **Figure 19-25.** The clutch pedal should be depressed and the clutch disengaged when changing the speed selector lever. The speed control pedal controls the speed in each range. The clutch pedal must be depressed when shifting the transaxle shift lever. If the transaxle is the single-speed type, the tractor speeds are controlled only by the variable-speed belt drive system. See **Figure 19-26.** The transaxle provides forward, neutral, and reverse direction only. The shift lever will have forward, neutral, and reverse positions.

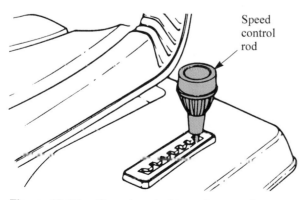

Figure 19-25. *Speed control lever is moved to desired location and held in place by notches. On many tractor models it is located on rear fender. (MTD Products, Inc.)*

Internal components of a heavy-duty, single-speed transaxle are illustrated in **Figure 19-27.**

A two-speed transaxle, coupled with a variable-speed drive, will provide a high-speed gear range and a low-speed gear range. See **Figure 19-28.** Within each range are a fixed number of variable-speeds. For example: the number of speeds may be seven high speeds plus seven low speeds. The clutch pedal must be depressed and the tractor not moving when shifting gears. The two-speed transaxle shift lever will have high, low, neutral, and reverse positions. See **Figure 19-29.**

Hydrostatic transmissions

Hydrostatic transmissions consist of a variable displacement hydraulic pump, a fixed displacement hydraulic motor, and a system of check valves contained within one housing. See **Figure 19-30.** Hydrostatic transmissions can be used in various types of applications where variable output speed is required. This type of transmission has many advantages over other variable speed drives and gear type transmissions. The advantages of the hydrostatic transmission are as follows:

- **Response.** Hydrostatic transmissions respond faster than any other type of power transmission system.
- **Precise speed.** Hydrostatic transmissions are capable of maintaining precise speed under varying load conditions.
- **Ease of operation.** One lever controls direction and speed smoothly without gear change.
- **Low maintenance.** Simple design keeps maintenance minimal.
- **Increased productivity and versatility.** Hydrostatic transmissions allow complete matching of power to load.
- **Completely self contained.** There are no external hydraulic lines, separate drive components, etc.
- **Simplified final product design.** Hydrostatic transmissions reduce the number of mechanical drive components.
- **Positive braking action.** The one lever that controls speed and direction also controls braking. As the control lever is moved toward neutral the output shaft slows until it finally stops at the neutral position.
- **Self lubricating.** The system is completely filled with oil at all times and all internal parts move in a bath of oil.

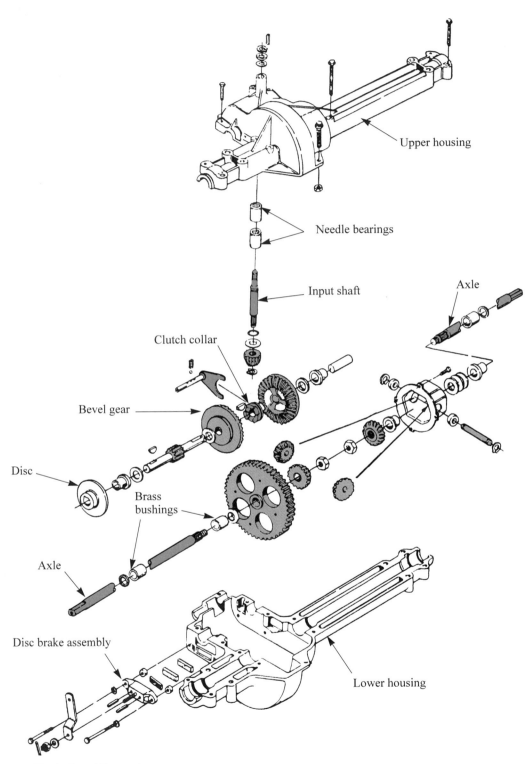

Upper housing

Needle bearings

Input shaft

Axle

Clutch collar

Axle

Bevel gear

Disc

Brass bushings

Axle

Disc brake assembly

Lower housing

Single-Speed Transaxle

Figure 19-26. *Inner components of a single-speed transaxle. Tractor speeds are controlled only by the variable-speed belt drive system. (MTD Products, Inc.)*

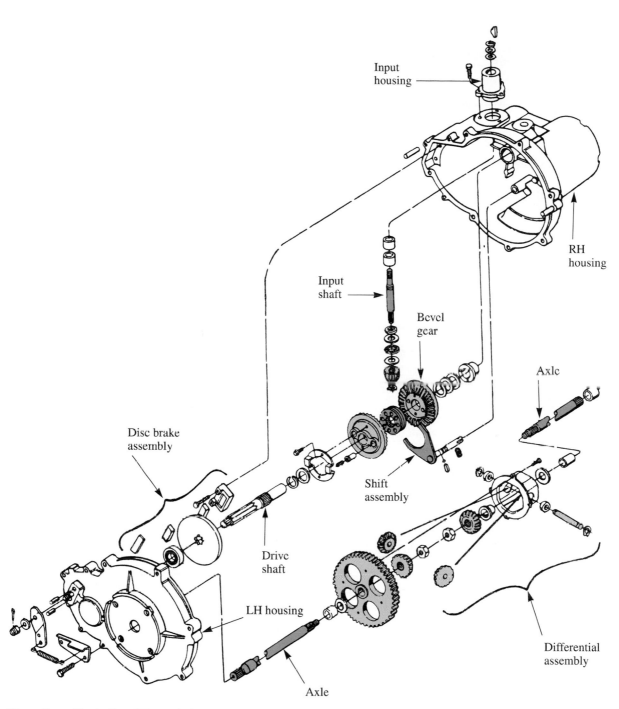

Input
housing

RH
housing

Input
shaft

Bevel
gear

Axle

Disc brake
assembly

Shift
assembly

Drive
shaft

LH housing

Differential
assembly

Axle

Heavy-Duty, Single-Speed Transmission

Figure 19-27. *Heavy-duty, single-speed transaxle. (MTD Products, Inc.)*

The difference in performance between the hydrostatic transmission and a speed gear transmission is shown with a graph in **Figure 19-31.** The smooth curve shows the uniform matching of torque and speed requirements by the hydrostatic transmission. As speed increases, torque decreases smoothly. The geared transmission has three peaks and valleys requiring manual shifting between each of them.

The pump section of the transmission controls

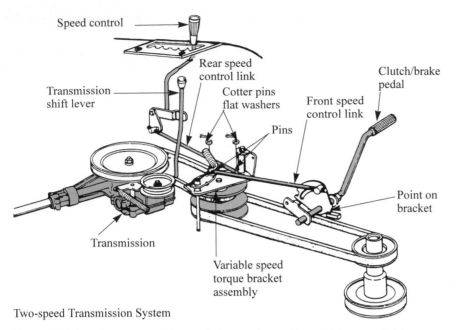

Two-speed Transmission System

Figure 19-28. *A two-speed transmission system with variable-speed drive. (MTD Products, Inc.)*

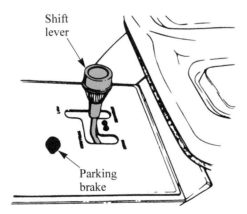

Figure 19-29. *A shift lever with* High, Low, Neutral, *and* Reverse *positions. (MTD Products, Inc.)*

speed and direction of the output shaft by a single lever. See **Figure 19-32.** By varying the displacement ratios between the pump and motor, infinite speed control is achieved. Speed increases as the lever is moved farther from neutral. From neutral to forward produces one direction of rotation of the output shaft. When the lever is in the neutral position, the output shaft stops producing an effective braking action. To produce opposite rotation of the output shaft the lever is moved from neutral to reverse.

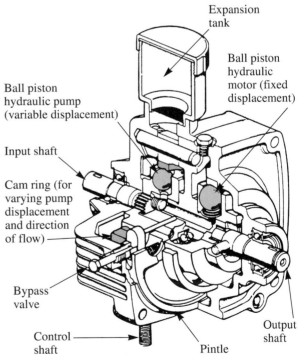

Hydrostatic Transmission

Figure 19-30. *Hydrostatic transmission is a variable displacement hydraulic pump driving a fixed displacement hydraulic motor. The input shaft is driven by the engine, the output shaft drives the axles. (MTD Products, Inc.)*

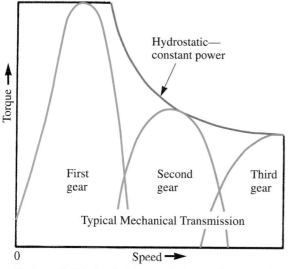

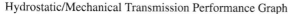
Hydrostatic/Mechanical Transmission Performance Graph

Figure 19-31. *Unlike the mechanical gear type transmissions, hydrostatic transmission provides constant smooth power. (MTD Products, Inc.)*

Hydrostatic transmission fluid flow

The variable displacement **ball piston pump** is driven by the engine at a constant speed. See **Figure 19-32.** The ball pistons rotate and roll within a close fitting circular housing. When the control lever places the cam ring in a position that causes the balls to reciprocate in their cylinders, they draw in oil from one port and force it out the opposite port to the motor (F). When the cam ring is centered so the balls do not reciprocate and only roll in a concentric circle (N), no oil is moved to the motor. This causes an effective braking action at the motor output shaft. When the control lever is moved to the reverse position (R), the cam ring causes the balls to reciprocate. This causes the pumping of fluid in the opposite direction to the motor, resulting in reversing of the output shaft.

During use, hydraulic oil gets warm and expands. A **reservoir** is necessary to accommodate the oil expansion without overflowing. See **Figure 19-33.** When the oil is cold, the pump and motor cavities must be full of oil. If they are not full, air occupies the space that should contain oil. Because air is compressible and oil is not, any air in the system will cause erratic running, noise, and eventual serious damage to the pump and motor. This condition is called *cavitation* and should not be tolerated for any lengthy period of time. It is important that the hydraulic oil level be checked frequently and oil added as needed. When cold, fill only to the cold level indicated on the reservoir.

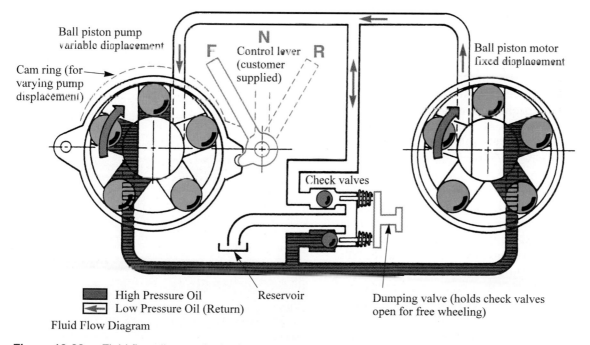

Figure 19-32. *Fluid flow diagram for hydrostatic transmission system. Notice function of check valves. (MTD Products, Inc.)*

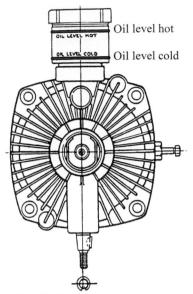

Oil level hot

Oil level cold

Reservoir/Expansion Tank

Figure 19-33. *Oil level should be checked and maintained when unit is cold. Reservoir allows oil to expand due to heat without overflowing. (MTD Products, Inc.)*

Overfilling will cause fluid to spill out at operating temperature.

Some transmissions utilize variable displacement pistons connected to a ***swash plate***. The plate, pistons, and cylinders rotate together. As the swash plate is tilted the pistons travel a greater distance in their cylinders and more oil is pumped. As the swash plate tilt is lessened the pistons travel less, and less fluid is drawn in and pumped out to the motor until no oil is pumped, and the motor output shaft stops. Both pump systems function in the same basic way but vary in their mechanical makeup.

Spring-loaded check valves are used in the hydrostatic fluid flow system to permit flow in one direction and prevent flow in the opposite direction. See **Figure 19-32.** Notice that when free wheeling is desired to manually push the tractor both check valves can be opened allowing fluid to flow freely either direction in the system. This overrides the normal neutral braking action. When the pump is in neutral, both check valves are closed preventing fluid flow in either direction in the system. The tractor will not move in this case.

To keep from overheating the system a cooling fan is placed on the input shaft of the transmission. See **Figure 19-34.** The input shaft and fan are driven from a V-belt between the engine and transmission. Cooling fins are pro-

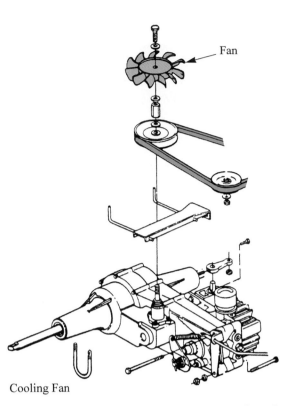

Fan

Cooling Fan

Figure 19-34. *A cooling fan on the input shaft cools transmission system. (MTD Products, Inc.)*

vided on the differential and transmission housing. See **Figure 19-35.** The ***differential gears*** permit each rear axle to turn at different speeds as is necessary when turning corners. See **Figure 19-36.**

There are many different makes and models of hydrostatic transmissions. The troubleshooting chart in **Figure 19-37** is general in its scope and may not be adequate for certain makes and models. Refer to the manufacturer's technical service manuals

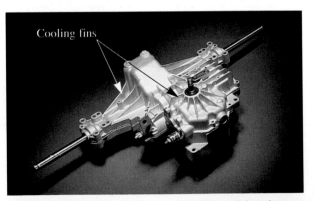

Cooling fins

Figure 19-35. *Cooling fins on the differential and transmission housing disperse heat. (MTD Products, Inc.)*

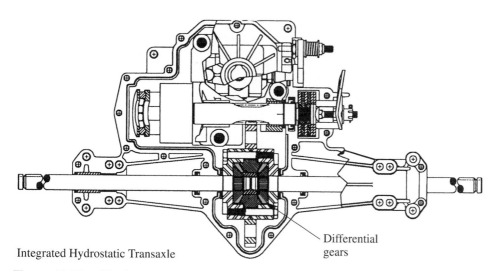

Integrated Hydrostatic Transaxle

Differential gears

Figure 19-36. *The integrated hydrostatic transaxle. Differential gears allow each rear axle to turn at different speeds when cornering. (MTD Products, Inc.)*

Hydrostatic Transmission Troubleshooting	
No output torque (power) in ether direction, cold start.	1. Recheck relief valve position, control linkage, input drive. 2. Oil level in reservoir. 3. Broken control shaft dowel pin. Transmission must be repaired or replaced.
Loss of output torque, continuous load.	1. Operating at conditions approaching hydraulic stall. The transmission fluid has exceeded 180°F. 2. Internal leakage due to wear. Transmission should be repaired or replaced. 3. Water in transmission fluid. Purge system of all fluid and replace with new transmission fluid. Replacement of the transmission is generally not necessary.
No output torque in one direction.	1. One of the directional valves is stuck. Transmission should be repaired or replaced. 2. Low oil level.
Riding mower cannot be pushed with engine off.	1. Relief valve control not set. 2. Relief valve travel not adjusted. 3. Motor piston or rotor seized. Transmission must be repaired or replaced.
No neutral.	1. Recheck linkage. Loose linkage creates an adjustment problem. Note: The hydraulic neutral band is very narrow. Deflection in the linkage may make it difficult to obtain neutral from both directions. It is recommended that neutral should be positive from forward drive.
Oil leakage at the relief valve.	1. Check O-ring for damage. Apply loctite PST5924 to threads and torque to 13 to 17 ft-lbs.
Oil leakage at the control shaft seat.	1. Spillage when fluid has been added to the reservoir. 2. Spillage at the vent in the reservoir at operating temperatures due to cold level being too high or water in the fluid. Reduce fluid level or replace fluid in the event there is water in it (milky color). 3. Loose oil reservoir. 4. Loose vent bolt. 5. Damaged control shaft seal. Transmission should be repaired.
Noisy operation.	1. Operating at part throttle. Hydrostatic transmission is designed to operate with the engine running at full throttle. 2. Water in transmission fluid. Replace transmission fluid. 3. Air in transmission fluid. Bleed air from vent.
Output shaft rotates in the opposite direction.	1. The transmission body is 180° out of position. Transmission has to be removed and reassembled correctly.

Figure 19-37. *Sample Hydrostatic Transmission Troubleshooting Chart. (MTD Products, Inc.)*

for the particular make and model in need of service or repair.

Electrical Safety Systems

Manufacturers have incorporated safety devices and interlock systems in tractors to protect against certain possible accidents. These systems are used to protect the operators and those people around them. Children and pets are often unpredictable and should never be allowed near tractors or other work equipment that may cause them injury. Even adults have been seriously injured by improper use of power equipment.

Safety interlocks

Safety interlocks are two or more devices that are connected in such a way that none of them can be operated independently. Safety interlock systems are used to provide added safety on lawn and garden tractors. This arrangement of devices may be mechanical, hydraulic, electronic, or any combination of the three. They are arranged so the functioning of one device is controlled by the functioning of another. Examples of safety interlocks are the reverse safety switch and the seat-activated switch.

Reverse safety switches are safety interlocks that are installed in the electrical system. The switches require the cutting deck (mower unit) to be disengaged before the unit can be shifted into reverse. For example, a lever may be used to lift and lower the cutting deck to establish cutting height. Pulling the lever all the way back and locking it will disengage the blades. The lift and disengage lever must be all the way back and the blades disengaged in order to start the engine, or shift into reverse.

As an added safety feature, **ANSI (American National Standards Institute)** requires that all lawn and garden tractors produced after July, 1987 have an operator present for the tractor to be operational. The engine must stop if the operator leaves the seat with the blades or PTO (Power-Take-Off) engaged. A **seat-activated switch** is the safety device used for this protection. The seat is mounted on a pivot bracket. As long as the operator's weight is upon the seat, the switch remains closed. When the operator gets off the seat a spring raises the seat, opens the switch, and shuts off the engine. See **Figure 19-38.** In addition, some models incorporate a new type of switch that

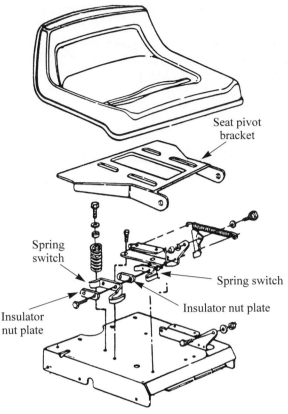

Operator Present Safety Seat

Figure 19-38. *Operator present safety seat. (MTD Products, Inc.)*

requires the parking brake be engaged before the operator leaves the seat. This can prevent a tractor from rolling away on an incline without an operator aboard.

Electric start systems

A typical **electric start system** including safety interlocks is shown in **Figure 19-39.** To start the engine with this system the ignition key must be turned on and both safety switches activated. The clutch safety switch is activated when the clutch is depressed. The blade safety switch is activated when the blade is disengaged. The circuit is complete between the battery and the coil primary of the **solenoid** when the safety switches are activated. The solenoid switch will close allowing the starter motor to crank the engine. In the circuit of the electric start tractor, the safety switches are wired in **series**. If any one switch is not activated, the tractor will not start.

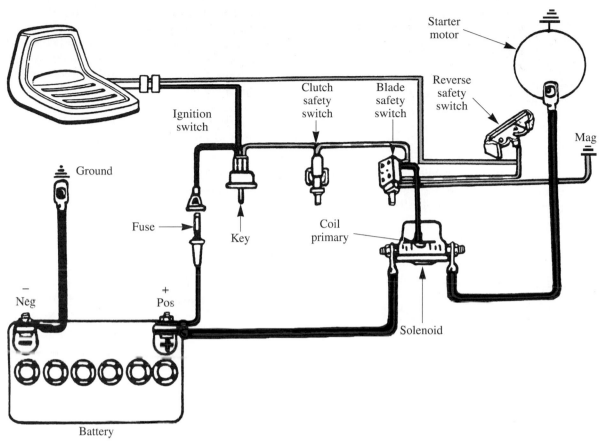

Typical Electric Start System

Figure 19-39. *Typical electric start system with safety interlock switches wired in series. (MTD Products, Inc.)*

Testing the interlock electrical system

The following testing procedures are provided by *MTD Products, Inc.* and correspond to the electrical interlock system shown in **Figure 19-39.**

1. Starting instructions.
   a. Disengage the blade or PTO.
   b. Depress the clutch pedal.
   c. Set the throttle (and choke if separate).
   d. Turn the ignition key to the *Start* position.

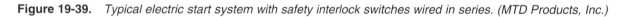

If the engine cranks but does not start, the problem *is not* with the interlock system. If the engine does not crank, use the following procedure to check out the system.

2. Check the two safety switches to see that the disengaging of the blade and the depressing of the clutch depresses the switch plunger a minimum of 1/8".

3. Check the fuse or circuit breaker between the positive terminal of the battery and the ignition switch. If the fuse or circuit breaker is blown the engine will not crank.

4. Check the following terminals to see that the wires are in place.
   a. The positive terminal of the battery. A large and a small wire should be fastened securely to this terminal. On some units both wires are cast into one clamp.
   b. The negative terminal on the battery and the ground to the frame.
   c. The ignition switch terminal.
   d The clutch safety switch.
   e. The blade safety switch.
   f. The solenoid terminals. A small wire is fastened to the coil primary and the two larger wires are fastened to each side.

5. Check the condition of the battery. Even if the battery is dead, a click should be heard at the

solenoid. Hearing a click at the solenoid verifies that the starting system is operating at least to that point. The specific gravity of the battery should be 1.265.

6. A *multimeter* (*continuity tester*) can be used to check the continuity between each component of the interlock system.

 To test the interlock system further, the safety switches will be bypassed. Make sure the clutch is disengaged and the blade engagement lever is in the disengaged position. If the clutch cannot be placed in the disengaged position, place the gear shift lever in the neutral (N) position. When using a jumper wire in the following tests the engine may crank over.

 Disconnect the spark plug wire and ground it against the engine block.

7. Use the jumper wire between the following points:

   a. The positive terminal on the battery to the terminal on the solenoid (coil primary). If the engine cranks, then test within this circuit to find the exact area of the problem. See steps b and c below.

   b. The positive terminal of the battery and the *S terminal* on the ignition switch. If the engine cranks, the problem is between the battery and the ignition switch.

   c. The *S terminal* on the ignition switch to the coil primary terminal on the solenoid. If the engine cranks, the problem is located between the ignition switch and the solenoid.

   d. Jump between the two large terminals on the solenoid. Only use wire that is at least as heavy as the wire from the solenoid to the starter. The wire is used with an alligator clip. If there is current up to the coil primary terminal of the solenoid and the starter will not crank, but you can crank the starter with the jumper wire, the problem is with the solenoid. Check to see if the base of the solenoid has a good ground to the frame of the unit. If it still fails to operate, replace the solenoid.

Testing the reverse safety switch

If the engine can be started, but it stalls when the blade is engaged, use the following procedure to determine if problem is in the reverse safety switch:

1. Disconnect wire going to magneto on engine.
2. Disconnect wire attached to spring switch.
3. Test continuity by attaching one lead of a multimeter to spring switch and other lead to ground. If there is continuity, the fiber washers could be damaged and should be replaced.

Testing the solenoid on electric start tractors

The following list contains possible solenoid problems. Replacement of the solenoid is required.

- **Solenoid is stuck.** Unit will start with ignition key in *Off* position.
- **Coil wire (inside solenoid) is bad.** Solenoid will not function.
- **Bad washer (inside solenoid).** Solenoid clicks, but starter motor does not turn.

Other problems that can appear to be a defective solenoid are as follows:

- Faulty ground.
- Defective safety switch.
- Discharged battery.
- Defective starter motor.
- Blown circuit breaker.
- Defective ignition switch.
- Defective wire harness.

Recoil Start System

The *recoil start system* is completely different from the electric start system. If the clutch is not depressed (disengaged), the blade is not disengaged, or the ignition key is not *On*, the ignition will be disabled and the engine cannot be started. In order for the blades and clutch to be engaged without stopping the engine, the recoil starter handle must be inserted into the dash panel and turned a quarter of a turn. This will disengage the wire that grounds the magneto. The safety switches are wired in parallel on the recoil start models. See **Figure 19-40.**

Testing the interlock system on the recoil start system

To test the interlock system on the recoil start system, start the engine as follows:

1. Disengage the blade.

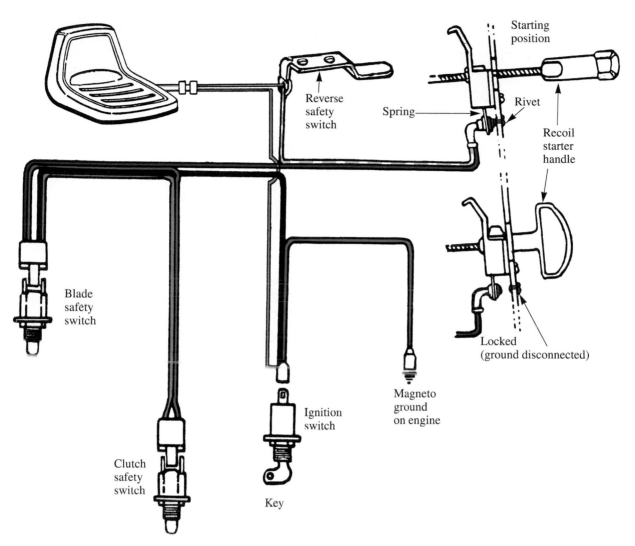

Typical Recoil Start System

Figure 19-40. *Typical recoil start system with safety interlock switches wired in parallel. (MTD Products, Inc.)*

2. Depress the clutch pedal and lock it in the disengaged position.
3. Set the throttle control.
4. Turn the ignition key to the *On* position.
5. Grasp the recoil starter handle and unlock it by twisting it 1/4 turn. Pull out sharply and hold it in the out position. See **Figure 19-41.**
6. Slowly let it rewind and pull it out again if the engine does not start.
7. After the engine starts, slowly let the recoil starter handle rewind and lock it into the dashboard by turning it a quarter turn. See **Figure 19-42.**

If the engine will not start and the gasoline shut-off valve is open, there is fuel in the gasoline

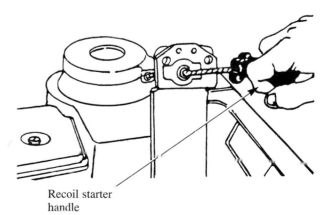

Recoil starter handle

Figure 19-41. *Recoil starter handle. (MTD Products, Inc.)*

Turn and lock

Locking Recoil Starter Handle

Figure 19-42. *Locking the recoil starter handle by turning it 1/4 turn. (MTD Products, Inc.)*

tank, and the spark plug wire is attached, then use the following procedure to determine if the problem is in the engine or the safety interlock system:

1. Check the two interlock switches to see that the disengaging of the blade and the depressing of the clutch depresses the red plunger a minimum of 1/8″.

 To determine if the problem is in the engine or interlock system, it is necessary to bypass the safety switches on the mower or tractor. This makes this procedure quite dangerous. Use extreme caution when performing these tests.

2. Disconnect yellow wire that connects ignition switch to primary wire of breaker assembly.

 The engine can no longer be shut off with the key.

3. If the engine starts, the problem is within the interlock system.
4. Check the grounding system behind the recoil starter handle. When the recoil starter handle is being pulled, the bolt on the spring should be grounded against the rivet. When the recoil starter handle is locked in place, the bolt on the spring should not touch the rivet.

 The recoil start system is not a fail-safe system. When a wire becomes unplugged from any component, it does not prevent starting as the electric start system does. If the engine can be engaged, or with the ignition key in the *Off* position, the unit should not be returned to the customer.

Testing the safety reverse switch on the recoil start system

To test the safety reverse switch on the recoil start system, proceed as follows:

The transmission lever must not be touching the reverse spring switch and the key must be in the *On* position.

1. Disconnect the yellow wire going to the magneto on the engine.
2. Disconnect the wire attached to the spring switch.
3. Attach one lead of a continuity tester to the spring switch and the other lead to ground. If there is continuity, the fiber washers could be damaged and should be replaced.

Testing procedure for operator present system (safety seat)

To check the operation of the safety seat, proceed as follows:

1. Start the unit as instructed in the owners guide.
2. Set the parking brake.
3. Place the shift lever in neutral gear.
4. Engage the PTO or blades.
5. Raise up off the seat (this will activate the seat kill mechanism).

At this point the engine should stop running. If unit continues to run, check wire lead and seat plug for proper connection. If this connection is satisfactory, then the seat switch mechanism and wire lead continuity should be inspected for shorts in the electrical system. See **Figure 19-43.**

Disassembly procedure

To disassemble the safety seat, proceed as follows:

1. Remove molding clip on lower front of seat. Remove molding.
2. Remove seat covering and foam padding.

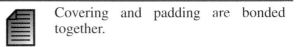 Covering and padding are bonded together.

3. Remove screws, metal plate, bushings and foam pad. See **Figure 19-44.**

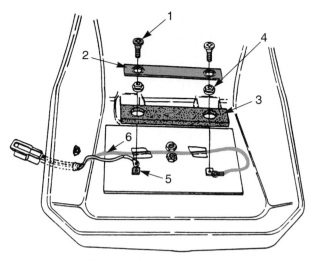

Safety Seat Electrical System

1. Shoulder screw
2. Seat switch base
3. Form seat switch

4. Shoulder spacer
5. Push-in nut
6. Wire harness

Figure 19-43. *Safety seat electrical system.* *(MTD Products, Inc.)*

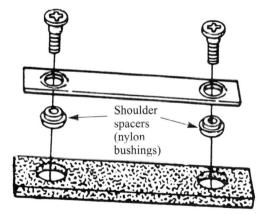

Safety Seat Disassembly

Figure 19-44. *Safety seat disassembly.* *(MTD Products, Inc.)*

> When reassembling, note the position of nylon bushings. The shoulder of the bushing must be placed upward through plate.

4. Check for broken terminal end or frayed plug wire. Tape to the bottom of seat pan.

Assembly procedure

Assemble seat in reverse order. Once assembled, check by pushing downward on metal plate. The distance between metal plate and screws must be maintained for proper switch operations.

> Some lawn riders and tractors may now have modified seat safety switch types and locations. One is illustrated in **Figure 19-45.**

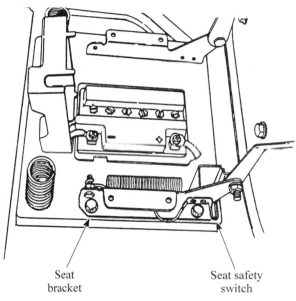

Safety Switch Location

Figure 19-45. *Safety switch location.* *(MTD Products, Inc.)*

Circuits for Study

The electrical circuits in **Figure 19-46** through **Figure 19-50** are for a 600 Series MTD tractors with a *Briggs & Stratton*, Vee, Twin-Cylinder, Vanguard, OHV engine. The electrical circuit in **Figure 19-51** is for a 800 Series MTD tractors with a *Briggs & Stratton*, Vee, Twin-Cylinder, Vanguard, OHV engine. They all are included for further study and practice. You should be able to identify and trace circuitry for components previously presented. On these circuits you will find additional circuitry for headlights and headlight switch, bulb monitors, PTO (Power-Take-Off) switch, and oil pressure switch.

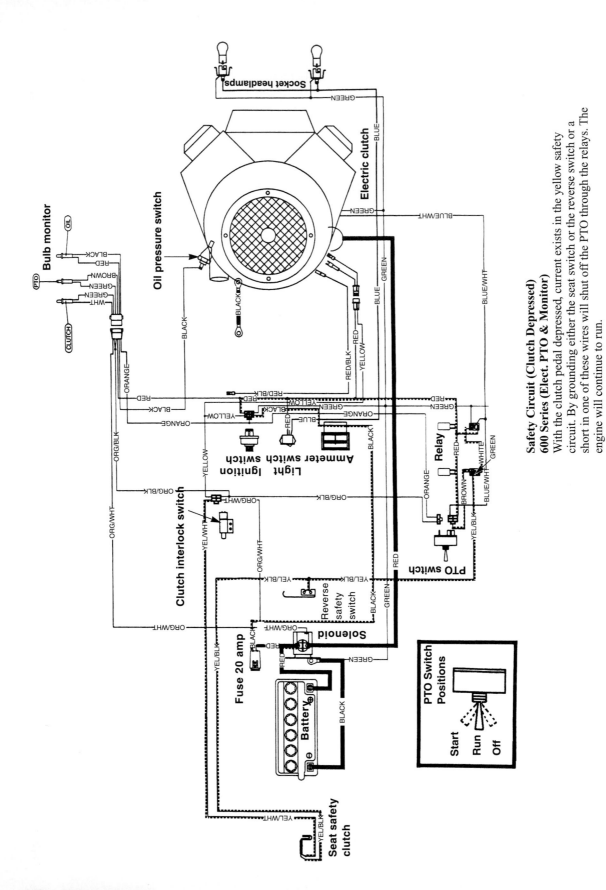

Figure 19-46. *Electrical circuit for 600 Series MTD tractors with Briggs & Stratton, Vee, Twin-Cylinder, Vanguard, OHV engines. (MTD Products, Inc.)*

**Safety Circuit (Clutch Depressed)
600 Series (Elect. PTO & Monitor)**
With the clutch pedal depressed, current exists in the yellow safety circuit. By grounding either the seat switch or the reverse switch or a short in one of these wires will shut off the PTO through the relays. The engine will continue to run.

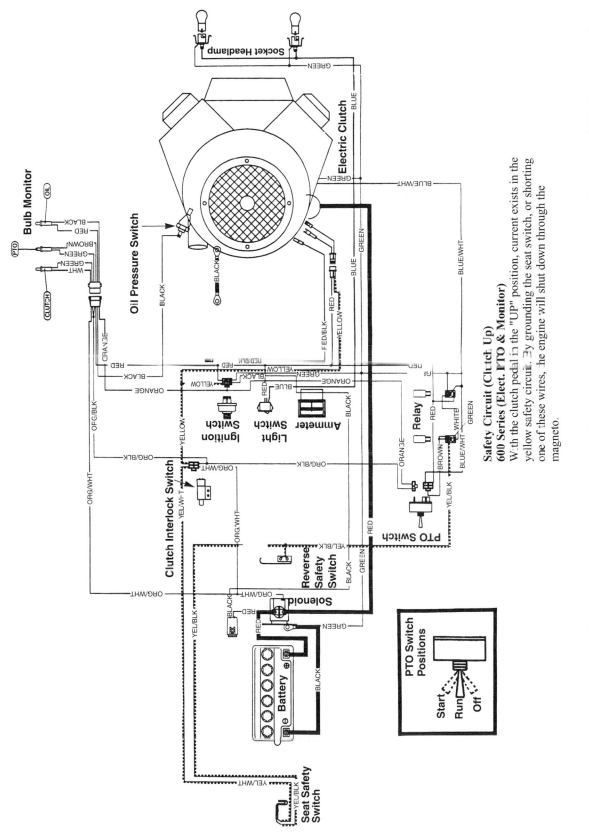

Safety Circuit (Clutch Up)
600 Series (Elect. PTO & Monitor)

With the clutch pedal in the "UP" position, current exists in the yellow safety circuit. By grounding the seat switch, or shorting one of these wires, the engine will shut down through the magneto.

Figure 19-47. *Electrical circuit for 600 Series MTD tractors with Briggs & Stratton, Vee, Twin-Cylinder, Vanguard, OHV engines. (MTD Products, Inc.)*

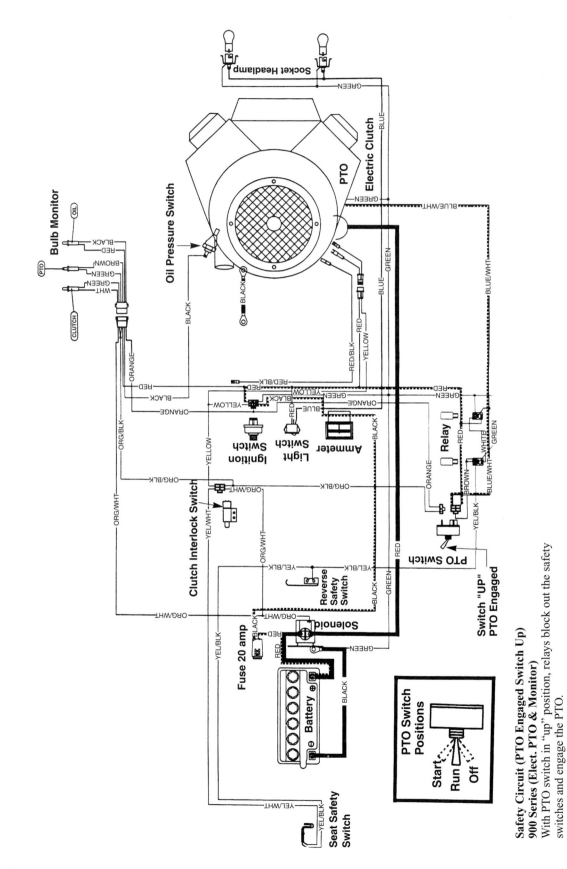

**Safety Circuit (PTO Engaged Switch Up)
900 Series (Elect. PTO & Monitor)**
With PTO switch in "up" position, relays block out the safety
switches and engage the PTO.

Figure 19-48. *Electrical circuit for 600 Series MTD tractors with Briggs & Stratton, Vee, Twin-Cylinder, Vanguard, OHV engines. (MTD Products, Inc.)*

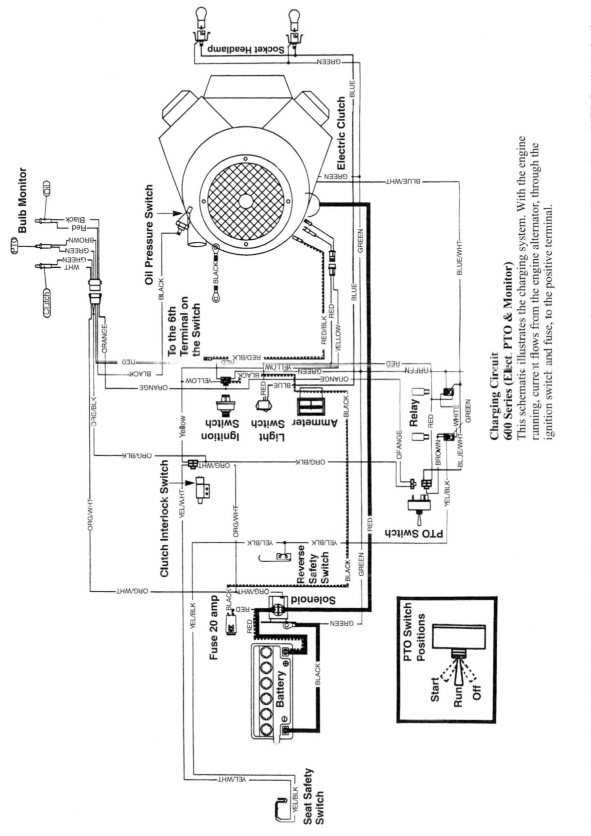

Figure 19-49. *Electrical circuit for 600 Series MTD tractors with Briggs & Stratton, Vee, Twin-Cylinder, Vanguard, OHV engines. (MTD Products, Inc.)*

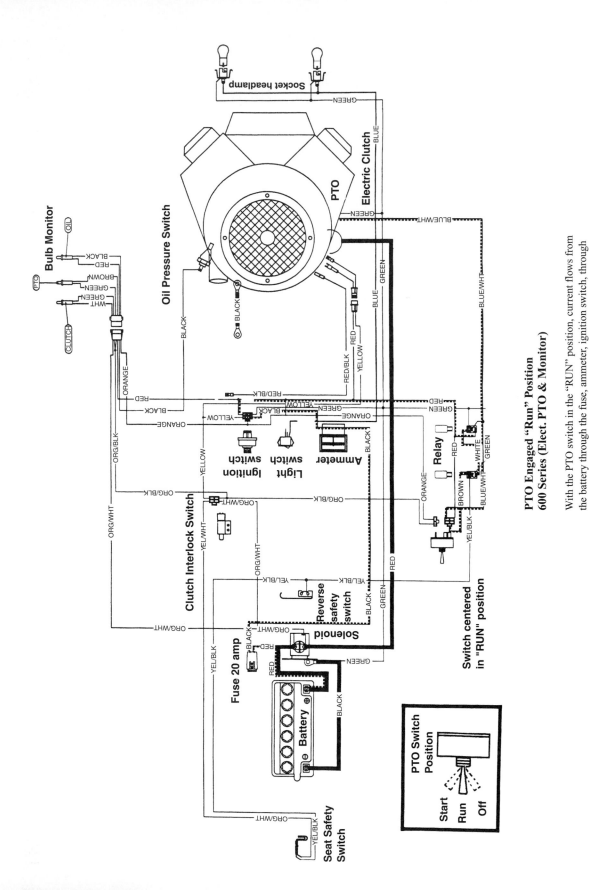

**PTO Engaged "Run" Position
600 Series (Elect. PTO & Monitor)**

With the PTO switch in the "RUN" position, current flows from the battery through the fuse, ammeter, ignition switch, through the PTO switch and relays, engaging the PTO.

Figure 19-50. *Electrical circuit for 600 Series MTD tractors with Briggs & Stratton, Vee, Twin-Cylinder, Vanguard, OHV engines. (MTD Products, Inc.)*

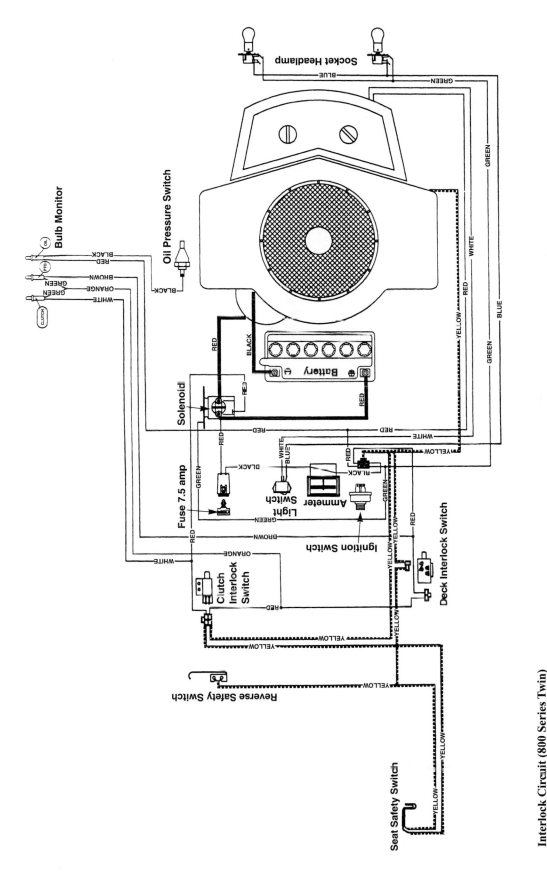

Interlock Circuit (800 Series Twin)
With the engine running, current exits in the safety circuit.
By grounding the reverse switch or the seat switch or a short
the yellow wires, the engine will shut down.

Figure 19-51. *Electrical circuit for 800 Series MTD tractors with Briggs & Stratton, Vee, Twin-Cylinder, Vanguard, OHV engines. (MTD Products, Inc.)*

The information on electrical systems in this chapter is presented for educational purposes only. It is not intended to be directly applicable to all products, though there may be some similarities. Manufacturer's technical service manuals for a particular make and model should be consulted when troubleshooting and servicing engine powered products.

General Maintenance

Maintenance is an important part of working with small gas engines and the machines they power. Lawn and garden tractor engines should be maintained regularly to keep them in good operating condition. A regular maintenance schedule will extend the life of the engine. The following is a list of considerations for performing tractor maintenance safety:

1. Keep maintenance area clean and dry.
2. *Never* lubricate or service machine while it is moving or engine is running.
3. Keep hands, feet, and clothing away from power driven parts.
4. Disengage all power and operate controls to relieve pressure. Lower equipment to the ground. Stop the engine. Remove the key. Allow machine to cool.
5. Securely support machine elements that must be raised for service work.
6. Disconnect battery ground, or negative, cable (–) before making adjustments on electrical systems or welding on machine.
7. Dispose of waste materials safely. Used oil, fuel, brake fluid, and batteries are harmful to the environment. Refer to local *recycle center* for disposal information.
8. Dispose of cleaning rags in a safe manner. Never leave used cleaning rags lying around. ***Spontaneous combustion*** can occur and result in a serious fire. Keep fire extinguishers available and fully charged. Consult with local fire department for information.

Refer to Chapter 1 of this text for more safety information.

Spark plugs

The spark plug should be removed for inspection and cleaning. Blow dirt away from the base of the spark plug before removing it with compressed air. Remove the high tension wire by pulling on the insulating boot only. Do not pull or jerk the wire. Install a deep spark plug socket socket on the plug and carefully turn the ratchet handle counterclockwise. Unscrew the spark plug and examine the electrode end of the plug. The electrodes and ceramic insulator should be dry. If it is wet with fuel or oil there may be more serious engine problems. The following are several reasons for a wet spark plug:
- Over choking.
- Excessively rich fuel mixture.
- Water in fuel.

If the insulator is dry and has a beige or gray/tan color, the plug is in good condition. Next, examine the electrodes for any erosion, burning or carbon fouling. Refer to *Spark Plug Condition* chart in the *Appendix* of this text. If the spark plug insulator and electrodes appear to be good, check the electrode gap with a wire type spark plug gapping tool. The engine service manual will specify the proper gap setting. If the gap is too large, carefully bend the outer electrode toward the center electrode until the gap is correct. Use the gap setting tool. See **Figure 19-52.** If the spark plug appears to be beyond simple servicing, replace it with a new one. Refer to the manufacturer's technical service manuals for the proper replacement spark plug.

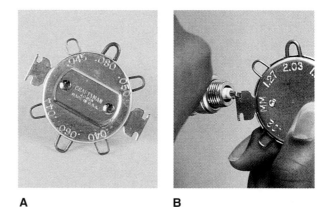

A   B

Figure 19-52. *A—A common spark plug gapping tool. B—Use a spark plug gapping tool to bend the outer electrode toward or away from the center electrode.*

Before replacing the spark plug clean the external ceramic insulator by wiping it with a clean cloth. Condensation can cause flashover (sparking externally) and erratic running if the insulator has a coating of dirt on it. If the metal base is rusty, clean it on a wire wheel or with a wire brush. If the threads are dirty, they should be wiped clean or wire brushed before inserting the plug back into the spark plug hole. The gasket should be in good condition or replaced with a new one.

Insert the spark plug in the spark plug hole and turn it clockwise by hand until it stops. Be careful not to cross thread the spark plug in the hole. If it will not turn, do not force it. Remove it and examine the threads in the hole. If they are damaged, a thread chaser (special thread tapping tool) may be needed to clean and correct the threads. If the spark plug turns smoothly by hand until it seats, tighten the plug with a torque wrench to 13 to 15 lb-ft. Do not overtighten.

Examine the high tension wire for deterioration before replacing it on the spark plug. Heat and dirt can make the insulation brittle and insulation cracking can occur. Cracked wire insulation can allow sparking to metallic parts of the engine and cause hard starting and erratic running. Oil or fuel soaked wires will leak current and cause weak firing of the plug. The insulating boot should be dry and in good condition.

Engine lubrication

Using the manufacturer recommended engine oil assures that the internal parts are kept lubricated during all running conditions. Changing oil and the oil filter cartridge at proper specified intervals will greatly extend the service life of the engine. Draining the oil removes the contaminated oil and any particles from the crankcase that might cause internal abrading. The manufacturer's technical service manual will specify the number of hours of engine operation before changing oil. A new engine must have the oil changed after 5 to 8 hours to remove any residual manufacturing contaminants and break-in material. If the engine uses an oil filter cartridge it should also be replaced at the first oil change. The oil should be drained after the engine has been run long enough to heat the oil. A drain plug is located at the lowest point in the crankcase. Drain the oil into a metal or plastic pan and inspect it for metal particles. Save the old oil in a sealed container and take it to a recycle center. Never pour oil into a drain or onto the ground.

Select the correct oil filter cartridge for replacement. Remove the old filter cartridge with an oil filter cartridge wrench. Apply a film of engine oil to the rubber gasket and thread the cartridge onto the engine until hand tight. Use the oil filter cartridge wrench to turn the cartridge another 1/2 to 3/4 turn to seal it. Do not over tighten the cartridge.

When replacing engine lubricating oil the correct SAE viscosity and API service type should be used. Refer to the engine manufacturer's recommendation. Replace the oil plug and install a new filter cartridge. You should become familiar with reading the oil dip stick. Remove the oil filler cap and dip stick and pour in the new oil. Do not overfill with oil. Excess oil can cause spark plug fouling and other internal problems. Wipe and check the dip stick until the oil level shows full. Run the engine for 30 seconds, stop the engine, and check the oil level again. Add oil if necessary. Check oil level before each use of the tractor.

Optional oil accessories

The engine may have an oil pressure switch with a warning light on the instrument panel. If the engine runs low on oil, the switch will activate the warning light or stop the engine. An oil level light may also be located on the instrument panel to indicate a low level condition. If the level is low, the light flickers when the engine is started. The engine should be stopped and oil added to the *Full* mark on the dip stick.

Air filter

Internal combustion engines consume a far greater volume of air than fuel. The air filter prevents dirt and abrasive particles in the air from entering the carburetor and engine. The air filter is installed on the intake side of the carburetor. The filter material is contained in a metal or plastic case. The air filter may be of the single or dual element, dry-type filter. To service the air filter, remove the cover and take out the filter element. Examine the element for dirt accumulation. Clean the element by tapping it on a flat surface to shake off any dirt particles. Do not use compressed air or solvents. Replace the element if it still appears to have dirt on it. Air filters should be serviced about every 25 hours of use. The time may vary depending upon the working conditions in which the tractor is used. Very dusty conditions require more frequent service.

Mufflers

The muffler is a sound deadening device, but it also affects engine efficiency. Mufflers are designed to control the amount of back pressure generated by the exhaust gasses. Check the muffler periodically to see that it is not rusting and leaking exhaust around the flanges or gaskets. If the inside baffles corrode away the sound emitted will change and become louder. This is an indication that a new muffler is needed. Sulfuric acid is one of the by products of combustion and causes corrosion inside the muffler. Mufflers get very hot when the engine is running. Let the engine cool down before touching or making any examination of the exhaust system. If an exhaust spark arrestor is required, it should also be removed and inspected.

Engine cleaning

Tractor engines can get very dirty and greatly reduce cooling efficiency. Remove any grass or other materials that have become lodged in the cooling fins. If necessary, wash the cold engine to remove soil—use water from a garden hose. If leaking oil has mixed with dirt and is caked on the engine, use a degreasing solvent. Spray the solvent on the cold engine and then wash it off. Follow the instructions on the solvent container

Radiator cleaning

Some tractors have water-cooled engines with a radiator like those found in automobiles. Clean radiator cooling fins by directing low pressure water from a garden hose through the fins. Do this from the back side of the radiator. The coolant inside of the radiator and engine block should be drained each year. Flush the cooling system with a flushing compound and install new antifreeze. The antifreeze should be a mixture of 50% water, and 50% Ethylene glycol to lower the freezing point to −34°F (−38°C). This mixture is used in warm climates also because the boiling point of the liquid is increased to 265°F (129°C). If the antifreeze does not contain a rust inhibitor, one should be

added to the cooling system. Rust inhibitors help keep the cooling system clean.

⚠ Dispose of used coolant properly. Do not allow it to enter drainage systems. Do not pour it into sink drains. Ethylene glycol is poisonous to animals and humans if ingested.

Battery maintenance

The battery may be maintenance free or conventional lead-acid type. If the battery is a conventional lead-acid type, the fluid (electrolyte) should be checked each time the tractor is used. Add only distilled water to bring the electrolyte to the proper level. The electrolyte should always be kept above the lead plates.

After the specified electrolyte has been installed at the initial service of a maintenance free battery, the sealed caps should never be removed. Checking electrolyte level or adding distilled water is not necessary when servicing a maintenance free batteries.

Clean the battery terminals for both the maintenance free and conventional lead-acid type. If a white, pasty, corrosive substance is found on the battery terminals and connectors wash it off with 1/4 cup of baking soda mixed with a quart of water. Pour the solution over the terminals and rinse with clear water. Then dry the terminals. Coat the terminals with Vaseline petroleum jelly.

⚠ Wash your hands thoroughly after touching or handling a lead acid battery to prevent acid burns. Protect your eyes with safety glasses.

If the tractor is not used for long periods keep the battery charged with a trickle charger. A trickle charger provides a very low charging current. It can be left connected to the battery indefinitely without harm.

If a battery no longer will hold a charge, it should be replaced. If this is the case, a maintenance free battery should never be replaced with a conventional lead-acid battery. The electrical system is designed exclusively for a maintenance free battery. The electrical system will not work properly with a conventional lead-acid battery substituted for a maintenance free battery.

Sharpening and balancing mower blades

A sharpened and balanced mower blade, results in an even cut. To sharpen and balance mower blades, proceed as follows:
1. Wear goggles and gloves when removing and sharpening blades.
2. Sharpen blades with a grinder, hand file, or electric blade sharpener. Maintain the original bevel angle of the blade.
3. Balance the blade as follows:
   a. Clean debris from blade.
   b. Put blade on nail in wall or vise. See **Figure 19-53.**
   c. Turn blade to horizontal position. Heavy end will drop.
   d. Remove metal from heavy end by grinding or filing cutting edge until blade remains horizontal.

> Other methods of balancing mower blades are covered in Chapter 18 of this text.

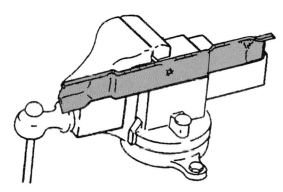

Figure 19-53. *Balancing mower blade. (Deere & Co.)*

Fueling and refueling

The following is the procedure for refueling a tractor safely:
1. *Do not* fuel or refuel machine:
   a. While smoking.
   b. When machine is near open flame or sparks.
   c. When engine is running. *Stop* engine.
2. Refuel outdoors.
3. Keep a first aid kit and fire extinguisher close by.
4. Have emergency phone number(s) readily accessible.

Storing a tractor

When storing a tractor, use a fuel stabilizer to prevent carburetor varnish from forming or partial plugging of carburetor jets. Either condition can cause engine to run lean and hot with possible piston seizure and engine failure. When storing a tractor, proceed as follows:
1. Run engine for at least 10 minutes to distribute stabilizer in carburetor.
2. Change engine oil while it is warm.
3. Repair any worn or damaged parts. Install new parts as necessary.
4. Service, or replace air cleaner.
5. Lubricate all grease fittings and wipe clean.
6. Wash tractor with mild soap and water. Wipe dry.
7. Start engine and engage PTO for a couple minutes to remove water. *Stop* engine.
8. Remove any oil, grease, or dirt from and around engine.
9. Apply paint to bare metal areas to prevent rust.
10. Remove battery:
    a. Clean battery.
    b. Check electrolyte level.
    c. Charge battery.
    d. Store battery in a cool, dry place where children cannot reach it.
11. Clean belts with a damp rag. Do not use petroleum based solvents on belts. *Lock* park brake to relieve transmission drive belt tension.
12. Close fuel shut-off valve.
13. Remove and clean mower unit.
14. Check the mower belts and relieve drive belt tension.
15. Remove, sharpen and balance mower blades.
16. Store tractor in a dry, protected location. If stored outside, place a waterproof cover over it.
17. Take the weight off the tires by placing blocks or stands under tractor. Let about one-third of air out of tires.

Removing a tractor from storage

To remove a tractor from storage, proceed as follows:
1. Inflate tires to correct pressures.
2. Remove any blocks or chocks from under tractor.

3. Clean and gap spark plug(s).
4. Check fluid level in battery. Install and charge if necessary.
5. Adjust transmission drive belt.
6. Lubricate all grease fittings.
7. All shields and guards should be in place.
8. Move mower drive belt tension lever *In* to tighten belt.
9. Check engine oil level.
10. Check coolant level.
11. Open fuel shut-off valve.
12. Fill fuel tank with fresh fuel.
13. Run engine 5 minutes with no load. Make sure the area is well ventilated.

Summary

Working with a tractor requires good operator safety understanding and attitude. Like other engine powered machines, there is inherent danger when used carelessly. All safety instructions provided by the manufacturer of the vehicle should be studied before attempting to use an unfamiliar machine and its attachments. Children should be kept out of the work area and under supervision. Mowing in reverse can be very dangerous and should always be avoided. If done, extreme caution and vigilance should be maintained. Remember, older machines may not have the safety interlock systems that are provided in modern equipment.

Lawn and garden tractors are very versatile machines but they are generally smaller with less horsepower than farm type tractors. Lawn and garden tractors can accomplish a wide variety of jobs with the appropriate accessories. Lawn mowing and mulching, leaf collection, plowing and throwing or blowing snow, soil preparation, trailer pulling, front-end loading, pulling loads, augering holes, rotary sweeping, and backhoeing are some of the types of work that can be done.

Nearly all engines used in lawn and garden tractors are of the four-cycle, electric start, with overhead valves and electronic ignition systems. They are trouble free if proper care and maintenance is followed. Engine horsepower may vary from tractor to tractor from about 14 hp to 22 hp. Some engines are of the single cylinder type and others have two opposed or vee configured cylinders. Some engines have a cast-iron block, others may have aluminum blocks with cast-iron inserted cylinders. Overhead valves with hydraulic valve

lifters are most efficient and provide greatest service life.

External components of the engine should be easy to access and service as needed. Fueling; oil checking, changing, and filling; air filter inspection and changing; and fuel filter inspection are items that need attention periodically. Battery fluid level and condition also need regular inspection and attention. Exhaust should be directed away from the operator at all times and conditions.

The chassis and steering portions of a tractor must be strong enough to withstand twisting and torsional load imposed upon it while holding all other parts in proper alignment. Steering should be smooth and positive in action even when running over rough ground. Zerk fittings are utilized at strategic points to enable greasing to be done with a hand operated grease gun. The steering wheel should be of such size, angle, and location that it is comfortable for the driver to operate without straining muscles. Four wheel steering turns the front and rear wheels for better maneuvering around trees and other sharp turns. This is available on some tractors.

The transmission, usually called a transaxle, is a hard working part of the tractor and is driven by the engine with either a V-belt or drive shaft. Variable-speed pulley systems are often used with a single, or two-speed, geared transmission. This allows smooth, gradual, speed changes within a particular selected gear ratio at the transmission. A clutch mechanism is used to disengage the engine from the transmission during gear changes.

Hydrostatic transmissions consist of a variable displacement hydraulic pump and a fixed displacement hydraulic motor. Both of these units, the differential and axles are all contained in one housing. This type of transmission has many advantages over other types of transmissions. Hydrostatic transmissions may be driven from the engine by belts or drive shaft. In the neutral mode the transmission provides braking which eliminates the need for a separate disc type braking system. Cooling is provided by fins on the housing and a fan attached to the input shaft.

Tractor electrical systems incorporate safety interlocks. One method requires mower decks to be disengaged before a tractor can be shifted into reverse. Another interlock switch shuts off the engine if the operator leaves the seat when the blades or PTO (power-take-off) is engaged. An interlock switch may be installed that requires the parking brake be engaged before the operator leaves the seat to prevent the tractor from rolling on an incline without a driver.

An electric start system may incorporate interlocks that require the clutch pedal to be depressed and the blade disengaged to start the engine with the key switch. The safety switches are wired in series so that any one switch can prevent starting.

Recoil (rope pull) starting systems are different from electric start systems. The interlock switches are wired in parallel circuits.

Know These Terms

mulching
compost
single-stage snow thrower
two-stage snow blower
rotary garden tilling
power-take-off
PTO
three-point hitch
overhead valves
chassis
grease (zerk) fitting
four wheel steering
transaxle
variable-speed pulley
moveable sheave
speed ranges
hydrostatic transmission

ball piston pump
reservoir
cavitation
variable displacement
swash plate
spring-loaded check valve
differential gears
safety interlock
reverse safety switch
ANSI
seat-activated switch
electric start system
solenoid
series
multimeter (continuity tester)
recoil start system
spontaneous combustion

Chapter 19 Review Questions

Answer the following questions on a separate sheet of paper.

1. List the ten kinds of work done with garden tractors.
2. What are the three techniques for mowing lawns?

3. What tractor accessory is available to improve traction for snow plowing?
4. When using a front end loader on a tractor what should be attached to the rear of the tractor?
5. A tractor outfitted for front end loading will have hydraulic _____ and hydraulic _____ to operate the bucket.
6. List six components that are common on nearly all garden tractor engines today.
7. _____ and _____ are the two forces that act upon the chassis during tractor operations.
8. _____ is an excellent material used for front axles.
9. The grease fittings are called _____ fittings.
10. On a four wheel steer tractor when the front wheels turn right, the rear wheels turn _____ after 45° of front wheel turn.
11. About a(n) _____° delay of turn between the front and rear wheels is intentional for safety reasons.
12. The transaxles of tractors may be driven from the engine by _____ or a(n) _____.
13. A low speed gear will have _____ pulling power.
14. The center portion of a variable-speed pulley is called a moveable _____.
15. The variable-speed pulley allows the operator to smoothly change pulley size _____ and thereby slow or speed-up the tractor.
16. The hydrostatic transaxle is a self contained unit consisting of what four primary components?
17. List nine advantages of the hydrostatic transmission over other types.
18. When does the hydrostatic transmission act as brakes for the tractor?
19. How is reverse obtained with the hydrostatic transmission?
20. Electrical interlocks are incorporated in tractors for _____ reasons.
21. ANSI requested that after _____ tractors have *operator present* interlock switches installed in electrical systems.
22. Electric start systems have interlock switches wired in _____.
23. Recoil (rope start) systems have interlock switches wired in _____.
24. Unless it is absolutely necessary, mowing with a tractor should not be done in _____.

Suggested Activities

1. Visit a store or tractor dealership and examine the various tractor models available. Ask questions about them.
2. Demonstrate proper procedures for normal tractor maintenance such as: oil, air filter, fuel filter, coolant (if used), battery, fuel, tire pressures.
3. Practice proper engine starting and stopping procedure.
4. Examine tractor drive belts. Replace belts if needed.
5. Lubricate a tractor chassis.
6. Show how a variable-speed drive pulley operates.
7. Practice some electrical tests following instructions in a tractor technical service manual for the make and model being tested.
8. Locate the safety interlock switches on a new model tractor.
9. Demonstrate safe procedure for attaching and removing a lawn cutting deck or other implement to a tractor.
10. Test drive a tractor to see if all systems are working properly. Note any problems and take corrective action.

Snow Throwers

After studying this chapter you will be able to:
▼ Safely operate and service snow throwing equipment.
▼ List important purchasing considerations for snow throwers.
▼ Identify major parts of walk behind snow throwers.
▼ Make adjustments to snow throwers.
▼ Properly maintain snow throwing machines.

Operating Safely

Safety is of primary importance when working with small gas engines. It is expected that persons servicing their own equipment, or the equipment of others, should at all times use good judgment. Before attempting to operate a snow thrower all safety instructions provided by the manufacturer should be read and understood. Safety signs and labels placed on the machine should be read and understood before use.

Because there have been accidents involving users of snow throwers, it is important to be aware of certain possible dangers inherent in these machines. In practically every case, accidents resulted from a disregard of safety instructions. Too often consumers neglect to read instructions provided by the manufacturer or take chances that result in serious injuries. Unfortunately, most the safety devices and guards provided by manufacturers can be circumvented by carelessness, ignorance, or a disregard for danger. Good judgment must always

be used when operating or working with any powered equipment. The specific make and model should be thoroughly understood by the operator or technician servicing a machine. Manufacturer's technical service manuals should be consulted whenever there is a question regarding safe repair, maintenance, or operating procedures.

Learn how to properly operate the machine and its controls. Keep the machine in proper operating condition. Do not let anyone operate the snow thrower who has not had proper instruction about its use. The following safety precautions are typical of those provided by manufacturers and cover most situations.

Protect children

When operating a snow thrower and approaching blind corners, shrubs, trees, or other objects that may block vision, be alert for children. *Never* let children operate the snow thrower.

Children are curious by nature, and are often attracted to operating machinery and associated activity. Children should be kept in the house under supervision, while a snow thrower is being operated. If children enter a work area, turn the snow thrower off. Take the children out of the area to a safe location where they will be supervised.

Preparation for using a snow thrower

Before using a snow thrower, there are a number of safety-related items that the operator should be aware of. The following is a list of these items:

1. Never attempt to make adjustments while the engine is running.
2. Disengage all clutches (release drive levers) before starting engine.
3. Wear proper winter clothing. See **Figure 20-1.** Wear footwear with soles that improve footing on slippery surfaces. Do not wear dangling items, such as scarves outside the coat, draw-strings, etc.

Figure 20-2. *Gasoline should be stored in an approved and clearly marked container.* Never *store gasoline, or other flammable liquids in glass containers. Refuel outdoors away from any flames.* (Jacobsen Manufacturing Co.)

Figure 20-1. *Gasoline powered snow throwers are an effective way of removing snow. Proper attire and well-maintained equipment can make the work safe and efficient. (Deere & Co.)*

4. Handle gasoline carefully, away from any open flame or sparks.
   a. Use an approved gasoline container. See **Figure 20-2.**
   b. Never remove fuel tank cap or attempt to refuel a running or hot engine.
   c. Never add fuel indoors.
   d. Wipe up spilled gasoline and dispose of rags promptly.
5. Adjust skid shoes to clear gravel or crushed rock surface. See **Figure 20-3.** See *Adjusting Skid Shoes* section in this chapter for this adjusting procedure.

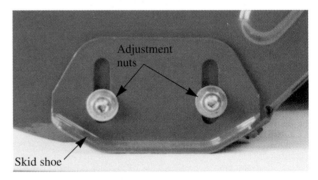

Figure 20-3. *The skid shoe attached to the auger housing is adjustable to establish intake height and reduce wear on the housing edges.*

Using a snow thrower

Before attempting to operate the snow thrower, read the manufacturer's technical service manuals. The following is a list of rules to adhere to when using a snow thrower:

1. Inspect the machine and be sure all hardware is tight. Damaged or badly worn parts should be repaired or replaced. Guards and shields should be in good condition and in place. If adjustments are needed, make them before attempting to operate the machine.
2. Examine and clear the work area of objects that may be thrown. People should be well away from the work area and pets should be safely confined. If a person or pet does enter the work area, stop the machine.

3. If an object is hit, stop and inspect the machine. Make any repairs before resuming operation. Properly maintain the machine. All shields and guards should be in place at all times.

4. Do not allow snow to be thrown toward people, cars, buildings, windows, or anything that may be damaged.

5. Do not use a snow thrower above ground level, such as on a roof of a building.

6. If clearing snow from a gravel path be extremely careful. Be aware that gravel may be picked up with the snow.

7. Do not leave a running machine unattended.

8. Operate the machine only during daylight or in well lighted areas. Some machines have a headlight for night operation.

9. Make sure to have good footing. Operate machinery slower on a slippery surface. It is important to be especially careful when pulling a running machine backwards.

10. Do not blow snow *across* a sloped surface. Avoid clearing snow from steep slopes and keep away from edges of ditches and banks. If slopes must be cleared, do so up and down the slopes. Never clear going across the face of the slopes. Be careful when changing direction on slopes.

11. Avoid sudden starts and stops. Keep a firm grip on the machine handles.

12. Do not run the engine in a confined space, such as a shop or garage. Dangerous carbon monoxide from exhaust fumes can accumulate in such places.

13. Keep hands and feet away from rotating parts. Stay clear of the discharge chute opening at all times.

14. Always use a grounded, 3-wire plug receptacle for electric starting.

15. Do not overload the machine capacity by clearing snow at too fast of a ground speed.

16. Disengage auger drive when transporting unit, or when it is not in use.

17. Do not change engine governor speed settings or overspeed the engine.

Maintenance and storage

Keeping snow throwers in good working order is an essential part of safe work habits. Storing the equipment properly will prevent any possible accidents. The following is a list of safety considerations for maintaining and storing snow throwers:

1. Keep all nuts, bolts, and screws tight. Check guards and shields.

2. Never store equipment with gasoline in the tank in a building where fumes may reach an open flame or spark. Remember, light switches may produce unseen sparks that could ignite gas fumes. Allow the engine to cool before storing in an enclosure.

3. For extended storage refer to the manufacturer's technical service manuals.

4. Run the auger drive after each use to clear out snow and prevent freeze up.

5. As a reminder to the operator of the most important precautions, safety and warning decals are placed at strategic locations on the snow thrower. They must be read and obeyed. If any decals are lost, damaged, or worn off, replace them. These decals are available at a local snow thrower dealer.

Purchasing Considerations

Walk-behind snow removal equipment is available for use in geographical areas where snowy conditions are an annual event. See **Figure 20-1.** Gasoline engines provide the power for the large majority of snow thrower machines. Because there are various sizes and kinds of snow throwing equipment, the selection of one should be researched thoroughly.

Single-stage snow throwers and two-stage snow throwers perform the same function of snow removal. However, there is some difference in the way they perform the function. For practical purposes the term snow thrower will generally be used except where specific reference to single-stage or two-stage type is made.

Machine size and type

Snow throwers are available in various sizes and work capacities to meet the needs of varying conditions of snow removal. Engines for snow throwers are designed for easy starting in low temperatures.

Small snow throwers

Small machines are available where snowfalls are generally light and the area to be cleaned is more limited, such as steps, walks, and small driveways. The cleaning width may be 16″ to 20″. Engine size is about 3 horsepower (hp), and two-cycle engines are most common. Almost all small machines are recoil start (Rope pull). See **Figure 20-4.** The handle on recoil start pull ropes should be large enough to grasp easily with a gloved hand. See **Figure 20-5.** Electric start is sometimes an option for this size of snow thrower. See **Figure 20-6.**

Small snow throwers are single-stage type. *Single-stage snow throwers* have steel or rubber blades that rotate like paddles and throw the snow upward and forward. See **Figure 20-7.** The chute can be rotated to the left, right, or straight ahead.

See **Figure 20-8.** Direction of snow throwing is controlled by a discharge chute located above the housing. Direction and distance of snow throwing may be more limited compared to larger machines. The maximum quantity of snow removal for small snow throwers is about 1000 lb per minute depending upon other factors.

Figure 20-4. *The recoil start rope handle on snow throwers should provide an easy, secure grasp with a gloved hand. The engine should start quickly with one or two pulls, even in cold weather. (The Toro Co.)*

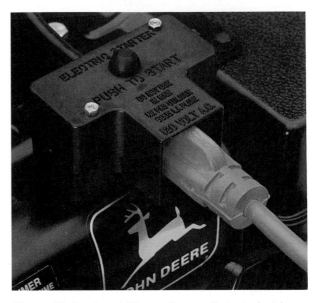

Figure 20-6. *In addition to the recoil start, electric starters may also be installed for convenience where an electric source is readily available from an extension cord. (Deere & Co.)*

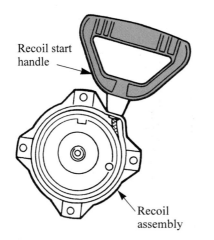

Recoil start handle

Recoil assembly

Figure 20-5. *A recoil start handle designed to accommodate a gloved hand for snow thrower engines. (The Toro Co.)*

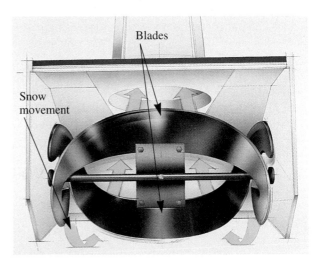

Blades

Snow movement

Figure 20-7. *Some smaller snow throwers have paddle-like impellers that throw the snow upward and out the discharge chute. Various designs exist. (The Toro Co.)*

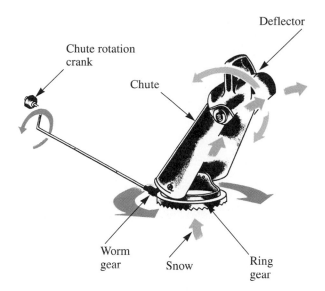

Figure 20-8. *By turning the chute rotation crank, the discharge chute can be rotated left, right, or straight ahead. The deflector can be raised or lowered to establish throwing distance. (Simplicity Manufacturing Inc.)*

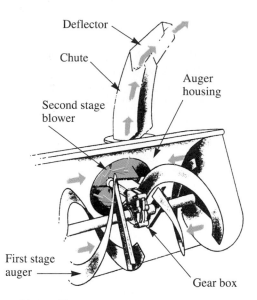

Figure 20-9. *Two-stage snow throwers have an auger to draw snow into a high-speed, second stage blower fan. Single-stage snow throwers do not have a blower fan. (Simplicity Manufacturing Inc.)*

Small snow throwers are pushed manually through the snow. They have small wheels with hard plastic tires. These tires should be checked every season. These tires can crack and make it difficult to push the equipment. If needed, replacement tires can be found at a local dealer.

Mid-size snow throwers

Mid-size snow throwers are heavier and have engines from 3 hp to 8 hp. The engine may be two-cycle or four-cycle. Mid-size machines may be either single-stage or two-stage type. **Figure 20-9** illustrates a two-stage type. Appropriate SAE oil weight commensurate with the temperature should be used. Remember, two-cycle engines must have oil mixed in proper proportions with the gasoline prior to fueling. Mid-size snow throwers have a chute and deflector to direct the snow and control the height and distance to which it is thrown. The chutes can be rotated as much as 230°. See **Figure 20-8.**

Some mid-size snow throwers are self-propelled. Three or four forward speeds, and one or two reverse speeds are available on most self-propelled snow throwers. Wheels are pneumatic rubber with heavy treads for traction. See **Figure 20-10.** With the traction engagement knob, the wheels can be adjusted to provide one or two wheel traction. A free wheeling feature allows for easy transporting. Tire pressure should be periodically checked. Proper air pressure in the tires help with traction, and eliminate improper wear of the tire tread. See **Figure 20-11.**

Figure 20-10. *Heavy treads on rubber pneumatic tires provide good traction in snow. Chains can be installed to increase traction on ice. The traction knob is used to provide traction to one, or both wheels. (Deere & Co.)*

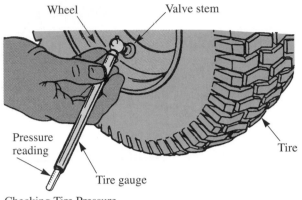

Figure 20-11. *Proper tire inflation should be maintained by checking with a tire gauge. (Simplicity Manufacturing Inc.)*

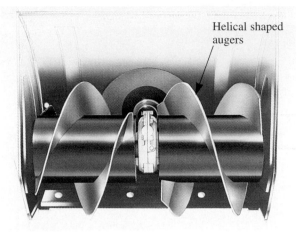

Figure 20-13. *A helical steel auger for pulling the snow into the machine. (The Toro Co.)*

Electric start may be a standard feature or an option for a mid-size unit. Ignition systems are electronic (CDI modules) and alternators are available for powering a sealed beam headlight for night snow removal. See **Figure 20-12.**

The augers are helical shaped steel that rotate and pull the snow into the machine. See **Figure 20-13.** Some augers have serrated edges to help break up ice and snow into finer particles before entering the second stage blower. See **Figure 20-14.**

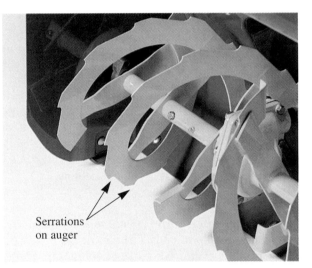

Figure 20-14. *An auger with serrated edges can break up ice and snow into finer particles before entering the second stage blower. (Deere & Co.)*

Heavy-duty snow throwers

Heavy-duty snow throwers range from 8 hp to 20 hp, and are exclusively four-cycle overhead valve engines. They have all the same features as mid-size machines, but are heavier and clean wider paths. The greater horsepower allows larger quantities of snow to be removed. They can remove up to about 2500 lb per minute. The clearing width may be as great as 32″.

Heavy-duty snow throwers are self-propelled, two-stage machines. Two-stage machines have an auger that pulls the snow into the auger housing. The snow is then picked up by a blower with blades

Figure 20-12. *This model has a sealed beam headlamp for night work. Unless the electrical system is voltage regulated, the light will brighten and dim as engine speed varies. (Deere & Co.)*

that force the snow at high velocity out through the chute above the machine. The distance snow is thrown may be as far as 40′, but this is controllable by a deflector at the top of the chute. Plastic (polymer) chutes are often used, so snow will not stick in the chute causing an obstruction.

Large pneumatic tires are common to mid-sized and heavy duty snow throwers. Correct tire inflation is important. Pneumatic tires can have chains installed to increase traction in icy conditions. Cleated tracks are found on the larger models and provide exceptional traction on slippery ice and snow. See **Figure 20-15.** The tracks can be engaged singularly for steering, or together for maximum traction. Clutch levers are provided on each of the handles to control direction of travel. See **Figure 20-16.** The tracks can be disengaged for ease of transporting. Six forward speeds and two reverse speeds are controlled by a shift lever located on the control panel.

Available on some models are features such as weight shifting by moving the wheels forward or backward; and raising, tilting, and lowering the auger housing. See **Figure 20-17.** These features should be considered, depending on the terrain and snow conditions to be encountered.

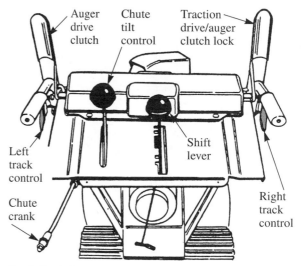

Track Controls

Figure 20-16. *Right and left track clutch levers permit responsive direction control. All controls are within easy reach of driver. (MTD Products, Inc.)*

Figure 20-17. *Weight shifting is accomplished by moving the wheels forward or backward. (The Toro Co.)*

Engine controls

Engine controls will vary from one make and model to another. **Figure 20-18** illustrates the basic controls that are quite common to four-cycle engines. Read the manufacturer's technical service manuals for specific details about engine controls.

It is common for snow thrower engines to have a recoil (rope pull) starter. Electric starters may be provided as standard, or optional equipment. Snow throwers with electric starts also have recoil starters. When supplemental electrical power is not available, the recoil starter can be used for starting.

Figure 20-15. *Rubber tracks provide exceptional traction on ice and snow and can be engaged singly, together, or disengaged entirely. (Deere & Co.)*

Engine Controls			
A	Electric start button (optional)	Activates electric starter.	
B	Fuel valve	Turns fuel supply on or off.	
C	Starter handle	Used to start engine.	
D	Primer button	Primes carburetor for faster cold starting.	
E	Throttle lever	Controls engine speed.	
F	Engine key	Prevents starting of engine without key. Stops engine when removed.	
G	Choke knob	Adjusts air/fuel mixture.	

Figure 20-18. *The basic engine controls vary from one make and model to another. (Simplicity Manufacturing Co.)*

Electric starting

Electric starting is convenient due to the size of the larger engines and cold conditions that may be encountered. The power is 110 volt (V) ac. The snow thrower is connected to a grounded outlet using a 3-wire extension cord. The ac current is reduced to 12V dc current for the starter motor. With some models, the cord is plugged into the receptacle on the snow thrower and a button on the receptacle is depressed to start the engine. See **Figure 20-19.** With other models, after the cord is plugged in, the engine is started by depressing the start button on the control panel. See **Figure 20-20.**

Drive systems

There are basically two driving systems. The first drive system is for driving the auger in single-stage machines, or for the auger and blower in two-stage machines. For those models that are self-propelled, the second drive system is for the wheels or tracks. There are many drive designs

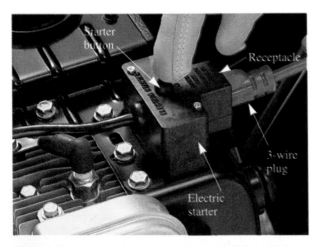

Figure 20-19. *The 3-wire extension cord is plugged into the electric starter receptacle on the snow thrower. A button on the electric starter is depressed to start the engine. (The Toro Co.)*

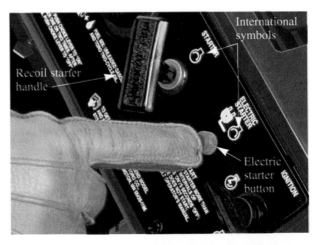

Figure 20-20. *Electric start button on a control panel. Notice the recoil start handle to the left of the start button and the international symbols for the various controls. (The Toro Co.)*

and each manufacturer may utilize different ones for different models. Transmissions may consist of gears, belts, chains, or combinations of gears, belts, and chains.

Figure 20-21 shows a portion of chains and belts used in a wheel driving system. **Figure 20-22**

shows the location of the reduction gear box transmission that drives the first and second stages from the engine. The internal parts of a typical gear box utilizing a ring gear and pinion gear are shown in **Figure 20-23.** Before beginning any service requiring disassembly and reassembly of parts, refer to the manufacturer's technical service manuals for the exact make and model.

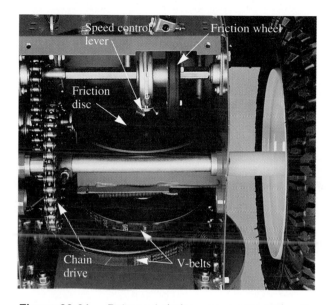

Figure 20-21. *Belts and chains are used to transmit motion from the engine to the wheels and auger. The friction wheel drives the friction disc and changes speeds when it is moved right or left by the operator. Moving to the right slows down the wheels and moving left speeds the wheels up. (Deere & Co.)*

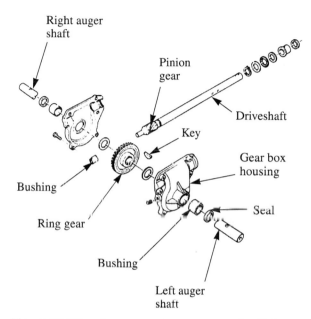

Figure 20-23. *Auger gear box components. Pinion gear and ring gear reduce rotation speed of auger and increase auger torque. (The Toro Co.)*

Track drive snow thrower

Self-propelled snow throwers with rubber tracks may need service. The tracks may become worn out or damaged. Also, the chain drive may need a new chain. Track adjusters allow loosening or tightening of the tracks by moving the idler wheel assembly forward or backward. This is done by loosening a nut or bolt. **Figure 20-24** is an exploded view of a track drive system for one model of snow thrower.

Track chain replacement

To replace the track chain, proceed as follows:
1. Block up the frame of the unit so the tracks are off the ground.
2. Loosen the nut on the track tension adjusters.
3. Roll the track off the idler and drive wheel assembly as shown in **Figure 20-25.**

Figure 20-22. *Auger reduction gear box reduces rotation speed of the auger. (Deere & Co.)*

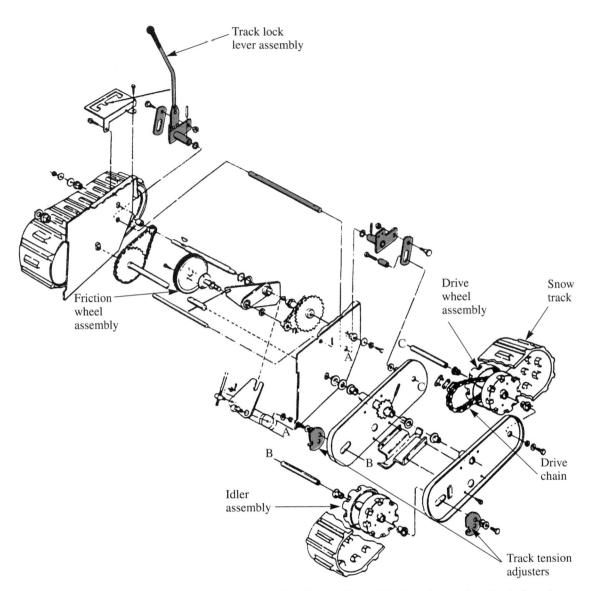

Figure 20-24. *The track lock lever is used to select the position of the housing and method of track operation. It moves tracks forward or backward relative to the housing. (MTD Products, Inc.)*

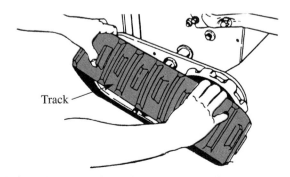

Figure 20-25. *Rolling the track off the idler and drive wheel assembly after releasing tension. (MTD Products, Inc.)*

4. Remove the chain by disconnecting the master link. See **Figure 20-26.** A master link and chain assembly is shown in **Figure 20-27**. The keeper is pried open with a screw driver and removed so the link can be removed. The open end of the keeper should trail the direction of rotation of the chain. It is best to replace the keeper on the outer side of the chain so it can be removed more easily.

5. Reassemble in reverse order with a new chain and/or track. When reassembling track, be certain the tread of the track is running in the proper direction. See **Figure 20-28.**

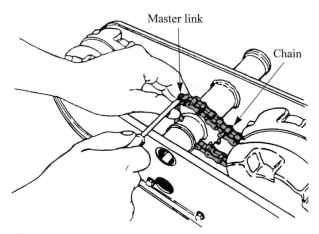

Figure 20-26. *Removing drive chain by disconnecting master link with screwdriver. (MTD Products, Inc.)*

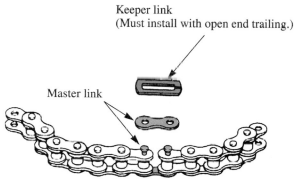

Figure 20-27. *Details of roller chain and master link with keeper. (Simplicity Manufacturing Co.)*

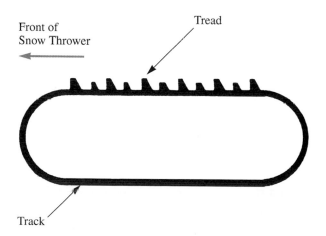

Figure 20-28. *Direction of track tread must be as shown when installing on idler and drive wheels. (MTD Products, Inc.)*

Track tension adjustment

It is important that the track tension on both sides be adjusted properly and equally. There are track adjusters for adjusting the tension. To adjust track tension, proceed as follows:

1. Disconnect the spark plug wire and ground it against the engine.
2. Drain the fuel from the fuel tank.
3. Tip the snow thrower forward so that it rests on the auger housing.
4. Loosen the track adjuster bolt.
5. Insert a screwdriver between tab on the track adjuster to tighten or loosen track. See **Figure 20-24.** The track is adjusted properly when the track can be deflected about 1/2″ (12.7mm).
6. Tighten track adjuster bolts.

Track lock lever

The track lock lever assembly shown in **Figure 20-24,** located on the right side of the snow thrower is used to select the position of the auger housing and the method of track operation. The lever can be moved to the right, then forward or backward to one of three positions. The lever positions are used to meet the following operating conditions:

- **Transport.** Raises front end of snow thrower for easy transporting, or to clear snow from gravel driveways without disturbing the gravel.
- **Normal snow.** Allows tracks to be suspended independently for continuous ground contact.
- **Packed snow.** Locks front end of snow thrower down to the ground for hard-packed or icy snow conditions.

Skid shoes

Skid shoes establish the height of the scraper bar above the surface. The shoes are located at the bottom, outer edges of the auger housing, and absorb the wear during operation. Fasteners fitted into slotted holes on the skid shoes, allow easy adjustment and replacement. See **Figure 20-29**. The scraper bar is a metal bar fastened along the bottom edge of the auger housing intake. The scraper bar adds stiffness to the sheetmetal housing and can be replaced when it becomes worn.

On smooth surfaces such as concrete or asphalt, the scraper bar should lightly scrape the surface or clear by about 1/8″ (3mm). On gravel surfaces the scraper bar should be high enough so

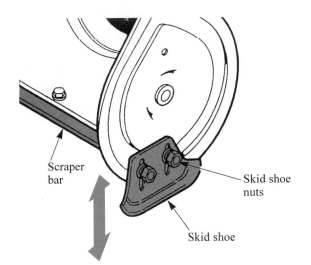

Figure 20-29. *Skid shoes establish height of scraper bar above the surface. (Deere & Co.)*

that it will not pick up gravel or debris; about 1 1/4″ (30mm). If the scraper bar is damaged or worn excessively, replace it before adjusting skid shoes.

Skid shoe adjustment

The skid shoes shown in **Figure 20-30** are adjusted in the following manner:
1. Park the snow thrower on a hard, smooth surface.
2. Stop the engine, remove key, wait for all moving parts to stop.

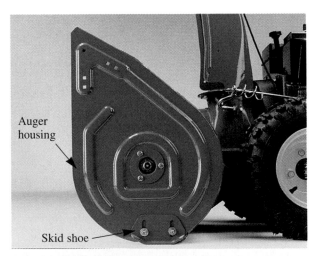

Figure 20-30. *On a level surface place two pieces of wood of desired thickness under scraper bar. Loosen skid shoe nuts, lower skids to surface, and tighten nuts; 1/8″ (3mm) for smooth surfaces and 1 1/4″ (30mm) for gravel surfaces. (Deere & Co.)*

3. Remove wire from spark plug and ground it to engine to prevent accidental start up.
4. Check tire pressure with a tire gauge:
   a. Air pressure should be equal in both tires. Check for a maximum of 20 psi (138 kPa) in each tire. Recommended tire pressure may be printed on the side wall of the tire or printed in the manufacturer's technical service manuals.
   b. Each tire must be resting on a hard surface, not a cross-link of tire chain.
5. Put a block of the desired height under the scraper bar of the auger housing.
6. Inspect the skid shoes. If shoes have excessive wear, replace them.
7. Loosen the adjustment nuts, lower the shoes to the surface, and tighten the nuts.

Operator presence

To provide additional safety, *operator presence controls* require that the auger automatically stops rotating when the drive (traction) lever is released. The auger (impeller) is engaged by depressing a spring-loaded *auger control* lever on top of the right handle. See **Figure 20-31.** When the *drive control* lever on the left handle is engaged, traction is established and the snow thrower moves forward. Due to a ratchet mechanism, when the drive control lever is engaged the impeller lever will stay engaged without holding it. When the drive control lever is released it also releases the ratchet mechanism holding the auger control lever and the auger stops. This allows one hand guidance of the snow thrower, but stops all movement when the drive control lever is released. In an emergency, all one has to do is let go of the drive control lever and all movement stops, except the engine continues to run.

Some snow throwers have two spring-loaded control handles. One handle (right handle) engages the auger when depressed. The second handle (left handle) engages the drive wheels when depressed. When both handles are released all motion stops, except the engine continues to run. See **Figure 20-32.** This requires the operator to hold both handles down while removing snow.

 Never reach into the chute or auger housing without stopping the engine, and removing and grounding the spark plug wire.

Snow Thrower Controls

1	Speed selector	Selects forward speeds 1-5, reverse speeds 1-2.
2	Drive control	Engages drive to wheels as it is depressed, disengages when released.
3	Auger control	Engages auger/impeller as it is depressed; disengages when released.
4	Chute direction control	Rotates discharge chute to desired direction.
5	Chute deflector	Controls vertical angle snow is thrown.
6	Chute deflector knob	Locks chute deflector at desired angle.
7	Skid shoes	Controls height of scraper bar.
8	Traction lock pins	Engages and disengages for drive or free-wheeling.

Figure 20-31. *Controls 2 and 3 require the operator to be present at the controls during operation. If the drive control lever 2, is released, auger control lever 3, also releases. The wheels and auger stop. (Simplicity Manufacturing Co.)*

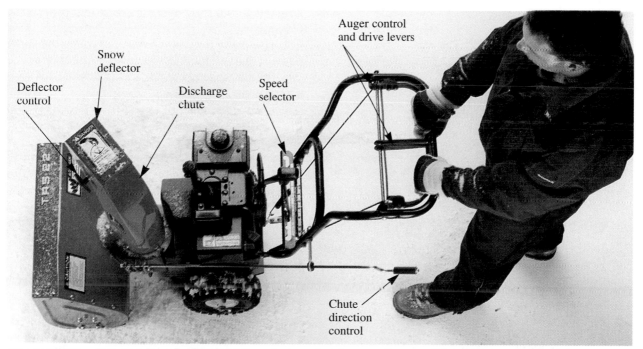

Figure 20-32. *Two handles must be held down to operate this snow thrower. When both are released the wheels and auger stop. The engine continues to run. (Deere & Co.)*

Snow Thrower Operation

There are a variety of models of snow throwers, but starting and using them is similar. Always refer to the manufacturer's technical service manuals for instruction for the specific model being used.

Engine starting

The procedure for starting a cold engine is slightly different than the procedure for warm engine restarting. See **Figure 20-33** for the location of parts and controls for a snow thrower. Since snow throwers models differ, refer to the manufacturer's technical service manual for the specific model. To start a cold engine (references to warm engine restarting included), proceed as follows:

1. Turn the fuel valve on. Fuel valves are located where the fuel exits the bottom of the tank.
2. Insert the engine ignition key into the switch and turn to the *Run*, or *Start*, position. This switch may be located on a control panel.
3. Move the throttle lever to the *Fast* position. The throttle lever may be located on a control panel.
4. Close the choke knob. Do not choke a warm engine to restart.
5. Press the primer button several times if engine is cold. Do not prime a warm engine for restarting.

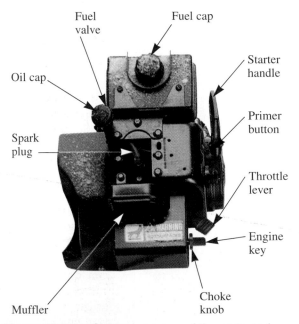

Figure 20-33. *Shown are many of the parts and controls for this snow thrower engine. (Deere & Co.)*

6. Pull starter handle rapidly, or push starter button if equipped with electric starter. Do not let the starter handle snap back. Let the starter rope rewind slowly, while keeping a firm grip on the starter handle.
7. As the engine starts and begins to operate evenly, turn the choke knob to the *Off* position, and set the throttle lever to *Slow*. If the engine begins to slow down and run roughly, turn the choke knob until the engine runs smoothly. Let the engine warm up more and return the choke knob setting to *Off*.

Allow the engine to warm up at *Slow* throttle before operating the snow thrower at full speed. The engine must reach operating temperature before it can develop full power. For best performance, when at operating temperature run the snow thrower engine at full operating speed. Use the throttle lever to set the engine speed.

Snow throwing

The procedure for snow throwing may vary from model to model. Refer to the manufacturer's technical service manuals. To start snow throwing with the model shown in **Figure 20-32,** proceed as follows:

1. Use chute direction control to rotate the snow discharge chute to the desired direction.
2. Set the speed selector to the desired forward speed.
3. Fully press and hold the auger control lever to begin auger rotation. To disengage the auger, completely release the lever.
4. Fully press and hold the traction drive control lever to engage the traction drive and begin moving the snow thrower. To disengage the traction drive, release the lever.
5. To select the forward or reverse speed as needed use the speed selector lever. Release the drive control lever, and change to the desired speed or direction.

General Maintenance

Maintenance is an important part of working with small gas engines and the machines they power. Small gasoline engines in snow throwers are similar to the engines in other kinds of implements. Snow thrower engines are designed to start easily and run under moderate to extremely cold working conditions. When working with snow throwers, there are a number of components that need checking, service, and periodic adjustments. Performing these on a regular basis, will result in a properly running snow thrower.

Spark plugs

The spark plug should be removed for inspection and cleaning. Before removing the spark plug, blow dirt away from its base with compressed air. Remove the high tension wire by pulling only on the insulating boot. Do not pull or jerk the wire. Install a deep spark plug socket on the plug and carefully turn the ratchet handle counterclockwise. Unscrew the spark plug and examine the electrode end of the plug. The electrodes and ceramic insulator should be dry. If it is wet with fuel or oil there may be more serious engine problems. The following are several reasons for a wet spark plug.

- Over choking.
- Excessively rich fuel mixture.
- Water in fuel.

If the insulator is dry and has a beige or gray/tan color, the plug is in good condition. Next, examine the electrodes for any erosion, burning or carbon fouling. Refer to *Spark Plug Condition* chart in the *Appendix* of this text. If the spark plug insulator and electrodes appear to be good, check the electrode gap with a wire type spark plug gapping tool. The engine service manual will specify the proper gap setting. If the gap is too large carefully bend the outer electrode toward the center electrode until the gap is correct. Use the gap setting tool. If the spark plug appears to be beyond simple servicing, replace it with a new one. Refer to the manufacturer's technical service manuals for the proper replacement spark plug.

Before replacing the spark plug clean the external ceramic insulator by wiping it with a clean cloth. Condensation can cause flashover (sparking externally) and erratic running if the insulator has a coating of dirt on it. If the metal base is rusty, clean it on a wire wheel or with a wire brush. If the threads are dirty, they should be wiped clean or wire brushed before inserting the plug back into the spark plug hole. The gasket should be in good condition or replaced with a new one.

Insert the spark plug in the spark plug hole and turn it clockwise by hand until it stops. Be careful not to cross thread the spark plug in the hole. If it will not turn, do not force it. Remove it and examine the threads in the hole. If they are damaged, a thread chaser (special thread tapping tool) may be needed to clean and correct the threads. If the spark plug turns smoothly by hand until it seats, tighten the plug with a torque wrench to 13 to 15 lb-ft. Do not overtighten.

Examine the high tension wire for deterioration before replacing it on the spark plug. Heat and dirt can make the insulation brittle and insulation cracking can occur. Cracked wire insulation can allow sparking to metallic parts of the engine and cause hard starting and erratic running. Oil or fuel soaked wires will leak current and cause weak firing of the plug. The insulating boot should be dry and in good condition.

Snow thrower engine fuel

Snow throwers that have a two-cycle engine must have the oil mixed with the fuel in proper proportions. Use only oil labeled for two-cycle engines. Refer to the manufacturer's technical service manuals for proper fuel-to-oil mixture ratio. Most two-cycle engines use a ratio of 40 or 50 parts fuel to 1 part of oil. Use a one gallon fuel container that has *Gasoline* clearly printed on it. It is very helpful to mark the fuel/oil ratio on the container for future mixing reference. To mix the fuel and oil, proceed as follows:

1. The procedure for mixing is to add about 1/4 gallon of clean, fresh, lead-free gasoline to the container.
2. Add the amount of premeasured oil. Close the container and shake the mixture well.
3. Add more gasoline until the container is about 3/4 full. Shake it again.
4. Fill the container to the one gallon mark and shake. The fuel is ready for use.

Before removing the fuel cap on the snow thrower wipe the cap and area around it clean and dry. On some machines if snow and ice is not cleared away they can accumulate and fall into the tank. Fill the fuel tank after each use of the snow thrower to prevent moisture from condensing in the fuel system. Water is heavier than gasoline and will settle in the lowest point of the fuel tank or fuel system. The water will freeze when the outside air temperature drops below 32°F (0°C). Fuel line freeze-up will prevent the engine from starting. Should this happen, the snow thrower will have to be moved to a warm area until the ice melts.

Four-cycle engines for snow throwers do not require mixing oil with fuel. The fuel tank can be filled with clean, fresh, lead-free gasoline. Some engine manufacturers caution against using fuels containing alcohol, such as gasahol. If gasoline with alcohol must be used, it should be limited to 10% Ethanol and must be removed before storage. To avoid damage to gaskets and other synthetic materials in the carburetor, gasoline with methanol must not be used. As with two-cycle engines, top off the tank with fuel after each use to prevent condensation. Close the fuel valve below the tank and leave a little air space for expansion of the fuel to avoid spilling over.

If the snow thrower has an in-line fuel filter, check it for contaminants. In-line filters are not cleanable. Remove it and replace it if needed. The normal change cycle for in-line filters is once each year before seasonal use.

Four-cycle engine lubrication

Unlike two-cycle engines, four-cycle engines receive their lubrication from oil in the crankcase. This oil must be the correct viscosity and service type. Do not use multi-viscosity oils above 40°F (4°C). Snow throwers use winter grade oil due to the cold temperatures in which they are used. Generally, SAE 5W-30 to 10W-30 detergent oil is recommended. The oil must be drained periodically and replaced with new oil. The manufacturer's technical service manuals will specify the number of hours of engine use before changing oil. A new engine must have the oil changed after 5 to 8 hours to remove any manufacturing particulates and break-in residue.

The engine will have an oil drain plug at the lowest point in the crankcase. The engine should be run before draining to heat the oil so it thins and drains completely. Place a metal or plastic pan under the drain plug and remove the plug.

Remove the oil filter cartridge with an oil filter wrench. Allow the oil to drain thoroughly. If the engine utilizes an oil filter cartridge it should be replaced during the first oil change. Discard the old filter cartridge. Select the correct, new, oil filter cartridge. Apply a film of engine oil to the rubber gasket and replace it on the engine, hand tight. Using the oil filter wrench, turn the filter cartridge another 1/2 to 3/4 turn. Do not over tighten the cartridge.

After installing the new filter cartridge, replace the plug and inspect the old oil for contamination of metal or other particles. Remove the oil filler cap and dip stick and pour in the new oil. Wipe and check the dip stick until the oil level shows full. Do not overfill. Excess oil will cause spark plug fouling and other internal problems. Run the engine for 30 seconds, stop the engine, and check the dip stick again. Add oil if necessary. Check the oil level before each use of the snow thrower. Add oil as needed.

Optional oil accessories

Some engines may have an oil pressure switch with a light on the instrument panel. If the engine runs low on oil, the pressure switch will activate the warning light or stop the engine. An oil level light may also be available to indicate a level condition. If the light flickers during starting, the oil level should be checked and filled to the *Full* mark on dip stick.

Air filter

The air filter prevents dirt and abrasive particles in the air from entering the carburetor and engine. An air filter is fastened to the intake side of the carburetor. The filter element may be in a metal or plastic case. Remove the cover and take out the filter element. Examine the element for dirt accumulation. Air filters may be of the cartridge, dual element, or dry-type filter. The dry-type element is cleaned by tapping it gently on a flat surface to shake off the dirt. Do not use compressed air or solvents. If it still appears to be dirty, replace it with a new element. Reinstall the filter element and cover.

Muffler

The purpose of the muffler is to reduce engine noise and produce the proper back pressure in the exhaust system. Mufflers get very hot when the engine is running and should be cool before touching or examining them. If there are signs of rust causing pin holes the muffler should be replaced. If the muffler sounds excessively loud the baffles inside may be corroded. This can cause overheating and warpage of exhaust valves. If the muffler shows serious deterioration it should be replaced immediately.

Preventing freeze-up

Normal use of a snow thrower may result in a build-up of snow packed in and around the starter cord housing and engine controls. Engine heat will prevent snow from freezing solid while the machine is running. After the engine is stopped and the engine cools, some snow may continue melting and later freeze around moving parts. To prevent freeze-up around engine controls and external parts, after each use it is good practice to do the following:

1. Before stopping the engine, pull the starter rope out several times and allow it to rewind slowly. This will help clear packed snow from the starter cord area.
2. Stop the engine by moving the throttle lever to the stop position, or by removing the ignition key.
3. Disconnect the spark plug wire and ground it on the engine away from the spark plug. Tie it away with string or tape if necessary.
4. Brush snow and ice from the snow thrower. Be sure to clear the engine and snow thrower controls, discharge chute, worm and chute rod gears, clutch rod areas, and anywhere snow has accumulated.
5. Remove and store engine key in a safe place to prevent unauthorized use.
6. If the snow thrower is kept in a cold shelter, fill the fuel tank to prevent condensation. Do not store near sparks or flames.

Auger and auger gear box lubrication

The auger shaft is driven from a gear box that requires lubrication. A filler plug may be located near the top of the case as in **Figure 20-34,** or at a lower level on the gear box housing. If the filler plug is at the top, it may be necessary to add the necessary quantity by first draining the oil and adding the correct weight in ounces. If the filler plug is at a level below the top of the gear box, it is only necessary to fill oil to the lower edge of the plug. See **Figure 20-35.** See manufacturer's technical service manuals for amount and type of oil to add. The following procedure can be used to add a winter weight worm gear oil to the auger gear box:

1. Place the snow thrower on a level surface.
2. Disconnect the spark plug wire and ground on the engine away from the spark plug. Tie it away with string or tape if necessary. Remove key from ignition.

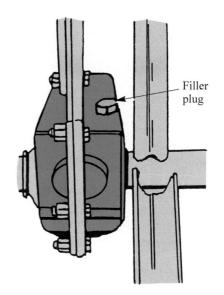

Figure 20-34. *Filler plug located at top on auger gear box requires oil to be measured out before filling. (MTD Products, Inc.)*

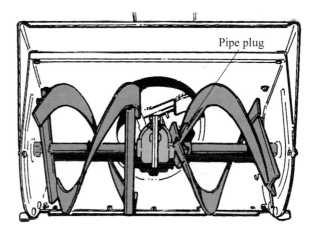

Figure 20-35. *Filler plug located on side of auger gear box. Remove plug and fill to lower edge of plug hole. (Simplicity Manufacturing Co.)*

3. Remove the filler plug.
4. Check the level. Add the proper amount as specified.
5. Replace the filler plug and tighten securely.

The auger blades are made in two halves. The center driving members are hollow tubes installed over the solid shafts on each side of the gear box. A **shear bolt** or **pin** is installed through each hollow tube and the solid shaft to connect them together. Should something become lodged in the auger housing and stop the auger, the bolt(s) or pin(s) will shear to protect the auger and gear box from more serious damage. See **Figure 20-36.** After removal of the obstruction and the sheared bolts or pins, new bolts or pins can be installed.

 Replace sheared bolts or pins with the exact same kind of bolts or pins. Replacements must have the proper shear strength. Stronger bolts or pins may not shear. This can cause serious damage to the gears in the auger gear box.

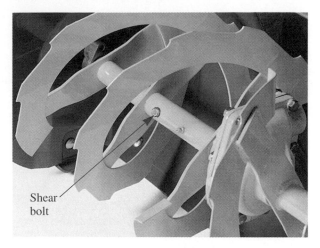

Figure 20-36. *Each auger blade has a shear bolt or pin that protects the auger gear box from damage resulting from an obstruction in the auger housing. (Deere & Co.)*

Lubrication

There are various places on snow throwers that need periodic lubrication to extend the useful life of the machine. Some parts require application of oil, while others require grease. Where contact between moving metal parts is made, oil should generally be applied. External parts can be oiled with an oil can filled with a medium weight (10W) oil. When oiling parts it is important not to over oil. Excess oil might drip on floors or on parts that should be kept dry, such as the traction drive or friction disc. Visible, excess oil generally provides no lubricating value. It mixes with dirt. **Figure 20-37** shows most of the parts that need oiling and greasing. Oil locations are shown by an orange oil

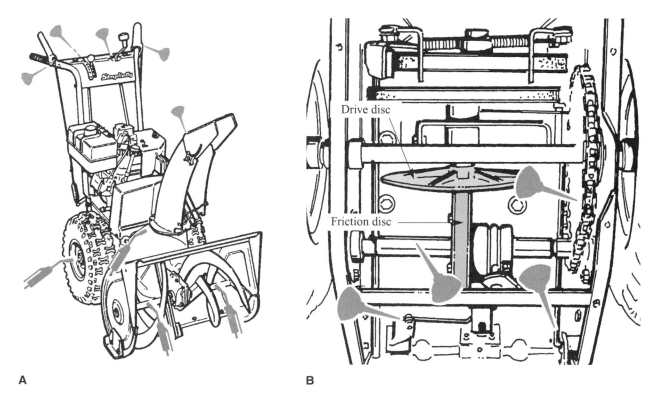

A   B

Figure 20-37. *A—General lubrication points on snow thrower. B—Lubrication points of a drive area with the bottom cover removed from snow thrower. (Simplicity Manufacturing Co.)*

can, and grease locations are shown by a green grease gun. Using a grease gun, grease should be applied directly to the grease (zerk) fittings. See **Figure 20-38**. When grease fittings are not present, parts must be disassembled to apply the grease and then reassembled.

It is very important that grease fittings on the auger shaft are lubricated regularly. If the auger rusts to the shaft, damage to the worm gear may occur if the shear bolts or pins do not break. There are two grease fittings on the auger shaft. See **Figure 20-37A**. Wipe the fittings clean and apply grease, using a grease gun. Wipe away any excess grease from fittings or seepage between parts. If there are no grease fittings, waterproof grease must be applied liberally to auger shafts during assembly to prevent corrosion and seizing of parts.

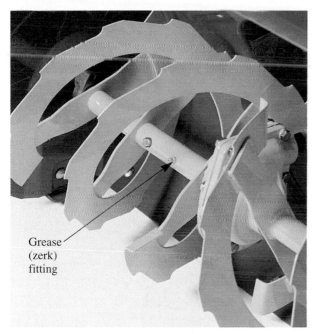

Grease (zerk) fitting

Figure 20-38. *Shown is the grease (zerk) fitting located on the auger shaft. (Deere & Co.)*

Off-Season Storage

When preparing to store a snow thrower for an extended period of time, refer to the manufacturer's technical service manuals for any *specific* instructions about the make and model being stored. If storing outside, a weather resistant cover may be placed over the snow thrower to protect it against severe weather conditions.

Gasoline may develop gummy deposits if left in the tank unused for more than 30 days. This can adversely affect the carburetor and filter causing the engine to malfunction. To avoid this condition, add a fuel stabilizer to the fuel tank, or drain all fuel from the system. Once drained, start the engine and let the engine run until it uses all of the remaining fuel.

Remove the spark plug and spray a fogging oil into the spark plug hole while cranking the engine slowly with the recoil starter. This will allow the fogging oil to enter between the valve seats and faces. This will prevent corrosion. Now replace the spark plug.

Clean the snow thrower thoroughly with a clean cloth. Protect any bare metal surfaces from corrosion by coating with paint, oil, or grease.

Store the snow thrower in it's normal position. This prevents any oil from the crankcase from entering into the cylinder head of the engine.

Starting after storage

After being stored for an extended period of time, there are some important steps to take when starting a snow thrower. To start a snow thrower after storage, proceed as follows:

1. Remove the spark plug, wipe it dry and inspect it for excess carbon, burned electrodes, and proper gap. Replace the plug with a new one if necessary.
2. If fuel tank was emptied before storing, fill the fuel tank with fresh gasoline. Skip this step if a fuel stabilizer was used before storing.
3. Make sure cooling fins are clean and no air flow obstructions exist.
4. Check engine oil level and lubricate the snow thrower. Change oil if condition warrants. If an oil filter is used on the engine, change it when changing engine oil.
5. Belts that are worn or loose should be adjusted, or replaced.
6. Start the engine outdoors. Warm up engine by running at *Slow* speed for a few minutes before running at *Fast* speed, or removing snow.
7. Check the operating condition of all controls. Make any adjustments that may be needed.
8. If the engine or snow thrower does not perform as expected refer to a troubleshooting chart such as the one in **Figure 20-39**.

Troubleshooting Guide— Snow Thrower		
Trouble	*Possible Cause(s)*	*Corrective Action*
Engine fails to start	1. Fuel tank empty, or stale fuel. 2. Blocked fuel line. 3. Key not in switch on engine. 4. Spark plug wire disconnected. 5. Faulty spark plug.	1. Fill tank with clean fresh gasoline. 2. Clean fuel line. 3. Insert key. 4. Connect wire to spark plug. 5. Clean, adjust gap or replace.
Engine run erratically	1. Unit running on CHOKE. 2. Blocked fuel line or stale fuel. 3. Water or dirt in fuel system. 4. Carburetor out of adjustment.	1. Turn choke knob to OFF position. 2. Clean fuel line; fill tank with clean fresh gasoline. 3. Use carburetor bowl drain to drain fuel tank. Refill with fresh fuel. 4. Adjust carburetor.
Loss of power	1. Spark plug wire loose. 2. Gas cap vent hole plugged.	1. Connect and tighten spark plug wire. 2. Remove ice and snow from cap. Be certain vent hole is clear.
Engine overheats	1. Engine oil level low. 2. Carburetor not adjusted properly.	1. Fill crankcase with proper oil. 2. Adjust carburetor.
Excessive vibration	Loose parts or damaged impeller.	Stop engine immediately and disconnect spark plug wire. Tighten all bolts and nuts. Make all necessary repairs. If vibration continues, have unit serviced by authorized service dealer.
Hard to shift, or will not shift	Shift rod misadjusted.	Adjust drive clutch.
Unit fails to propel itself	1. Incorrect adjustment of drive clutch. 2. Drive belt loose or damaged.	1. Adjust drive clutch. 2. Replace drive belt.
Unit fails to discharge snow	1. Auger shear bolt broken. 2. Discharge chute clogged. 3. Foreign object lodged in auger. 4. Incorrect adjustment of auger drive clutch. 5. Auger drive belt loose or damaged.	1. Replace auger shear bolt. Refer to Maintenance section. 2. Stop engine immediately and disconnect spark plug wire. Clean discharge chute and inside of auger housing. 3. Stop engine immediately and disconnect spark plug wire. Remove object from auger. 4. Adjust auger clutch. 5. Replace auger drive belt.

Figure 20-39. *Troubleshooting Guide—Snow Thrower. (MTD Products, Inc.)*

Know These Terms

single-stage snow throwers
operator presence controls

auger control
drive control
shear bolt
shear pin

Chapter 20 Review Questions

Answer the following questions on a separate sheet of paper.

1. List four safety considerations to protect children from snow thrower accidents.

2. When removing snow on sloped surfaces, it is safest to travel _____.
   a. across the face of the slope
   b. up and down the face of the slope
   c. at a 45° angle to the slope
3. List five safe operating practices when operating a snow thrower.
4. Before making any mechanical adjustments to a snow thrower, always _____ first.
5. Snow throwers are of two basic types. They are either _____-stage or _____-stage.
6. The more efficient type of snow thrower is the _____-stage type.
7. Heavy duty snow thrower engines are _____.
   a. exclusively two-cycle
   b. exclusively four-cycle
   c. Neither A or B
8. Snow is drawn into a snow thrower by a device called a(n) _____.
9. On heavy duty machines, wheels or tracks can be engaged independently for easier _____.
10. When operating a snow thrower the engine should run at _____.
    a. *High* speed
    b. *Low* speed
    c. *Medium* speed
11. Heavy duty snow throwers can remove as much as _____ lb per minute.
    a. 1100
    b. 1600
    c. 2300
12. To prevent damage and excessive wear to the auger housing _____ are fastened at the outer edges.

13. The auger housing height is established for various terrain and snow conditions by adjusting the _____.
14. List four steps to prevent freeze-up of a snow thrower.
15. How is the auger gear box protected from having something accidentally jammed into the impeller?

Suggested Activities

1. Identify and point out all the external parts of a snow thrower.
2. Check a snow thrower for proper condition of all visible parts.
3. Demonstrate and/or describe lubrication procedures for a snow thrower.
4. Check tire condition and pressure. If there are tracks, check condition and tension. Adjust track tension, or show how it is adjusted.
5. Demonstrate skid shoe adjustment.
6. Check engine condition and prepare for starting.
7. Demonstrate safe and proper engine starting procedures, and start the engine.
8. If snow is available, demonstrate proper use of snow thrower to clear snow. If snow is not available, briefly perform a simulated snow clearing operation on a clean, dry, outdoor surface.

Example of a two-stage snow thrower in use. (MTD Products, Inc.)

21

Personal Watercraft

After studying this chapter you will be able to:

▼ Describe personal watercraft safety and established boating regulations.

▼ Understand the major components of personal watercraft.

▼ Explain proper personal watercraft operation.

▼ Understand personal watercraft engine systems.

▼ Maintain and make adjustments to personal watercraft.

▼ Properly store and remove from storage a personal watercraft.

Introduction

The *personal watercraft* described in this chapter will refer only to a type of popular small boat, which is propelled and guided by a high velocity jet of water. For clarity, the abbreviation *PWC* will be used to refer to these small jet driven personal watercraft. Many manufacturers produce small jet driven PWC. The term *Jet Ski,* though used frequently for all types of PWC, is a registered trade mark of the Kawasaki Motors Corp., U.S.A. and relates only to their brand of PWC. See **Figure 21-1.**

The U.S. Coast Guard considers all personal watercraft, regardless of size, to be power boats. No matter how simple they are to ride, all power boats have the same requirements under the law for registration and regulation. The PWC in this chapter are termed *Class A Inboard Boats* by the U.S. Coast Guard; these are boats less than 16' in

length. Over 45% of the recreational boats in use today are less than 16' (4.87m) long.

Some states have adopted other definitions of PWC. An operator should become aware of these definitions and the laws that relate to them if operating in these states. It is common practice to transport watercraft by trailer from one state to another for recreational purposes. It is the boater's responsibility to be aware of any variations in regulations.

The difference in the PWC that will be discussed in this chapter from other more traditional watercraft is the innovative jet drive propulsion system. See **Figure 21-2.** These craft do not have

Figure 21-1. *This is a personal watercraft (PWC). This particular one is a Jet Ski made by Kawasaki. (Kawasaki Motors Corp., U.S.A.)*

Figure 21-2. *The jet drive and steerable nozzle on the PWC replaces the propeller and rudder common to other boats. (Polaris)*

ferent from other propeller driven boats. They can operate in shallower water and are quickly and easily maneuvered. Acceleration is rapid, but the boat is more affected by waves, turbulence, and obstructions than larger boats. It is important to take extra precautions when operating PWC in boat traffic conditions, because operators of other boats may not understand the capabilities of the smaller jet-type PWC.

This chapter covers the *do's* and *don'ts* relating to operating personal watercraft safely. Those who maintain, as well as those who operate PWC, should be aware of the serious nature of the equipment. PWC operation requires persons to be responsible and knowledgeable about the unique rules and safety requirements that govern all watercraft.

In some of the figures in this chapter the colors used do not correspond to the color code in the front of this text. Orange and green will still be used for components of interest and secondary components, respectively. Other colors used in this chapter are used to colorize the figures.

an exposed propeller and rudder like most other boats used on the water.

PWC of the kind discussed in this chapter are small boats powered by an inboard engine that has an axial jet pump mechanism. See **Figure 21-3.** The capabilities and limitations are somewhat dif-

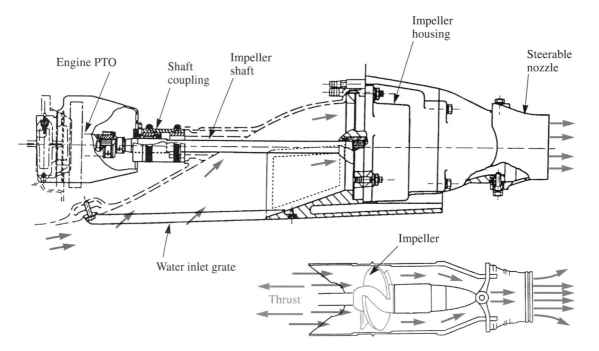

Figure 21-3. *An axial jet pump can operate in shallow water and can maneuver the craft quickly and easily. The engine driven impeller draws water through the inlet and pumps it out the nozzle at high velocity. The reaction creates thrust to push the boat ahead. (Sea-Doo)*

Operating Safely

One of the most important safety considerations for operation, is to know the basic terms used while boating. Operators should know the specific parts and limitations of the PWC. Before attempting to operate a PWC, the operator should know all boating navigational rules and all safe boating practices.

General boating terms

A number of general boating terms will be used throughout this chapter. See **Figure 21-4.** The following list are terms that should be learned before operating a PWC:

- **Port.** The left side of a PWC when you are aboard and facing the bow (front).
- **Starboard.** The right side of the PWC when you are aboard and facing the bow.
- **Bow.** The front of the PWC.
- **Stern.** The rear of the PWC.
- **Hull.** The body of the PWC from bow to stern.
- **Forward.** Toward the bow of the PWC.
- **Aft.** Toward the rear of the PWC.
- **Deck.** Any permanent horizontal covering over the hull.
- **Steering control.** The device designed for controlling the direction of a personal watercraft.

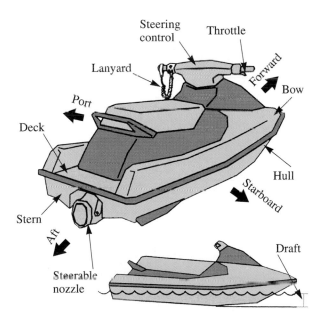

Figure 21-4. *Boating terminology is used by PWC operators to communicate. (Kawasaki Motors Corp., U.S.A.)*

- **Steerable nozzle.** A device for directing the stream of water from the jet pump to the left or right at the stern of a PWC.
- **Throttle.** Controls speed by regulating the amount of fuel delivered to the engine.
- **Lanyard.** Engine shut-off cord on PWC attached to the operator and the stop switch.
- **Draft.** The depth of the PWC below the waterline.

Operation precautions

Statistics show that inexperience and lack of education are the chief reasons why PWC were involved in accidents. In one year, a state reported that nearly 50% of its 689 boating accidents were PWC-related. In that year, only 5% of the 930,000 registered boats in that state were PWC. The ***Department of Natural Resources (DNR),*** PWC manufacturers, boating organizations, and boating law administrators are working together to establish classes for all operators of PWC. Often a PWC is the first watercraft an individual owns. If the operator is older than 16 years of age, no boating education is required. The result is inexperienced operators, which quickly get into unexpected difficulties on the water. The following are suggestions for PWC operation and warnings about conditions inherently associated with operating a PWC:

1. It is recommended that all operators complete a boating safety course. The Coast Guard office or state boating authority can furnish applicable age limit and operating rules.
2. It should be understood that a PWC is not a toy. It is a high-performance Class A powerboat. It requires learned and practiced skills. All operators and passengers should become familiar with the necessary maneuvering techniques before riding. The watercraft should always be run at a speed appropriate for the water conditions and level of experience of the operator.
3. These PWC *do not* have brakes. The PWC is stopped by releasing the throttle and gliding to a stop by the natural resistance of the water. Jet thrust is required to steer and turn the vehicle. Releasing the throttle eliminates the ability to steer the watercraft. Always use enough throttle to provide power to maneuver the watercraft; especially around obstacles. This is particularly dangerous for inexperienced

operators. Operating skills are developed through experience and practice. These skills must be accomplished before carrying a passenger. From full throttle, coasting to a stop will take a minimum of 300′ (90m). See **Figure 21-5.** Try to imagine the length of a football field which is 300′ from goal to goal.

4. Know and observe all local, state, and federal boating regulations and speed limits. Boating laws and navigational rules are for the benefit of everyone sharing the waterways.

5. Beginners should not tow water-skiers, kneeboards, other watercraft, or any object or person behind the PWC. Towing can cause loss of steering control and will create a hazardous condition which could result in severe injury or death.

6. Never go over a ski jump or attempt to jump waves or other objects in the water. The PWC can be severely damaged and the operator can be injured due to loss of visibility, watercraft control, and reduced reaction time.

7. The operator and passenger must wear a U.S. Coast Guard approved *personal flotation device (PFD)* at all times because of the drowning hazards associated with boating. The three common PFD types are shown in **Figure 21-6.**

   a. The seat of a PWC is not a flotation device.

   b. A helmet is not required. It may be helpful in some situations and detrimental in others. Good judgment must prevail in all cases.

Operating Do's and Don'ts	
	***Type 1—Offshore Lifejacket*** This PFD is designed for extended survival in rough, open water. It has 22 lb of buoyancy and usually will turn an unconscious person face-up. This is the best PFD to keep you afloat in remote regions where rescue may be slow in coming.
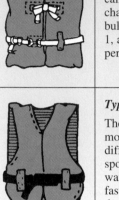	***Type 2—Near Shore Life Vest*** This *classic* PFD is designed for calm inland water where there is a chance of fast rescue. It is less bulky and less expensive than Type 1, and may turn an unconscious person face-up in the water
	Type 3—Floatation Aid These lifejackets are considered most comfortable, with styles for different boating activities and sports. They are for use in calm water where there is a chance of fast rescue since they are not designed to turn a person face-up.

Figure 21-6. *A personal flotation device (PFD) is a required part of PWC equipment. Each onboard passenger is required by the U.S. Coast Guard to wear an approved, good wearable condition Type I, Type II, or Type III PFD.*

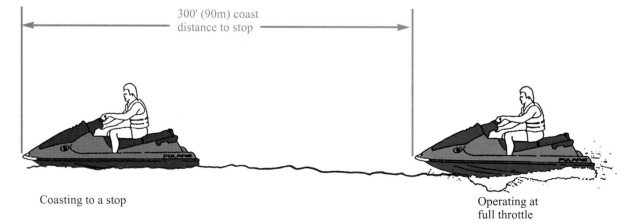

Coasting to a stop   300′ (90m) coast distance to stop   Operating at full throttle

Figure 21-5. *PWC do not have brakes. By releasing the throttle the PWC glides to a stop. It also eliminates the ability to steer. A minimum of 300′ (90m) should be allowed to coast to a stop. Inexperienced riders must gain stopping and steering skills. (Polaris)*

c. It is recommended that all riders wear additional personal protection including deck shoes and a wet/dry suit. These items can protect riders from exposure and potential hazards in the water such as debris and hidden objects.

d. Eye protection (goggles) should be worn at all times to protect against water spray.

8. Before starting the PWC the operator must make sure the lanyard is connected to the wrist or PFD. See **Figure 21-7.** The lanyard should be free and not wrapped around the handlebars or controls. The lanyard should always be disconnected from the stop switch when the PWC is not in service.

9. The PWC should never be overloaded. Overloading significantly reduces stability and control. When two or three persons are aboard, the PWC handles differently. The operator must have adequate, prior experience to safely handle the PWC with passengers. The maximum safe load should be known by the operator. Refer to the manufacturer's technical service manuals for maximum safe load specifications.

10. A federal regulation requires that an approved operable fire extinguisher be on aboard. The operator should know how to quickly reach and use the extinguisher. If there is any doubt about the ability to extinguish a fire, swim away from the PWC as quickly as possible and seek help from other boats or from individuals ashore.

11. Be aware of severe weather conditions. Observe weather forecasts and conditions before venturing out. Do not operate a PWC in poor visibility. Operating PWC in bad weather can result in severe injury or death due to exposure (hypothermia, or subnormal body temperature). This can also result in accidents due to rough water and poor visibility.

12. Be aware of the danger of hypothermia, which can result in severe injury or death in a very short time. Hypothermia can begin in water as warm as 80°F (27°C). When going into remote areas or large areas of open water, always ride with another PWC. Take along a flare gun to signal for help if necessary.

13. The stream of water from the jet pump, and falling into the water at high speed can cause severe personal injury; especially to body orifices such as eyes, mouth, ears, rectum, etc. Normal swimming attire may not provide adequate protection. The jet pump output is especially dangerous to a passenger that falls to the rear of a PWC, or to a person behind a moving PWC. If a passenger falls from a PWC, the throttle should be released immediately to avoid injury. Do not exceed idle speed if any person is within 50′ (15m) of the rear of the craft.

14. Watch for dangerous obstacles above and below the water at all times; especially in shallow water. The PWC must be at least 2′ (60cm) off the bottom when starting. This will prevent the jet pump from being damaged by objects sucked up from the bottom. See **Figure 21-8.** Never ride in water that is less than 2′ (60cm) deep. Do not operate at more than idle speed in water that is less than 6′ deep. If an operator or passenger is thrown

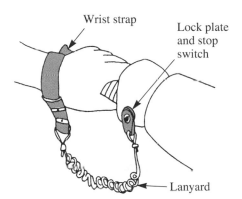

Figure 21-7. *Before starting a PWC, the operator must make sure the lanyard is securely fastened to the wrist. It should be free and not wrapped around the handlebars or other controls. (Polaris)*

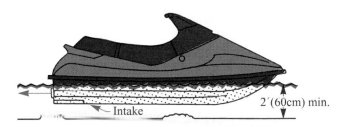

Figure 21-8. *The minimum depth of water to ride in is 2′ (60cm). Debris can be picked up by the impeller and cause serious damage to the impeller blades. Idle speed should not be exceeded in water that is less than 6′ (1.8m) deep. (Kawasaki Motors Corp., U.S.A.)*

from a PWC in shallow water, an object below the water could be struck. Severe injury or death could result.

15. Always perform the preride check before starting and riding the watercraft. The preride checklist is discussed later in this chapter.

16. Riding a PWC is strenuous. All operators should be competent swimmers and in good physical condition. Never travel beyond swimming distance from shore.

17. It is illegal to operate a PWC between sunset and sunrise. Many accidents happen at dusk, after sunset, when the unlighted craft cannot be seen well by others.

18. Quick turns or acceleration can throw an unaware passenger. The operator should alert passengers before making sudden maneuvers.

19. Do not allow hands, feet, ropes, clothing, straps, or long hair to come in contact with the water inlet grate located on the bottom of the PWC. Never operate a PWC with the inlet grate removed or allow any object to enter the inlet port. Severe injury or death could result from becoming tangled in the jet pump drive-line components of a PWC.

20. To avoid injury from shock, never touch or remove electrical parts when starting or during operation of a PWC.

21. Do not give a ride to someone whose feet cannot firmly touch the floorboards. Operator and passenger should keep their feet firmly on the floorboards while the PWC is in motion. It is possible to lose balance, fall overboard, or possibly injure feet from objects in the water. The passenger should face forward and firmly hold onto the operator's waist or seat strap.

22. If the PWC becomes capsized it must be uprighted in accordance to the procedure described in the manufacturer's technical service manuals. See **Figure 21-9.** Rolling the craft in the wrong direction to right it may allow water to enter the engine through the exhaust pipe. This can cause major engine damage. If water is taken in through the exhaust, follow engine draining procedure described in the manufacturer's technical service manuals as quickly as possible. *Do not* try to restart engine before draining the engine.

23. Manufacturers warn that safe operation of their craft requires good judgment and physical skills. Persons with cognitive or physical

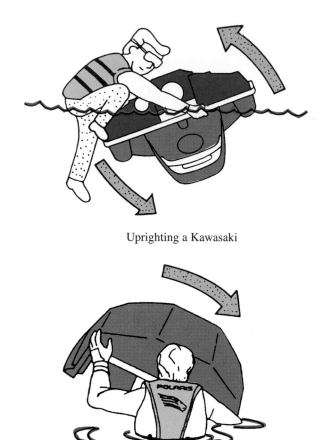

Uprighting a Kawasaki

Uprighting a Polaris

Figure 21-9. *When attempting to right an overturned PWC it is essential to know the correct direction to roll the craft to prevent water from getting into the engine. Do not attempt to restart the PWC if water has been taken in by the engine. Have the PWC towed to shore and follow special engine draining procedures to avoid serious engine damage.*

disabilities who operate a PWC have an increased risk of overturns and loss of control which may result in injury or death.

24. Alcohol is involved in over half of all boating accidents and fatalities. No one should ever operate a PWC while under the influence of drugs or alcohol!

Preride checklist

The following items should be checked *each day* before starting to use the watercraft. Use of a typed, laminated checklist kept aboard the PWC is good operating practice to avoid forgetting items.

Outside check

- **Clean pump.** Clear the water inlet, jet pump, and drive shaft of any foreign objects. See **Figure 21-10.**
- **Pump cover tight.** Check the jet pump cover and inlet grate for looseness. If needed, tighten the mounting bolts.
- **Hull damage.** Inspect the hull for any cracks or other damage.
- **Drain screw.** Make sure the drain screw or plug in the stern is securely installed. See **Figure 21-11.**
- **Steering.** Inspect the operation of the steering for any binding. See **Figure 21-12.** If needed, adjust the cable. Refer to control cable adjust-

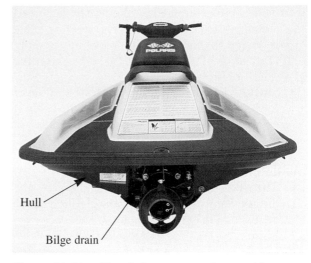

Figure 21-11. *The drain screw or plug must be secured to avoid taking on water during riding. (Polaris)*

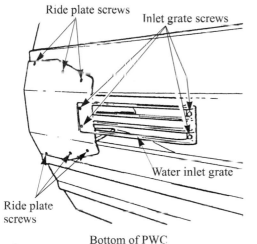

A

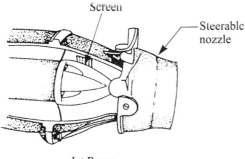

B

Figure 21-10. *The inlet grate and jet pump must be frequently checked for any foreign objects. The drive shaft, as shown in **Figure 21-3**, should also be checked frequently for foreign objects. (Sea-Doo)*

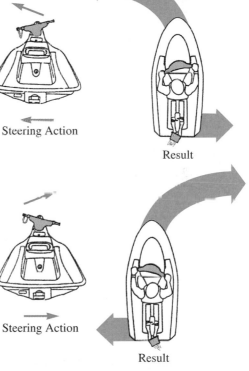

Figure 21-12. *Steering controls must be checked for freedom and proper operation. Turning craft to right swings bow to starboard and stern to port. If the engine is not running at least medium speed, the boat will not be steerable.*

ment procedures in the manufacturer's technical service manuals. The steering cable is sealed at both ends and does not require lubrication. The cable must be replaced if the seals are broken.

- **Shift lever.** Check operation of the shift lever for binding or rough spots. Adjust the cable if necessary (see adjustment procedures in maintenance manual).

Inside check

- **Throttle control.** Check operation of the throttle control lever for binding, rough spots, or excessive play. Adjust if needed. See **Figure 21-13.**

 If the throttle does not return freely and completely, it may cause loss of control.

- **Ventilate engine compartment.** Open the storage compartment lid, take out the storage box, remove the seat, and keep open for at least several minutes to purge gasoline fumes from the engine compartment.

 A concentration of gasoline fumes in the engine compartment can cause a fire or explosion.

- **Fire extinguisher.** Fire extinguisher should have a full charge and be secured properly in place. Lock pin in place.
- **Fuel level.** Check the fuel level. Refill if necessary and turn fuel selector valve to the *On* position. See **Figure 21-14.**
- **Engine oil level.** Check oil level in oil tank. Add oil as needed. Bleed air from line if needed.
- **Fuel leaks.** Check the engine compartment for any fuel leaks.
- **Oil leaks.** Check engine compartment for any oil leaks.
- **Sediment bowl.** Drain any water out of the sediment bowl and clean it. See **Figure 21-15.**
- **Fasteners.** Check and tighten any loose bolts, nuts, or clamps.
- **Hose connections.** Check security of all hose connections and clamps. Check condition of hoses. Look for cracks and deterioration of rubber. Replace hoses and/or clamps if necessary. See **Figure 21-16.**

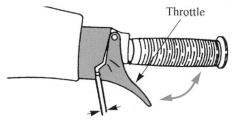

.020″-.060″ (.55mm-1.5mm)

Figure 21-13. *The throttle control lever should move freely and not have excessive play. (Polaris)*

Figure 21-14. *The fuel selector valve should be turned to the* On *position before starting engine. (Polaris)*

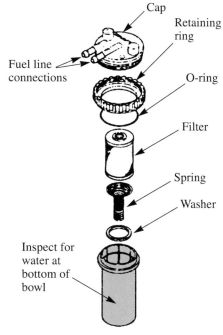

Assembly of Sediment Bowl

Figure 21-15. *The sediment bowl retaining ring must be unscrewed to remove the bowl. Check the filter and clean the bowl. O-ring must be seated properly when assembling to avoid leakage. (Polaris)*

Clamps

Nonreusable

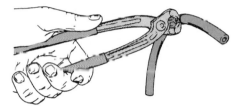

Cutting Clamp

Securing Clamp

Figure 21-16. *Check hose connections and clamps before riding. Look for cracks and deterioration of hoses. Replace bad hoses. Non-reusable clamps require a special tool. (Sea-Doo)*

- **Drain bilge.** Drain any water out of the engine compartment by removing drain plug. Replace drain plug securely after draining. See **Figure 21-11.**
- **Engine shut-off lanyard key.** Start the engine and run it for a few seconds. Pull the engine lanyard key off the engine stop button to see if engine stops immediately. See **Figure 21-17.**

⚠ Do not run the engine in a closed area. Exhaust gases contain carbon monoxide: a colorless, odorless, poisonous gas. Breathing exhaust gas leads to carbon monoxide poisoning, asphyxiation, and death.

- **Stop button.** Restart the engine, run it for a few seconds and check that the engine *Stop* button works. See **Figure 21-17.**

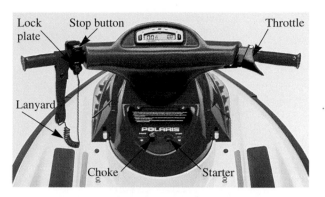

Figure 21-17. *The engine shutoff lanyard key should be tested before riding. Start engine, pull lanyard key off the engine stop button to see if engine stops. Check it two times. If this system fails when an operator falls off, the PWC could continue to run leaving the operator stranded in the water. (Polaris)*

⚠ Do not run the engine with the PWC out of the water for more than 15 seconds at a time. Overheating will cause serious engine and exhaust system damage.

- **Seat.** Check that the seat latch works and is secure. See **Figure 21-18.**
- **Rider protection.** Make sure proper flotation and protective gear are available for riders. See **Figure 21-19.**

Because of the number of items, it should be obvious that a printed checklist is necessary to maintain consistency and thoroughness. If the checklist is plastic laminated both sides it can be kept in the watercraft storage compartment without being destroyed by water. A well maintained

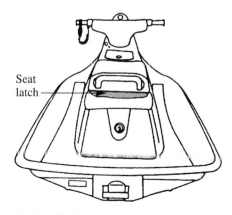

Figure 21-18. *Before riding, make sure the seat latch works and is secured. (Polaris)*

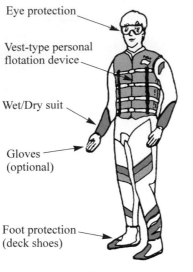

Eye protection

Vest-type personal
flotation device

Wet/Dry suit

Gloves
(optional)

Foot protection
(deck shoes)

Protective Gear

Figure 21-19. *All flotation and protective gear should be in good condition. This gear should be available for the operator and each rider. (Polaris)*

watercraft will provide many hours of enjoyable and trouble free recreation.

Launch ramp etiquette

Each operator should be considerate and efficient when launching a PWC. The craft and all preride safety checks should be performed beforehand. It is best to make all preparations and checks to the PWC before getting into the launching area. When it is time to launch the PWC, do so as quickly as possible.

When removing the PWC from the water and loading it onto a trailer, stop short of the launch ramp. Be patient for others who may be loading or unloading. Give adequate room and take your turn when the ramp is clear. Load quickly and pull away promptly to an area where you can attend to other details. Make sure all tie downs are secured and the PWC is resting properly on the cradle.

Identification numbers and placards

The engine and hull identification numbers are used to register a PWC. These numbers are the only way of positively identifying one PWC machine from another. They are unique numbers and may also be needed when ordering parts. If a PWC is stolen, investigating authorities require

both numbers in addition to the make and model number with other unique features that could identify it. The owner should record these numbers for ready access should they ever be needed. The records should be stored safely, but not in the boat.

The hull identification number is placed on the hull by the manufacturer. It is usually located near the stern (back) of the boat. The engine identification number will be stamped somewhere on the engine block by the manufacturer. See the manufacturer's technical service manuals for the exact location of the numbers.

Warning label decals are located in appropriate and readily readable locations. All labels should be read, understood, and followed carefully by anyone owning and/or operating a PWC. If any label becomes illegible, or comes off, it should be replaced as soon as possible. The PWC dealer will provide the proper decals, usually at no cost, and will also replace them if requested. Special procedures are required to remove and replace old decals. Some of the typical warning labels are shown in **Figure 21-20.**

Registration numbers must be located visibly on the hull. When registering a PWC the directions for placement will be provided and may vary from states to state. See **Figure 21-21A.** Registration numbers must be block characters no less than 3″ (76.2mm) in height. They should be a color that contrasts greatly with the background color on the hull. The spaces between the numerals and the prefix/suffix letters must be equal to the width of any letter except I, or any number except 1. **See Figure 21-21B.** When a PWC is purchased and registered, a specified number of days is allowed in most states to install the numbers. However, the registered PWC may be legally operated during this time period without the numbers displayed.

Signs, symbols, buoys, and markers

United States waters and state waterways are marked for safe navigation through the use of signs, symbols, buoys, and markers. These are made up of various shapes, colors, numbers, and lights to show proper course, speed limits, and other restrictions. See **Figure 21-22.** Markings may vary by geographic location. One of the most important parts of operating a PWC is to know the signs, symbols, buoys, and markers that will be encountered and then abide by them.

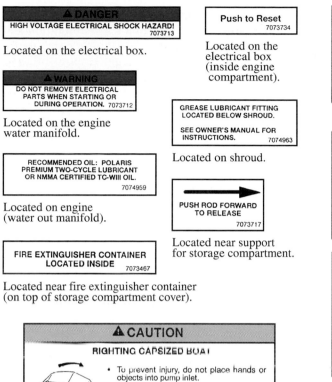

▲ DANGER
HIGH VOLTAGE ELECTRICAL SHOCK HAZARD!
7073713

Located on the electrical box.

▲ WARNING
DO NOT REMOVE ELECTRICAL PARTS WHEN STARTING OR DURING OPERATION. 7073712

Located on the engine water manifold.

RECOMMENDED OIL: POLARIS PREMIUM TWO-CYCLE LUBRICANT OR NMMA CERTIFIED TC-WIII OIL. 7074959

Located on engine (water out manifold).

FIRE EXTINGUISHER CONTAINER LOCATED INSIDE 7073467

Located near fire extinguisher container (on top of storage compartment cover).

Push to Reset
7073734

Located on the electrical box (inside engine compartment).

GREASE LUBRICANT FITTING LOCATED BELOW SHROUD.
SEE OWNER'S MANUAL FOR INSTRUCTIONS. 7074963

Located on shroud.

PUSH ROD FORWARD TO RELEASE
7073717

Located near support for storage compartment.

▲ CAUTION
RIGHTING CAPSIZED BOAT

• To prevent injury, do not place hands or objects into pump inlet.
• To prevent major engine damage: Make sure engine is stopped by pulling lanyard from engine stop switch and turn boat to upright position in a clockwise direction. 7073865

Located at rear of watercraft and positioned upside down allowing the operator to read it when the boat is in the capsized position.

Figure 21-20. *Decal labels such as those shown are positioned to warn and instruct riders about important information. Labels should be replaced if they are removed or unreadable. (Polaris)*

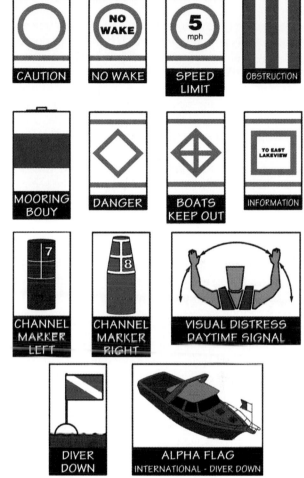

Figure 21-22. *These are some signs, symbols, markers, and buoys that a PWC operator will be expected to know. (Kawasaki Motors Corp., U.S.A.)*

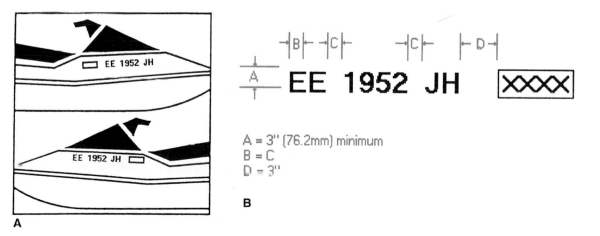

A = 3" (76.2mm) minimum
B = C
D = 3"

Figure 21-21. *A—Registration numbers must be located visibly on the hull and contrast greatly with the background color. B—All letters and numbers must be block letters no less than 3" (76.2mm) in height.*

Courtesy and common sense

Courtesy toward other users of the water and shoreline inhabitants can be extremely important to being welcome with a PWC. People not associated with PWC can become irritated when their privacy seems to be invaded by excessive wake and noise in their adjacent water. They may have their own boats tied at their docks. These boats can be damaged due to turbulence created by other boats passing at a fast speed. Other people may not appreciate the fun one is having because all they see and hear is the noise, and what may be seen as reckless riding. Courtesy and common sense can go a long way toward keeping peace and reducing restrictions upon all boaters.

Boating courtesy

The following is a list of courtesy rules that a PWC operator should follow:

1. As much as possible, stay away from residential areas, docks, moorings, etc. Respect the rights of shoreline property owners. PWC operators are responsible for any damage caused by their crafts wake. PWC operators must have prior permission from the owner to launch, land, or moor a boat on private property.
2. Pay attention to posted speed limits. Excessive speed can be dangerous and can create large wakes. Be particularly aware of *No Wake* areas usually marked with signs or buoys. See **Figure 21-22.** These areas should be traveled at 5 miles per hour (mph) or less. In many places, the required speed limit is posted with signs or buoys.
3. Excessive noise from PWC is the fastest way to make them unpopular with other water users and shoreline inhabitants. Noise carries further on water. Early morning and evening hours are when most people want to enjoy peace and quiet. Avoid riding for extended periods in front of private residences, condominiums, or apartments.
4. Keep a sharp lookout for swimmers at all times. Glare and sun can make swimmers and other objects hard to see. Avoid swimming areas.
5. Give sailboats and sailboarders plenty of room. High-speed wakes can cause them to veer off course and lose their source of power, the wind.
6. You should be familiar with the *Diver Down* flag. See **Figure 21-22.** Displayed on a boat or float it indicates that divers are present in the area. Avoid the area surrounding the flag by at least 300′.
7. A PWC operator has legal responsibility to stay clear of other nonpowered, less maneuverable vessels and watercraft.
8. *Never* jump the wake of a boat towing a skier. This is dangerous for everyone involved.
9. People fishing should be given plenty of room. On boats or ashore, people fishing usually have lines or nets out. These can be damaged by a PWC, a PWC can be damaged by them, or the PWC operator can be injured. Noise and turbulence may be perceived by some to scare fish away. This can cause strained relations between the people fishing and the PWC operator.
10. Stay away from dams. Areas immediately above and below dams are extremely dangerous. Observe and obey all signs and instructions about any dams in the area.

Navigational rules

Operating a PWC must be done in accordance with all rules and regulations governing it and the waterway on which it is being used. These rules are used and enforced internationally, as well as by the *U.S. Coast Guard* and local law enforcement agencies. A complete set of rules can be obtained from a *U.S. Coast Guard Auxiliary* or *Department of Motor Vehicles.* The complete set may also be obtained when registering ownership of a PWC.

The rules discussed in the following sections are not a complete set of rules. The rules covered are in a condensed format provided for convenience only.

Right-of-way and give-way

In nautical terms the ***stand-on vessel (privileged)*** has the right-of-way. The ***give-way vessel (burdened)*** must give way, or yield, to the stand-on vessel.

Whenever coming near another boat, be cautious and use common sense. Do not rely on other boaters to know or follow these rules.

Stand-on vessel

The stand-on vessel has the right-of-way and the duty to continue its course and speed, except to avoid an immediate collision. By maintaining course and speed the give-way vessels should be able to determine how best to maneuver and avoid the stand-on vessel.

Give-way vessel

The give-way vessel, which does not have the right-of-way, is responsible to take positive action to stay out of the way of the stand-on vessel. Normally, the give-way vessel should not cross in front of the stand-on vessel. The give-way vessel should slow down or change direction briefly and pass behind the stand-on vessel. The actions of the give-way vessel should be clear and understandable by the stand-on vessel.

Rule 2

Rule 2 is *The General Prudential Rule* in the International Rule. It tells the operator to follow standard *stand-on/give-way vessel* procedure except when a collision will occur. If both vessels try to avoid the collision, then both vessels become give-way vessels.

Encountering vessels

There are three main situations in which a PWC operator may encounter other vessels. These situations are *overtaking* (passing), *meeting* (approaching another vessel head-on), and *crossing* (traveling across another vessel's path).

Overtaking. If a PWC operator is passing another vessel, the PWC is the give-way vessel. The other vessel is expected to maintain its course and speed. The PWC must stay out of its way until it is past it.

The same would be true if the PWC was being overtaken by another vessel. The PWC would be the stand-on vessel, and should maintain course and speed until the other vessel has passed. See **Figure 21-23A.**

Meeting. If a PWC operator is meeting another vessel head-on, and the PWC is close enough to possibly collide, neither vessel has the right-of-way. Both vessels must alter course to avoid an accident. The PWC should keep the other vessel to the port side (left side). This rule does not apply if the PWC will be clear of the other vessel by maintaining course and speed. See **Figure 21-23B.**

Crossing. When a PWC encounters another power vessel that will cross paths close enough to run the risk of collision, the vessel having the other on the starboard side (right side) must avoid the other. If the other vessel is on the starboard side (right side), then the PWC is the give-way vessel and must keep out of way of the other vessel.

If the other vessel is on the port side (left side), the PWC is the stand-on vessel and should maintain course and speed. This is providing that the give-way vessel gives the PWC the proper right-of-way. See **Figure 21-23C.**

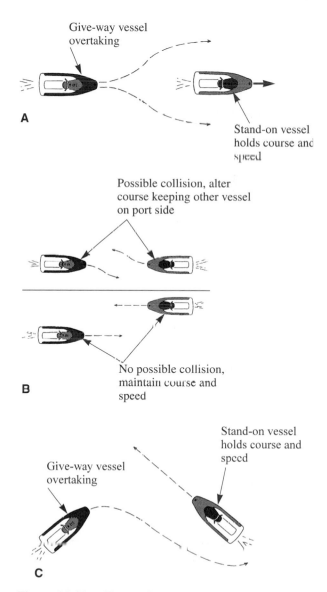

Figure 21-23. *The navigational rules for avoidance should be understood and applied properly when in the vicinity of other watercraft.*

Nonmotorized watercraft

Nonmotorized craft are normally given the right-of-way, except in the following situations:

- When a nonmotorized craft is overtaking a power vessel the power vessel has the right-of-way.
- Nonmotorized craft should stay clear of fishing vessels.
- In a narrow channel, a nonmotorized craft should not interfere with the safe passage of a power vessel.

Fishing vessel right-of-way

Under International Rules, a *fishing vessel* is any vessel that is fishing with nets, lines, or trawls. Vessels with trolling lines are not considered fishing vessels. Fishing vessels have the right-of-way regardless of position. Fishing vessels cannot interfere with the passage of other vessels in narrow channels.

Purchasing Considerations

PWC are available in several types. They vary in performance, stability, and the amount of skill necessary to operate them. Some are ridden in a sitting position and others are ridden in a kneeling or standing position. See **Figure 21-24.** Passenger carrying capacity can vary from one person boats to those that can carry up to three persons.

All PWC are designed to allow the operator to fall safely overboard. The reduced risk is due to the jet propulsion system that replaces the rudder

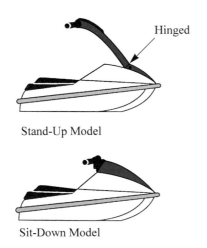

Figure 21-24. *The stand-up model is ridden in a kneeling or standing position. The sit-down model is ridden with the rider sitting down.*

and propeller on the outside of the hull common to other power boats. Also, the lanyard on the PWC stops the engine immediately if the operator falls overboard. This prevents the PWC from running away out of control. See **Figure 21-25.** If safety and manufacturer's instructions are followed, reboarding the PWC can be done with minimal risk.

Hull shapes and lengths vary from one type of PWC to another. The variations depend upon the purpose for which the PWC was designed, such as racing, water stunts, or pleasure cruising.

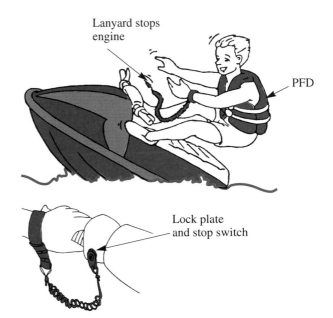

Figure 21-25. *A lanyard is attached to the rider's wrist and the engine switch. If the operator falls from the craft, the lanyard stops the engine, the PWC stops, and the rider can reboard from the rear of the craft.*

PWC main components and parts

Storage compartments are convenient for stowing personal gear, tools, and food. An approved, fully-charged fire extinguisher is required on the PWC. The extinguisher should be properly fastened in the storage compartment. High-performance racing or stunt-type PWC may have more limited storage space compared to larger cruising-type PWC. Fuel capacity and cruising range is also more limited in high-performance racing or stunt-type PWC.

Before purchasing a PWC, become familiar with the components and parts that are common to most all PWC. Once a specific make and model is purchased, the operator should become familiar with the specific components and parts of their PWC. This will enable the operator to take full advantage of the capabilities of the PWC. It will also give the operator confidence that the PWC is in proper working order and that complete control of the machine is possible.

Figure 21-26 shows the components found on most PWC. The following is a list of those components:

- **Seat strap**. A seat strap is available on PWC designed to carry more than one person. It provides the passenger with a handle to hold while riding.
- **Seat/engine compartment.** The seat is molded to a comfortable shape and should be made of a material that resists chaffing. Removing the seat provides access to the engine, oil tank, battery, electrical box, muffler, and other mechanical components.
- **Handlebars.** Handlebars are used to steer and control a PWC by turning the steerable jet nozzle at the stern. Turning the handlebar left turns the nozzle left and the boat turns left. A starter button, stop button with lanyard key, and trim adjust switch are located on the left handlebar. Before starting the PWC the operator must always attach the lanyard cord to their left wrist or personal floatation device (PFD), making sure it is secure.
- **Fire extinguisher.** The fire extinguisher should be securely held in the storage container. The U.S. Coast Guard requires that a class B-1 fire extinguisher in working condition be readily accessible aboard boats. Class B-1 extinguishers are charged with carbon dioxide for extinguishing class B fires, or flammable liquid fires. Refer to **Figure 1-10.** The *1* indicates the size of fire that the extinguisher can be used against. The larger the number the larger the fire the extinguisher can be used for. Locking pins and sealing wires should be in place.
- **Storage compartment door.** Opens to provide access to fire extinguisher and other items being carried.
- **Air intake openings.** These openings allow air to enter the engine compartment to supply engine with air and ventilate compartment.
- **Fuel tank cap.** Turn cap counterclockwise to open and clockwise to replace. Cap must be tight before operating PWC.

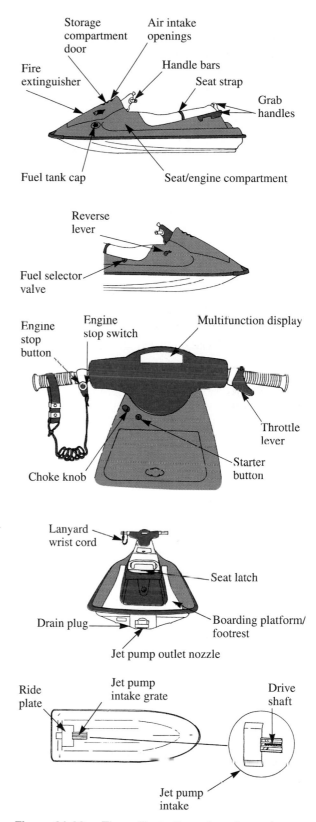

Figure 21-26. *These illustrations show the main components that are typically found on PWC.*

- **Grab handles.** Assist riders when boarding from water.
- **Fuel selector valve.** The fuel valve knob or lever has three positions: *On*, *Off*, and *Res* (reserve). The fuel must be turned *On* before starting the engine, and turned *Off* after stopping the engine. If the PWC runs out of fuel with the fuel selector in the *On* position, the low fuel warning light will come on. Switching to *Res* will allow a specific amount of additional running time to get to shore. The rider should know the amount of time and approximate distance that can be covered after switching to *Res*. After refueling, the fuel selector should always be set to the *On* position, not *Res*, for the next ride. Refueling should be done as soon as possible to avoid needing a tow, or being stranded in open water. Being stranded in open water is very dangerous.
- **Reverse lever.** A reverse shifting lever is used to engage a reverse bucket over the jet nozzle. The reverse lever has a lock button that must be pressed before the lever can be turned to the reverse position. The reverse bucket causes the jet stream to flow toward the bow of the boat. This causes the PWC to back up. This is very useful in docking or close quarter maneuvering situations. The boat should always be allowed to slow down before attempting to shift into reverse. The reverse bucket should *never* be used as a brake. This action can cause the PWC to dive into the water throwing the occupants forward, thus causing possible injury.
- **Multifunction display.** The multifunction display is similar to the instrument panel on an automobile. The display may contain such things as oil level warning light, low fuel warning light, fuel level lights, cooling water temperature warning light, engine tachometer, and trim position indicator.
- **Starter button.** To start the engine the starter button is pressed until the engine starts. *Never* press the button longer than 5 seconds. If the engine does not start wait 15 seconds and try again. If the engine does not start after several attempts, refer to the troubleshooting guide in the manufacturer's technical service manuals. The button should be released as soon as the engine starts.

The lanyard and lock plate must be attached to the engine stop switch or the engine will not crank or start.

- **Choke knob.** When starting a cold engine turn on the choke knob to help the engine obtain a rich mixture of fuel. This will make the engine start easier. As soon as the engine has started and runs a few seconds, the choke knob should be slowly turned off. Choking the engine should not be necessary if the engine is warm.
- **Throttle lever.** The throttle lever is located on the right hand side of the handlebar. The engine increases speed when the throttle handle is squeezed. When it is released, spring pressure returns the engine to idle speed. The throttle lever should move freely and return to idle quickly when released. This should always be checked before starting the engine.
- **Engine stop button.** The stop button is red for easy identification. It is located on the left hand side of the handlebar. The stop button will stop the engine when pressed.
- **Engine stop switch.** The lock plate end of the lanyard cord attaches to the stop switch on the left hand side of the handlebar. Should the rider be dislodged from the PWC the lanyard connected to the operator's wrist will pull the lock plate out and stop the engine.

On some PWC, disconnection of the lanyard slows the engine to idle speed and automatically turns the craft in a left circle. The PWC will return to the operator. If this feature is available it should be known before riding.

- **Jet pump outlet nozzle.** High-velocity water is pumped out through the nozzle by an axial-flow jet pump driven from the engine shaft. An impeller blade, like the propeller on an outboard engine, is housed inside a tubular shaped pump housing. The rotating impeller draws water in through the grate on the hull bottom. It forces the water out through the jet nozzle at high velocity. This drives the craft ahead similar to the operation of an aircraft jet engine. The jet pump outlet nozzle is hinged to allow it to control the direction of the PWC. The nozzle is directed by turning the handlebars right or left. Control cables connect the handlebars to the nozzle.

- **Lanyard wrist cord.** The lockplate end is fastened to the stop switch on the left hand side of the handlebar. The wrist end is fastened to the operator's wrist or PFD. If the rider is dislodged, the lanyard lock plate is pulled from the stop switch and the engine automatically stops, or slows to idle speed.
- **Seat latch.** Holds seat in place. When released the seat can be raised to provide access to engine compartment.
- **Boarding platform/footrest.** The place for riders feet while riding and assists boarding the PWC.
- **Drain plug.** When water gets into the bilge (bottom of hull) it can be drained by removing the PWC from the water and removing the drain plug. Some watercraft have an automatic bilge system that removes water from the engine compartment while the craft is being ridden. A venturi vacuum is created by the jet pump. This vacuum draws water from the engine compartment through a hose to a nozzle in the jet pump. A siphon breaker (check valve) is installed in the hose to prevent water from entering the hull when the engine is not running.
- **Jet pump intake grate.** The intake grate prevents large debris from entering into the jet pump thus damaging the impeller and drive shaft. Riders, who may suddenly become swimmers, should stay clear of the intake grate when the engine is running. Long, loose items such as hair can be drawn into the shaft and impeller. Loose articles of clothing or jewelry should be kept safely inside a wet/dry suit. PWC should never be used in water less than 2' (60cm) deep. This will prevent small debris from entering the pump. This could severely damage impeller blades.
- **Driveshaft.** The driveshaft transmits power from the engine to the impeller. The drive shaft is located under the intake grate.
- **Jet pump intake.** The impeller draws water through the grate and into this opening.
- **Ride plate.** This plate covers and protects the jet pump. It provides leveling control for the PWC. Some PWC have electric trim control. Pressing a trim switch on the left hand side of the handlebar starts an electric motor that raises or lowers the jet pump nozzle. Pressing

the switch down directs the nozzle down and lowers the bow of the PWC. Pressing the switch up raises the nozzle and raises the bow of the PWC. A trim indicator on the panel shows the position of the nozzle. The electric trim motor provides subtle changes in the boats attitude for a more efficient and comfortable ride.

PWC Engines

Engines used in PWC are precision built machines. These engines are two or three cylinder, two-cycle, water-cooled, high-speed engines. See **Figure 21-27.** An operator should become familiar

A

B

Figure 21-27. *PWC engines are two- or three-cylinder, two-cycle, water-cooled, high-speed precision machines. The engine in A has two cylinders and the one in B has three cylinders. These engines are specifically designed for jet propelled watercraft. (Kawasaki Motors Corp., U.S.A.)*

with the parts of these engines and their function, as wells as the cooling and jet propulsion systems. New PWC engines require a break-in period to prevent damage.

Basic engine parts

Figure 21-28 identifies many of the basic engine components of a two-cylinder engine. The cylinders contain intake, exhaust, and transfer ports. They direct the passage of air/fuel mixture from the crankcase to the combustion chamber. The cylinders accommodate the pistons and piston rings. The cylinder head is an aluminum alloy casting, bolted to the top of the cylinder head. See **Figure 21-29.** It contains the combustion chamber and threaded holes for the spark plugs. The engine cylinders and cylinder heads are jacketed to circulate cooling water. See **Figure 21-30.**

Pistons

The piston serves four purposes. It transmits combustion expansion forces to the crankshaft. It acts as a valve for the opening and closing of intake, exhaust, and transfer ports. (Two-cycle engines do not have poppet valves like four-cycle engines). The piston retains piston rings that seal the cylinder

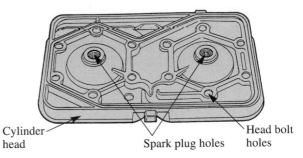

Figure 21-29. *The cylinder head is an aluminum alloy casting. (Sea-Doo)*

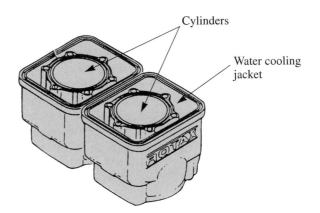

Figure 21-30. *Cylinders are surrounded by a water jacket for cooling. (Sea-Doo)*

bores and transmit heat to the cylinders. Its dome shape provides proper mixing of fuel and air.

The pistons are cast aluminum alloyed with manganese and copper or nickel. See **Figure 21-31.** This increases durability and heat resistance. To improve the pourability of the molten metal and reduce thermal expansion in the finished piston, 10% to 25% silicon is added to the alloy. The use

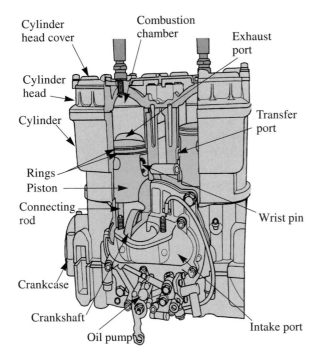

Figure 21-28. *Shown are the major components of a two-cylinder PWC engine. (Sea-Doo)*

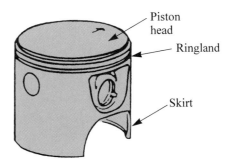

Figure 21-31. *Pistons are special cast alloy aluminum. This reduces the weight and mass to allow high speeds and power output.*

of aluminum in the piston reduces the weight of the reciprocating mass. This allows the high-rotational speeds and high power output of modern engines.

The pistons are connected to the crankshaft by a connecting rod and wrist pin. Pistons are manufactured out of round. See **Figure 21-32.** The reason is the wrist pin bosses have a greater volume of metal than other sections of the piston, and consequently absorb more heat. The greater heat absorption, the greater the area will expand. Under normal operating temperatures, the piston expands and becomes a more rounded shape to fit the cylinder more precisely.

Piston rings

Piston rings are placed in grooves immediately below the piston dome to seal the piston and cylinder wall. Within each ring groove is a small pin to provide a locating point for each ring end. See **Figure 21-33.** This pin prevents the rings from rotating in the grooves. Without it the ring ends would catch on the port edges and would result in the ring breaking. When installing a new piston and/or rings, it is critically important that piston ring locating pin be matched with the gap in piston ring. This should be checked as the cylinders are being installed.

Selection of the proper type of piston ring is relative to the kind of use of a particular engine. Rotax engine pistons use two types of compression rings. See **Figure 21-34.** For medium revolution engines rectangular rings provide adequate piston sealing. High-performance engines that develop higher revolutions, require better than average sealing from rings, such as the L-type trapezoid rings.

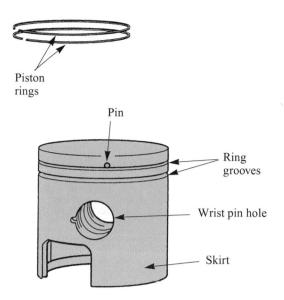

Figure 21-33. *Piston rings seal between the piston and cylinder wall. A small pin prevents the rings from rotating around the piston. (Sea-Doo)*

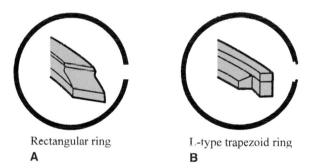

Rectangular ring
A

L-type trapezoid ring
B

Figure 21-34. *A—Rectangular rings are suitable for normal medium speed engines. B—L-type trapezoidal rings are for use on high-performance engines that require greater sealing ability. (Sea-Doo)*

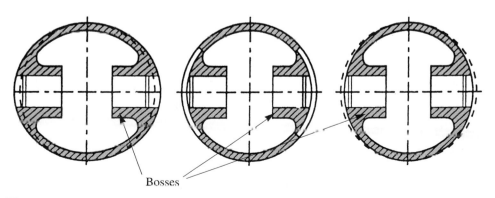

Figure 21-32. *Pistons are made out-of-round. Carefully calculated thermal expansion causes the piston to become round to fit the cylinder properly. (Sea-Doo)*

Crankcase

Crankcases for PWC are made as horizontally split, matched halves. See **Figure 21-35.** To reduce engine weight, the crankcase is made of light-weight aluminum alloy. To assist with the fuel pumping cycle, a pulse nipple is incorporated into the crankcase. A tube connects the pulse nipple to the carburetor fuel pump.

The crankcase serves three purposes. It supports the crankshaft in perfect alignment, serves as a primary compression chamber for incoming air/fuel mixture, and is the supporting structure for the cylinders, armature plate, coils, etc.

When assembling a crankcase it is important that the proper torque amount and sequence of tightening crankcase bolts be followed. **Figure 21-36** shows proper torquing sequence of crankcase bolts.

Crankshaft

The backbone of the engine is the crankshaft. It is the driving member that converts reciprocating movement of the pistons to rotary motion. See **Figure 21-37.** The crankshaft is supported by ball bearings mounted inside the crankcase. It is machined to obtain precision journals and perfect alignment. Any misalignment, distortion, or out of roundness would impair the necessary free, smooth rotation. This will shorten engine life.

Connecting rod

The link between the piston and the crankshaft is the connecting rod. See **Figure 21-37.** Needle bearings are installed within each bored end of the rod. Connecting rods may be an integral part of the crankshaft assembly and cannot be

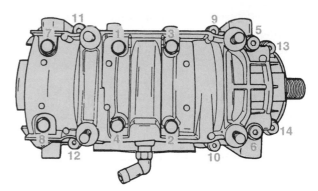

Crankcase Torquing Sequence

Figure 21-36. *Proper sequence and torque are important when assembling crankcase halves. (Sea-Doo)*

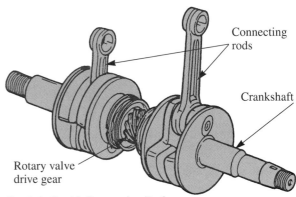

Crankshaft with Connecting Rods

Figure 21-37. *The crankshaft and connecting rods are assembled as a unit and installed in the crankcase. The crankshaft is supported in bearings so that it will turn freely and withstand the loads imposed upon it by the pistons and rods. (Sea-Doo)*

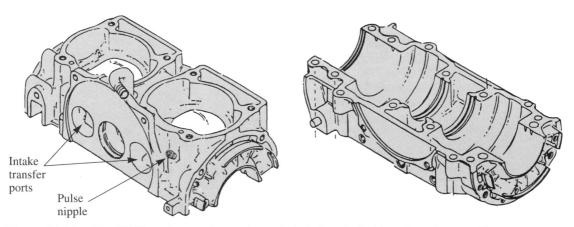

Figure 21-35. *The PWC engine crankcase is made in halves bolted together at assembly. Notice intake transfer ports. (Sea-Doo)*

removed unless the crankshaft is disassembled. Special tooling is required for this procedure and should not be attempted unless the proper tooling is available.

Some engines have the crankshaft and connecting rods mounted in sleeve type bearings instead of roller or needle bearings. The roller and needle bearings are more friction free and produce longer engine life.

Wrist pins

The wrist pin links the connecting rod to the piston. It is a precision machined steel pin inserted through a needle bearing in the upper (smaller) end of the connecting rod. It also is placed in the piston bosses and held in position with internal snap rings. Refer to Chapter 16 of this text.

Two-stroke cycle engine

The two-stroke cycle engine performs the same operation as a four-stroke cycle engine. However, the two stroke cycle is completed in 360° of crankshaft rotation instead of 720° of rotation in the four-stroke cycle. Refer to Chapter 5 of this text.

The air/fuel transfer may be accomplished by one of three methods. These transfer methods are through a crankcase reed valve, the piston skirt opening and closing a transfer port, or a rotary valve driven by the crankshaft. See **Figure 21-38**. As the piston starts its upward stroke, the air/fuel mixture is drawn into the crankcase from the carburetor. The drawing of the mixture is done through the intake port uncovered by one of the three transfer methods.

Rotary valve engines

The major differences between a piston port engine and a rotary valve engine are:
1. The intake port is directly positioned in the crankcase.
2. The opening and closing of the intake port is controlled by a rotary valve instead of the piston, allowing asymmetrical timing. Symmetrical timing reduces volumetric efficiency.

A rotary valve enables a very short air/fuel mixture inlet track. The design introduces the mixture in a very suitable position without obstruction to the gas flow. The intake position also enhances the lubrication of the lower connecting rod bearings. With rotary valves, the opening duration of

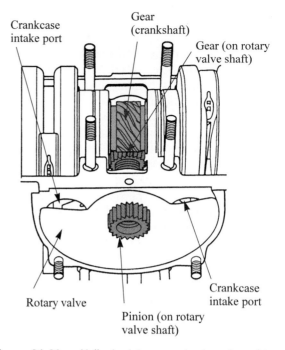

Figure 21-38. *Air/fuel mixture can be transferred to the crankcase by a rotary valve driven from the crankshaft gear. Atmospheric pressure moves air/fuel mixture into crankcase when ports are aligned. (Sea-Doo)*

the intake port is precisely controlled by the valve. This allows the manufacturer to determine and design the maximum possible intake with benefit to crankcase filling. **Figure 21-39** compares the piston port engine to the rotary valve engine.

The rotary valve intake duration is asymmetrical. In piston port engines, intake duration is symmetrical. Complete control of intake timing by using rotary valves produces greater engine torque at lower RPM, more peak power and easier starting.

Intake	*Piston Port Engine*	*Rotary Valve Engine*
Total duration	150°	195°
Opening (BTDC)	75°	130°
Closing (ATDC)	75°	65°

Figure 21-39. *As shown for the rotary valve engine, the total duration of the intake is greater and the opening starts earlier. This produces a greater volumetric efficiency. Also, in the rotary valve engine, the intake closes earlier to avoid fresh charge spitback. (Sea-Doo)*

Reed valve engines

Reed valve engines are utilized in some watercraft to increase performance over piston port engines. However, reed valve engines have some disadvantages when compared to rotary valves engines. The disadvantages of reed valve engines are as follows:

1. The intake port is restricted by the reeds and cage (reed stop). Refer to Chapter 5 of this text.
2. The reeds tend to separate air from fuel.
3. Because the crankcase *vacuum* must first open the reed to permit intake, this initial force is not fully applied to the intake operation. Therefore, there is a partial loss of intake potential.
4. At high speeds, the delay in closing the reed affects the reopening of the reed. Potential volumetric efficiency is adversely affected (the ability of air/fuel mixture to fill the combustion chamber).

5. A continual rebounding of the reed can cause further intake restrictions.

Engine cooling system

There are a number of cooling systems used in PWC engines. They all work by water circulating around engine parts. The cooling system shown in **Figure 21-40** is just one example. Engine cooling is obtained by intake water circulated through the engine. The engine cylinders and cylinder heads are jacketed to circulate cooling water. Universally the water is taken from a pressurized area between the impeller and venturi in the jet pump. The water passes through a tee, where a small amount is diverted directly into the exhaust gas for noise reduction and performance improvement.

The majority of water passes into the exhaust system and warmed. The water enters the cooling

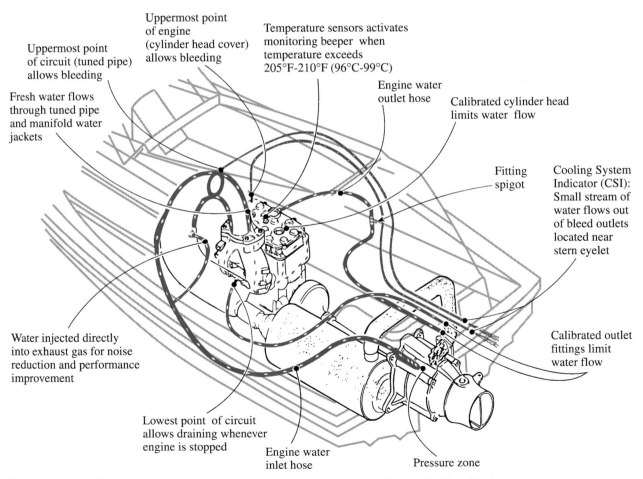

Uppermost point of engine (cylinder head cover) allows bleeding

Temperature sensors activates monitoring beeper when temperature exceeds 205°F-210°F (96°C-99°C)

Uppermost point of circuit (tuned pipe) allows bleeding

Fresh water flows through tuned pipe and manifold water jackets

Engine water outlet hose

Calibrated cylinder head limits water flow

Fitting spigot

Cooling System Indicator (CSI): Small stream of water flows out of bleed outlets located near stern eyelet

Water injected directly into exhaust gas for noise reduction and performance improvement

Calibrated outlet fittings limit water flow

Lowest point of circuit allows draining whenever engine is stopped

Engine water inlet hose

Pressure zone

Figure 21-40. *This is one example of an engine cooling system used in PWC. (Sea-Doo)*

jacket on the tuned exhaust pipe and travels into the cooling jacket around the exhaust manifold. Next, it enters the cooling jackets of the cylinders through small passages under the exhaust ports. The water surrounding the cylinders moves upward through calibrated holes in the cylinder head exiting the engine at the intake side of the cylinder head. The hot water is returned to its source through a hose and calibrated, limited water flow fittings at the stern of the PWC.

Bleed valves and lines prevent air entrapment in uppermost parts of the engine. If bleeding is not done, hot spots can occur.

Jet pump propulsion system

The jet pump is an axial flow device, which means it has a single impeller driven by a centrally located shaft. See **Figure 21-41.** Other than the engine, it is the major component that drives the PWC. It provides thrust in water similar to the way a jet engine produces thrust in air. The impeller draws water in through the grate screen on the bottom of the hull and forces it out through a discharge nozzle at great velocity. The reaction of the water being forced out the stern forces the boat forward. Essentially, the engine provides the rotational force transmitted to the shaft and impeller. The motive force is provided by the difference in pressures between the front and rear of the impeller blades. See **Figure 21-42.** The low-pressure side drawing water in, the high-pressure side forcing water out.

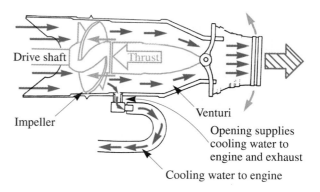

Figure 21-42. *Impeller draws water in and forces water out through the venturi where velocity is increased before water jets out through the steering nozzle. The reactionary force is opposite of the water direction, therefore forces the craft forward. (Kawasaki Motors Corp., U.S.A.)*

Performance of the PWC is closely related to the amount of pitch in the impeller blades and the horsepower of the engine The *pitch* of the impeller is the angular relationship of the blades to a line perpendicular to the shaft it is mounted on. See **Figure 21-43.** Racing craft have higher pitch impellers with high-horsepower engines to rotate them. Blade area and shape is also a factor in attaining proper efficiency. To achieve both acceleration and top speed, blades may have low pitch leading edges and high pitch trailing edges. Impellers are precision machined and foreign materials passing through them can cause minor to severe damage to their blades. See **Figure 21-44.**

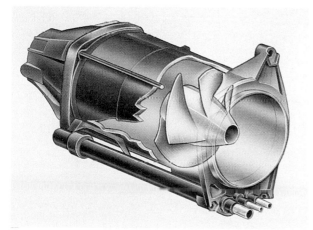

Figure 21-41. *An axial flow jet pump with single impeller that is driven by a centrally located shaft is the major driving unit of the PWC. (Polaris)*

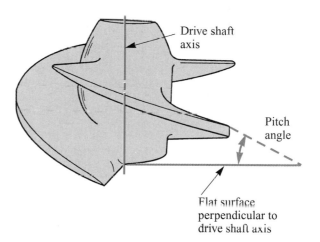

Figure 21-43. *The pitch of the impeller blades is the angle measured from the blade to a line perpendicular to the axis of the drive shaft. (Sea-Doo)*

Figure 21-44. *Impellers are precision machined and can be damaged if abrasive materials such as sand or stones pass through them. (Polaris and Sea-Doo)*

Figure 20-45 shows the internal components of an axial flow pump. *Stator vanes* reduce the tendency for the water to revolve as it enters the venturi. The venturi restriction causes the water to increase in velocity as it exits through the steering nozzle. The blade ends of the impeller rotate close to, but do not touch the wear ring. Clearance is about .020″ (.5mm). This clearance should be checked with a long feeler gauge at assembly.

In addition to providing propulsion for the PWC, the pump provides cooling water circula-tion for the engine as shown in **Figure 21-42.** Likewise, on some PWC a bilge draining system is installed by creating a venturi vacuum to draw water out with water in the discharge nozzle. See **Figure 21-46.**

> The general relationship of parts for a PWC jet pump is shown in Figure 21-47. Study this figure to become familiar with the jet pump parts.

Figure 21-45. *The components of a jet pump are precisely fitted to each other and must be in exact alignment. Stator vanes in the housing reduce the rotational tendency of the water as it leaves the impeller. Lubrication is essential. Leaks cannot be tolerated in the shaft area. (Sea-Doo)*

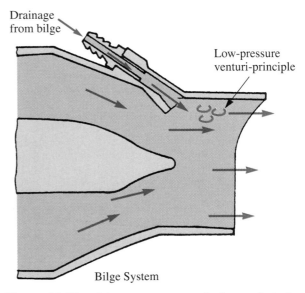
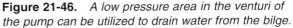

Figure 21-46. *A low pressure area in the venturi of the pump can be utilized to drain water from the bilge.*

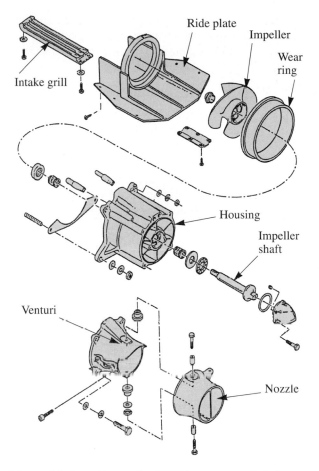

Ride plate

Impeller

Wear ring

Intake grill

Housing

Impeller shaft

Venturi

Nozzle

Figure 21-47. *Parts of a PWC jet pump and their general alignment positions to one another. Engine drive shaft and coupling are not shown. (Sea-Doo)*

New engine break-in

New PWC need a careful break in period for the engine parts to wear in to each other to produce smooth, long wearing surfaces. Overheating during this period due to improper use can do extreme damage to engine parts and shorten its useful life.

The break-in period for a specific engine will be outlined in the operating instructions of the manufacturer's technical service manuals. It may vary from one full tank of fuel to a specified number of hours of operation. During this period a fuel/oil mixture in the fuel tank is used in addition to the recommended oil in the oil injection tank. The fuel/oil mixture ratio is usually 50:1; 1 pint (.5 liter) of oil to 6 gallons (23 liters) of gasoline. This mixture provides additional lubrication during the break-in period. It is important to use the oil specified or engine warranty may be voided if engine damage is experienced.

For the best results, vary the throttle speeds during the break-in period without subjecting the boat to heavy loads or full-throttle settings. It is recommended that a passenger not be carried during the break-in period.

After the break-in period mixing oil with the gasoline is no longer necessary. The oil injection system will now provide the necessary engine lubrication.

Engines vary and require the operator to follow the manufacturer's recommendations for the specific make and model engine in the watercraft. Break-in information is provided in the manufacturer's technical service manuals

General Maintenance

The PWC engine drives a jet pump that draws water in and forces it out the back of the craft at high velocity to propel the boat forward. Like any gasoline engine, the PWC require proper maintenance and care to keep them in good operating condition. Even under ideal conditions, a certain amount of wear and deterioration of parts and eventually service of some kind can be expected. This section will introduce more engine parts, their functions, and their maintenance considerations. Earlier chapters of this text explain engine theory and many details that are common to PWC engines.

If extensive engine work is to be done, it will be necessary to have or obtain special tools designed by the manufacturer to properly do the job. Some of the tools are mandatory, others are recommended or more common to the average shop. Such tools as a degree wheel for setting timing of the rotary valve, special pullers, drivers, installers, and gauges may be necessary. Of course, the kind and extent of the work will determine the tools needed. If engine disassembly is needed, the engine will have to be removed from the PWC. This means all hoses and electrical connections must be removed, as well as mechanical separation from the jet pump. The engine should be removed from the PWC and placed in a cradle in the shop. Never attempt engine removal while the PWC is in the water. Lifting the engine from the hull may require an engine hoist. Eye bolts or lifting lugs on the engine may be needed. Trying to manually lift an engine out of the hull could be dangerous. If the engine is dropped, damage may be caused to the engine, the hull, and the person lifting.

The most important documents to have at hand to do any kind of engine work are the manufacturer's technical service manuals for the engine make and model. They will provide invaluable information about general PWC information, engine components, fuel and carburetion, electrical systems, lubrication system, cooling system, propulsion and drive system, steering system, hull and body maintenance, storage, and inspection.

Spark plugs

Spark plugs should be removed and examined for their condition periodically. Refer to *Spark Plug Condition Chart* in the *Appendix* section of this text. When checking the spark plugs, make sure the engine is cold. Blow any dirt away from the base of the plugs with compressed air before removing them to prevent dirt from falling into the spark plug holes. Use a deep spark plug socket being careful not to chip or crack the porcelain insulator. See **Figure 21-48.**

Spark plugs that do not show any signs of serious deterioration may be cleaned, gapped, and reinstalled. If reinstalling the used plugs, wipe off the threads and apply antiseize dielectric grease. When it is necessary to replace old sparks plugs with new ones, always use the spark plugs specified for the engine. When installing the plugs, do not overtighten. Use a torque wrench to tighten plugs to 13 lb-ft to 15 lb-ft. Refer to the *Spark plugs* section in Chapter 14 of this text.

Fuel filter

The fuel filter for a PWC is usually an in-line type. The location of the fuel filter may vary from one PWC to another. However, it will be installed in the fuel line between the fuel tank and the carburetor. Before removing the fuel filter turn off the fuel valve at the fuel tank to avoid spillage. See the manufacturer's technical service manuals.

The fuel filter is designed to prevent water or dirt particles from entering the carburetor and engine. If foreign particles and/or water accumulate in the filter, it should be removed and replaced with a new one.

> In-line fuel filters are sealed and are not serviceable.

Fuel filter screens

There may be fuel filter screens located at the fuel outlet of the fuel tank. The screens can be removed and cleaned or replaced. Always shut off the fuel tank valve before removing screens for inspection.

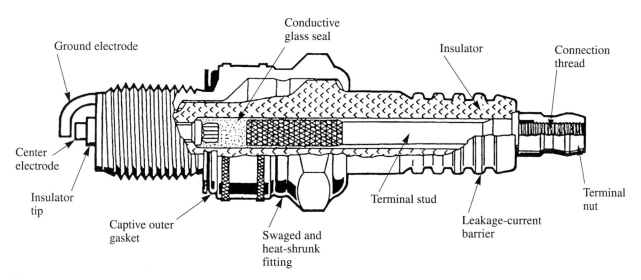

Figure 21-48. *Spark plugs can affect the running efficiency of an engine and should be in good condition. Aside from electrode wear and erosion the porcelain insulator can be cracked by using an improper tool or removal technique. (Sea-Doo)*

Fuel vent check valve

A rubber vent hose is connected from the fuel tank to a sediment bowl. A check valve mounted in the vent hose allows air to enter the fuel tank, but minimizes fuel spillage if the PWC is overturned.

To examine the check valve, remove it from the system and test it by blowing through each end. Air should pass easily in one direction, and none should pass through in the opposite direction. If the valve fails either of these tests, it should be replaced. See **Figure 21-49.** When installing the check valve in the line, the direction of air flow should point toward the tank. An arrow on the valve should indicate the direction of air flow.

 The fuel filler cap should be loosened to relieve tank pressure before disconnecting any lines in the system.

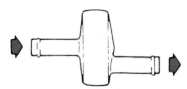

Air Should Pass This Direction

Air Should Not Pass This Direction

Figure 21-49. *Test the fuel vent check valve by blowing into it from each end. Air should pass through one direction only. If it fails, replace the valve. Install with arrow pointing toward the fuel tank. (Kawasaki Motors Corp., U.S.A.)*

Sediment bowl

The sediment bowl, located in the vent hose, prevents water from entering the fuel tank. The sediment bowl can be disassembled, drained of any water, and cleaned. The sediment bowl retaining ring is unscrewed and the bowl removed. The

bowl is sealed with an O-ring. The O-ring must be in place when the bowl is reassembled and the retaining ring tightened. See **Figure 21-15.**

Oil filter and oil filter screen

The oil filter is located in-line between the oil tank and the engine. See **Figure 21-50.** It is usually made of clear plastic, so visual inspection is performed without problem. The filter should be checked periodically. These filters cannot be cleaned. If the filter is dirty, restricted, or clogged, replace it immediately.

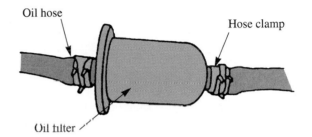

Figure 21-50. *An in-line oil filter should be changed annually during the preseason inspection. This filter should not be cleaned and reused. (Polaris)*

The oil tank has an oil filter screen set into the oil filler neck. This filter screen should be checked every time oil is added to the tank. If any dirt particles are seen on the filter screen remove, clean, and reinstall it before adding oil. The filter screen can be cleaned by washing it in a nonflammable or high flash-point solvent. A brush can be used to remove any trapped material. Be careful not to allow bristles from the brush to enter the oil tank or adhere to the filter screen.

 Do not use gasoline or a low flash-point solvent to clean the filter. This is dangerous fire hazard and may cause an explosion.

Oil pump bleeding

When either of the oil pump hoses is removed, air may become trapped inside the pump. Oil flow can be obstructed by the trapped air in the line. The following procedure should be used to *bleed* the air from the line.

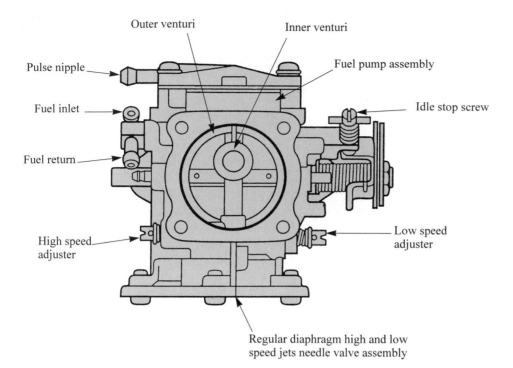

Outer venturi

Inner venturi

Pulse nipple

Fuel pump assembly

Fuel inlet

Idle stop screw

Fuel return

High speed adjuster

Low speed adjuster

Regular diaphragm high and low
speed jets needle valve assembly

Figure 21-51. *Shown are the external parts of one type of PWC carburetor. (Sea-Doo)*

1. Fill the oil tank with engine oil.
2. Place a rag under the oil pump.
3. Loosen the air bleed screw on the oil pump until oil flows out, then tighten the bleeder screw.
4. Provide engine cooling by running water through the cooling hose.
5. Start the engine and run it at idle speed. Observe the flow of oil passing through the transparent outlet hose.
6. Continue running the engine until all air bubbles in the outlet hose disappear.

> The engine must be running before the cooling water is turned on. The water should be turned off before the engine is stopped. Do not run the engine without cooling water for more than 15 seconds.

Carburetors

Idle speed is the lowest speed at which the engine will run slow and smoothly. An idle adjust screw is used to adjust the idle speed. If the screw is turned clockwise the idle speed increases, and if the screw is turned counterclockwise the idle speed decreases. A good idle speed is about 250 rpm in the water and 700 rpm out of the water. To read the rpm use a tachometer. Most PWC are equipped with tachometer.

Carburetors for PWC engines are adjusted at the factory for best performance under most sea level conditions. It is recommended by the manufacturers that these setting not be changed. If the PWC is to be used at high altitudes the air/fuel mixture will become richer due to the less dense air in the atmosphere. If the watercraft is to be used at altitudes above 3000′ (914m) mean sea level (MSL), the carburetor should be adjusted accordingly. For this operation, refer to the manufacturer's technical service manuals or take the PWC to an authorized dealer.

Figure 20-51 shows the external components of a PWC carburetor. There are a number of carburetor types. The type depends on the size and kind of engine the carburetor is used with. **Figure 20-52** shows the internal features of a carburetor.

When compared to the carburetors in Chapter 8 of this text, the carburetors illustrated in **Figure 21-52** have some interesting and unique features.

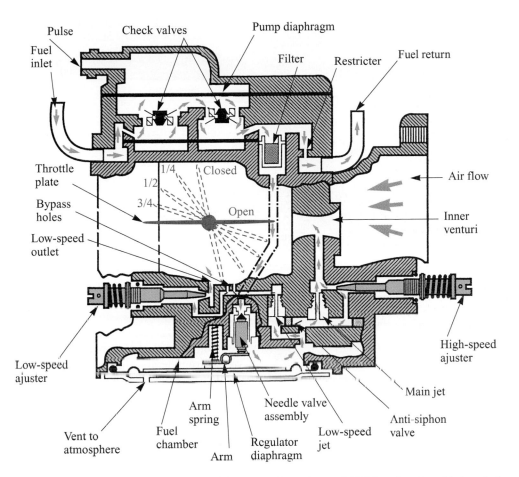

Figure 21-52. *Shown are the internal features of one type of PWC carburetor. Air flow is from right to left. Internal features of carburetor showing fuel flow at idle speed setting. Notice bypass holes in relationship to the position of throttle plate; closed through full throttle settings. (Sea-Doo)*

If the fuel flow in the high-speed circuit is followed, it becomes apparent the fuel flow to the high-speed jet is divided. Part of the fuel is metered by the low-speed screw. The fuel flow to the low-speed outlet takes an additional route by the by-pass holes (transition circuit). Fuel is drawn through the by-pass holes as the throttle is opened and the throttle plate (butterfly valve) exposes the holes to the air flow. The ability for fuel to pass through the low-speed circuit and transition is controlled by *pop-off pressure* (lifting of the needle from its seat) at the needle valve assembly, and then by the size of the low-speed jet. If the low-speed jet is changed, it affects the mixture at idle and speeds other than idle.

The fuel flow is also divided at the high-speed jet. This means even if the high-speed screw is fully closed, fuel would still pass through the jet.

There is a relationship between the arm spring and the needle in the needle valve assembly. The spring exerts a pressure through the arm onto the needle. The size of the hole in the valve seat contributes to the performance of the carburetor in several ways. First, and most importantly, the hole size helps determine pop-off pressure. Four things combine to create the pop-off pressure:

- Vacuum created within the carburetor body (manifold pressure)
- Atmospheric pressure
- Arm spring pressure
- Fuel pressure acting on the needle valve

Fuel pressure acts against the exposed end of the needle valve. For example, with a 2.0 size needle valve, the fuel pressure pushes against the needle with a certain force. If a larger 2.5 needle is used with the same spring, the fuel has a larger surface to push against and can move the needle more easily. Therefore, increasing needle valve size effectively decreases pop-off pressure; decreasing needle valve size will increase pop-off pressure.

 The principles of hydraulics apply when figuring the changes in needle size and its relationship to the change in pop-off pressure.

The arm spring can also be changed to accommodate adjustment. The manufacturer of the carburetor in **Figure 21-52** provides four springs with different gram ratings of stiffness. It also provides a reference chart to obtain approximate pop-off pressures with spring pressure and needle valve combinations.

Study the carburetor diagrams in **Figure 21-52.** Examine how the fuel flow is affected by the varying throttle settings. Air flow through the carburetor is from right to left.

Battery care

The battery in a PWC may be either a maintenance free or a conventional lead-acid battery. Check the manufacturer's technical service manuals to determine which type is being used.

After the specified electrolyte has been installed at the initial service of a maintenance free battery, the sealed caps should never be removed. The maintenance free battery should not be replaced with a conventional lead-acid battery. The electrical system is designed exclusively for a maintenance free battery and the electrical system will not work properly with a conventional lead-acid battery. Checking electrolyte level or adding distilled water is not necessary in maintenance free batteries.

If the PWC is designed for a conventional lead-acid battery, only distilled water should be used to bring the electrolyte to the proper level. See **Figure 21-53.** The electrolyte should always be kept above the lead plates.

Battery safety

Removing, servicing, charging, or installing a battery must be done carefully to avoid spilling any electrolyte. If a battery is to be stored for an extended period of time, store it in a cool, dry place, out of direct sunlight.

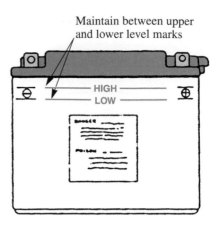

Figure 21-53. *The electrolyte level in a conventional battery of this type is kept between the lines on the case by adding distilled water as needed. Electrolyte level should not be allowed to drop below the plate tops. (Polaris)*

⚠ Keep children away from batteries. Conventional lead-acid storage batteries contain poisonous electrolyte. They contain sulfuric acid that can cause serious burns to the skin, eyes, or clothing if contact is made. The following are antidotal measures to be taken if the battery acid is contacted:
- **External.** Flush with water.
- **Internal.** Drink large quantities of water or milk. Follow with milk of magnesia, beaten egg or vegetable oil. Call a physician immediately.
- **Eyes.** Flush with water for 15 minutes and get prompt medical attention.

Batteries produce hydrogen gas which is very explosive. Keep batteries away from sparks, flame, cigarettes, etc. Charging and using a battery should be done in a well ventilated space. Always wear eye protection when working with batteries.

Battery removal

Batteries used in PWC are secured in a battery box, tray, or similar device fastened to the hull. The battery is strapped in position to resist being thrown about during operation. See **Figure 21-54A.** A loose battery would quickly result in a cracked case and electrolyte leaking into the hull. This could cause serious damage to the hull and possible injury to riders.

Follow the proper procedure recommended in the manufacturer's technical service manuals for removing straps and battery. The following are general safe procedures:

1. Maintenance free battery. See **Figure 20-54A.**
   a. Disconnect the black (negative –) cable first.
   b. Disconnect the red (positive +) cable last.
   c. Lift the battery out of the hull carefully. Do not tip the battery.
2. Conventional lead-acid battery: See **Figure 20-54B.**
   a. Remove battery vent tube from battery.
   b. Disconnect the black (negative –) cable first.
   c. Disconnect the red (positive +) cable last.

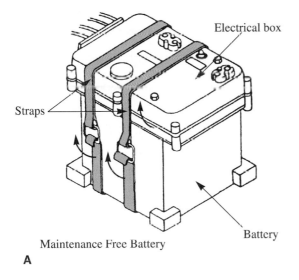

Maintenance Free Battery

A

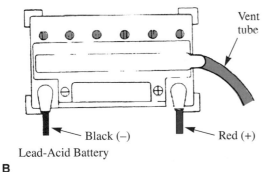

Lead-Acid Battery

B

Figure 21-54. *A—Shown is a maintenance free battery. B—Shown is a conventional lead/acid battery. All batteries must be strapped in mounting to prevent cracking of case. When removing either type, handle with care, follow proper methods, and do not tip battery or spill acid. (Polaris)*

d. Lift the battery out of the PWC carefully. Do not tip the battery to avoid spilling electrolyte.

Conventional lead-acid batteries need occasional additions of distilled water to keep the electrolyte level above the plates. Fill to the level shown on the case of the battery. See **Figure 21-53.** Use a bulb syringe to add water and do not overfill. Maintenance free batteries *do not* need to have fluid added and should not have the sealed caps removed.

Battery terminals

Corrosion at the battery terminals and cable connectors is not uncommon. The corrosion should be cleaned from the terminals. To clean the terminals, proceed as follows:

1. Keep the battery caps in place.
2. Use a stiff wire brush to remove the corrosion from the cable connectors and battery terminals.
3. Use a solution of baking soda and water to wash the cable connectors and battery terminals. (One teaspoon of baking soda to one cup of water).
4. Rinse with clear water and dry the battery case and terminals with absorbent cloths.
5. Coat the terminals with a waterproof dielectric grease to prevent further corrosion.

Battery charging

Only charge a battery when it needs it. Normally, the alternator driven by the engine keeps the battery charged. If the battery is old and does not crank the engine the next day after using it, it should be tested and possibly replaced with a new one. A battery that is low on fluid will not take a full charge. This battery will eventually overheat the plates and fail completely. Sulfated plates will cause battery failure also. Refer to the *Batteries* section in Chapter 14 of this text.

When using a battery charger, the battery should be connected to the charger before the charger is turned on. If the charger is turned on first, sparks will occur at the terminals and possibly cause an explosion. If it is a conventional lead-acid battery, remove the caps from each battery cell and add distilled water before charging. Fill each cell to the correct level.

⚠️ A battery that is low on charge may freeze in cold weather. *Never* attempt to charge a frozen battery. It may explode.

Installing the battery

If an old battery is to be replaced, always replace it with the same battery type. Refer to the manufacturer's technical service manuals for the proper type. To install a battery proceed as follows:

1. Set the battery into its holder.
2. Replace the battery vent tube on conventional lead-acid batteries.
3. Connect and tighten the red (positive) cable first.
4. Connect and tighten the black (negative) cable last.
5. Coat the cable connectors and terminals with waterproof dielectric grease.
6. Install battery cover and secure hold down straps.
7. Inspect cables and vent tube.

Cooling system flushing

The cooling system of a PWC must be protected from accumulating salt deposits and sand. To prevent this, the system must be thoroughly flushed out with fresh water. This should be done after *every use* in salt water. If used in fresh water, the system should be flushed *after each day's use* and whenever there appears to be reduced water flow from the bleed outlets. See **Figure 21-40.** These outlets are located on the hull of the craft at the stern. A PWC cooling system should be flushed clean before any extended storage period. The procedure for flushing is also used for providing auxiliary cooling water during out of water maintenance checks, such as oil pump air bleeding.

Operating in shallow or dirty water, will require more frequent flushing than in deep, clear freshwater. The water utilized for cooling the entire system is from the same water that the PWC is operated in. Beaching a PWC with the engine running should always be avoided. If the PWC is beached, the cooling system must be flushed before the engine is run again or severe engine damage will result.

To flush the cooling system of a PWC, proceed as follows:

 Always follow the manufacturer recommended flushing procedure when flushing the watercraft cooling system to avoid engine damage. Never flush a hot engine. Severe engine damage could result.

 Do not touch any electrical part when the engine is running. Severe injury or death could result.

1. Remove seat.
2. Locate the inlet for auxiliary cooling water. Refer to manufacturer's technical service manuals for this location.
3. If a special adapter is required, install it and fasten a garden hose as directed in the manufacturer's technical service manuals. If a fitting is already installed for attaching a garden hose, follow the prescribed procedure in the manufacturer's technical service manuals.
4. Start the engine allowing it to idle *before* turning on the water.

The engine must be turned on *before* the water is turned on, or water may flow back through the exhaust pipe into the engine resulting in severe engine damage.

5. Turn the water on immediately (within 10 seconds) after the engine is started. Remember, the engine should never be run longer than 15 seconds without cooling water.
6. Adjust the water flow so that a small trickle of water discharges out of the bypass outlet of the hull.
7. Slightly *rev* the engine (increase engine speed) intermittently for one or two minutes to flush the system completely.
8. Turn off the water. Turn off the engine within 10 seconds.
9. Remove the garden hose and reassemble hoses and clamps, or install the plug in accordance with the instructions in the manufacturer's technical service manuals.

Flushing the bilge

The bilge is the inside bottom of the hull of a PWC. Some PWC have a drain plug in the stern to drain water that may be accumulated in the bilge. See **Figure 21-11.** To flush and drain the bilge the PWC should be removed from the water, the plug removed, and bilge water allowed to drain completely. Flush the bilge with fresh water and continue to drain remaining water. The bilge should be wiped dry with towels or shop cloths. Reinstall the plug securely before launching the PWC.

Some PWC have automatic bilge draining systems. These systems also need flushing with fresh water and will require certain hoses and fittings to be disconnected. A garden hose is connected to the hoses and fresh water turned on for flushing of the system. During this procedure, water may flow into the engine compartment. Do not allow a large amount of water to enter the engine compartment. Remove the drain plug in the stern of the watercraft and drain *all* the water from the engine compartment.

Jet pump water inlet screen

The purpose of the water inlet screen is to prevent grass and debris from entering the engine cooling system. See **Figure 21-55.** The water inlet screen is located inside the stationary nozzle of the jet pump. After using the PWC, the screen should be visually inspected for buildup of any contaminants and debris. Clean the screen as needed, by flushing engine and screen. If the water inlet screen becomes plugged with contaminants the engine will overheat. The PWC should be stopped immediately to avoid serious internal engine damage.

External lubrication points

Because of exposure to corrosive elements in water and atmospheric conditions, proper lubrication of all moving parts is essential to maintain good performance and extend the useful life of the PWC. Grease used for PWC is special marine grease that is water and salt resistant. At points where grease should be applied, always use the type specified for the particular PWC being treated. The following are common locations on the PWC that require periodic lubrication:

- **Throttle cable and choke cable.** Special cable lube is used to lubricate the inner cables. The throttle lever should be depressed and cable lube squirted onto the cable. The lubricant is worked down the cable by pushing and releasing the throttle several times. The choke cable is lubricated similarly. See **Figure 21-56.**
- **Steering cable joints and inner wire.** Lubricate same as throttle cable and choke cable. Lubricate steering nozzle end and handlebar ends. Seals can be moved to allow oil to enter into the cables. Replace seals after oiling.
- **Steering nozzle pivot shaft.** Lubricate the pivot connections.
- **Steering handle pivot shaft.** Lubricate shaft and bushing. Tighten shaft if it has loosened.
- **Seat latch and hooks.** Grease the locking mechanism.

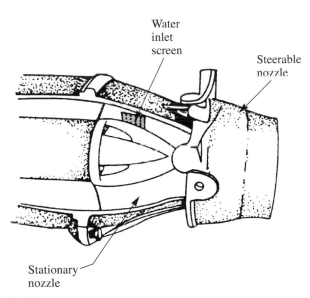

Water inlet screen

Steerable nozzle

Stationary nozzle

Jet Pump

Figure 21-55. *The jet pump water inlet screen must be inspected for foreign materials. If cooling water to the engine cannot pass through this screen the engine will overheat and seize causing severe damage. (Polaris)*

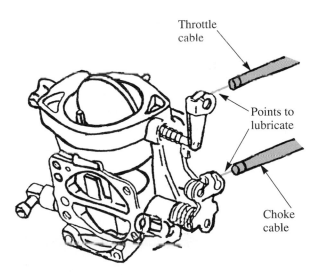

Throttle cable

Points to lubricate

Choke cable

Figure 21-56. *Grease springs and exposed portions of cable and shafts at the carburetor. Use specified waterproof grease. (Polaris)*

- **Carburetor and oil injection pump.** Grease springs, exposed portions of cable and shafts at the carburetor. If the PWC is used in salt water, grease more often.
- **Electrical connections.** Apply dielectric grease to battery posts and exposed cable connections.

Depending upon the make and model of the PWC, in addition to components mentioned others may need lubrication. Refer to the manufacturer's technical service manuals for detailed information about procedures and specific kinds of lubricants to use.

Transporting PWC

It is common practice to transport PWC on trailers. Trailers are commercially available to handle one PWC, or two PWC side by side. See **Figure 21-57.** All aluminum trailers are corrosion resistant, light, and easy to handle. The trailer should be equipped with greasable, sealed wheel bearings to prevent water from entering the bearings. See **Figure 21-58.** If the wheel bearings are not sealed and water enters them they will corrode and eventually fail. Grease fittings on the axle hub caps permit pressure greasing with a hand grease gun without removing the wheels and wheel bearings from the axles. A waterproof wheel-bearing grease should always be used.

PWC trailers are often backed down a launch ramp and the wheels are submerged until the PWC floats off the trailer. Likewise, the trailer is lowered down the ramp, the PWC floated up to it, the bow of the PWC is hooked to the trailer, and the PWC is pulled onto the trailer. The trailer and PWC can then be pulled out of the water.

The watercraft should be securely fastened to the trailer. A bow line, usually a cable or line (rope), attached to the trailer or winch, holds the bow in place. See **Figure 21-59.** Manufacturers provide a tie down at the bow of the boat for this purpose. The stern of the craft should also be tied down to the trailer. Tie downs, or boarding handles at the stern may be used to fasten a line through, and to, the trailer frame. Additional tie down lines may be used between the bow and stern for additional security. Study the manufacturer's technical service manuals for the recommended method to secure the craft to a trailer. It is important that the trailer cradle supports be located properly so the weight of the craft and road bumps do not damage

Figure 21-57. *Trailers are available for towing one or two PWC. The trailers should have proper supports and tie down equipment. Running lights and license illumination are required. (Featherlite Trailers)*

Figure 21-58. *The trailer should be equipped with greasable, sealed wheel bearings to prevent water from entering the bearings. Grease fittings on the axle hub caps permit pressure greasing with a waterproof wheel-bearing grease. (Dutton-Lainson Co.)*

the hull. It may be important to avoid supporting the hull at particular places to avoid exceeding strength limits. A cracked hull can surely spoil a long anticipated day of recreation for the owner. To protect the overall appearance and keep the craft clean while trailering over long distances an opaque fitted cover can be used.

When transporting a PWC, the fuel valve should always be turned to the *Off* position. The seat should be latched and compartment doors secured. The trailer should be matched to the PWC

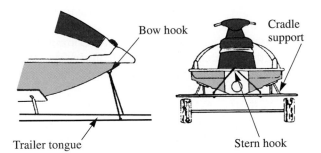

Figure 21-59. *Fore and aft tie downs secure the PWC for towing, while the hull rests on cradle pads. (Polaris)*

weight and hull configuration. Adjustments to the trailer cradle, rollers, and/or pads are possible to conform to the hull shape and length. Abide by the trailer laws and regulations for the area in which it will be used. Proper trailer licensing and lights are mandatory.

Running lights, stop lights, and turn signals are required. These must be functioning anytime the trailer is being towed on the road. A locking trailer hitch on the trailer tongue must conform to the weight of the trailer and include safety chains. The ball hitch on the towing vehicle must also be securely attached and conform to the weight of the trailer and PWC being towed. Electrical connections for the trailer lights should have a quick disconnect receptacle with a ground wire to the vehicle ground. Grounding through the ball hitch is not proper due to grease, corrosion, and constant movement between the ball and the hitch. If this is done lights will be dim, intermittent, and may be entirely out of service at times.

Towing PWC to shore

For some unforeseen reason, it is entirely possible that a PWC might need to be towed to shore by another watercraft. Running out of reserve fuel, engine problems, dead battery, or other complications could necessitate such towing. To accomplish this, tie one end of a 20′ (6m) line to the bow of the dead PWC. Attach the other end to the tow boat. Towing must be done *slowly* at idle speed. Do not go over 5 mph (8 km/h).

 It would be wise to carry a 20′ tow rope in the storage compartment of the PWC for such an occasion.

PWC storage

When preparing a PWC for off-season storage some preventive maintenance is recommended to preserve the PWC and its components. By taking a few precautions at the time of storage, the PWC will be ready to ride the next season with minimum preparation time. The following are the steps for preparing the PWC for storage:

1. Flush the entire cooling system with fresh water and dry. See *Cooling system flushing* section in this chapter.
2. Make sure the exhaust system is drained and dry. Do this by starting the engine and *briefly rev* the engine. Not more than 10 seconds.

⚠ Gasoline fumes are highly flammable and can be explosive under certain conditions. Do not smoke. Make sure the area is well ventilated and free from any source of flame or sparks; including appliance pilot lights.

3. If it is desired to leave gasoline in the fuel tank during storage, a fuel stabilizer should be added to the fuel tank. Fill the tank and replace the cap. The stabilizer will prevent varnish from forming. Filling the fuel tank will prevent condensation of water in the tank. When using a fuel stabilizer in the fuel, run the engine briefly to circulate the stabilizer through the carburetor.

An alternate method is to starve the engine of gas. To do this, add the stabilizer to the fuel tank and then shut off the fuel from the tank with the shut off valve. Now, run the engine at low speed until it starves of fuel. Because it may take longer than 15 seconds to starve the engine of fuel, this procedure *must be* done with cooling water circulating in the engine. If cooling water is not used, stop the engine after 10 seconds, let it cool 5 minutes, repeat the process until the engine starves and stops by itself. When the engine has stopped, leave the fuel shut off valve in the *Off* position. This method removes all fuel from inside the carburetor(s) and avoids varnish build up in them during storage.

If it is desired that the fuel tank be completely emptied for storage, drain the tank with a fuel siphon, or hand operated pump. Leave the fuel tank cap loose. This will allow air to circulate in the tank and prevent water condensation. Fogging the engine with rust-preventive oil and drying out the carburetors can be done simultaneously. Fog the engine with a rust-preventive oil by following the procedures on the can. Fogging oil is available from the PWC dealer or an automotive service store. The fogging oil will protect the internal components of the engine such as cylinders and rings, bearings, crankshaft, etc. Run the engine at part throttle until the engine starves of fuel. Because it may take longer than 15 seconds to starve the engine of fuel, this procedure *must be* done with cooling water circulating in the engine. Do not run the engine more than 15 seconds without cooling water. If cooling water is not used, stop the engine after 10 seconds, let it cool 5 minutes, repeat the process until the engine starves and stops by itself. Under any circumstances, do not allow engine to become overheated.

Lubrication

Before storing a PWC for the season, there are many components that need to be lubricated. To lubricate the PWC for storage, proceed as follows:

1. Remove the spark plugs and pour about one tablespoon of two-cycle engine oil into each cylinder.
2. Inspect the spark plug condition. Spread a small quantity of antiseize grease to spark plug threads and reinstall old plugs or replace with new plugs as needed. Tighten plugs with a torque wrench to 13 lb-ft to 15 lb-ft. Do not overtighten.
3. Lubricate the choke, throttle, and steering cables. See *External lubrication points* section in this chapter.
4. Lubricate all other areas described in *External lubrication points* section in this chapter.

Cleaning

Before storing a PWC for the season, it must be cleaned. To clean the PWC before storage, proceed as follows:

1. Drain the bilge and engine area by removing the drain plug. Cleaning may be done with hot water and mild detergent (dish soap). Rinse and drain. Dry remaining water with clean, absorbent towels. Leave the drain open and prop up the seat 1″ to 2″ during storage.

This prevents condensation from forming in the engine compartment.
2. Coat the engine and engine compartment with a protective spray and lubricant. These are available for use on the engine and in the engine compartment.
3. Wash the exterior of the PWC with fresh water and mild detergent. Rinse thoroughly.
4. Inspect and clean the jet pump intake, outlet, and impeller area. These components should receive service at this time if any damage is observed.

> ⚠ Never use strong detergents, abrasives, degreasers, paint thinner, acetone, window cleaners, ammonia or products containing alcohol for cleaning the PWC. They can cause damage to finishes, decals, vinyl and plastics and accelerate UV breakdown. This breakdown can cause color change and early deterioration of parts.

5. Protect and shine the exterior with a nonabrasive silicone wax. The seat and other vinyl surfaces can be protected with a vinyl protector.
7. Place an opaque cover over the PWC during storage to keep it clean. Store it in a cool, dry place.

Know These Terms

personal watercraft (PWC)	personal flotation device (PFD)
port	stand-on vessel (privileged)
starboard	give-way vessel (burdened)
bow	
stern	
hull	overtaking
forward	meeting
aft	crossing
deck	fishing vessel
steering control	pitch
steerable nozzle	stator vanes
throttle	idle speed
lanyard	pop-off pressure
draft	
Department of Natural Resources (DNR)	

Chapter 21
Review Questions

Answer the following questions on a separate sheet of paper.

1. The Coast Guard considers PWC of the kind in this chapter as Class_____ boats.
2. Name eight general boating terms.
3. List six items that should be on a preride checklist.
4. PWC of the type in this chapter may be designed to carry from _____ to _____ passengers.
5. The major propulsion system of a PWC is a _____.
6. There are two styles of PWC. They are the _____ and _____ models.
7. The Coast Guard recommends that all operators of PWC complete a _____.
8. When the engine of a PWC is slowed to idle speed from high speed the danger that exists is a lack of _____ and _____.
9. Coasting from high speed to complete stop may take as much as _____'.
10. It is illegal to ride a PWC between _____ and _____.
11. If a PWC is overturned and uprighted in the wrong direction the result may be engine damage caused by _____.
12. No one should ever operate a PWC while under the influence of _____ or _____.
13. It is a requirement that *all* riders of PWC wear a Coast Guard approved _____.
14. Personal watercraft are identified by the manufacturer with a number on the _____ and the _____.
15. Name twelve common components of PWC.
16. PWC engines are _____ cycle, water cooled engines.
17. What is the most important thing to have if any extensive engine work is to be done?
18. Name the three means that air/fuel mixtures can be introduced into the crankcase.
19. The PWC engine's main job is to drive the _____.
20. In addition to providing propulsion for the watercraft, the jet pump also provides _____ water for the engine.
21. In addition to providing propulsion, in some PWC the jet pump will also _____ from the bilge.
22. If water gets into the engine, what must be done as soon as possible?
23. What parts of an engine will, by inspection, give an indication of the engine condition?
    a. spark plugs
    b. piston dome
    c. Neither A nor B.
    d. Both A and B.
24. A PWC engine should never be run without cooling water for more than _____ seconds.
    a. 5   c. 3
    b. 15   d. 17
25. To avoid severe engine damage when flushing the cooling system, the engine must be turned on _____ seconds before the flushing water is turned on.
26. If a PWC is used in salt water, the cooling system should be thoroughly flushed clean with fresh water every _____.
    a. week it is used
    b. season
    c. time it is used
    d. couple of days

Suggested Activities

1. Demonstrate the proper clothes to wear when operating a PWC.
2. Make a presentation on safe procedures when operating a PWC.
3. Make a display of right-of-way and give-way rules.
4. Make a display of signs, symbols, buoys, and markers to be familiar with when operating a PWC.
5. Demonstrate proper cooling system flushing procedures.
6. Demonstrate proper pre-ride inspection procedures.
7. Demonstrate proper engine and systems maintenance.
8. Demonstrate how to launch or remove a PWC from the water and secure it and a trailer for towing.

This Kawasaki Jet Ski is one example of a personal watercraft. (Kawasaki Motors Corp., U.S.A.)

Career Opportunities and Certification

After studying this chapter, you will be able to:
- ▼ Identify several career opportunities in the small gas engine field.
- ▼ List qualities that are essential for anyone pursuing a career in small engines.
- ▼ List the advantages and disadvantages of entrepreneurship.
- ▼ Identify the benefits of outdoor power equipment certification.

Career Opportunities in the Small Gas Engine Field

The small gas engine field offers career opportunities in three different areas: service, sales, and manufacturing. Through training, study, and work experience, you can become an engine service technician, service manager, sales manager, or general manager of a small engine service center. You can also be a manufacturer's technician, service representative, or engineer.

Engine service technician

Implement sales facilities and equipment rental centers need *engine service technicians* to do tune-ups, service equipment, and make repairs. Often, the quality of workmanship and reliability of the technician who services the customers' equipment directly affects the reputation and sales volume of the business.

Small gas engine service technicians must be able to diagnose engine troubles and make appro-

priate repairs and/or part replacements. See **Figure 22-1.** They should be able to analyze the mechanical condition and performance of an engine and make proper recommendations to the owner. A service technician should also be competent in the use of test equipment and thoroughly familiar with manufacturers' technical service manuals.

Good service technicians keep their tools and equipment in first-class condition and organized for convenient use. They must have specialty tools

Figure 22-1. *A small gas engine service technician must know engine construction and principles of operation. Technicians must be proficient in troubleshooting, maintenance, service, and repair. Notice the certified OPE technician arm patch.*

(pullers, drivers, etc.), which are available from the engine manufacturers or tool manufacturers. Small gas engine technicians must know how to use micrometers and dial indicators. Basic machining and welding experience is also desirable. Engine service technicians generally receive their knowledge and skill at technical career centers or vocational schools. Most engine manufacturers have their own training programs for service personnel. Certification tests are available. See *Certification* Section in this chapter.

Service manager

There are numerous opportunities as service managers in small gas engine shops with more than one service technician. *Service managers* are responsible for quality workmanship and satisfactory shop operation. They must plan and supervise the activities of all service department employees.

Service managers discuss service problems with customers, make recommendations, write job tickets, and assign work to the technicians. They handle customer complaints, are responsible for the training of apprentices, and, generally, inspect all finished repair work. They report directly to the general manager or the owner of the service facility.

Sales manager

Sales managers are needed to sell or rent implements and vehicles that utilize small gasoline engines. Some sales managers handle a variety of products, such as yard and farm equipment, marine and sports vehicles, construction equipment, and emergency repair or rescue equipment. Others specialize in one field or sell a few closely related products such as personal watercraft, motorcycles, all-terrain vehicles (ATVs), and snowmobiles.

The geographical location of a sales or rental business often determines the type of vehicle or equipment that is most in demand by the consumer. Gasoline engine applications are numerous, and new uses are being developed each year.

General manager

Managing a successful small engine sales and service business requires experience and education. Keeping accurate sales and service records, budgeting, promoting sales, and maintaining adequate tools, parts, supplies, and accessories are just a few of the responsibilities of the *general manager.*

The general manager must have a sincere, personal commitment to provide fair, quality service to customers. Often, a mechanic or salesperson works up to this position through years of work experience and college courses in business and management.

Manufacturer's Technician

Small gas engine manufacturers need people to develop prototype engines or engine parts and test new design theories. See **Figure 22-2.** *Manufacturer's technicians* need to be skilled in the use of tools, materials, and machine processes in order to produce a special part or engine unit. They are usually required to run exhaustive tests using dynamometers and other specialized testing equipment, observing and recording test results.

Technicians are generally involved with experiments, tests, and analysis of various engine systems and designs in the plant and in actual field use. When testing is completed, they present the test results and recommend changes to engineers and others involved in the project. Their observations and recommendations may be presented orally or in writing. Therefore, they must be able to communicate in clear, technical language. This

Figure 22-2. *Manufacturer's technicians help small gas engine manufacturers develop prototype engines or engine parts and test new design theories.* (Briggs & Stratton Corp.)

requires adequate communicative skills combined with technical talents.

Many colleges, technical institutes, and universities offer programs for technicians. Engine mechanics can become technicians if they have the desire to further their education through evening courses and service training programs.

Service representative

Small gas engine manufacturers may train certain employees with broad service experience to become *service representatives*. These representatives are required to work closely with service managers and mechanics in the field to catch and correct chronic service problems. In some cases, service representatives write and distribute service bulletins concerning these problems. They also meet with and report findings to company engineers involved in engine design.

Engineer

Manufacturers need *engineers* to design engines that will perform satisfactorily under specific environmental conditions. For example, a small gas engine designed for use in a garden tractor is quite different from one intended for use in a chain saw. Engineers must use their knowledge of scientific principles to design and create engines that will meet all of the specified operating requirements.

Engineers usually have college engineering degrees. Sometimes, however, an engineer's license can be obtained by passing special examinations. Engineers must have a strong background in science, mathematics, and many specialized technical subjects, such as electronics, drafting, or fluid power. They must be analytical and creative, with a practical knowledge of manufacturing processes and materials.

Executive

Any of the careers outlined in this chapter can serve as stepping stones to high-level management positions in the small gas engine field. Many of the successful *executives* in this field began their careers in engine production, design, sales, or service. Almost invariably, the key to their success is learning to do the job at hand to the best of their ability.

Entrepreneur

Many people in the small engine field start their own business. These people are called *entrepreneurs*. See **Figure 22-3.** Entrepreneurs must have a total understanding of the managerial, financial, and technical aspects of the small engine business.

Figure 22-3. *The entrepreneurs of this lawn and garden shop sell, service, and repair small gas engine equipment. In addition to technical knowledge about small gas engines, they must have a complete understanding of the managerial and financial aspects of the business.*

There are several advantages associated with owning a business. As the owner, you have total control over the way a business grows and develops. You have the opportunity to hire and train people as you desire. Additionally, your income is only limited by the success of your business.

On the other hand, owning a business can be extremely difficult. Entrepreneurs work very long hours trying to establish and maintain a profitable business. They also take many risks to get the new business started. Most entrepreneurs spend years repaying loans that were taken to go into business. Entrepreneurs are responsible for these loans even if their business fails.

Teacher

Teaching is a rewarding career choice for many individuals. The industrial technology and vocational education teaching field can use qualified teachers with the knowledge of small gas engines.

To teach at the high school or vocational school level, a college degree with a specialization in teaching is required. Teaching certification is obtained in the state in which one wishes to teach at one of these levels. This type of teaching is a rewarding career for individuals who enjoy working with young people.

In addition to hiring certified teachers, many community colleges hire teachers that have extensive industry-related experiences, as well as the aptitude for teaching. Manufacturers also hire teachers to come to their facility and teach service courses to mechanics and technicians.

Certification

The *Engine & Equipment Training Council (EETC)* is an organization that creates the voluntary technicians certification tests for outdoor power equipment technicians, mechanics, and service managers. The *Engine Service Association (ESA)* works in conjunction with EETC to promote the improvement of the outdoor power equipment industry. With the *Outdoor Power Equipment (OPE)* tests that EETC creates, ESA coordinates the certification process on a national level.

The technicians certification tests are available for technicians or service managers wishing to enhance their training, employment opportunities, and personal credibility. Employers having certified technicians are recognized by their customers as having qualified service personnel who are competent and will produce quality work.

Why get certified?

This certification can provide personal and professional benefits. Personal prestige and credibility is gained by individuals who have demonstrated interest by studying and meeting certification requirements to advance their professional qualifications.

Those individuals that pass one of the certification tests will receive an arm patch bearing the certified OPE technician emblem and an $8'' \times 10''$ certificate. Wearing the official emblem, obtained by passing a certification test, shows customers that service will be in the hands of a competent person who can proficiently diagnose and repair equipment. See **Figure 22-1.** Passing a certification test indicates that an individual has met industry standards of professionalism and has studied to

obtain certification. This accomplishment is a sign of professionalism and can increase one's chances for advancement and monetary awards.

Dealerships that advertise they employ certified technicians can increase their sales of equipment. The service dealer who employs certified technicians can assure customers that any service problems during and after equipment warranties will be taken care of in a fair and equitable manner.

What is on a certification test?

OPE certification tests are carefully constructed to measure knowledge in basic skills, interpersonal relationship skills, engine fundamentals, theory, servicing, failure analysis, troubleshooting, and repair. Certification can be obtained in one or more of the following test areas:
- Four-Stroke Engines
- Two-Stroke Engines
- Diesel Engines
- Electrical
- Drive Systems
- Generators

Each of the test areas has its own certification test. Sample certification tests for Basic Four-Stroke Engines and Basic Two-Stroke Engines can be found in the *Workbook for Small Gas Engines*, which can be ordered directly from Goodheart-Willcox. A study guide published by ESA is available for each test area and summarizes about 85% of the topics covered.

Each certification test consists of 150 to 200 multiple choice questions. The questions are divided into four main categories:
- Fundamentals
- Servicing Engine Systems
- Failure Analysis
- Troubleshooting

There is a two hour time limit for taking a certification test. A minimum passing score is 70%.

Who may take the test, and where?

Anyone may take the test at any scheduled test session. Tests are administered at ESA-approved locations throughout the United States. ESA promotes the certification and schedules all testing locations and dates. ESA handles the registration, collection of fees, and certification-related questions. The registration fee is kept minimal, but it must be paid 30 days prior to the test date.

ESA provides registered participants all pertinent details prior to testing and notifies them of scores after testing. ESA handles the dissemination of arm patches, certificates, and maintains individual records on all applicants. Certification is valid for three years in each test area. ESA notifies all applicants when recertification is necessary:

For information on testing locations, testing dates, registration, obtaining a study guide, or any other inquiries, contact:

Engine Service Association
210 Allen
Exton, PA 19341
Phone: (610) 363-3844
Fax: (610) 363-3817
Internet: www.opecert.com

Summary

The small gas engine field offers career opportunities in several areas. The engine service technician diagnoses engine trouble and makes appropriate repairs. Service managers plan and supervise the activities of all service department employees. Sales managers sell implements and vehicles that utilize small gas engines. General managers oversee both the service and sales aspects of the business.

Small engine manufacturer's technicians develop prototype engines and test new design theories. Manufacturers often train employees with broad service experience to become service representatives. Service representatives work to solve chronic service problems. Manufacturers also need engineers to design new engines that will perform satisfactorily under specific conditions.

Any career in small engines can lead to high-level management positions. Most successful executives began their careers in production, design, sales, or service. Many individuals in the small engine field start their own business. These people are entrepreneurs.

The industrial technology and vocational education teaching field can use qualified teachers with the knowledge of small gas engines. Teaching can be a rewarding career choice.

The Engine Service Association (ESA) Outdoor Power Equipment technicians certification tests are available for anyone interested in improving their employment potential and credibility in the workplace. Passing the certification tests can bring personal and professional benefits.

Know These Terms

engine service technician
service manager
sales manager
general manager
manufacturer's technician

service representative
engineer
executive
entrepreneur
EETC
ESA
OPE

Chapter 22
Review Questions

Answer the following questions on a separate sheet of paper.

1. Small engine service technicians generally receive their training at _____ _____.
2. The service manager _____ the activities of all service department employees.
3. General managers are often required to keep accurate _____ and _____ records.
4. Small engine manufacturer's technicians must have good oral and written communications skills. True or False?
5. The _____ _____ works closely with service managers and service technicians to catch and correct chronic service problems.
6. Many successful executives in the small engine field began their careers in production, sales, or service. True or False?
7. An entrepreneur is an individual who starts his or her own business. True or False?
8. What is the name of the organization that creates the OPE certification tests?
9. What organization administers the OPE certification tests?
10. What are the six test areas available for OPE certification?
11. How many questions are on each of the OPE tests?
12. How much time is allotted to take each of the OPE tests, and what is the minimum passing score?
13. How long is OPE certification valid?

Appendix

Check Sheet for Four-Cycle Engine Reconditioning

Engine: Make _____ Model No. _____

Hp _____ Serial No. _____

Check List for Disassembly

_____ 1. Remove all gasoline from engine.
_____ 2. Inspect engine for broken or missing parts.
_____ 3. Record all important data on an information sheet.
_____ 4. Remove spark plug. Check gap and condition of electrodes.
_____ 5. Take compression reading. Record on information sheet.
_____ 6. Check ignition output with spark tester.
_____ 7. Drain oil from crankcase.
_____ 8. Disconnect all linkage from remote throttle assembly to engine.
_____ 9. Remove engine from mountings.
_____ 10. Clean engine housing or mounting area.
_____ 11. Remove blower housing from engine.
_____ 12. Remove carburetor and carburetor linkage.
_____ 13. Remove governor air vane and governor linkage.
_____ 14. Remove muffler.
_____ 15. Remove valve chamber cover.
_____ 16. Remove cylinder head and head gasket.
_____ 17. Measure bore and stroke. Record measurements.
_____ 18. Check and record valve clearance.
_____ 19. Remove air deflector shields.
_____ 20. Check and record armature air gap.
_____ 21. Remove starter clutch and flywheel nut.
_____ 22. Remove flywheel.
_____ 23. Remove ignition breaker point cover.
_____ 24. Check and record ignition point gap setting.
_____ 25. Remove ignition points, condenser, and ignition cam (or solid state ignition unit).
_____ 26. Remove magneto assembly.
_____ 27. Remove all rust and burrs from end of crankshaft.
_____ 28. Remove mounting flange, if any.
_____ 29. Check timing marks.
_____ 30. Remove camshaft and oil pump.
_____ 31. Remove valve tappets.
_____ 32. Remove piston and rod assembly. Note: Mark rod first.
_____ 33. Remove piston rings from piston.
_____ 34. Check and record ring end gap.
_____ 35. Clean ring grooves in piston.
_____ 36. Check and record piston ring-to-land clearance.
_____ 37. Remove crankshaft and inspect it.
_____ 38. Remove intake and exhaust valves.
_____ 39. Wash and clean all parts that will not be damaged by solvent.
_____ 40. Inspect engine block for scores or imperfections.
_____ 41. Check all bearings and oil seals for possible damage.

_____ 42. Recondition or replace necessary engine components.

Check List for Reassembly

_____ 1. Clean valve seats with wire wheel and brush.
_____ 2. Have instructor check valve parts after they are cleaned.
_____ 3. Lap valves against valve seats, using lapping compound.
_____ 4. Install valve assemblies.
_____ 5. Install crankshaft.
_____ 6. Fit rings on piston in proper order.
_____ 7. Oil cylinder wall. Install piston and rod assembly in proper direction.
_____ 8. Torque rod bolts to specifications. Bend up retainer clips.
_____ 9. Install tappets, camshaft, and oil pump.
_____ 10. Align timing marks on camshaft and crankshaft.
_____ 11. Bolt mounting flange (crankcase cover) on engine. Check for proper fit.
_____ 12. Check and record valve clearance measurements.
_____ 13. Assemble valve cover and breather and bolt to engine block.
_____ 14. Install cylinder head and head gasket. Torque to specifications.
_____ 15. Reassemble and install ignition system. If necessary, set point gap to specifications.
_____ 16. Install ignition point cover (if necessary).
_____ 17. Install flywheel key and flywheel. Torque flywheel nut.
_____ 18. Set armature air gap to specifications.
_____ 19. Fasten governor air vane to engine block.
_____ 20. Mount carburetor on engine block.
_____ 21. Install muffler.
_____ 22. Install blower housing.
_____ 23. Connect fuel lines and valve breather tube.
_____ 24. Mount engine in implement or equipment.
_____ 25. Connect engine to drive train.
_____ 26. Connect all linkage between remote throttle and engine.
_____ 27. Tighten oil plug and fill engine with proper oil.
_____ 28. Clean and install air cleaner.
_____ 29. Check engine compression.
_____ 30. Clean and set gap of spark plug electrodes. Install spark plug.
_____ 31. Check to be sure all components are tight and properly adjusted.
_____ 32. Fill fuel tank with clean gasoline.
_____ 33. If indoors, turn on exhaust fan and wear goggles.
_____ 34. Engage carburetor choke and start engine.
_____ 35. Adjust carburetor.

General Torque Specifications in Consideration of Fastener Quality

The following rules apply to the chart:
1. Consult manufacturers' specific recommendations when available.
2. The chart may be used directly when any of the following lubricants are used:
   a. Antiseize compounds, Molykote, Fel-Pro C-5, graphite and oil, or similar mixtures.
3. Increase the torque by 20% when using engine oil or chassis grease as a lubricant. (these lubricants are not generally recommended for fasteners)
4. Reduce torque by 20% when plated bolts are used.
5. Increase torque by 20% when multiple tapered tooth lock washers are used.

SAE Standard/Foot-Pounds							Metric Standard					
Grade of Bolt	SAE 1 & 2	SAE 5	SAE 6	SAE 8			Grade of Bolt		5D	.8G	10K	12K	
Min. Ten. Strength	64,000 P.S.I.	105,000 P.S.I.	133,000 P.S.I.	150,000 P.S.I.			Min. Ten. Strength		71,160 P.S.I.	113,800 P.S.I.	142,200 P.S.I.	170,679 P.S.I.	
Markings on Head	⬢	⬢	⬢	⬢	Size of Socket or Wrench Opening		Markings on Head		5D	.8G	10K	12K	Size of Socket or Wrench Opening
U.S. Standard	Foot Pounds				U.S. Regular		Metric		Foot Pounds				Metric
Bolt Dia.					Bolt Head	Nut	Bolt Dia.	U.S. Dec. Equiv.					Bolt Head
1/4	5	7	10	10.5	3/8	7/16	6mm	.2362	5	6	8	10	10mm
5/16	9	14	19	22	1/2	9/16	8mm	.3150	10	16	22	27	14mm
3/8	15	25	34	37	9/16	5/8	10mm	.3937	19	31	40	49	17mm
7/16	24	40	55	60	5/8	3/4	12mm	.4720	34	54	70	86	19mm
1/2	37	60	85	92	3/4	13/16	14mm	.5512	55	89	117	137	22mm
9/16	53	88	120	132	7/8	7/8	16mm	.6299	83	132	175	208	24mm
5/8	74	120	167	180	15/16	1	18mm	.7090	111	182	236	283	27mm
3/4	120	200	280	296	1-1/8	1-1/8	22mm	.8661	182	284	394	464	32mm

NOTE: To convert pound-feet to Newton-meter, multiply pound-feet by .7376. The torque specifications are given in pound-feet. The pound-inch equivalent may be obtained by multiplying the specification by 12.

V-Belt Failure and Correction Chart

Cause of Failure	Correction	Cause of Failure	Correction
1. Normal wear.	Replace belt.	6. Damaged or worn pulleys.	Align pulleys (except where an offset system is used with special pulleys).
2. Poor operating habits.	Do not engage and disengage clutch excessively.	7. Incorrect tensions.	Replace belt. Check idler spring tension. Lubricate idler brackets.
3. Damaged or worn idler pulleys.	Replace idler, frozen bearings, and belt.	8. Oil and grease damage.	Replace belts. Eliminate oil leakage. Use oil resistant belts if possible.
4. Incorrectly positioned belt guards.	Realign guards. Replace damaged guards.	9. Heat damage (140° or higher).	Use heat resistant belt. Avoid polyester belts. Shield belts from heat source.
5. Misaligned pulleys.	Replace pulleys.	10. Incorrect installation.	Install with care. Never use force. Recheck belts after 48 hours of use.

Units of Measure

U.S. CUSTOMARY	METRIC

LENGTH

U.S. CUSTOMARY	METRIC
12 inches = 1 foot	1 kilometer = 1000 meters
36 inches = 1 yard	1 hectometer = 100 meters
3 feet = 1 yard	1 dekameter = 10 meters
5,280 feet = 1 mile	1 meter = 1 meter
16.5 feet = 1 rod	1 decimeter = 0.1 meter
320 rods = 1 mile	1 centimeter = 0.01 meter
6 feet = 1 fathom	1 millimeter = 0.001 meter

WEIGHT

U.S. CUSTOMARY	METRIC
27.34 grains = 1 dram	1 tonne = 1,000,000 grams
438 grains = 1 ounce	1 kilogram = 1000 grams
16 drams = 1 ounce	1 hectogram = 100 grams
16 ounces = 1 pound	1 dekagram = 10 grams
2000 pounds = 1 short ton	1 gram = 1 gram
2240 pounds = 1 long ton	1 decigram = 0.1 gram
25 pounds = 1 quarter	1 centigram = 0.01 gram
4 quarters = 1 cwt	1 milligram = 0.001 gram

VOLUME

U.S. CUSTOMARY	METRIC
8 ounces = 1 cup	1 hectoliter = 100 liters
16 ounces = 1 pint	1 dekaliter = 10 liters
32 ounces = 1 quart	1 liter = 1 liter
2 cups = 1 pint	1 deciliter = 0.1 liter
2 pints = 1 quart	1 centiliter = 0.01 liter
4 quarts = 1 gallon	1 milliliter = 0.001 liter
8 pints = 1 gallon	1000 milliliter = 1 liter

AREA

U.S. CUSTOMARY	METRIC
144 sq. inches = 1 sq. foot	100 sq. millimeters = 1 sq. centimeter
9 sq. feet = 1 sq. yard	100 sq. centimeters = 1 sq. decimeter
43,560 sq. ft. = 160 sq. rods	100 sq. decimeters = 1 sq. meter
160 sq. rods = 1 acre	10,000 sq. meters = 1 hectare
640 acres = 1 sq. mile	

TEMPERATURE

FAHRENHEIT		CELSIUS
32° F	Water freezes	0° C
68° F	Reasonable room temperature	20° C
98.6° F	Normal body temperature	37° C
173° F	Alcohol boils	78.34° C
212° F	Water boils	100° C

Useful Conversions

WHEN YOU KNOW:	MULTIPLY BY:	TO FIND:
TORQUE		
pound-inch pound-foot	0.11298 1.3558	newton-meters (N•m) newton-meters
LIGHT		
foot candles	1.0764	lumens/meters^2 (lm/m^2)
FUEL PERFORMANCE		
miles/gallon	0.4251	kilometers/liter (km/L)
SPEED		
miles/hour	1.6093	kilometers/hr (km/h)
FORCE		
kilogram ounce pound	9.807 0.278 4.448	newtons (n) newtons newtons
POWER		
horsepower	0.746	kilowatts (kw)
PRESSURE OR STRESS		
inches of water pounds/sq. in.	0.2491 6.895	kilopascals (kPa) kilopascals
ENERGY OR WORK		
btu foot-pound kilowatt-hour	1055.0 1.3558 3600000.0	joules (J) joules joules

Conversion Table
Metric to U.S. Customary

WHEN YOU KNOW ▼	MULTIPLY BY: * = Exact		TO FIND ▼
	VERY ACCURATE	APPROXIMATE	
LENGTH			
millimeters	0.0393701	0.04	inches
centimeters	0.3937008	0.4	inches
meters	3.280840	3.3	feet
meters	1.093613	1.1	yards
kilometers	0.621371	0.6	miles
WEIGHT			
grains	0.00228571	0.0023	ounces
grams	0.03527396	0.035	ounces
kilograms	2.204623	2.2	pounds
tonnes	1.1023113	1.1	short tons
VOLUME			
milliliters	0.20001	0.2	teaspoons
milliliters	0.06667	0.067	tablespoons
milliliters	0.03381402	0.03	fluid ounces
liters	61.02374	61.024	cubic inches
liters	2.113376	2.1	pints
liters	1.056688	1.06	quarts
liters	0.26417205	0.26	gallons
liters	0.03531467	0.035	cubic feet
cubic meters	61023.74	61023.7	cubic inches
cubic meters	35.31467	35.0	cubic feet
cubic meters	1.3079506	1.3	cubic yards
cubic meters	264.17205	264.0	gallons
AREA			
square centimeters	0.1550003	0.16	square inches
square centimeters	0.00107639	0.001	square feet
square meters	10.76391	10.8	square feet
square meters	1.195990	1.2	square yards
square kilometers		0.4	square miles
hectares	2.471054	2.5	acres
TEMPERATURE			
Celsius	*9/5 (then add 32)		Fahrenheit

Conversion Table
U.S. Customary to Metric

WHEN YOU KNOW ⬇	MULTIPLY BY: * = Exact		TO FIND ⬇
	VERY ACCURATE	APPROXIMATE	
LENGTH			
inches	* 25.4		millimeters
inches	* 2.54		centimeters
feet	* 0.3048		meters
feet	* 30.48		centimeters
yards	* 0.9144	0.9	meters
miles	* 1.609344	1.6	kilometers
WEIGHT			
grains	15.43236	15.4	grams
ounces	* 28.349523125	28.0	grams
ounces	* 0.028349523125	.028	kilograms
pounds	* 0.45359237	0.45	kilograms
ton	* 0.90718474	0.9	tonnes
VOLUME			
teaspoons	* 4.97512	5.0	milliliters
tablespoons	* 14.92537	15.0	milliliters
fluid ounces	29.57353	30.0	millilitres
cups	* 0.236588240	0.24	liters
pints	* 0.473176473	0.47	liters
quarts	* 0.946352946	0.95	liters
gallons	* 3.785411784	3.8	liters
cubic inches	* 0.016387064	0.02	liters
cubic feet	* 0.028316846592	0.03	cubic meters
cubic yards	* 0.764554857984	0.76	cubic meters
AREA			
square inches	* 6.4516	6.5	square centimeters
square feet	* 0.09290304	0.09	square meters
square yards	* 0.83612736	0.8	square meters
square miles	* 2.589989	2.6	square kilometers
acres	* 0.40468564224	0.4	hectares
TEMPERATURE			
Fahrenheit	* 5/9 (after subtracting 32)		Celsius

Millimeter Conversion Chart

mm In.	15 = .5905	30 =1.1811	45 =1.7716	60 =2 3622	75 =2.9527	90 =3.5433	105 =4.1338	120 =4.7244
.25=.0098	15.25=.6004	30.25=1.1909	45.25=1.7815	60.25=2.3720	75.25=2.9626	90.25=3.5531	105.25=4.1437	120.25=4.7342
.50=.0197	15.50=.6102	30.50=1.2008	45.50=1.7913	60.50=2.3819	75.50=2.9724	90.50=3.5630	105.50=4.1535	120.50=4.7441
.75=.0295	15.75=.6201	30.75=1.2106	45.75=1.8012	60.75=2.3917	75.75=2.9823	90.75=3.5728	105.75=4.1634	120.75=4.7539
1=.0394	16=.6299	31=1.2205	46=1.8110	61=2.4016	76=2.9921	91=3.5827	106=4.1732	121=4.7638
1.25=.0492	16.25=.6398	31.25=1.2303	46.25=1.8209	61.25=2.4114	76.25=3.0020	91.25=3.5925	106.25=4.1831	121.25=4.7736
1.50=.0591	16.50=.6496	31.50=1.2402	46.50=1.8307	61.50=2.4213	76.50=3.0118	91.50=3.6024	106.50=4.1929	121.50=4.7885
1.75=.0689	16.75=.6594	31.75=1.2500	46.75=1.8405	61.75=2.4311	76.75=3.0216	91.75=3.6122	106.75=4.2027	121.75=4.7933
2=.0787	17=.6693	32=1.2598	47=1.8504	62=2.4409	77=3.0315	92=3.6220	107=4.2126	122=4.8031
2.25=.0886	17.25=.6791	32.25=1.2697	47.25=1.8602	62.25=2.4508	77.25=3.0413	92.25=3.6319	107.25=4.2224	122.25=4.8130
2.50=.0984	17.50=.6890	32.50=1.2795	47.50=1.8701	62.50=2.4606	77.50=3.0512	92.50=3.6417	107.50=4.2323	122.50=4.8228
2.75=.1083	17.75=.6988	32.75=1.2894	47.75=1.8799	62.75=2.4705	77.75=3.0610	92.75=3.6516	107.75=4.2421	122.75=4.8327
3=.1181	18=.7087	33=1.2992	48=1.8898	63=2.4803	78=3.0709	93=3.6614	108=4.2520	123=4.8425
3.25=.1280	18.25=.7185	33.25=1.3091	48.25=1.8996	63.25=2.4901	78.25=3.0807	93.25=3.6713	108.25=4.2618	123.25=4.8524
3.50=.1378	18.50=.7283	33.50=1.3189	48.50=1.9094	63.50=2.5000	78.50=3.0905	93.50=3.6811	108.50=4.2716	123.50=4.8622
3.75=.1476	18.75=.7382	33.75=1.3287	48.75=1.9193	63.75=2.5098	78.75=3.1004	93.75=3.6909	108.75=4.2815	123.75=4.8720
4=.1575	19=.7480	34=1.3386	49=1.9291	64=2.5197	79=3.1102	94=3.7008	109=4.2913	124=4.8819
4.25=.1673	19.25=.7579	34.25=1.3484	49.25=1.9390	64.25=2.5295	79.25=3.1201	94.25=3.7106	109.25=4.3012	124.25=4.8917
4.50=.1772	19.50=.7677	34.50=1.3583	49.50=1.9488	64.50=2.5394	79.50=3.1299	94.50=3.7205	109.50=4.3110	124.50=4.9016
4.75=.1870	19.75=.7776	34.75=1.3681	49.75=1.9587	64.75=2.5492	79.75=3.1398	94.75=3.7303	109.75=4.3209	124.75=4.9114
5=.1968	20=.7874	35=1.3779	50=1.9685	65=2.5590	80=3.1496	95=3.7401	110=4.3307	125=4.9212
5.25=.2067	20.25=.7972	35.25=1.3878	50.25=1.9783	65.25=2.5689	80.25=3.1594	95.25=3.7500	110.25=4.3405	125.25=4.9311
5.50=.2165	20.50=.8071	35.50=1.3976	50.50=1.9882	65.50=2.5787	80.50=3.1693	95.50=3.7598	110.50=4.3504	125.50=4.9409
5.75=.2264	20.75=.8169	35.75=1.4075	50.75=1.9980	65.75=2.5886	80.75=3.1791	95.75=3.7697	110.75=4.3602	125.75=4.9508
6=.2362	21=.8268	36=1.4173	51=2.0079	66=2.5984	81=3.1890	96=3.7795	111=4.3701	126=4.9606
6.25=.2461	21.25=.8366	36.25=1.4272	51.25=2.0177	66.25=2.6083	81.25=3.1988	96.25=3.7894	111.25=4.3799	126.25=4.9705
6.50=.2559	21.50=.8465	36.50=1.4370	51.50=2.0276	66.50=2.6181	81.50=3.2087	96.50=3.7992	111.50=4.3898	126.50=4.9803
6.75=.2657	21.75=.8563	36.75=1.4468	51.75=2.0374	66.75=2.6279	81.75=3.2185	96.75=3.8090	111.75=4.3996	126.75=4.9901
7=.2756	22=.8661	37=1.4567	52=2.0472	67=2.6378	82=3.2283	97=3.8189	112=4.4094	127=5.0000
7.25=.2854	22.25=.8760	37.25=1 4665	52.25=2.0571	67.25=2.6476	82.25=3.2382	97.25=3.8287	112.25=4.4193	
7.50=.2953	22.50=.8858	37.50=1.4764	52.50=2.0669	67.50=2.6575	82.50=3.2480	97.50=3.8386	112.50=4.4291	
7.75=.3051	22.75=.8957	37.75=1.4862	52.75=2.0768	67.75=2.6673	82.75=3.2579	97.75=3.8484	112.75=4.4390	
8=.3150	23=.9055	38=1.4961	53=2.0866	68=2.6772	83=3.2677	98=3.8583	113=4.4488	
8.25=.3248	23.25=.9153	38.25=1.5059	53.25=2.0965	68.25=2.6870	83.25=3.2776	98.25=3.8681	113.25=4.4587	
8.50=.3346	23.50=9252	38.50=1.5157	53.50=2.1063	68.50=2.6968	83.50=3.2874	98.50=3.8779	113.50=4.4685	
8.75=.3445	23.75=.9350	38.75=1.5256	53.75=2.1161	68.75=2.7067	83.75=3.2972	98.75=3.8878	113.75=4.4783	
9=.3543	24=.9449	39=1.5354	54=2.1260	69=2.7165	84=3.3071	99=3.8976	114=4.4882	
9.25=.3642	24.25=.9547	39.25=1.5453	54.25=2.1358	69.25=2.7264	84.25=3.3169	99.25=3.9075	114.25=4.4980	
9.50=.3740	24.50=.9646	39.50=1.5551	54.50=2.1457	69.50=2.7362	84.50=3.3268	99.50=3.9173	114.50=4.5079	
9.75=.3839	24.75=.9744	39.75=1.5650	54.75=2.1555	69.75=2.7461	84.75=3.3366	99.75=3.9272	114.75=4.5177	
10=.3937	25=.9842	40=1.5748	55=2.1653	70=2.7559	85=3.3464	100=3.9370	115=4.5275	
10.25=.4035	25.25=.9941	40.25=1.5846	55.25=2.1752	70.25=2.7657	85.25=3.3563	100.25=3.9468	115.25=4.5374	
10.50=.4134	25.50=1.0039	40.50=1.5945	55.50=2.1850	70.50=2.7756	85.50=3.3661	100.50=3.9567	115.50=4.5472	
10.75=.4232	25.75=1.0138	40.75=1.6043	55.75=2.1949	70.75=2.7854	85.75=3.3760	100.75=3.9665	115.75=4.5571	
11=.4331	26=1.0236	41=1.6142	56=2.2047	71=2.7953	86=3.3858	101=3.9764	116=4.5669	
11.25=.4429	26.25=1.0335	41.25=1.6240	56.25=2.2146	71.25=2.8051	86.25=3.3957	101.25=3.9862	116.25=4.5768	
11.50=.4528	26.50=1.0433	41.50=1.6339	56.50=2.2244	71.50=2.8150	86.50=3.4055	101.50=3.9961	116.50=4.5866	
11.75=.4626	26.75=1.0531	41.75=1.6437	56.75=2.2342	71.75=2.8248	86.75=3.4153	101.75=4.0059	116.75=4.5964	
12=.4724	27=1.0630	42=1.6535	57=2.2441	72=2.8346	87=3.4252	102=4.0157	117=4.6063	
12.25=.4823	27.25=1.0728	42.25=1.6634	57.25=2.2539	72.25=2.8445	87.25=3.4350	102.25=4.0256	117.25=4.6161	
12.50=.4921	27.50=1.0827	42.50=1.6732	57.50=2.2638	72.50=2.8543	87.50=3.4449	102.50=4.0354	117.50=4.6260	
12.75=.5020	27.75=1.0925	42.75=1.6831	57.75=2.2736	72.75=2.8642	87.75=3.4547	102.75=4.0453	117.75=4.6358	
13=.5118	28=1.1024	43=1.6929	58=2.2835	73=2.8740	88=3.4646	103=4.0551	118=4.6457	
13.25=.5217	28.25=1.1122	43.25=1.7028	58.25=2.2933	73.25=2.8839	88.25=3.4744	103.25=4.0650	118.25=4.6555	
13.50=.5315	28.50=1.1220	43.50=1.7126	58.50=2.3031	73.50=2.8937	88.50=3.4842	103.50=4.0748	118.50=4.6653	
13.75=.5413	28.75=1.1319	43.75=1.7224	58.75=2.3130	73.75=2.9035	88.75=3.4941	103.75=4.0846	118.75=4.6752	
14=.5512	29=1.1417	44=1.7323	59=2.3228	74=2.9134	89=3.5039	104=4.0945	119=4.6850	
14.25=.5610	29.25=1.1516	44.25=1.7421	59.25=2.3327	74.25=2.9232	89.25=3.5138	104.25=4.1043	119.25=4.6949	
14.50=.5709	29.50=1.1614	44.50=1.7520	59.50=2.3425	74.50=2.9331	89.50=3.5236	104.50=4.1142	119.50=4.7047	
14.75=.5807	29.75=1.1713	44.75=1.7618	59.75=2.3524	74.74=2.9429	89.75=3.5335	104.75=4.1240	119.75=4.7146	

Decimal Equivalents of 8ths, 16ths, 32nds, 64ths

8ths	32nds	64ths	64ths
1/8 = .125	1/32 = .03125	1/64 = .015625	33/64 = .515625
1/4 = .250	3/32 = .09375	3/64 = .046875	35/64 = .546875
3/8 = .375	5/32 = .15625	5/64 = .078125	37/64 = .578125
1/2 = .500	7/32 = .21875	7/64 = .109375	39/64 = .609375
5/8 = .625	9/32 = .28125	9/64 = .140625	41/64 = .640625
3/4 = .750	11/32 = .34375	11/64 = .171875	43/64 = .703125
7/8 = .875	13/32 = .40625	13/64 = .203125	45/64 = .671875
16ths	15/32 = .46875	15/64 = .234375	47/64 = .734375
1/16 = .0625	17/32 = .53125	17/64 = .265625	49/64 = .765625
3/16 = .1875	19/32 = .59375	19/64 = .296875	51/64 = 796875
5/16 = .3125	21/32 = .65625	21/64 = .328125	53/64 = .828125
7/16 = .4375	23/32 = .71875	23/64 = .359375	55/64 = .859375
9/16 = .5625	25/32 = .78125	25/64 = .390625	57/64 = .890625
11/16 = .6875	27/32 = .84375	27/64 = .421875	59/64 = .921875
13/16 = .8125	29/32 = .90625	29/64 = .453125	61/64 = .953125
15/16 = .9375	31/32 = .96875	31/64 = .484375	63/64 = .984375

Rules Relative to the Circle

To find circumference—
Multiply diameter by 3.1416 . . . . Or divide diameter by .03183

To find diameter—
Multiply circumference by 0.3183 . . . . Or divide circumference by 3.1416

To find radius—
Multiply circumference by 0.15915 . . . . Or divide circumference by 6.28318

To find side of an inscribed square—
Multiply diameter by 0.7071
Or multiply circumference by 0.2251 . . . . Or divide circumference by 4.4428

To find side of an equal square—
Multiply diameter by 0.8862 . . . . Or divide diameter by 1.1284
Or multiply circumference by 0.2821 . . . . Or divide circumference by 3.545

Square—
A side multiplied by 1.4142 equals diameter of its circumscribing circle.
A side multiplied by 4.443 equals circumference of its circumscribing circle.
A side multiplied by 1.128 equals diameter of an equal circle.
A side multiplied by 3.547 equals circumference of an equal circle.
A side multiplied by 1.273 equals circle inches of an equal circle.

To find the area of a circle—
Multiply circumference by one-quarter of the diameter
Or multiply the square of diameter by 0.7854
Or multiply the square of circumference by .07958
Or multiply the square of 1/2 diameter by 3.1416

Tap Drill Sizes for Machine Screw Threads (75% Depth of Thread)

A bolt inserted in an ordinary nut, which has only one-half of a full depth of thread, will break before stripping the thread. Also a full depth of thread, while very difficult to obtain, is only about 5% stronger than a 75% depth.

These tables give the exact size of the hole, expressed in decimals, that will produce a 75% depth of thread, and also the nearest regular stock drill to this size. Holes produced by these drills are considered close enough for any commercial tapping.

$$\text{Diameter of Tap, Minus } \frac{.974}{\text{No. threads per Inch}} = \text{Diameter of Hole}$$

Tap Size	Threads per Inch	Diameter Hole	Drill
0	80	.048	3/64
1	72	.060	53
1	64	.058	53
2	64	.071	50
2	56	.069	50
3	56	.082	45
3	48	.079	47
4	48	.092	42
4	40	.088	43
4	36	.085	44
5	44	.103	37
5	40	.101	38
5	36	.098	40
6	40	.114	33
6	36	.111	34
6	32	.108	36
7	36	.124	1/8
7	32	.121	31
7	30	.119	31
8	36	.137	29
8	32	.134	29
8	30	.132	30
9	32	.147	26
9	30	.145	27
9	24	.136	29
10	32	.160	21
10	30	.158	22
10	28	.155	23
10	24	.149	25
12	28	.181	14
12	24	.175	16
14	24	.201	7
14	20	.193	10
16	22	.224	2
16	20	.219	7/32
16	18	.214	3
18	20	.245	D
18	18	.240	B
20	20	.271	I
20	18	.266	17/64
22	18	.292	L
22	16	.285	9/32
24	18	.318	O
24	16	.311	5/16
26	16	.337	R
26	14	.328	21/64
28	16	.363	23/64
28	14	.354	T
30	16	.389	25/64
30	14	.380	V

Tap Drill Sizes for Fractional Size Threads (75% Depth of Thread)

Tap Size	Threads per Inch	Diam. Hole	Drill
1/16	72	.049	3/64
1/16	64	.047	3/64
1/16	60	.046	56
5/64	72	.065	52
5/64	64	.063	1/16
5/64	60	.062	1/16
5/64	56	.061	53
3/32	60	.077	5/64
3/32	56	.076	48
3/32	50	.074	49
3/32	48	.073	49
7/64	56	.092	42
7/64	48	.090	43
7/64	40	.089	43
1/8	48	.105	36
1/8	40	.101	38
1/8	36	.098	40
1/8	32	.095	3/32
9/64	40	.116	32
9/64	36	.114	33
9/64	32	.110	35
5/32	36	.132	30
5/32	32	.129	30
5/32	30	.126	1/8
11/64	36	.145	27
11/64	32	.141	9/64
3/16	36	.161	20
3/16	32	.157	22
3/16	30	.155	23
3/16	24	.147	26
13/64	32	.173	17
13/64	28	.171	11/64
13/64	24	.163	20
7/32	28	.188	12
7/32	24	.184	13
7/32	20	.178	16
15/64	28	.204	6
15/64	24	.200	8
15/64	20	.194	10
1/4	32	.220	7/32
1/4	28	.215	3
1/4	24	.214	3
1/4	27	.209	4
1/4	20	.201	7
5/16	32	.282	9/32
5/16	27	.276	J
5/16	24	.272	I
5/16	20	.264	17/64
5/16	18	.258	F
3/8	27	.339	R
3/8	24	.334	Q
3/8	20	.326	21/64
3/8	16	.314	5/16
7/16	27	.401	Y
7/16	24	.397	X
7/16	20	.389	25/64
7/16	14	.368	U
1/2	27	.464	15/32
1/2	24	.460	29/64
1/2	20	.451	29/64
1/2	13	.425	27/64
1/2	12	.419	27/64
9/16	27	.526	17/32
9/16	18	.508	33/64
9/16	12	.481	31/64
5/8	27	.589	19/32
5/8	18	.571	37/64
5/8	12	.544	35/64
5/8	11	.536	17/32
11/16	16	.627	5/8
11/16	11	.599	19/32
3/4	27	.714	23/32
3/4	16	.689	11/16
3/4	12	.669	43/64
3/4	10	.653	21/32
13/16	12	.731	47/64
13/16	10	.715	23/32
7/8	27	.839	27/32
7/8	18	.821	53/64
7/8	14	.805	13/16
7/8	12	.794	51/64
7/8	9	.767	49/64
15/16	12	.856	55/64
15/16	9	.829	53/64
1	27	.964	31/32
1	14	.930	15/16
1	12	.919	59/64
1	8	.878	7/8
1 1/16	8	.941	15/16
1 1/8	12	1.044	1 3/64
1 1/8	7	.986	63/64
1 3/16	7	1.048	1 3/64
1 1/4	12	1.169	1 11/64
1 1/4	7	1.111	1 7/64
1 5/16	7	1.173	1 11/64
1 3/8	12	1.294	1 19/64
1 3/8	6	1.213	1 7/32
1 1/2	12	1.419	1 27/64
1 1/2	6	1.338	1 11/32
1 5/8	5 1/2	1.448	1 29/64
1 3/4	5	1.555	1 9/16
1 7/8	5	1.680	1 11/16
2	4 1/2	1.783	1 25/32
2 1/8	4 1/2	1.909	1 29/32
2 1/4	4 1/2	2.034	2 1/32
2 3/8	4	2.131	2 1/8
2 1/2	4	2.256	2 1/4
2 5/8	4	2.381	2 3/8
2 3/4	4	2.506	2 1/2
2 7/8	3 1/2	2.597	2 19/32
3	3 1/2	2.722	2 23/32
3 1/8	3 1/2	2.847	2 27/32
3 1/4	3 1/4	2.972	2 31/32
3 3/8	3 1/4	3.075	3 1/16
3 1/2	3 1/4	3.200	3 3/16
3 5/8	3 1/4	3.325	3 5/16
3 3/4	3	3.425	3 7/16
4	3	3.675	3 11/16

Typical Ignition Component Tests

The following pages illustrate typical ignition component tests using three common testers. These examples are for illustrative purposes only and may not apply to all makes and models. Always refer to an appropriate service manual for specific hookup information before performing ignition component tests.

Condenser Continuity Test Using a Merc-O-Tronic Model 79 Tester:

1. Insert the test leads into the *Test Leads* jacks.
2. Connect the black test clip to the base of the condenser.
3. Connect the red test clip to the condenser lead.
4. Turn the selector switch to the *Cont./Cond. Test* position.
5. Press the *Cont./Cond. Test* push switch. The L.E.D. will glow brightly and then slowly dim as the condenser becomes charged.
6. If the L.E.D. remains bright, the condenser is defective (due to leakage or short).

Coil Continuity Test Using a Merc-O-Tronic Model 79 Tester:

1. Insert the test leads into the *Test Leads* jacks. Turn the selector switch to *Cont./Cond. Test.*
2. By connecting the red and black test clips to the ignition coil primary or secondary wires (or across any wire), continuity can be checked.
3. Press the *Cont./Cond. Test* push switch. The L.E.D. will light if there is continuity.

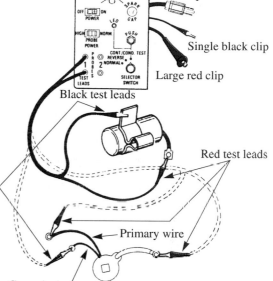

Solid State Pulse Transformer Test Using a Merc-O-Tronic Model 79 Tester:

1. Insert the test leads into the *Test Leads* jacks.
2. Connect the black test lead to the coil primary ground.
3. Connect the red test lead to the coil positive primary wire.
4. Connect the large red test lead to the high tension wire (use adapter if needed).
5. Connect the single black test lead from the analyzer to the coil primary ground wire. Normally, both black test leads will be connected in the same place.
6. Place the tester's selector switch in the *Normal* position.
7. Turn the power on.
8. Push the probe power switch to *Normal.*
9. A strong, steady spark should occur across the spark gap.
10. If the spark is faint, intermittent, or does not occur at all, the coil is defective and must be replaced.

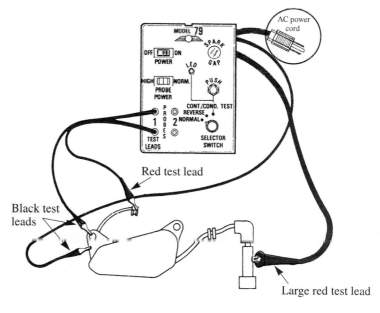

Condenser Continuity Test Using a Graham-Lee Model 31-SM (31-SM X-H) Tester:

 This is a test for continuity only—Microfarad values are not used here. The condenser may be tested on the stator, but it must be disconnected at the terminals. *Short* the condenser lead to the condenser body to remove any charge.

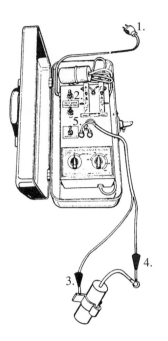

1. Plug the tester into a 110/120 volt outlet.
2. Turn the toggle *Main Switch* to the *On* position. Allow the tester to warm up for a minute.
3. Connect the black lead (alligator clip) of the tester to the condenser body.
4. Connect the red lead (alligator clip) of the tester to the condenser terminal.
5. The continuity light will glow briefly if the condenser is good. If the continuity light remains on, the condenser is leaking.
6. To repeat this test, disconnect the red and black leads, ground the condenser to the terminal lead, and repeat the steps to achieve the results in step 5.

Coil Firing Test Using a Graham-Lee Model 31-SM (31-SM X-H) Tester:

 This coil test can be made on the engine, but the condenser and the coil leads must be disconnected at the terminal. (If the condenser and coil leads are clamped together, they must be separated.)

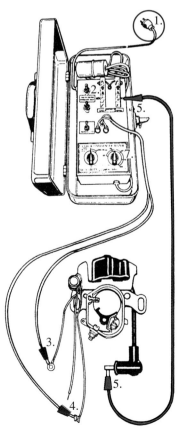

1. Plug the tester into a 110/120 volt outlet.
2. Turn the toggle *Main Switch* to the *On* position. Allow the tester to warm up for a minute.
3. Connect the black lead (alligator clip) of the tester to the ground lead of the coil.
4. Connect the red lead (alligator clip) of the tester to the primary lead of the coil.
5. Connect a lead from the spark plug gap labeled *Normal* to the secondary terminal (high tension lead) of coil.
6. Turn the *Coil Index* knob counterclockwise as far as it will go.
7. Turn the *Coil Index* knob clockwise until a spark arcs across the air gap. The setting should not exceed 70.

Solid State Ignition System Test (Ignition Units) Using a Graham-Lee Model 31-SM (31-SM X-H) Tester:

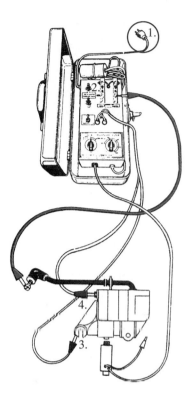

This solid state unit must be tested off of the engine.

1. Plug the tester into a 110/120 volt outlet.
2. Turn the toggle *Main Switch* to the *On* position. Allow the tester to warm up.
3. Connect the black lead (alligator clip) to the ignition unit ground (frame) or the trigger pack ground lead.
4. Connect the red lead (alligator clip) to the ignition unit terminal.
5. Connect the spark plug terminal to the spark gap labeled *Normal* on the tester.
6. Set the *Coil Index* to 80. Hold the *Coil Tests* switch to the *On* position.
7. With the red end of the trigger coil up to the trigger magnet on the ignition unit, a low frequency arcing should occur. When the trigger pack is pulled away, firing should stop.
8. Release the spring switch while the trigger pack is pulled away from the ignition unit and disconnect the red lead.
9. Operate the spring switch and bring the trigger coil red end to the trigger magnet on the ignition unit as before. A single spark should jump the spark gap, indicating that the capacitor in the ignition unit is holding a charge.

Condenser Capacity Test Using a Merc-O-Tronic Model 9800 (98, 98A) Tester:

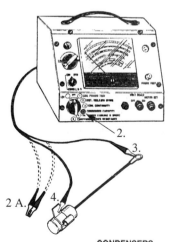

To determine if a condenser meets its rated value, check the appropriate specification table. See the example below. A condenser can be tested on the stator but must be disconnected at the terminals.

1. If a Merc-O-Tronic tester 98 or 98A is used, plug the tester into a 110/120 volt outlet. (Model 9800 does not require 110/120 volts).
2. Position the *Selector Switch* to position 4, *Condenser Capacity*. To Re-calibrate Meter:
   a. Clip the black and red test leads together.
   b. For Merc-O-Tronic testers 98 and 98A, depress the red button to set the meter to the top of the scale. For a 9800 model, turn the *Meter Set* knob to set line on scale 4.
   c. Unclip the test leads.
3. Connect the red test lead (alligator clip) to the condenser terminal.
4. Connect the black test lead (alligator clip) to the body of the condenser.
5. Read the value on scale 4. (On 98 and 98A models, the red button must be depressed.)
6. If the value found in the test is not the same as listed in the specification table, replace the condenser.

CONDENSERS

29164	.16 - .20
29177	.15 - .18
30548A	.16 - .18
30548B	.16 - .18
610370	.15 - .19
610467	.12 - .16
610588	.18 - .22
610642	.16 - .18
610767A	.16 - .18

Coil Power Test Using a Merc-O-Tronic Model 9800 (98, 98A) Tester:

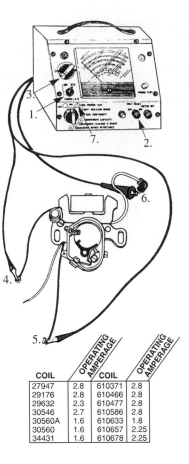

 Coil test can be made on the engine. Isolation of the coil leads can be made by placing a piece of cardboard between the points or by separating the coil primary lead, ground lead, and condenser lead as shown.

⚠ Never perform this test without the spark plug high tension lead attached to the tester's large red lead.

1. The ignition selector switch on the 9800 tester must be in the *Std.* position. (If 98 or 98A testers are used, a 55-980 adapter must be used.)
2. The *Volt Scale* must be in the *Off* position.
3. The *Lo-Hi* control knob must be in the lowest possible position.
4. Connect the black test lead (alligator clip) to the coil primary ground lead or the stator plate.
5. Connect the red test lead (alligator clip) to the primary coil lead.
6. Connect the large red test lead (alligator clip) to the terminal of the spark plug wire.
7. Turn the selector switch to coil power test No. 1.
8. Slowly turn the *Lo-Hi* knob clockwise and note the value on scale 1. When the meter reads the operating amperage for a particular winding (refer to coil manufacturer's specifications), stop turning the knob. The 5mm gap should fire steadily.

COIL	OPERATING AMPERAGE	COIL	OPERATING AMPERAGE
27947	2.8	610371	2.8
29176	2.8	610466	2.8
29632	2.3	610477	2.8
30546	2.7	610586	2.8
30560A	1.6	610633	1.8
30560	1.6	610657	2.25
34431	1.6	610678	2.25

Solid State Ignition System Test Using a Merc-O-Tronic Model 9800 (98, 98A) Tester and a CD 55-700 Adapter:

If using a model 9800 tester, turn the ignition selector switch to the *Std.* position.

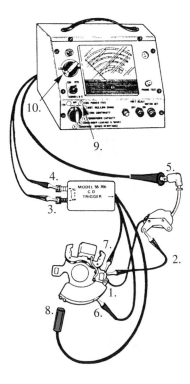

1. Attach a lead from the ignition coil terminal to the terminal on the ignition unit.
2. Attach a jumper lead from the ignition mounting plate to the ignition coil mounting plate.
3. Connect the black test lead (alligator clip) to the black terminal on the CD 55-700 unit.
4. Connect the red test lead (alligator clip) to the red terminal on the CD 55-700 unit.
5. Connect the large red test lead from the tester to the spark plug terminal on the high tension lead.
6. Attach the CD 55-700 adapter red lead to the ground cut-off terminal on the ignition unit (not pictured, located on the backside of the ignition unit).
7. Attach the CD 55-700 adapter black lead to the ignition unit mounting plate.
8. Place the electro-magnetic triggering device near the trigger terminal of the ignition unit.
9. Position the selector switch to *Coil Power Test.*
10. Slowly turn the *Lo-Hi* current knob clockwise until the ignition unit fires when viewed through the spark gap window.

📋 While the test is in progress, it may be necessary to move the triggering device to achieve proper alignment with the trigger on the ignition unit.

(Ignition component tests courtesy of Tecumseh Products Co.)

Spark Plug Condition Chart

Normal Appearance
A spark plug in a sound engine operating at the proper temperature will have deposits that range from tan to gray in color. If LPG is used, the deposits will be brown. Under normal conditions, the electrode should wear slightly, but there should be no evidence of burning.

Carbon Fouling
Carbon fouling (dry, black, sooty carbon deposits) can be caused by plugs that are too cold for the engine, an over-rich fuel mixture, a clogged air cleaner, a faulty choke, or sticking valves. Installing a hotter plug will temporarily solve this problem.

Oil Fouling
Oil fouling (wet, black deposits) is caused by excessive oil in the combustion chamber. Worn rings, valve guides, valve seals, and cylinder walls can cause oil fouling. Switching to a hotter spark plug may temporarily relieve the symptoms, but will not correct the problem.

Ash Fouling
Ash fouling is caused by the buildup of heavy combustion deposits. These deposits are formed by burning oil and/or fuel additives. Although ash fouling is not conductive, excessive deposits can cause a spark plug to misfire.

Splashed Fouling
Splashed fouling can occur after a long-delayed tune-up. When a new plug is installed in an engine with excessive piston and combustion chamber deposits, the plug will restore regular firing impulses and raise the combustion temperature. When this occurs, accumulated engine deposits may flake off and stick to the hot plug insulator.

Gap Bridging
Gap bridging (combustion deposit bridging the center and ground electrodes) is caused by a sudden burst of high speed operation following excessive idling. It can also be caused by improper fuel additives, obstructed exhaust ports (two-cycle engines), and excessive carbon in the cylinder.

High Speed Glazing
High speed glazing (hard, shiny, electrically conductive deposits) can be caused by a sudden increase in plug temperature during hard acceleration or loading. High speed glazing can cause the engine to misfire at high speeds. If high speed glazing recurs, a cooler plug should be used.

Preignition
Preignition (fuel charge ignited by a glowing combustion chamber deposit or a hot valve edge before the spark plug fires) can cause extensive plug damage. When plugs show evidence of preignition, check the heat range of the plugs, the condition of the plug wires, and the condition of the cooling system.

Detonation
Detonation can cause the insulator nose of a spark plug to fracture and chip away. The explosions that occur during heavy detonation produce extreme pressure in the cylinder. Detonation can be caused by low octane fuel, advanced ignition timing, or an excessively lean fuel mixture.

Overheating
Overheating (dull, white insulator and eroded electrodes) can occur when a spark plug is too hot for the engine. Advanced ignition timing, cooling system problems, detonation, sticking valves, and excessive high-speed operation can also cause spark plug overheating.

Mechanical Damage
Mechanical damage can be caused by a foreign object in the combustion chamber. It can also occur if the piston hits the firing tip of a spark plug with improper reach. When working on an engine, keep spark plug hole(s) and carburetor throat covered to prevent foreign objects from entering the combustion chamber.

Worn Out
Extended use will cause the spark plug's center electrode to erode. When the electrode is too worn to be filed flat, the plug must be replaced. Typical symptoms of worn spark plugs include excessive fuel consumption and poor engine performance.

Small Nonroad Spark-Ignition Engines (Phase 1 Information)

- Final Rule Published *July 3, 1995;* Federal Register Volume 60 pp. 34582–34657; available electronically
- Sets emissions standards for *new* nonroad small spark-ignition engines at or below 19 kilowatts (25 horsepower) used in lawn, garden, and utility equipment.
- Equipment types fall into two categories: non-hand-held (such as lawnmowers, tillers, chippers, generators, pumps, augers, lawn and garden tractors, forklifts, and golf carts) and hand-held (such as string trimmers, hedge clippers, edgers, leaf blowers, and chain saws).
- This category of engines accounts for 5 % of summertime HC in ozone nonattainment areas and 5 % of wintertime CO in CO nonattainment areas.
- Rule will achieve 32% reduction in HC and 7% reduction in CO from these sources upon complete fleet turnover in 2020, for $280 per ton of HC removed.
- Resembles *California's Tier 1* rule for small engines; minor differences in the federal rule include:
  - looser CO standards
  - looser standards for two-stroke lawnmowers (phasing to the tighter standards over several years)
  - engine classification for some equipment types (and, hence, their applicable standards)
  - later effective date
  - EPA-performed production line auditing
  - streamlined certification process (1-pager)
  - includes farm and construction equipment (preempted from being covered in California)
- Elements of the rule include:
  - new engine standards only (no in-use program)
  - exhaust standards only (no evaporative emissions standards)
  - HC, CO, and NO_x standards
  - five engine classes defined by size and equipment type (see below) effective with model year 1997 (model year is January 2, 1996–December 31, 1997; manufactures define annual production periods for each engine family within that window; engines built in annual production periods starting September 1, 1996 or later must be certified; annual production periods not defined are assumed to start January 1, l996) (exception; class V engines preempted in California have an effective date of January 1, 1998)

Engine Failure Analysis*

Most major failures are usually caused by improper/infrequent air cleaner service, insufficient amount of lubrication, overheating, overspeeding or breakage. In some instances, these failures can produce similar damaging effects. For this reason, always evaluate all engine parts. The following cause and effect relationships can be used as a guideline when analyzing the probable cause of engine failure.

Common Effects of Improper/Infrequent Air Cleaner Service

Abrasives Found:
- On threads of air cleaner stud.
- On inside surface of air cleaner cover.
- In air cleaner cup and underneath curled edge of cup.
- In crossover tube, intake elbow, or manifold.
- On carburetor choke and throttle plates.
- In crankcase oil, cylinder, and oil sump/cover.

Parts Worn:
- Carburetor choke and throttle shafts.
- Intake valve's face, seat, guide, and stem.
- Cylinder bore—no cross hatch—bore ridge at top of ring travel.
- Piston rings and piston skirt.
- Crankshaft journals and bearings.

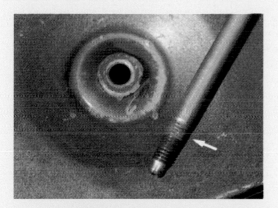

Improper/Infrequent Air Cleaner Service

When abrasive grit passes through an improperly /infrequently serviced air cleaner, it is typical to find the following sequence of events:

A. Grit accumulates on threads of air cleaner stud.

B. Grit accumulates in air cleaner cup and on inside surface of air cleaner cover.

(Continued)

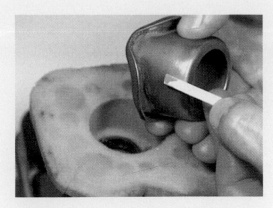

C. Grit accumulates underneath curled edge of air cleaner cup.
NOTE: If the air cleaner assembly is properly maintained and correctly assembled, an accumulation of grit should not be found within the air cleaner assembly as described.

D. Bisecting an improperly/infrequently serviced oil foam air cleaner element usually reveals the penetration of grit through the full thickness of the element. The dark streaks running through the element represent entry paths the grit chose as it passed through the element.

E. Grit accumulates in carburetor. As a result, choke and throttle shafts are commonly found excessively worn.

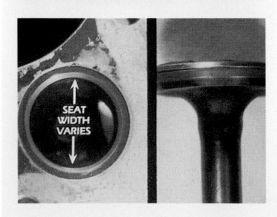

F. Worn intake valve seat and face. Each time the valve closes, some of the incoming grit is caught between the valve face and seat, producing a most unique wear pattern. Wear on the valve seat is indicated by the varying width of the seat. Generally, the greatest amount of seat wear (widest width) is found nearest the cylinder bore. Wear on the valve face is represented by the extreme groove. Wear in the valve guide will also usually occur from the grit being dragged down the guide each time the valve closes.

(Continued)

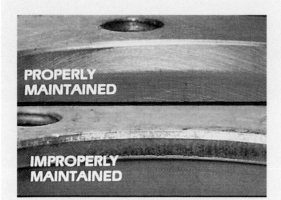

G. Worn cylinder bore. Grit working its way in between the piston ring face the bore results in excessive ring wear and bore wear, and cross hatch removal. A bore ridge develops at top of ring travel. Excessive oil consumption and smoking occur.
NOTE: An engine properly maintained will not show excessive bore wear even after hundreds of hours of rugged use.

H. Worn piston skirt—many fine scratches with *etched* appearance.

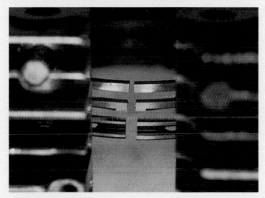

I. Worn piston rings. At 3600 RPM, the piston assembly changes its direction of travel 120 times every second. Incoming grit causes rapid premature ring wear, resulting in excessive ring end gaps.
NOTE: An engine properly maintained will not show excessive ring wear even after hundreds of hours of rugged use.

J. Worn oil ring. The two thin oil wiping rails worn flat is caused by grit, and is also an indication of a worn ring. The extent of wear varies. Ring wear can be as excessive as shown.

(Continued)

K. Worn compression rings. The varying width of the compression ring is also caused by grit, and is a further indication of a worn ring. The extent of wear varies. Ring wear can be as excessive as shown.

L. Worn internal parts. Incoming grit mixes with crankcase oil. All internal moving parts are *worked on* by the grit, resulting in worn crankshaft journals, main bearings, etc.

Common Causes of Abrasive Grit Entering Engine Causing Premature Wear

- Improperly/infrequently serviced air cleaner assembly.
- Entrance when crankcase oil level is checked or oil added.
- Dry oil foam air cleaner element.
- Bent/loose air cleaner stud.
- Incorrect assembling of air cleaner assembly.
- Deteriorated oil foam air cleaner element/ sealing lip torn.
- Operating without an air cleaner assembly.
- Torn paper air cleaner cartridge.
- Damaged/missing air cleaner assembly parts.
- Ruptured…breather tube, grommet, oil seal, etc.
- Torn gasket…breather, oil sump, intake elbow, etc.
- Loose intake elbow, crossover tube or manifold.
- Missing welch plug (float-type carburetor).

Common Effects of Insufficient Amount of Lubrication (Oil)

- Discoloration of parts:
  - Connecting rod
  - Crankshaft journals
  - Aluminum main bearings
  - Cylinder bore
  - Piston pin
  - Piston skirt
- Scoring of parts:
  - Two or more crankshaft journal bearing surfaces

(Continued)

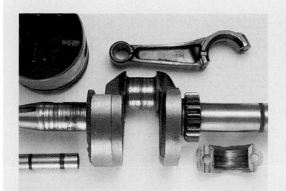

Insufficient Amount of Lubrication (Oil)

When run with an insufficient amount of lubrication (oil), it is typical to find the following conditions:

A. Discoloration of internal parts. Without enough lubricating oil, or having too much oil, engine operating temperatures increase beyond acceptable levels. These excessive operating temperatures cause discoloration. Parts which typically discolor are the connecting rod, crankshaft journals, aluminum main bearings, cylinder bore, piston pin, and piston skirt.

B. Scoring of two or more bearing surfaces. Scoring is the transfer of aluminum bearing material to its mating crankshaft journal. When run with an insufficient amount of lubrication (oil) it is typical for two or more crankshaft journal-bearing surfaces to score.

The following describes additional information on scoring:

A. Scoring can be less severe on one or two of the crankshaft journal-bearing surfaces. Even surfaces starting to score, with discoloration of parts, can generally be attributed to running with an insufficient amount of lubrication (oil).

B. When a dual ball bearing equipped engine runs with an insufficient amount of lubrication (oil), the outer crankshaft journals do not score like previously described. However, discoloration still occurs, and generally is found on crankpin journal, connecting rod bearing, cylinder bore, piston pin, and piston skirt. Scoring is also usually evident at the connecting rod bearing

(Continued)

C. When scoring is isolated to a single crankshaft journal-bearing surface, the cause can usually be attributed to either a manufacturing defect (such a failure would occur quickly), or to the manner in which the engine has been mounted. Example: Excessive belt tension, misalignment between engine and equipment, etc. Discoloration will typically occur, but it will be limited only to that scored bearing surface.

Oil Recommendation

Use a high quality detergent oil classified *For Service SF, SE, SD, or SC*. Detergent oils keep the engine cleaner and retard the formation of gum and varnish deposits.

Nothing should be added to the recommended oil. Oil additives merely increase the operating cost of the engine.

Common Causes of Discolored and Scored Parts
- Running with an insufficient amount of lubrication (oil).
- Overfilling crankcase with oil.
- Oil viscosity not matched to season.
- Excessive oil consumption die to engine wear, loss of crankcase vacuum or plugged cylinder cooling fins.
- Operating at excessive angles.
- Leaking oil seal/oil sump gasket.
- Crankcase oil dilution with gasoline.
- Bearing overload (PTO).

Common Effects of Overheating

- Vapor lock (air bubble form in fuel restricting fuel flow).
- Loose exhaust valve seat.
- Distorted cylinder head gasket mounting surface.
- Cylinder bore *hot* spots.
- Excessive oil consumption.
- Crankcase oil loses viscosity (thickens).

(Continued)

Overheating

A. Air-cooled engines require unobstructed flow of cooling air, especially throughout the cylinder fin area. Engine operating temperatures increase beyond acceptable levels when the fins become plugged with grass clippings, leaves, dirt, etc. Even a partial restriction of the fins is sufficient to raise the operating temperatures above normal.

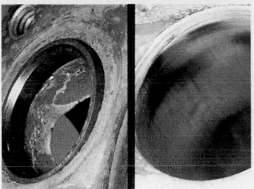

B. One of the possible consequences of excessive operating temperatures in the valve area is the *loosening* of the exhaust valve seat. The excessive heat also commonly produces *hot* spots within the cylinder bore.

Common Causes of Overheating

- Plugged cylinder cooling fins.
- Operating engine in too confined environment (restricts incoming cooling air and outgoing hot exhaust gases…causes recirculation of heated air and/or exhaust gases).
- Too lean carburetor mixture adjustment.

Common Effects of Overspeeding

- Connecting rod breaking close to piston pin.
- Likelihood of seizure.
- Liability risk/personal injury and property damage.

(Continued)

Overspeeding

A connecting rod breaking as the result of overspeeding may not necessarily show signs of overheating (discoloration) or seizures. It will, however, break in a unique place. It usually breaks very close to the piston pin. While the connecting rod can also break elsewhere, it is the break close to the piston pin which distinguishes this failure.

Top No Load Governed Speed

Refer to the Briggs & Stratton Service Engine Sales Manual, MS-4052, or the Yellow Bar Microfiche card for the correct top no load governed engine speed.*

CAUTION: Always use an accurate tachometer when adjusting idle speed, and top no load governed speed.

Common Causes of Overspeeding

- Plugged cylinder cooling fins on air vane governor equipped engine.
- Misadjusted mechanical governor.
- Bypassing the governor system.
- Binding, modified, broken, missing, non-original governor parts.

Major Engine Failures Cause and Effect Relationships

Cause	Effect
Improper/Infrequent Air Cleaner Service Damaged Air Cleaner Mounting Gasket, Bent Air Cleaner Stud. ● ◆	● Premature wear. ▲ Seizure. ■ Breakage. ○ Scoring. ▼ Overspeeding. ❑ Overheating (discoloration). ◆ Excessive Oil Consumption and Smoking.
Insufficient Amount of Lubrication (Oil). ▲ ■ ○ ❑	
Misadjusted Governor/Bypassing Governor. ▼ ■ ▲	
Plugged Cooling Fins. ❑ ▼ ■ ◆	
Loss of Crankcase Vacuum. ◆	

*Information courtesy of Briggs & Stratton Corp.

Summary of Power & Recreational Manufacturers' Positions on Oxygenated Fuel Use

The following recap is based on a review of each manufacturer's equipment owner's manuals. Working may vary slightly across a manufacturers' product line but it is generally similar if not identical. Position and wording for a manufacturers two-stroke verses four-stroke models may vary. Also some manufacturers use several engine suppliers (Briggs & Stratton, Tecumseh, and Kohler) and may utilize the applicable engine manufacturer's fuel recommendations for models with those engines. Finally it should be noted that these recommendations are for new or late model equipment and may or may not apply to earlier models.

Manufacturer	Ethanol	MTBE	Manufacturer	Ethanol	MTBE
Power Equipment			*Motorcycle*		
Am. yard Prd/Roper/Rally	yes*	NM	Harley Davidson	yes	yes
Ariens	yes[1]	NM	Honda	yes	yes
Bolens/Troy-Bilt	yes*	yes	Kawasaki	yes	yes
Briggs & Stratton	yes	yes	Suzuki	yes*	yes
Coleman	yes*	NM	Yamaha	yes	NM
Cub Cadet	NM	NM			
Dixon	yes	yes	*Recreational*		
Echo	yes	yes	ArticCat (Arctco)	yes*	yes*
Grasshopper	NM	NM	Honda	yes	yes
Homelite	yes	yes	Kawasaki	yes	yes
Honda Power Eq.	yes	yes	Polaris	yes*	yes*
John Deere (four-stroke)	yes	yes	SkiDoo/Bombardier	yes	NM
Kawasaki	yes	yes	Suzuki	yes*	yes
Kohler	yes	yes	Yamaha	yes	NM
Kubota	NM	NM			
McColloch	yes*	yes*	*Boats/Marine*		
MTD	yes	yes	Honda	yes	yes
Onan	yes*	yes*	Kawasaki	yes	yes
Poulan/Weedeater	NM	NM	Mercury	yes*	yes*
Ryobi	yes*	yes*	OMC (Johnson/Evinrude)	yes*	yes*
Sears	yes*	yes*	Pleasurecraft	yes*	yes
Shindaiwa	NM	NM	Tigershark (Arctco)	yes*	yes*
Simplicity	yes	NM	Tracker	yes*	NM
Snapper	NM	NM	Yamaha	yes*	NM
Stihl Inc.	NM	NM			
Tecumseh	yes*	yes*			
Toro/Lawnboy	yes	yes			

Legend

yes	= permitted/approved	
yes*	= indicates approved but may contain precautionary language or require modification	
NM	= not mentioned in owner's manual	
[1]	= engine manufacturer indicates approval but equipment manufacturer does not.	
Ethanol	= oxygenate (an alcohol)	
MTBE	= oxygenate (an ether)	

Additional Manufacturer Recommendations*

Arctic Cat

- Recommended gasoline 87 minimum octane regular unleaded.
- Oxygenated gasolines up to 10% ethanol or up to 15% MTBE are acceptable with carburetor jet one size larger than main jet required for unleaded gasoline. Ethanol fuel does not require gasoline antifreeze additive.
- Oxygenated gasoline additives: Ethanol (grain alcohol: *Gasohol*) and methyl tertiary butyl ether (MTBE). Oxygenated fuels lean air fuel mixture when compared to nonoxygenated fuels. This may cause hotter running engines.

Briggs & Stratton

- Purchase unleaded fuel that can be used in 30 days to assure freshness. Recommend Briggs & Stratton Gasoline additive. Leaded gasoline may be used outside U.S. if unleaded is unavailable. Excessive amounts of oxygenated fuels can damage fuel system or cause performance problems. Do not use gasoline containing Methanol.

Harley Davidson

- Gasolines containing up to 15% MTBE can be used in Harley Davidson motorcycles.
- Gasoline ethanol blends, up to 10% ethanol, 90% unleaded gasoline, can be used in Harley Davidson motorcycles.
- Harley Davidson recommends use of reformulated gasolines (RFG) or oxygenated gasolines whenever possible.

Honda

- Gasoline with up to 10% ethanol by volume may be used. Gasoline with up to 15% MTBE by volume may be used.

Kawasaki

- Gasolines with up to 10% ethanol are approved. Gasolines with up to 15% MTBE are approved.

Kohler Company

- Gasolines with up to 10% ethanol, 90% unleaded gasoline are approved. Unleaded gasoline blends up to 15% MTBE are approved.

Mercury Marine

- Mercury Marine products produced since 1980 are designed to be used with any commercially available regular grade lead free gasoline, including those containing oxygenates. Refer to: Mercury Outboards/Mariner Outboards Service Bulletin No. 95-5. Any major brand of lead free automotive gasoline with a minimum pump posted octane rating (R+M)/2 of 87 is satisfactory for these outboard engines. Engines may use up to 10% ethanol, but the addition of a Quicksilver Water Separating Fuel Filter is recommended.
- 1979 and older engines should have additional inspections for deterioration of fuel system components caused by alcohol and acids in gasolines.

OMC

- Use of alcohol extended fuels is acceptable only if the alcohol content does not exceed:
  - 10% ethanol by volume.
  - 5% methanol with 5% cosolvents by volume.
  - There are no concerns or performance problems related to the use of reformulated and oxygenated gasolines.

Suzuki

- Unleaded gasoline containing up to 15% MTBE may be used in Suzuki motorcycles.
- Unleaded gasoline containing up to 10% ethanol may be used in Suzuki motorcycles.

Tecumseh

- Leaded gasoline, unleaded or leaded premium gasoline, gasohol containing no more than 10% ethanol, or unleaded gasoline containing no more than 15% MTBE can be used if unleaded regular gasoline is not available.

Toro/Lawnboy

- Use lead free gasoline, including oxygenated or reformulated gasoline, with a rating of 85 octane or higher. Use of premium grade gasoline is not recommended. Use gasoline containing up to 10% ethanol or 15% MTBE. NEVER use gasoline containing methanol.

*Information courtesy of Downstream Alternatives, Inc.

Conversion from conventional magneto to solid state magneto

Conversion of a conventional magneto to solid state magneto is possible. The Omega Electronic Ignition Module shown below, works well on most single cylinder, air cooled engines. It is designed for use in a magneto coil ignition system. A good coil is essential. Weak coils will fire mechanical ignition but will not work properly with solid state or transistorized modules. This module will not work on engines with battery ignition, twin cylinder, or flywheels with multiple magnets.

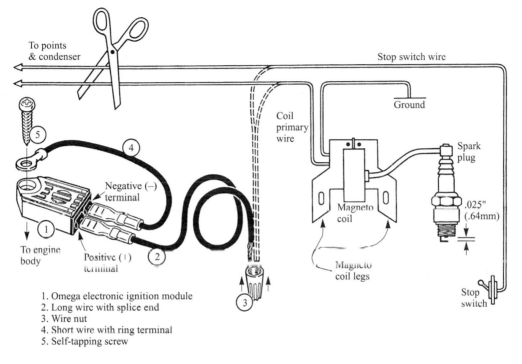

1. Omega electronic ignition module
2. Long wire with splice end
3. Wire nut
4. Short wire with ring terminal
5. Self-tapping screw

To install conversion unit:
1. Remove flywheel cover from engine. If the magneto coil is under the flywheel, remove flywheel also. Separate the primary wire of the magneto coil from the points and condenser. Separate the stop switch wire, if attached, from the points and condenser.
2. Attach long wire (2) to the positive (+) terminal of the module (1) with female connector. Splice other end of long wire (2) to coil primary wire from magneto coil and wire from stop switch. Secure splice with wire nut (3), or solder if desired.
3. Reinstall flywheel, if necessary, and flywheel cover.
4. Attach short wire (4) to the negative (–) terminal of the module (1) with female connector.
5. Insert self-tapping screw (5) through ring connector on other end of short wire (4) and through mounting hole in module. Screw module to metal flywheel cover or other desirable metal location on engine. This will properly ground the module. Pick a spot with good airflow so the module will not get hot. Check to be sure that the module is securely grounded.
6. To get good results using this module, the engine must be in good working condition. (Example—carburetor and other components must be in proper working condition.) The air gap between the flywheel and magneto legs should be between .011″ and .015″ (.3mm and .4mm). Adjust gap if necessary.
7. If spark plug is dirty, clean or replace it. Inspect the gap. It should be .025″ (.64mm). Adjust if necessary.
8. Using a spark plug tester, check for spark. If no spark is visible, it is probable the engine has reversed polarity. To get a spark, reverse the wire connections at the module terminals. Connect the long wire (2) to the negative (–) terminal and the short wire (4) to the positive (+) terminal.
9. Check again for spark.

Information courtesy of Power Lawnmower Parts.

Glossary

A

Abrasion: Wearing or rubbing away.

Additive: A material that is added to the oil to give it certain properties. Example: an additive is blended with engine oil to lessen its tendency to congeal or thicken at low temperatures.

Aft: Toward the rear of the PWC.

Air: A gas containing approximately 4/5 nitrogen, 1/5 oxygen, and some carbonic gas.

Air bleed: A tube in a carburetor through which air can pass into fuel moving through a fuel passage.

Air cleaner: A device for filtering, cleaning, and removing dust from the air admitted to an engine.

Air gap: The space between the flywheel and magneto core poles.

Air horn: Part of air passage in carburetor that is on the atmospheric side of the venturi. The choke valve is located in the air horn.

Air lock: A bubble of air trapped in a fluid circuit that interferes with normal circulation of the fluid.

Air-cooled engine: An engine cooled by air.

Air-fuel ratio: Ratio, by weight, of fuel as compared to air in carburetor mixture.

Alignment: An adjustment to bring related components into a line.

Allen wrench: A hexagonal (six-sided) wrench that fits into a recessed hexagonal hole. Used commonly with set screws.

Alloy: A mixture of different metals. For example, solder is an alloy of lead and tin.

Alternating current: An electrical current alternating flow back and forth in a circuit. Abbreviation is *ac*.

Alternator: A generator that produces alternating current.

Aluminum: A metal noted for its lightness. Aluminum is often alloyed with small quantities of other metals.

American Petroleum Institute: Abbreviation is API.

American Society of Mechanical Engineers: Abreviation is ASME.

Ammeter: An instrument for measuring the flow of electric current. See *Ampere* and *Multimeter*.

Ampere: Unit of measurement for flow of electric current.

Ampere-hour capacity: A term to indicate the capacity of a storage battery.

Anaerobic sealant: Similar to RTV, but can cure in the absence of air. This type of sealant can be used as a thread locking material or between two machined surfaces.

Antifreeze: A material, such as ethylene glycol, that is added to water to lower its freezing point.

Antifriction bearing: A bearing constructed with balls or rollers between journal and bearing surfaces to provide rolling instead of sliding friction.

Antiseize compound: A lubricant applied to threaded fastener and metal components that are exposed to constant heat. It prevents the metal material from being cold welded together.

Arc: A discharge of electric current across a gap, such as between electrodes.

Armature: Part of an electrical motor or generator that includes the main current-carrying winding.

Atmospheric pressure: The weight of air at sea level (about 14.7 psi; less at higher altitudes).

Automotive Electric Association: Abbreviation is AEA.

B

Babbitt: An alloy of tin, copper, and antimony having good antifriction properties. Used as a facing for bearings.

Back pressure: A resistance to free flow, such as a restriction in the exhaust line.

Backfire: Ignition of the mixture in the intake manifold by a flame from the cylinder, such as might occur from a leaking or open intake valve.

Backlash: The clearance or *play* between two parts, such as meshed gears.

Baffle or baffle plate: An obstruction for checking or deflecting the flow of gases or sound.

Ball bearing: An anti-friction bearing consisting of a hardened inner and outer race with hardened steel balls set between the two races.

Battery: Any number of complete electrical cells assembled in one housing or case.

Battery Council International: Abbreviation is BCI.

Bearing: A part in which a journal, pivot, or similar object turns or moves.

Bevel: The angle that one surface makes with another when they are not at right angles.

Blow-by: A leakage or loss of pressure, often used in reference to leakage of compression past piston ring between piston and cylinder.

Boiling point: The temperature at atmospheric pressure at which bubbles or vapors rise to the surface of a liquid and escape.

Bolts: Threaded fasteners that hold parts together by squeezing them between the head on one end and a nut on the other end.

Bore: The diameter of a hole, such as a cylinder; also to enlarge a hole as opposed to making a hole with a drill.

Boss: An extension or strengthened section that supports piston pin or piston pin bushings.

Bottom dead center: Abbreviation is BDC.

Bounce: As applied to engine valves, a condition where the valve is not held tightly to its seat when the cam is not lifting it. In an ignition distributor, a condition where breaker points make and break contact when they should remain closed.

Bow: The front of the PWC.

Brake horsepower: A measurement of power developed by an engine in actual operation. Abbreviation is BHP.

Break-in: The process of wearing into a desirable fit between surfaces of two new or reconditioned parts.

Breaker arm: Movable part of a pair of contact points in an ignition distributor or magneto.

Breaker points: Two separable points that interrupt primary circuit in a distributor or magneto for the purpose of inducing a high tension current in the ignition system. Breaker points are usually faced with silver, platinum, or tungsten.

British Thermal Unit: A measurement of the amount of heat required to raise the temperature of 1 lb of water, 1°F. Abbreviation is BTU.

Brushes: The bars of carbon or other conducting material that contact the commutator of an electric motor or generator.

Bushing: A removable liner for a bearing.

Bypass: An alternate path for a flowing substance.

C

Calibrate: To determine or adjust the graduation or scale of any instrument giving quantitative measurements.

Cam angle: The number of degrees of distributor shaft rotation during which contact points are closed.

Cam or breaker cam: The lobed cam rotating in the ignition system that interrupts the primary circuit to induce a high tension spark for ignition.

Cam-ground piston: A piston ground to a slightly oval shape, which, under the heat of operation, becomes round.

Camshaft: Shaft containing lobes or cams that operate engine valves.

Capacitance: The property that opposes any change in voltage in all electrical circuits.

Capacitor: A device that possesses capacitance (stores electricity). In simple state, consists of two metal plates separated by an insulator.

Carbon: A common nonmetallic element that is an excellent conductor of electricity. It also forms in the combustion chamber of an engine during the burning of fuel and oil.

Carbon monoxide: Gas formed by incomplete combustion. Colorless, odorless, and very poisonous. Chemical formula is CO.

Carburetor: A device for automatically mixing fuel in proper proportion with air to produce a combustible gas.

Carburetor icing: A term used to describe formation of ice on a carburetor throttle plate during certain atmospheric conditions.

Case harden: To harden the surface of steel.

Capacitive discharge ignition: Abbreviation is CDI.

Cell: Part of a battery containing a group of positive and negative plates along with electrolyte.

Cell connector: Lead bar or link connecting the pole of one battery cell to the pole of another.

Celsius: A measurement of temperature. Zero on Celsius scale is 32° on Fahrenheit scale.

Centrifugal force: A force that tends to move a body away from its center of rotation. Example: a whirling weight attached to a string.

Chamfer: A bevel or taper at the edge of a hole.

Charge (or recharge): Passing an electrical current through a battery to restore it to activity.

Chasing: The procedure of recutting a thread with a threading tap.

Check valve: A gate or valve that allows passage of gas or fluid in one direction only.

Choke: A reduced passage, such as a valve placed in a carburetor air inlet to restrict the volume of air admitted.

Circuit: The path of electrical current, fluids or gases. Examples: for electricity, a wire; for fluids and gases, a pipe.

Clearance: The space allowed between two parts, such as between a journal and a bearing.

Clevis pin: A pin that functions as an axle so a part can swivel on it.

Clockwise rotation: Rotation in same direction as the hands of a clock.

Coefficient of friction: A measurement of the amount of friction developed between two surfaces that are rubbed together.

Coil: An electrical device made up of a number of concentric coils in helical form to provide electrical resistance.

Combustion: The process of burning.

Combustion space or chamber: Volume of cylinder above piston with piston at top center.

Commutator: A ring of copper bars in a generator or electric motor providing connections between armature coils and brushes.

Components: The parts that constitute a whole.

Compound: A mixture of two or more ingredients.

Compression: The reduction in volume or the *squeezing* of a gas. As applied to metal, such as a coil spring, compression is the opposite of tension.

Compression ratio: Volume of combustion chamber at end of compression stroke as compared to volume of cylinder and chamber with piston on bottom center.

Concentric: Two or more circles having a common center.

Condensation: The process of vapor becoming a liquid; the reverse of evaporation.

Condenser: A device for temporarily collecting and storing a surge of electrical current for later discharge.

Conduction: The flow of electricity through a conducting body.

Conductor: A material along or through which electricity will flow with slight resistance; silver, copper, and carbon are good conductors.

Connecting rod: Rod that connects the piston to the crankshaft.

Contact breaker: A device for interrupting an electrical circuit; often automatic and may be known as a *circuit breaker, interrupter, cut-out,* or *relay*.

Contact points: Also called breaker points. Two separable points usually faced with silver, platinum, or tungsten, which interrupt the primary circuit in the distributor or magneto for the purpose of inducing a high tension current in the ignition system.

Contraction: A reduction in mass or dimension; the opposite of expansion.

Convection: A transfer of heat by circulating heated air.

Converter: As used in connection with liquefied petroleum gas, it is a device that converts or changes LPG from liquid to vapor state for use in the engine. See *Liquid petroleum gas.*

Corrode: To eat away gradually, especially by chemical action.

Cotter pin: A pin used to lock castle nuts and secure clevis pins. They may be made of copper, brass, aluminum, or stainless steel.

Counterclockwise rotation: Rotating in the opposite direction of the hands on a clock. Anti-clockwise rotation.

Coupling: A connecting means for transferring movement from one part to another; may be mechanical, hydraulic, or electrical.

Crankcase: The housing for the crankshaft and other related internal parts.

Crankcase dilution: Under certain conditions of operation, unburned fuel finds its way past the piston rings and into the crankcase and oil reservoir, where it dilutes or *thins* the engine lubricating oil.

Crankshaft: The main shaft of an engine that, working with the connecting rods, changes the reciprocating motion of the pistons into rotary motion.

Crankshaft counter-balance: A series of weights attached to or forged integrally with the crankshaft and placed to offset the reciprocating weight of each piston and rod assembly.

Crude oil: Oil as it comes from the ground. Unrefined.

Cubic inch: Abbreviation is *cu in.*

Current: The flow of electricity.

Cylinder: A round chamber of some depth bored to receive a piston; also sometimes referred to as *bore* or *barrel.*

Cylinder block: The largest single part of an engine. The basic or main mass of metal in which the cylinders are bored or placed.

Cylinder head: A detachable portion of an engine fastened securely to the top of the cylinder block that contains all or a part of the combustion chamber.

Cylinder sleeve: A liner or tube placed between the piston and the cylinder wall or cylinder block to provide a readily renewable wearing surface of the cylinder.

D

Dead center: The extreme upper or lower position of the crankshaft throw at which the piston is not moving in either direction.

Deck: Any permanent horizontal covering over the hull.

Degree: Abbreviation is deg., or indicated by a ° symbol placed alongside of a figure; may be used to designate temperature readings or angularity, one degree being 1/360 part of a circle.

Demagnetize: To remove the magnetization of a pole that has previously been magnetized.

Density: Compactness; relative mass of matter in a given volume.

Detergent: A compound used in engine oil to remove engine deposits and hold them in suspension in the oil.

Detonation: A excessively rapid burning or explosion of the mixture in the engine cylinders. It becomes audible through a vibration of the combustion chamber walls and is sometimes confused with a *ping* or spark knock.

Diaphragm: A flexible partition or wall separating two cavities.

Die: A cutting tool used for cutting (threading) external threads on a rod, bolt, shaft, or pin. See *Die-stock.*

Die casting: An accurate and smooth casting made by pouring molten metal or composition into a metal mold or die under pressure.

Die-stock: Used to hold die while cutting external threads. See *Die.*

Differential pressure test: A test thatchecks the compression of an engine by measuring leakage from the cylinder to other parts of the engine. This test device and procedure can identify a specific worn or damaged component in the engine that may, or may not, be directly related to the cylinder condition or rings. The device is designed so that specific leakages can be detected and isolated before disassembling parts of the engine.

Diluent: A fluid that thins or weakens another fluid. Example: gasoline dilutes oil.

Dilution: See *Crankcase Dilution.*

Diode: A semiconductor device that allows current to flow in only one direction.

Direct current: Electric current that flows continuously in one direction. Example: a storage battery. Abbreviation is dc.

Discharge: The flow of electric current from the battery; the opposite of charge.

Displacement: See *Piston Displacement.*

Distortion: A warpage or change in form from the original shape.

Dowel pin: A pin inserted in matching holes in two parts to maintain those parts in fixed relation to each other. They are heat treated and hardened. The dowel pin is pressed into a hole with an interference fit.

Down-draft: Used to describe a carburetor type in which the mixture flows downward to the engine.

Draft: The depth of the PWC below the waterline.

Drop forging: A piece of steel shaped between dies while hot.

Dwell period: The number of degrees the breaker cam rotates from the time the breaker points close until they open again. Also known as *cam angle.*

Dynamometer: A machine for measuring the actual power produced by an internal combustion engine.

E

Eccentric: One circle within another circle, each having a different center. Example: a cam on a camshaft.

Electrode: Usually refers to the insulated center rod of a spark plug. It is also sometimes used to refer to the rods attached to the shell of the spark plug.

Electrolyte: A mixture of sulphuric acid and distilled water used in conventional lead-acid type storage batteries.

Element: One set of positive plates and one set of negative plates complete with separators assembled together.

Electromotive force or voltage: Abbreviation is EMF.

Energy: The capacity for doing work.

Engine: The term applies to the prime source of power generation.

Engine displacement: The sum of the displacement of all the engine cylinders. See *Piston Displacement.*

Environmental Protection Agency: A government funded agency that regulates policies that deal with environmental issues. Abbreviation is EPA.

Evaporation: The process of changing from a liquid to a vapor, such as boiling water to produce steam. Evaporation is the opposite of condensation.

Exhaust: The spent fuel after combustion takes place in an internal combustion engine.

Exhaust pipe: The pipe connecting the engine to the muffler to conduct the exhausted or spent gases away from the engine.

Expansion: An increase in size. Example: when a metal rod is heated, it increases in length and diameter. Expansion is the opposite of contraction.

F

Fahrenheit (F): A scale of temperature measurement ordinarily used in English-speaking countries. The boiling point of water is 212° Fahrenheit, as compared to 100° Celsius.

Federal Clean Air Act: An act aimed at ridding the atmosphere of harmful road vehicle emissions. Since inception, there has been an increase of awareness of the hazards of harmful emission from small gas engines.

Feeler gauge: A metal strip or blade finished accurately with regard to thickness and used for measuring the clearance between two parts. Feeler gauges ordinarily come in a set of different blades graduated in thickness by increments of .001 in.

Ferrous metal: Metals that contain iron or steel and are, therefore, subject to rust

Field. In a generator or electric motor, the area in which magnetic flow occurs.

Field coil: A coil of insulated wire surrounding the field pole.

Fillet: A rounded filling, such as a weld between two parts joined at an angle.

Filter (oil, water, gasoline, etc.): A unit containing an element, such as a screen of varying degrees of fineness. The screen or filtering element is made of various materials depending upon the size of the foreign particles to be eliminated from the fluid being filtered.

Fit: A kind of contact between two machined surfaces.

Flange: A projecting rim or collar on an object for keeping it in place.

Flash point: The temperature at which an oil will flash and burn.

Flashover: Tendency of current to travel down the outside of a spark plug instead of through the center electrode.

Float: A hollow tank that is lighter than the fluid in which it rests. A float is ordinarily used to automatically operate a valve controlling the entrance of a fluid.

Float level: The predetermined height of the fuel in the carburetor bowl, usually regulated by means of a suitable valve.

Floating piston pin: A piston pin that is not locked in the connecting rod or the piston, but is free to turn or oscillate in both the connecting rod and the piston.

Flutter or bounce: A condition arising from a valve not being held tightly on its seat when the cam is not lifting it.

Flyweights: Special weights that react to centrifugal force to provide automatic control of other mechanisms, such as accelerators or valves.

Flywheel: A heavy wheel in which energy is absorbed and stored by means of momentum.

Foot-pound: This is a measure of the amount of energy or work required to lift 1 lb a distance of 1 ft. Abbreviation is ft-lb.

Force-fit: Also known as a press-fit, interference-fit or drive-fit. This term is used when a shaft is slightly larger than a hole and must be forced in place.

Forge: To shape plastic or hot metal by hammering.

Form-in-place sealant: A sealant that can be used in place of conventional gaskets when the exact replacement gasket is not available. See *Gasket*.

Forward: Toward the bow of the PWC.

Four-cycle engine: Also known as Otto cycle. In a four-cycle engine, an explosion occurs every other revolution of the crankshaft. A cycle is considered as a half revolution of the crankshaft. The cycles (strokes) are (1) intake; (2) compression; (3) power and; (4) exhaust stroke.

Friction: Resistance to motion created when one surface rubs against another.

Fuel knock: Same as detonation.

Fulcrum: The support on which a lever turns in moving a body.

G

Gas: A substance that can be changed in volume and shape according to the temperature and pressure applied to it. Example: air is a gas that can be compressed into smaller volume and into any shape desired by pressure. It can also be expanded by the application of heat.

Gasket: A soft, pliable material, such as fiber, rubber, neoprene (synthetic rubber), cork, treated paper, thin steel, or laminated material, used between engine parts to seal and prevent leakage of engine oil, coolant, compression, and vacuum; acts as a seal.

Gassing: The bubbling of the battery electrolyte that occurs during the process of charging a battery.

Gear ratio: The number of revolutions made by a driving gear as compared to the number of revolutions made by a driven gear of a different size. Example: if one gear makes three revolutions while the other gear makes one revolution, the gear ratio would be 3 to 1.

Generator: A device consisting of an armature, field coils, and other parts, which when rotated, generates electricity. A generator is usually driven by a belt from the engine crankshaft.

Glaze: An extremely smooth or glossy surface finish, such as a cylinder wall that is highly polished over a long period of time by the friction of the piston rings.

Glaze breaker: A tool for removing the glossy surface finish in an engine cylinder.

Glow plug: A device with a fine wire connected in series to an electrical circuit for the purpose of creating enough resistance and heat to ignite fuel in a combustion chamber. Used in place of spark plugs on some engines.

Governor: A mechanical, hydraulic, or electrical device that controls and regulates speed.

Grass discharge chute guard: An important safety device on lawn mowers and tractors that prevent objects from being thrown out the discharge chute.

Grid: The metal framework of an individual battery plate in which the active material is placed.

Grind: To finish or polish a surface by means of an abrasive material.

Grooved pin: A pin driven into an interference hole. The groove cuts into the wall of the hole and secures the pin.

Gum: Oxidized petroleum products that accumulate in the fuel system, carburetor, or engine parts.

H

Heat treatment: A combination of heating and cooling operations timed and applied to a metal in a solid state in a way that will produce desired properties.

High tension: The secondary or induced high-voltage electrical current. High tension is also used in reference to the wiring from the distributor cap to the coil and the spark plugs.

Hone: An abrasive tool for correcting small irregularities or differences in the diameter of a cylinder.

Horsepower: The energy required to lift 550 lb a distance of 1 ft in 1 second. Abbreviation is HP or hp.

Hot spot: Refers to a comparatively thin section or area of the wall between the inlet and exhaust manifold of an engine, the purpose being to allow the hot exhaust gases to heat the comparatively cool incoming mixture. Also used to designate local areas of the cooling system that have above average temperatures.

HP or hp: See *Horsepower.*

Hull: The body of the PWC from bow to stern.

Hydrocarbon: Any compound composed entirely of carbon and hydrogen, such as petroleum products.

Hydrocarbon engine: An engine using petroleum products, such as gas, liquefied gas, gasoline, kerosene, or fuel oil as a fuel.

Hydrometer: An instrument for determining the state of charge in a battery by finding the specific gravity of the electrolyte.

I

Idle: Refers to the engine operating at its slowest practical speed.

Ignition distributor: An electrical unit containing the circuit breaker for the primary circuit and providing a means for conveying the secondary or high tension current to the spark plug wires as required.

Ignition system: The means for igniting the fuel in the cylinders; includes spark plugs, wiring, ignition distributor, ignition coil, and source of electrical current supply.

Impeller: A rotor or wheel with blades to pump water or propel objects through water or other fluids.

Indicated horsepower: Horsepower developed by an engine, which is a measurement of the pressure of the explosion within the cylinder expressed in pounds per square inch. Abbreviation is IHP.

Induction: The influence of different strength magnetic fields that are not electrically connected to one another.

Induction coil: Essentially a transformer that, through the action of induction, creates a high tension current by means of an increase in voltage.

Inertia: A physical law that tends to keep a motionless body at rest and also tends to keep a moving body in motion. Therefore, effort required to start a mass moving or to retard or stop it once it is in motion.

Inhibitor: A material to restrain or hinder some unwanted action, such as a rust inhibitor, which is a chemical added to a cooling system to retard rust formation.

Inlet valve or intake valve: A valve that permits a fluid or gas to enter a chamber and seals against exit.

Inside diameter: Abbreviation is ID.

Insulation: Any material that does not conduct electricity; used to prevent the flow or leakage of current from a conductor. Also, used to describe a material that does not conduct heat readily.

Insulator: An electrical conductor covered or shielded with a nonconducting material, such as a copper wire within a rubber tube.

Intake manifold or inlet pipe: The tube used to conduct the gasoline and air mixture from the carburetor to the engine cylinders.

Integral: Formed as a unit with another part.

Intensify: To increase or concentrate, such as to increase the voltage of an electrical current.

Interference angle: Difference in angle between mating surfaces of a valve and a valve seat.

Intermittent: Motion or action that is not constant but occurs at intervals.

Internal combustion: The burning of a fuel within an enclosed space.

Internal Combustion Engine Institute, Inc: Abbreviation is ICEI.

J

Journal: The part of a shaft or crank that rotates inside of a bearing.

Jump spark: A high tension electrical current that jumps through the air from one terminal to another.

K

Key: A small block inserted between a shaft and hub to fasten a pulley or gear to the shaft.

Keyseat The recess in which the key rests. See *Key* and *Keyway.*

Keyway. The groove in the pulley, gear, or collar where the key and the keyseat match. See *Key* and *Keyseat.*

Knock: A general term used to describe various noises occurring in an engine; may be used to describe noises made by loose or worn mechanical parts, preignition, or detonation.

Knurl: To indent or roughen a finished surface.

L

Lacquer: A solution of solids in solvents that evaporate with great rapidity.

Laminate: To build up or construct out of a number of thin sheets. Example: the laminated core in a magneto coil.

Lanyard: Engine shut-off cord on PWC attached to the operator and the stop switch.

Lapping: The process of fitting one surface to another by rubbing them together with an abrasive material between the two surfaces.

Lead: A short connecting wire that makes electrical contact between two points.

Liner: Usually a thin section placed between two parts, such as a replaceable cylinder liner in an engine.

Linkage: Any series of rods, yokes, and levers, etc., used to transmit motion from one unit to another.

Liquid: Any substance that assumes the shape of the vessel in which it is placed without changing volume.

Liquid petroleum gas: Made usable as a fuel for internal combustion engines by compressing volatile petroleum gases to liquid form. When so used, must be kept under pressure or at low temperature in order to remain in liquid form. Abbreviation is LPG.

Lobe: An off-center or eccentric enlargement on a shaft that converts rotary motion to reciprocating motion. Also called a *cam*.

Lock nut: A nut designed to create friction to reduce the tendency for vibration or motion to rotate and loosen the nut.

Lost motion: Motion between a driving part and a driven part that does not move the driven part. Also see *Backlash*.

LPG: See *Liquid petroleum gas*.

M

Magnet (permanent): A piece of hard steel that can be charged with and retain magnetic power. Magnets are often bent into a *U* shape so as to have opposite poles.

Magnetic field: The flow of magnetic force or magnetism between the opposite poles of a magnet.

Magneto: An electrical device that generates current when rotated by an outside source of power; may be used for the generation of either low tension or high tension current.

Manifold: A pipe with multiple openings used to connect various cylinders to one inlet or outlet.

Mechanical efficiency: The ratio between the indicated horsepower and the brake horsepower of an engine.

Micrometer: A measuring instrument for either external or internal measurement in thousandths or ten thousandths of an inch.

Mill: To cut or machine with rotating, toothed cutters.

Misfiring: Failure of an explosion to occur in one or more cylinders while the engine is running; may be continuous or intermittent failure.

Module: A packaged functional assembly of wired electronic components for use with other such assemblies.

Motor: This term should be used in connection with an electric motor and should not be used when referring to the engine.

Motor and Equipment Manufacturer's Association: Abbreviation is MEMA.

Motor and Equipment Wholesaler's Association: Abbreviation is MEWA.

Muffler: A chamber attached to the end of the exhaust pipe that allows the exhaust gases to expand and cool. It is usually fitted with baffles or porous plates and serves to subdue much of the noise created by the exhaust.

Multimeter: A measuring instrument that is a combination ammeter, ohmmeter, and voltmeter. See *Ammeter, Ohmmeter,* and *Voltmeter*.

N

National Coarse: Coarse thread series designation. Abbreviation is NC.

National Fine: Fine thread series designation. Abbreviation is NF.

National Standard Parts Association: Abbreviaton is NSPA.

Needle bearing: An anti-friction bearing using a great number of small, cylindrical rollers; also known as a quill-type bearing.

Negative pole: The point from which an electrical current flows through the circuit. It is designated by a minus sign (–).

Neon tube: An electric *bulb* or tube filled with a rare gas. Neon tubes are often used on ignition test instruments.

Nonferrous metals: This designation includes practically all metals that do not contain iron (or contain very little iron) and, therefore, are not subject to rusting.

North pole: The pole of a magnet where the lines of force start; the opposite of south pole.

Nut: A nut holds parts together by squeezing them between the nut on one end and the head of a bolt on the other. See *Bolt*.

O

Ohm: A measurement of the resistance to the flow of an electrical current through a conductor.

Ohmmeter: An instrument that measures the resistance to the flow of electrical current through a conductor. See *Ohm* and *Multimeter*.

Oil pumping: A term used to describe an engine that is using an excessive amount of lubricating oil.

Open circuit: A break or opening in an electrical circuit that interferes with the passage of the current.

Oscillate: To swing back and forth like a pendulum.

Otto cycle: Also called four-stroke cycle. Named after the man who adopted the principle of four cycles of operation for each explosion in an engine cylinder. The four cycles of operation are: (1) intake, (2) compression, (3) power, (4) exhaust.

Outside diameter: Abbreviation is OD.

Oxidize: To combine an element with oxygen or convert it into its oxide. The process is often accomplished by a combination. For example, when carbon burns, it combines with oxygen to form carbon dioxide or carbon monoxide. When iron rusts, the iron has combined with oxygen from the air to form the rust (oxide of iron).

P

Personal watercraft: A type of popular small boat that is propelled and guided by a high velocity jet of water. Abbreviation is PWC.

Petroleum: A group of liquid and gaseous compounds composed of carbon and hydrogen, which are removed from the earth.

Phillips screw: A screw head having a cross instead of a slot for a corresponding type of screwdriver.

Pins: A piece of material used to either retain parts in a fixed position or to preserve alignment of parts. They are usually cylindrical in shape.

Piston: A cylindrical part that is closed at one end and connected to the crankshaft by the connecting rod. The force of the explosion in the cylinder is exerted against the closed end of the piston, causing the connecting rod to move the crankshaft.

Piston collapse: An abnormal reduction in the diameter of the piston skirt due to heat or stress.

Piston displacement: The volume of air moved or displaced by moving the piston from one end of its stroke to the other.

Piston head: The part of the piston above the rings.

Piston lands: Those parts of a piston between the piston rings.

Piston pin (wrist pin): The journal for the bearing in the small end of an engine connecting rod. The piston pin also passes through the piston walls.

Piston ring: An expanding ring placed in the grooves of the piston to seal it against the passage of fluid or gas.

Piston ring expander: A spring placed behind the piston ring in the groove to increase the pressure of the ring against the cylinder wall.

Piston ring gap: The clearance between the ends of the piston ring.

Piston ring groove: The channel or slots in the piston in which the piston rings are placed.

Piston skirt: The part of the piston below the rings and the bosses.

Piston slap: Rocking of loose fitting piston in a cylinder, making a hollow bell-like sound.

Pivot: A pin or short shaft upon which another part rests, turns, rotates, or oscillates.

Platinum: An expensive metal having an extremely high melting point and good electrical conductivity. Often used in magneto breaker points.

Polarity: The positive or negative terminal of a battery or an electric circuit; also the north or south poles of a magnet.

Polarize: To give polarity to an electric circuit so current will flow in the proper direction.

Poppet valve: A valve structure consisting of a circular head with an elongated stem attached in the center. A poppet valve is designed to open and close a circular hole or port.

Porcelain: General term applied to the material or element used for insulating the center electrode of a spark plug.

Port: 1. The left side of a PWC when you are aboard and facing the bow (front). 2. The openings in the cylinder block for valves, exhaust and inlet pipes, or water connections. In two cycle engines, the openings for inlet and exhaust purposes.

Positive pole: The point to which an electrical current returns after passing through the circuit. This is designated by a plus sign (+).

Post: The heavy circular part to which the group of plates is attached. The post extends through the cell cover to provide a means of attachment to the adjacent cell or battery cable.

Potential: An indication of the amount of energy available.

Potential difference: A difference of electrical pressure that sets up a flow of electric current.

Pounds per square inch: A pressure measurement. Abbreviation is PSI.

Preignition: Ignition occurring earlier than intended. Example: an explosive mixture could be fired in a cylinder by a flake of incandescent carbon before the electric spark occurs.

Press-fit: Also known as a force-fit or drive-fit. Fit accomplished by forcing a shaft into a hole that is slightly smaller in diameter than the shaft.

Primary winding: In an ignition coil or magneto armature, a wire that conducts the low tension current, which is to be transformed by induction into high tension current in the secondary winding.

Primary wires: The wiring circuit used for conducting the low tension or primary current to the points, where it is to be used.

Prony brake: A machine for testing the power of an engine while it is running against a friction brake.

Propane: A petroleum hydrocarbon compound that has a boiling point about 44°F and is used as an engine fuel. It is loosely referred to as liquid petroleum gas (LPG) and is often combined with Butane.

Pushrod: A connecting link in an operating mechanism, such as the rod interposed between the valve lifter and rocker arm on an overhead valve engine.

R

Ratio: The relation or proportion that one number bears to another.

Ream: To finish a hole accurately with a rotating, fluted tool.

Reciprocating: A back and forth movement, such as the action of a piston in a cylinder.

Rectifier: Device used to convert alternating current to pulsating direct current.

Reed valve: A flat, springy valve covering the ports between the carburetor and the crank chamber in a two-cycle engine.

Reel-type mower: A lawnmower that has helical blades that rotate around a horizontal shaft.

Regulator: An automatic pressure reducing or regulating valve.

Resistance: Quality of an electric circuit, or any component in it, to oppose the flow of electrical current.

Retaining ring: A ring made of circular spring steel that fits externally or internally into a groove of a part. It is used to restrict movement. Example: an external retaining ring prevents movement of a shaft beyond a point through a hole such as in a bearing; an internal retaining rings prevent a shaft from traveling beyond the retaining ring located in its groove in the cylindrical part.

Retard: To cause the spark to occur at a later time in the cycle of engine operation. Opposite of spark advance.

Revolutions per minute: Abbreviation is RPM.

Rocker arm: Device used in overhead valve system to transfer the upward motion of the pushrod to a downward force on the valve.

Roller bearing: An inner and outer race upon which hardened steel rollers operate.

Room temperature vulcanizing sealant: A form-in-place sealant that is also referred to as silicon sealant. Abbreviation is RTV.

Rotary mower: A lawnmower that has horizontally rotating, large diameter blades and heavy engines and housings.

Rotary valve: A valve construction in which ported holes move in and out of register with each other to allow fluids or gases to enter and exit.

Rotor: A rotating valve or conductor for carrying fluid or electrical current from a central source to the individual outlets as required.

Rubber: An elastic, vibration-absorbing material of either natural or synthetic origin.

Running-fit: Sufficient clearance allowed between a shaft and journal to allow free running without overheating.

S

Scale: A flaky deposit occurring on steel and iron or the mineral and metal build-up in a cooling system.

Score: A scratch, ridge, or groove marring a finished surface.

Screws: Threaded fasteners that hold parts together by passing through one part and threading into another.

Seat: A surface, usually machined, upon which another part rests or seats. Example: the surface upon which a valve face rests.

Secondary winding: In an ignition coil or magneto armature, a wire in which a secondary or high tension current is created by induction due to the interruption of current in the adjacent primary winding.

Sediment: Active material of the battery plates, which is gradually shed and accumulates in a space provided below the plates.

Seize: When one surface moving on another binds and sticks. Example: a piston will seize in a cylinder because of lack of lubrication or excessive heat-related expansion.

Semiconductor: A substance that shares the characteristics of both a conductor and an insulator.

Separators: Sheets of rubber or wood inserted between the positive and negative plates of a cell to keep them out of contact with each other.

Set screw: A screw that is heat-treated, hardened-alloy steel fasteners, which is used to secure such things as pulleys, gears, and shafts.

Shim: Thin sheets used as spacers between two parts, such as the two halves of a journal bearing.

Short circuit: To provide a shorter path; often used to indicate an accidental ground in an electrical device or conductor.

Shrink-fit: An exceptionally tight fit achieved by the heating and/or cooling of parts. The outer part is heated above its normal operating temperature or the inner part chilled below its normal operating temperature and assembled in this condition.

Shroud: A light cover over the flywheel that shields the flywheel and helps direct airflow over the engine to carry away heat.

Shunt: To bypass or turn aside. In electrical apparatus, an alternate path for the current.

Shunt winding: An electric winding or coil of wire that forms a bypass or alternate path for electric current. When applied to electric generators or motors, each end of the field winding is connected to an armature brush.

Side seals: The spring-loaded seals located on the sides of the triangular rotor in the Wankel engine.

Sillment seal: Compacted powder that helps ensure permanent assembly of a spark plug and eliminates compression leakage under operating conditions.

Sliding-fit: Clearance between a shaft and journal that is sufficient to allow free running without overheating.

Sludge: A composition of oxidized petroleum products along with an emulsion formed by the mixture of oil and water. This forms a pasty substance, clogs oil lines and passages, and interferes with engine lubrication.

Society of Automotive Engineers: Abbreviation is SAE.

Solenoid: An iron core surrounded by a coil of wire that moves due to magnetic attraction when an electrical current is fed to the coil. Solenoids are often used to actuate mechanisms by electrical means.

Solvent: A solution that dissolves some other material. Example: water is a solvent for sugar.

South pole: The pole of a magnet to which the lines of force flow; the opposite of north pole.

Spark: An electrical current possessing sufficient pressure to jump an air gap from one conductor to another.

Spark advance: Causing the spark to occur at an earlier time in the cycle of engine operation; opposite of retard.

Spark gap: The space between the electrodes of a spark plug across which the spark jumps. Also, a safety device in a magneto to provide an alternate path for the current when it exceeds a safe value.

Spark knock: See *Preignition*.

Spark plug: A device inserted into the combustion chamber of an engine. Spark plugs contain an insulated central electrode for conducting the high tension current from the ignition distributor or magneto. This insulated electrode is spaced a predetermined distance from the shell or side electrode in order to control the dimensions of the gap that the spark must jump.

Specific gravity: The relative weight of a substance as compared to water. Example: if a cubic inch of acid weighs twice as much as a cubic inch of water, the specific gravity of the acid would be 2.

Spiral bevel gears: A gear and pinion wherein the mating teeth are curved and placed at an angle with the pinion shaft.

Spline: A long keyway.

Spurt-hole: A hole drilled through a connecting rod and bearing that allows oil under pressure to be squirted out of the bearing for additional lubrication of the cylinder walls.

Square foot. Abbreviation is sq ft.

Square inch. Abbreviation is sq in.

Stamping: A piece of sheet metal cut and formed into the desired shape with the use of dies.

Starboard: The right side of the PWC when you are aboard and facing the bow.

Starter: An electric motor attached by gearing to an engine to provide power to turn it over for starting.

Stator: Stationary coils of an alternating current generator.

Steerable nozzle: A device for directing the stream of water from the jet pump to the left or right at the stern of a PWC.

Steering control: The device designed for controlling the direction of a PWC.

Stern: The rear of the PWC.

Straight pin: A pin used for alignment. They fit closely, but are not usually an interference fit like dowel pins. See *Dowel pin.*

Stress: The force or strain to which a material is subjected.

Stroboscope: A term applied to an ignition timing light, which, by being attached to the distributor points, gives the effect of making a mark on a rapidly rotating wheel, such as a flywheel, appear to stand still for observation.

Stroke: The distance moved by the piston.

Stud: A rod with threads cut on both ends, such as a cylinder stud, which screws into the cylinder block on one end and has a nut placed on the other end to hold the cylinder head in place.

Stud bolt. A bolt that is threaded on both ends. Example: the head on a gasoline engine may be held to the engine block with these.

Suction: Suction exists in a vessel when the pressure is lower than the atmospheric pressure. Also see *Vacuum.*

Sulfated: When a battery is improperly charged or allowed to remain in a discharged condition for some length of time, an abnormal amount of lead sulfate will collect on the plates. The battery is then said to be *sulfated.*

Sump: The part of the block in a small four-stroke engine that holds and collects the lubricating oil.

Supercharger: A device for increasing the volume of the air charge for an internal combustion engine.

Synchronize: To cause two events to occur in unison or at the same time.

T

Tachometer: A device for measuring and indicating the rotational speed of an engine.

Tap: To cut threads in a hole with a tapered, fluted, threaded tool.

Taper pin: A pin that has a uniform taper of .250 per foot over the length of the pin and each end is rounded slightly. They are generally used to fasten pulleys and gears to shafts to prevent rotation on the shaft.

Taper tap: A tap that has a slender taper at the beginning of the tap that makes it start easier in the threads. Used for tapping through holes.

Tappet: The adjusting screw for varying the clearance between the valve stem and the cam. May be built into the valve lifter or into the rocker arm on an overhead valve engine.

Tapping: The process of cutting threads in a hole.

Tension: Effort devoted toward elongation or *stretching* of a material.

Terminal: In electrical work, a junction point where connections are made, such as the terminal fitting on the end of a wire.

Thermal efficiency: A gallon of fuel contains a certain amount of potential energy in the form of heat when burned in the combustion chamber. Some of this heat is lost and some is converted into power. The thermal efficiency is the ratio of work accomplished compared to the total quantity of heat contained in the fuel.

Thermostat: A heat-controlled valve used in the cooling system of an engine to regulate the flow of water or used in the electrical circuit to control the current.

Thread: The helical portion of a screw or bolt, or the helix in a hole that it fastens into.

Thread adhesive: Adhesive applied to the threads of nuts, bolts, or screws to prevent them from loosening during service.

Threading: The process of making external threads on an external cylindrical surface.

Threading tap: A tool used to cut or recut the thread.

Throttle: Controls speed by regulating the amount of fuel delivered to the engine.

Through hole: A threaded hole that goes all the way through material. Example: a nut.

Throw: With reference to an engine, usually the distance from the center of the crankshaft main bearing to the center of the connecting rod journal.

Timing chain: Chain used to drive camshaft and accessory shafts of an engine.

Timing gears: Any group of gears that are driven from the engine crankshaft to cause the valves, ignition, and other engine-driven apparatus to operate at the desired time during the engine cycle.

Tolerance: A permissible variation between the two extremes of specified dimensions.

Top dead center: Abbreveation is TDC.

Torque: An effort devoted toward twisting or turning.

Torque wrench: A special wrench with a built-in indicator to measure the applied turning force.

Transfer port: In two-cycle engines, an opening in the cylinder wall permitting fuel to enter from the crankcase.

Transistor: A semiconductor device that is often used for switching applications. A transistor is used in place of breaker points in a TCI system.

Transistor controlled ignition: Abbreviation is TCI.

Troubleshooting: Refers to a process of diagnosing or determining the source of trouble or troubles from observation and testing.

Tune-up: The process of accurate and careful adjustments to obtain maximum engine performance.

Turbulence: A disturbed or disordered, irregular motion of fluids or gases.

Two-cycle engine: An engine design permitting one power stroke for each revolution of the crankshaft.

U

Unified National Coarse: Coarse thread series designation. Abbreviation is UNC.

Unified National Fine: Fine thread series designation. Abbreviation is UNF.

Up-draft: Used to describe a carburetor type where the mixture flows upward to the engine.

Upper cylinder lubrication: A method of introducing a lubricant into the fuel or intake manifold in order to permit lubrication of the upper cylinder, valve guides, and other parts.

V

Vacuum: A perfect vacuum has not been created as this would involve an absolute lack of pressure. The term is ordinarily used to describe a partial vacuum, that is, a pressure less than atmospheric pressure. See *Suction.*

Vacuum gauge: An instrument designed to measure the amount of vacuum existing in a chamber.

Valve: A device for alternately opening and sealing an aperture.

Valve clearance: Air gap allowed between end of valve stem and valve lifter or rocker arm to compensate for heat expansion.

Valve face: Part of a valve that mates with and rests upon a seating surface.

Valve grinding: A process of mating the valve seat and valve face performed with the aid of an abrasive.

Valve head: The portion of the valve upon which the valve face is machined.

Valve key or valve lock: The key, washer, or other device that holds the valve spring cup or washer in place on the valve stem.

Valve lifter: A rod or plunger that transfers motion from the cam and the other valve train components. A lifter is often adjustable to vary the length of the unit.

Valve margin: The space or rim between the surface of the head and the surface of the valve face.

Valve overlap: An interval expressed in degrees where both valves of an engine cylinder are open at the same time.

Valve seat: The matched surface upon which the valve face rests.

Valve spring: A spring attached to a valve to return it to the seat after it has been released from the lifting or opening operation.

Valve stem: That portion of a valve that rests within a guide.

Valve stem guide: A bushing or hole in which the valve stem is placed. A valve stem guide allows only two-way motion.

Vanes: Any plate or blade attached to an axis and moved by air or liquid.

Vapor lock: A condition wherein fuel boils in the fuel system, forming bubbles that retard or stop the flow of liquid fuel to the carburetor.

Vaporizer: A device for transforming or helping to transform a liquid into a vapor; often includes the application of heat.

Venturi: Two tapering streamlined tubes joined at their small ends so as to reduce the internal diameter.

Vibration damper: A device to reduce the torsional or twisting vibration that occurs along the length of the crankshaft used in multiple cylinder engines; also known as a harmonic balancer.

Viscosity: The resistance to flow or adhesive characteristics of an oil.

Volatility: The tendency for a fluid to evaporate rapidly or pass off in the form of a vapor. Example: gasoline is more volatile than kerosene because it evaporates at a lower temperature.

Volt: A unit of electrical force that will cause a current of one ampere to flow through a resistance of one ohm.

Voltage regulator: An electrical device for controlling or regulating voltage.

Voltmeter: An instrument for measuring the voltage in an electrical circuit. See *Volt* and *Multimeter*.

Volume: The measure of space expressed as cubic inches, cubic feet, or other units of linear measure.

Volumetric efficiency: A combination between the ideal and actual efficiency of an internal combustion engine. If the engine completely filled each cylinder on each intake stroke, the volumetric efficiency of the engine would be 100%. In actual operation, however, volumetric efficiency is lowered by the inertia of the gases, the friction between the gases and the manifolds, the temperature of the gases, and the pressure of the air entering the carburetor. Volumetric efficiency is ordinarily increased by the use of large valves, ports, and manifolds. It can be further increased with the aid of a supercharger.

W

Washer: A washer provides a wider bearing surface for a bolt or screw head and/or nut.

Water column: A reference term used in connection with a manometer.

Watt: A measuring unit of electrical power obtained by multiplying amperes by volts.

Wiring diagram: A detailed drawing of all wiring, connections, and components in an electrical circuit.

Wrist pin: See *Piston Pin.*

Index

A

Absolute vacuum, 124
Acceleration system, 128
Acceleration well, 128, 129
Acorn nuts, 47
Adhesives and sealants, 58, 59
Adhesives, thread, 58
Adjustable wrench, 22
Air cleaner, 204
    and air filters, 140-144
    filter, chain saw, 348
    oil-wetted, 140, 141
    removing foam element, 141
    service, 141, 204
Air filters, and air cleaners, 140-144
Air fuel mixture, 123, 124
    requirements, 124
Air vane, 135
Air vane governor, 135-138
Air-cleaner service, 204
Allen wrench, 22
Alnico, 159
Alternator system troubleshooting, 264
Alternators, 262-265
    maintaining, 262
    output tests, 262-265
Aluminum guides, replacing worn,
    309, 310
American National Standards Institute
    (ANSI), 384
American Petroleum Institute (API),
    180
Ammeters, 258
Ampere (A), 149
Anaerobic sealant, 59
Anco, locking nut, 48
Antifoaming additives, 179
Antifreeze, 193, 206
Antifriction bearings, 300
Antiseize compounds, 59
API engine oil service classification,
    180, 181
    symbol, 181
Atmospheric pressure, 124
Atomizing, 65
Auger,
    controls, 414
    gear box components, 411
    gear box lubrication, 419, 420
    reduction gear box, 411
Auto-transformer type ignition coil,
    169, 170
Automatic compression release, 326, 327
Axial clearance, 47

B

Babbitt, 176
Ball peen hammers, 25, 26
Ball piston pump, 381
Barrel pump, 183
Batteries, 253-258, 454-456
    battery starting circuits, 253
    hydrometer test, 254, 255
    installation 255, 256
    PWC maintenance, 454-456
Battery,
    lead-acid, 170
    undercharging, 256
Battery construction, 170
Battery failure,
    causes for failure, 256
    frozen electrolyte, 256
    lack of water, 256
    loose holddowns, 256
    overcharging, 256
Battery ignition systems, 167-169, 252,
    253, 259
    servicing, 252, 253
Battery installation, PWC mainte-
    nance, 456
Battery operated ignition systems, 251
Battery recharging, 256-258
    maintenance and service, 258
Battery removal, PWC maintenance,
    454, 455
Battery safety, PWC maintenance, 454
Battery starting circuits, 253
Battery terminals, PWC maintenance,
    455
Battery testing, visual inspection, 253
Battery voltage, 170
Bearing,
    clearance, 303
    crush, 299, 300
    splitter, 27
    spread, 298, 299
Bearing, crankshaft, valve, and
    camshaft service, 297-330
    friction-type rod bearings, 298-300
    introduction, 297
    rod bolt locking devices, 300, 301
Bilge,
    flushing, 456, 457
    system, 448
Blade guard, 361
Blind holes, 52
Boating courtesy, PWC, 436
Boating terms,
    aft, 427

bow, 427
draft, 427
forward, 427
hull, 427
lanyard, 427
port, 427
starboard, 427
steerable nozzle, 427
steering control, 427
stern, 427
throttle, 427
Bolts, 44, 46-49
    common types, 44
    grades, 48, 49
    head size, 48
    length, 48
    size, 48
    terminology, 48
Bolt torque chart, general, 49
Bore, 92
Boring machine, 277
Bottom dead center, 79
Bottom tap, 52
Bottom tapping tool, 51
Boundary layer, 191
Boundary lubrication, 177
Box-end wrench, 21
Brake horsepower (bhp), 98
    measuring, 99, 100
Breaker point ignition systems,
    adjusting, 247, 248
Breaker points, 157, 158
Breaking, 251
Brush cutting, 351-353
Brushcutters, 350
Brushes, 262

C

California Air Resources Board
    (CARB), 118
Cam follower (wear block), 248
Cam lobes, 73
Cam-ground piston, 282
Camshaft, 72-74, 326
Camshafts and gears, 326
Capacitive discharge ignition (CDI)
    system, operation, 163-165
Carbon monoxide, 15
Carburetion, 67, 123-146
    air-fuel mixture, 123, 124
    principles, 123-126
Carburetion principles, carburetor
    pressure differences, 124

Carburetor adjustment, high speed and idle mixture adjustment, 218, 219
Carburetor disassembly, 221, 222
Carburetor kits, 220
Carburetor overhaul, 220-235
    engine priming, 225
    inspection of parts, 222, 223
    removal, 220, 221
Carburetor pressure differences, 124
    vacuum, 124
Carburetors, 124
    float-type, 127
    primer, 131
    PWC maintenance, 452-454
    types of carburetors, 126-144
Career opportunities and certification, 463-467
    engineer, 465
    entrepreneur, 465
    executive, 465
    general manager, 464
    manufacturer's technician, 464, 465
    sales manager, 464
    service manager, 464
    service representative, 465
    service technician, 463, 464
    teacher, 465, 466
Casting, 69
Castle nut, 47
Cavitation, 381
CDI module laminations, 164
CDI system, 163-165
Center electrode, 153
Center punch, 26
Centrifugal, 135
Centrifugal force, 195
Centrifugal governors, hunting, 139
Centrifugal or mechanical governor, 135-140
Certification, 466, 467
    certification test, 466
    test location, 466, 467
    why get certified?, 466
Chain guard (scabbard), 346
Chain saw fuel and carburetor, 347, 348
Chain saw maintenance, 347-350
    cylinder fins, 348
    fuel filter, 348, 349
    lubrication, 349
    guide bar, 350
    muffler, 349, 350
    rules for safe operation, 347
    spark plug, 350
Chain saws, 344-350
    air cleaner (filter), 348
    purchasing considerations, 344
    safety features, 344, 346, 347
    storage, 350
Chasing threads, 51, 52
Chassis, 372-374
    and steering, 372-374
Chemical energy, 97

Choke, 127
    system, 127, 128
Class 1 Fit, 50
Class 2 Fit, 50
Class 3 Fit, 50
Clearances, 214
Clevis pins, 56
Coil specifications, 247, 249
Cold chisels, 26, 27
Cold running, 179
Cold welded, 59
Combination slip-joint pliers, 24
Combination square, 39
Combination wrench, 22
Combustion, 63
Commutators and brushes, 261, 262
Compressed air, safety, 16
Compression,
    ratio, 79, 93
    ring shapes, 285
    rings, 285, 286
    stroke, 77, 78
    test, 208
Condensation, 204
Condenser, 158
Conduction, 192
Conductors, 149
Connecting rod, 70, 297
    and bearings, 297-301
    cap installation, 300
    needle bearings, 300
Constant level splash system, 184
Continuity, 245
    test (multimeter), 251, 386
Convection, 192
Cooling, 67
Cooling fins, 67, 192, 201
Cooling system flushing, PWC maintenance, 456, 457
Cooling systems, 191-200
    how it works, 192
    principles, 191-195
Corrected horsepower, 102
Correction factor, 102-104
Cotter pins, 56, 57
Countersunk, 44
Crank offset, 92
Crankcase,
    breather, 204
    breather service, 204, 205
    seals, 304, 305
    torquing sequence, 444
    valving, 87
Crankshaft,
    and crankcase, 67-69, 301-305
    balance, 302
    clearances, 302, 303
    crankcase seals, 304, 305
    crankshaft clearances, 302, 303
    main bearings, 302
    measuring bearing clearance, 303, 304

    throw, 302
    with connecting rods, PWC, 444
Cross-scavenged, 82
Crossing, 437
Cutout relay service, 266
Cutout switches, 259
Cylinder block, 67
Cylinder fins, chain saw, 348
Cylinder gauge, 275
Cylinder head, 252, 323
    installing, 323
    reinstalling, 252
Cylinder reconditioning, 273-279
    honing, 278, 279
    inspection, 274, 275
    measurement, 275, 276
    reboring, 277, 278
Cylinder sleeve, 277
Cylinder taper, 275
Cylinder walls, and rings, 289, 290

D

Dampening coils, 322
Dc starter-generator circuit, checking, 259-261
Dead man switch, 16
Decimal equivalents and tap drill sizes, 53, 54
Department of Natural Resources (DNR), 427
Depth micrometer, 249
Detergent/dispersants, 178, 183
    additives, 178
Dethatcher blade, 335
Detonation, 110
Diagonal side-cutting pliers, 24
Dial indicator, 39, 40, 250
Diaphragm-type carburetors,
    operation, 132-134
    repair, assembling, 230-232
    troubleshooting charts, 232-235
Die, 52
Die-cast cylinder, 67
Die-stock, 52
Diesel fuel, 111
Dieseling, 110
Differential gears, 382
Differential pressure test, 209, 210
Digital micrometers, 33
Diode, 265
Dipper, 183
Dipstick, 202
Discharging battery, 170
Distributors, 266, 267
    ignition timing, 266, 267
    lubrication, 267
Dowel pins, 56
Downdraft carburetors, 126, 130
Drift punch, 26
Drive control, 414

Drive systems, snow thrower, 410, 411
Dry types, 140, 204
Dry-type air cleaner, 142, 143
Dry-type dual element, 143, 144
Dry-type filter element, 142
Dual-element, 140, 204
Dual-purpose alternator, 263
Dwell and cam angle, 162, 163
Dynamometer, 99

E

Earphone-type protectors, 15
Earplugs, 15
Economizer circuit, 129
Economizer system, 129
Edgers/trimmers, 359-363
    lubrication, 362
    maintenance, 362
    purchasing considerations, 359
    rules for safe operation, 361, 362
    safety features and adjustments,
        359-361
    storage, 362
Ejection and barrel pumps, 184
Elastic limit, 48
Electric start, 253, 338, 384-386
    systems, 384-386
    testing interlock electrical system,
        385, 386
    testing reverse safety switch, 386
    testing solenoid on electric start
        tractors, 386
Electric start tractors, testing solenoid,
    386
Electrical circuits, for MTD tractors,
    390-395
Electrical principles, fundamentals,
    148-152
Electrical safety systems, 384-386
    electric start systems, 384-386
    safety interlocks, 384
Electrical units of measurement, 149, 150
Electrical wiring/grounding
    procedures, 17
Electricity, and magnets, 151, 152
Electrodes, types, 156, 157
Electrolytes, freezing points, 257
Electron theory, 148, 149
Electronic ignition, 163
Electronic switching devices, 158
Emergencies, shop safety, 17
Emissions,
    small engine, 116-120
    standards, 117-120
End gap, 288
Energy, 97, 98
    types of, 97
Energy conserving II oil, 181
Engine & Equipment Training Council
    (EETC), 466

Engine bore and stroke, 92
Engine components, 372
Engine construction and principles of
    operation, 63-76
    basis for an engine, 66-75
    gasoline engines, 63-66
Engine controls, snow thrower,
    409, 410
Engine cooling system, PWC, 446, 447
Engine cooling, exhaust cooling,
    192, 193
Engine design and construction, 67
Engine displacement, 92, 93
Engine exhaust emissions, 117
Engine fuel/oil ratio chart, 181
Engine fuels, 109-116
Engine fuels, gasoline, 110
    kerosene and diesel fuels,
        110, 111
    two-cycle fuel mixtures, 111
Engine governor,
    adjustments, 235-237
    functions, 135
Engine identification, small, 77
Engine inspection, 272, 273
    disassembly, and cylinder
        reconditioning, 271-280
    introduction, 271
    organizing the job, 273
Engine lubrication, lubricating oil,
    181, 182
Engine operation, simple, 63, 64
Engine parts,
    clean, 178
    cools, 178
Engine performance,
    measuring, 91-108
    terminology, 92-107
    the engine, 91, 92
Engine safety, operation, 17
Engine Service Association (ESA), 466
Engine service technician, 463, 464
Engine starting,
    blade brakes, 338, 339
    lawn mowers, 337-341
    procedure, 339-341
Engine torque, 104, 105
Engine troubleshooting charts,
    211-213, 345
Engine, basis, 66-75
Engineers, 465
Entrepreneur, 465
Environmental Protection Agency
    (EPA), 119
Esna self-locking nut, 48
Executives, 465
Exhaust emissions, 117
Exhaust stroke, 77, 79
Exploded views, 210
Explosive force, 65, 66
Extended rope starter, 338

External threads, terminology, 49

F

Fasteners, sealants, and gaskets, 43-62
    adhesives and sealants, 58, 59
    gaskets, 59, 60
    keys, 57, 58
    pins, 56, 57
    retaining rings, 57
    screws, 44-46
    threaded fasteners, 43-54
    washers, 55, 56
Federal Clean Air Act, 117
Feeler gauges, 37-39, 248
Filler plug, 202
Fire extinguisher classifications, 18
Fire extinguishers, 17
Fires,
    Class A, 18
    Class B, 18
    Class C, 18
First aid kits, 17
Flammable, 14
Flashover, 155
Flashpoint, 15
Flat washer, 55
Flexloc self-locking nut, 48
Float, 127
Float bowl ventilation, 127
Float-type carburetor, 127, 220, 224-230
    assembling, 224, 225
    repair, 224-230
    troubleshooting, 225-230
Floating rings, 286
Flywheel, 74, 75, 244, 245
    inspecting, 245
    magnets, 164
    operation, 164
    reinstalling, 251, 252
    removing, 244, 245
Force, 94-96
Form-in-place sealants, 59
Four cycle torque specifications, 23
Four wheel steering, 373
Four-cycle engine,
    and two-cycle engines, 77-90
    bearings, 297
    compression stroke, 77, 78
    exhaust stroke, 77, 79
    intake stroke, 77, 78
    lubrication, 80
    lubrication, snow thrower, 418
    lubrication systems, 182, 183
    power stroke, 77-79
    vs two-cycle engine, 87
Friction, 175, 176
    preventing wear, 176
Friction bearings, 297
Friction-type connecting rod bearing,
    299

Friction-type crankshaft main bearings, 302
Friction-type rod bearings, 298-300
　bearing crush, 299, 300
　bearing spread, 298, 299
　inserts are matched, 300
Frictional horsepower (fhp), 101
Front end loading, 371
Fuel, atomized, 65
Fuel emission control systems, 109-122
　emission controls, 116-120
　engine fuels, 109-116
Fuel filler caps, 112
Fuel filter, 112, 113
　chain saw, 348, 349
　PWC maintenance, 450
　screens, PWC maintenance, 450
Fuel mixtures, two-cycle, 111
Fuel pick-up line, 112
Fuel pumps, 113-116, 217, 218
　hand primer, 114
　impulse diaphragm, 115
　mechanical, 114
　operation, 114
　without filter system, 114, 115
Fuel system service, 217-238
　carburetor overhaul, 220-235
　engine governor adjustments, 235-237
　troubleshooting fuel system, 217-219
Fuel transfer, 83
Fuel vent check valve, PWC maintenance, 451
Full-flow filter system, 186
Full-pressure lubrication system, 184, 185
Full-throttle sequence, part-throttle, full-throttle sequence, 130, 131

G

Gapping, 241
　tool, 241
Gasket, 43-62
　fasteners, and sealants, 43-62
　rules, 60
Gasoline, 64, 110
　ignited, 64, 65
　liquefied petroleum gas and natural gas, 110
Gasoline engine-powered chain saw, 346
Gasoline engines, 63-66
Gear puller, 27
General manager, 464
Give-away vessel (burdened), 436
Goggles, 15
Governor features, 140
Governor throttle controls, 135

variable speed, 137, 138
Governors, 135-140, 235
　air vane, 135-138
　centrifugal, 135-140
　mechanical, 135-140
　pneumatic, 135
　speed setting, change, 139
　types, 135-140
　vacuum, 135, 139, 140
Grass cutting, 351
Grass discharge chute guard, 333
Gravity-fed system, 217
Grease (zerk) fitting, 373
Grip-length, 46
Grooved pins, 56
Grounding out, 252
Guide bar, 350
Guide bushing installation, 310

H

Hammers, 25, 26
　ball peen, 25, 26
　soft-faced, 26
Hand primer, 114
Hand tapping, rules, 52
Hand tool use, 15, 16
Hazardous materials, handling, 14, 15
Hazards, 13
Heat dissipation, 67
Heat engine category, 91
Heat ranges, 153
Heavy-duty, single-speed transaxle, 379
Helix, 43
Hexagon nut, 47
Hexagon socket, 45
High speed adjust, 218
High tension lead, 153
High tension lead connectors, 155
High voltage secondary current, 169
Honing, 278
Honing the cylinder, 278, 279
Horsepower, 97-102
　corrected horsepower, 102
　formula, 98
　kinds, 98-102
　measuring engine brake horsepower, 99, 100
Horseshoe valve, 72
Hot spots, 192, 205
Humidity correction factor, 103
Hunt, 235
Hunting, 139
Hydrodynamic lubrication, 177
Hydrogen, 256
Hydrogen gas, 15
Hydrometer, 254
　test, 254, 255
Hydrostatic transmissions, 377-384
　hydrostatic transmission fluid flow, 381-384

troubleshooting chart, 383
performance graph, 381

I

Identification numbers and placards, PWC, 434
Idle mixture adjustment, 218
Idle speed stop screw, 218
Idling circuit, 129, 130
Ignition, 67
Ignition advance systems, 161, 162
Ignition coil, 152
Ignition key switches, 259
Ignition spark, timing, 251
Ignition system components, testing, 245-249
Ignition system operation, 147-180, 239-270
　fundamental electrical principles, 148-152
　service, 239-270
　tune-up, 239
Ignition systems, 147-174
　basic ignition system operation, 147-170
Ignition power, 83
Ignition testing procedure, 242, 243
Independent rope starter, 338
Indicated horsepower (ihp), 100, 101
Inertia, 92
Inertia starters, 338
Inside micrometer, 275
Insulator, 153
Intake and exhaust ports, 71
Intake into crankcase, 83
Intake stroke, 77, 78
Integrated hydrostatic transaxle, 383
Interference angle, 311
Internal threads, terminology, 49

J

Jam nut, 47
Jet pump propulsion system, 447-449
Jet pump water inlet screen, PWC maintenance, 457

K

Kantlink washer, 55
Kerosene, 110, 111
　and diesel fuels, 110, 111
Keys, 57, 58
Keyseat, 57
Keyway, 57
Kickback, chain saw, 347
Kinetic energy, 97
Knock-off tool, 244
Knowing Mowing program, 333

L

Lamson self-locking nut, 48
Lands, 281
Lapping, 312
    valves, 312
Launch ramp etiquette, PWC, 434
Lawn and garden tractors, 365-402
    chassis and steering, 372-374
    circuits for study, 389-395
    electrical safety systems, 384-386
    engine components, 372
    general maintenance, 396-400
    operating safely, 365-367
    purchasing considerations,
        367-371
    recoil start system, 386-389
    transmissions, 374-384
Lawn equipment, 331-364
    chain saws, 344-350
    edgers/trimmers, 359-363
    lawn mowers, 333-344
    string trimmers and brushcutters,
        350-359
    working safely, 331-333
Lawn mowers, 333-344
    engine starting, 337-341
    general maintenance, 341-344
    Knowing Mowing program, 333
    minor checks, 341
    purchasing considerations, 333-337
Lead-acid battery, 170
Leaf collecting, 369
Leaf-type feeler gauges, 241, 242
Liquefied petroleum gas (LPG), 110
    advantages, 110
    combustion, 110
    disadvantages, 110
Load adjusting needle, 128
Load adjustment, 128
Loaded oil, 202
Lock nuts, 47, 48
Lock washers, 55
Loop-scavenged, 82
LPG fuel system, 111
Lubricating oil, 177-179
    aids fuel economy, 179
    cooling, 178
    easy starting, 177
    engine lubrication, 181, 182
    engine parts clean, 178
    lubricates and prevents wear, 177
    oil selection, 179
    prevents foaming, 179
    protects against rust/corrosion,
        177, 178
    seals combustion pressures,
        178, 179
Lubrication, 67, 267, 349
    chain saw, 349
Lubrication systems, 175-190
    principles, 175-186

M

Machinist's vise, 28
Magnetism, 150, 151, 245
Magneto breaker point assembly, 157
Magneto ignition systems, comparison,
    166, 167
Magneto service, 243, 244
Magneto systems, 148, 239, 243
Magnets and electricity, 151, 152
Main bearings, 302
Maintenance and troubleshooting,
    201-216
    preventive, 201-216
    service information, 210-214
    systematic troubleshooting,
        206-210
Maintenance for lawn/garden tractors,
    396-400
    air filter, 397, 398
    battery maintenance, 398
    engine cleaning, 398
    engine lubrication, 397
    fueling and refueling, 399
    optional oil accessories, 397
    radiator cleaning, 398
    removing from storage, 399, 400
    sharpening and balancing blades,
        399
    spark plugs, 396
    storing tractor, 399
Maintenance free (sealed batteries),
    254
Maintenance procedures, lawn
    mowers, 341-344
    air cleaners, 342
    blade sharpening, 341, 342
    changing oil, 343
    muffler replacement, 343
    removal from storage, 344
    spark plugs, 342
    storage, 343, 344
    V-belts, 343
Major (largest) diameter, 49
Manual throttle control, 134, 135
Manufacturer's technicians, 464, 465
MBI magneto cycle, 159, 160
MBI magneto system, 158-160
Mean atmospheric pressure, 124
Mean effective pressure (mep), 101
Measuring bearing clearance, 303, 304
Measuring instruments, and tools,
    21-42
    measuring instruments, 29-40
    micrometers, 29-34
    small hole gauge, 37
    telescoping gauges, 35, 36
    thickness gauges, 37-39
Mechanical breaker point ignition
    (MBI), 158
Mechanical breaker points, 157
Mechanical efficiency, 106

Mechanical energy, 97
Meeting, 437
Metric (M), 49
Metric micrometer, 33
Metric tap drill chart, 55
Metric Vernier caliper,
    25 division scale, 35
    50 division scale, 35
Micrometer depth gauges, 33
Micrometers, 29-34
    cleaning and calibrating, 30
    depth gauges, 33, 34
    digital, 33
    reading metric, 33
    reading standard, 31
    reading Vernier, 31-33
    standard and Vernier, 31
    using, 30, 31
Moveable sheave, 375
Mower deck and blade adjustment, 369
Mowing and mulching, 367-368
Mowing, proper, 368, 369
Muffler, 205, 349, 350
    chain saw, 349, 350
    service, 205
Mulching, 368, 369
    proper, 369
Multi-piece crankshaft, 302
Multigrade oil, 180
Multimeter, 245, 386
    continuity tester, 386
Multiple contact switches, 259
Multiviscosity oil, 180

N

Natural draft carburetor, 126, 130
Needle nose pliers, 24
Neoprene, 61, 304
    oil seal, 304
Nonroad sources, 118
Nuts, 47, 48
    castle, 47
    hexagon, 47
    jam, 47
    lock, 47, 48
    self-locking, 48
    square, 47
    wing, 47

O

Occupational Safety and Health
    Administration (OSHA), 17
Octane number, 64
Ohm's law, 150
Ohms (Ω), 149, 150
Oil, 202-204
    changing, 203
    mixing oil and fuel, 203, 204
    when to change, 202, 203
Oil control rings, 286

Oil cooling, operation, 198
Oil filter and oil filter screen, PWC
    maintenance, 451
Oil filter systems, 185, 186
    bypass systems, 185-187
    full-flow, 186, 187
    shunt, 186
Oil-wetted air cleaner, 140, 141, 204
Open-end wrench, 21
Operation precautions, PWC, 427-434
Operator presence controls, 414
Optical tachometer, 208
OSHA requirements, 17, 18
Out-of-roundness, 275
Outboard water circulation systems,
    195-198
Outdoor Power Equipment (OPE), 466
Over square, 92
Overcharging, battery failure, 256
Overhaul, 220
Overhead valve system disassembly,
    320-322
    removing valves, 321, 322
Overhead valve systems, 320-325, 372
    adjusting valve clearance, 324, 325
    disassembly, 320-322
    installing, 322, 323
    installing cylinder head, 323
    installing rocker arms, 323, 324
    servicing, 322
Overtaking, 437
Oxygen, 256
Oxygenates, 64

P

Palnut, single-thread lock nut, 48
Part-throttle, full-throttle sequence,
    130, 131
Performance, 92
Personal flotation device (PFD), 428
Personal watercraft (PWC), 425-462
    boating terms, 427
    courtesy and common sense, 436
    encountering vessels, 437
    engines, 441-449
    fishing vessel right-of-way, 438
    general maintenance, 449-458
    give-way vessel, 437
    identification numbers and
        placards, 434
    introduction, 425, 426
    launch ramp etiquette, 434
    navigational rules, 436
    nonmotorized watercraft, 438
    operating safely, 427-438
    operation precautions, 427-430
    preride checklist, 430-434
    purchasing considerations, 438-441
    Rule 2, 437
    signs, symbols, buoys, and markers,
        434, 435

stand-on vessel, 437
    storage, 459, 460
    transporting PWC, 458, 459
Petroleum, 64
Phase 1 Standards for Small Spark-
    Ignition Engines, 120
Phase 2 Standards for Small Spark-
    Ignition Engines, 119-120
Phillips screwdrivers, 25
Pin boss, 282
Pin punch, 26
Pin-to-boss clearance, 294
Pinned rings, 286
Pins, 56, 57
    clevis, 56
    cotter, 56, 57
    dowel, 56
    grooved, 56
    straight, 56
    taper, 56, 57
Piston and piston ring service, 281
Piston construction, 281-283
    cam-ground pistons, 282, 283
    piston fit, 282
    piston head shape, 283
    piston head size, 283
    piston thrust surfaces, 283
Piston damage, 290-292
    scoring and scuffing, 290
Piston fit, 282
Piston head shape, 283
Piston head size, 283
Piston height, 248-250
Piston pins, 70, 292-294
    measuring piston pins and bosses,
        294
    removing, 293, 294
Piston ring, 70, 284-292
    and cylinder walls, 289, 290
    construction, 284, 285
    design, 284-292
    end gap, 288, 289
    movement, 286, 287
    piston damage, 290-292
    side clearance, 287, 288
    tension, 286
    types, 285, 286
    wear-in, 292
Piston ring types, 285, 286
    compression rings, 285, 286
    oil control rings, 286
Piston ring wear-in, 292
Piston rings, 70, 284-292
Piston skirt, 282
Piston thrust surfaces, 283
Piston to crankshaft, connecting,
    70, 71
Pistons and piston rings, 281-296
    piston construction, 281-283
    piston pins, 292-294
    piston ring design, 284-292
    service, 281

Pitch, 447
Plastigage, 303
Pliers, 23, 24
    combination slip-joint pliers, 24
    diagonal side-cutting, 24
    needle nose, 24
    vise-grip, 23
Plowing, throwing, and blowing snow,
    369, 370
Plug tap, 52
Plug tapping tool, 51
Plunger pump, 194
Pneumatic, 135
Point gap and dwell, relationship, 163
Pollution, consumer role, 120
Polyurethane foam filter, 141
Pop-off pressure, 453
Poppet valves, 71, 305
Porting, 325
Ports, 71
Ports, reeds, and rotary valves, 325, 326
Positive displacement oil pumps, 184
Potential energy, 97
Pounds per square inch (psi), 95
Power, 97
Power stroke, 77-79
Power tools, safety, 16
Power take-off (PTO), 366
Practical efficiency, 106
Preset torque wrench, 22
Pressure, 95
Pressure pulse exhaust tuning, 85
Pressure switches, 259
Pressure-vacuum water flow system, 195
Pressurized fuel system, 116
Preventive maintenance, 201-206
    checking oil level and condition,
        202
    general, 201-206
Prick punch, 26
Primer, 131, 225
Primer plunger, 131
Probe and pickup tools, 27, 28
Prony brake, 99
Proper clothing, safety, 15
Punches, 26
    center, 26
    drift, 26
    pin, 26
    prick, 26
Purchasing considerations,
    lawn mowers, 333-337
    conveniences, 336, 337
    optional features, 335
    push-type and self-propelled
        mowers, 334
    two-cycle or four-cycle engines,
        334, 335
Purchasing considerations, PWC,
    438-441
    main components and parts,
        438-441

Push button switches, 259
Push-type and self-propelled mowers, 334
Pushrods, 320
PWC engine crankcase, 444
PWC engines, 441-449
    basic engine parts, 442-445
    connecting rod, 444, 445
    crankcase, 444
    crankshaft, 444
    parts, 442-445
    piston rings, 443
    pistons, 442, 443
    wrist pins, 445
PWC maintenance, 449-458
    battery care, 454
    carburetors, 452-454
    cooling system flushing, 456, 457
    external lubrication points, 457, 458
    fuel filter, 450
    fuel filter screens, 450
    fuel vent check valve, 451
    jet pump water inlet screen, 457
    oil filter and oil filter screen, 451
    oil pump bleeding, 451, 452
    sediment bowl, 451
    spark plugs, 450
PWC purchasing considerations, 438-441
    main components and parts, 438-441
PWC storage, 459, 460
    cleaning, 460
    lubrication, 460
PWC, transporting, 458, 459

R

Radiator core, 197
Radiators, 197, 198
Rate of power, 98
Rated horsepower, 101, 102
Reach, 153
Reboring the cylinder, 277, 278
Reciprocating engines, 92
Recoil start system, 386-389
    assemble seat, 389
    disassemble safety seat, 388, 389
    testing interlock system, 386-388
    testing procedure of safety seat, 388, 389
    testing safety reverse switch, 388
Recoil starter, 337
Reed valve action, 88
Reed valve engine, 85-87
Reed valves, 88, 325
    four reed, 88
    multiple reed, 88
    single reed, closed position, 88
    single reed, open position, 88
    triple reed, 88
    twin reed, 88

Reeds, 219
Reel-type mower, 334
Reservoir, 381, 382
    expansion tank, 382
Retaining ring pliers, 24
Retaining rings, 57
Reverse flushing, 205
Reverse safety switch, 384, 386
    testing, 386
Ridge reaming tools, 274
Ring end gap, 288, 289
Ring movement, 286, 287
Ring side clearance, 287, 288
Ring tension, 286
Rings and cylinder walls, 289, 290
Rocker arms, 320, 323, 324
    installing, 323, 324
Rod bolt locking devices, 300, 301
Rod cap, 300
Rods and bearings, connecting, 297-301
Room temperature vulcanizing sealant (RTV), 59
Rotary broom sweeping, 371
Rotary disc valve, 84
    engine, 84, 85
Rotary garden tillers, 370
Rotary mowers, 334
Rotational speeds, 137
Rotor-type pump, 194
Rust inhibitor, 205

S

SAE viscosity grade, 179, 180
Safety, shop, 13-19
    small gas engine shop, 13-19
Safety glasses, 15
Safety interlocks, 384
Sales managers, 464
Scabbard (chain guard), 346
Scavenging, 79, 84
    and tuning, 84
Screw pitch gauge, 40, 41
Screwdrivers, 24, 25
    offset, 25
    Phillips, 25
Screws, 44-46
    self-tapping, 45, 46
    set screws, 45
Sealants,
    and adhesives, 58, 59
    gaskets, and fasteners, 43-62
Sealed batteries (maintenance free), 254
Seat-activated switch, 384
Sediment bowl, OWC maintenance, 451
Self-aligning reamer guide, 310
Self-locking nuts, 48
Self-retaining, 48
Self-tapping screws, 45, 46

Sensitivity, 140
Series, 384
Service information, 210-214
    troubleshooting charts, 211-213
Service managers, 464
Service manual, 210, 271
Service procedures, 210
Service representatives, 465
Set gauge, 275
Set screw heads, 45
Set screws, 45
Shear bolt, 419
Shop safety, 13-17
    adequate ventilation, 15
    compressed air, 16
    electrical wiring/grounding procedures, 17
    emergencies, 17
    engine operation, 17
    hand tool use, 15, 16
    lifting, 16, 17
    power tool use, 16
Shop tools, 21-40
    cleaning tank, 28
    cold chisels, 26, 27
    gear pullers, 27
    hammers, 25, 26
    pliers, 23
    probe and pickup tools, 27, 28
    punches, 26
    retaining ring pliers, 24
    screwdrivers, 24, 25
    vise, 28
    wrenches, 21-23
Shunt filter system, 186
Side clearance, 287
Signs, symbols, buoys, and markers, PWC, 434, 435
Single-piece crankshafts, 301
Single-speed transaxle, 378
Single-stage snow throwers, 370, 406
Slap, 283
Sliding vane pump, 194
Small engine,
    identification, 77-80
    rods and bearings, 298
    torque charts, 23
Small gas engine shop, 13-19
    adequate ventilation, 15
    handle hazardous materials properly, 14, 15
    proper clothing, 15
    safety, 13-19
    shop safety, 13-19
Small hole gauge, 37, 38
Snap rings, 293
Snow thrower,
    controls, 415
Snow thrower maintenance,
    air filter, 418
    auger and auger gear box lubrication, 419, 420

Snow thrower maintenance (*continued*)
  engine fuel, 417, 418
  four-cycle engine lubrication, 418
  general, 416-421
  lubrication, 420, 421
  muffler, 418
  optional oil accessories, 418
  preventing freeze-up, 419
  spark plugs, 417
Snow thrower off-season storage, 421, 422
  starting, 421
Snow thrower operation, 416
  engine starting, 416
  snow throwing, 416
Snow throwers, 403-424
  general maintenance, 416-421
  off-season storage, 421, 422
  operating safely, 403-405
  purchasing considerations, 405-415
  snow thrower operation, 416
Snow thrower safety, 403-405
  maintenance and storage, 405
  preparation for use, 403, 404
  protect children, 403
  usage, 404, 405
Snow thrower selection, 405-415
  drive systems, 410, 411
  electric starting, 410
  engine controls, 409, 410
  heavy-duty snow thrower, 408, 409
  machine size and type, 405
  mid-size snow thrower, 407, 408
  operator presence, 414, 415
  skid shoe adjustment, 414
  skid shoes, 413, 414
  small snow throwers, 405-407
  track chain replacement, 411-413
  track drive snow thrower, 411
  track lock lever, 413
  track tension adjustment, 413
Snow thrower troubleshooting chart, 422
Socket sets, 22
Soil, preparing for planting, 370
Solenoid, 259, 384
Solid state, 163, 247
  ignition systems, advantages, 163
  systems, 247
Solvent, 14
Spark arrestor, 347
Spark plugs, 152-156, 239-242, 350, 396, 417, 450
  analysis of used spark plugs, 241
  chain saw, 350
  cleaning, 241
  gap, 239
  gapping, 241, 242
  heat transfer, 153, 156
  identification, 154
  installation, 242
  reach, 155

  removal, 249
  temperature, measuring, 156
Special-purpose tools, 21
Speed ranges, 377
Spin-lock nut, 48
Splash lubrication system, 183
Spontaneous combustion, 396
Spring-loaded check valves, 382
Square, 92
Square nuts, 47
Stability, 140
Stand-on vessel (privileged), 436
Standard micrometer, 31
  reading, 31
State of tune, 102
Stator adjustment bolt, 251
Stator assembly, 262
Stator plate, 248
Stator vanes, 448
Stop switch, 160, 161
Straight pins, 56
String trimmers, 350-359
  air filter, 359
  and brushcutters, 350-359
  brush cutting, 351-353
  carburetor adjustment, 357
  grass cutting, 351
  handles and cutters, 351
  lubrication, 357, 358
  maintenance, 357
  muffler, 359
  operator safety, 353
  purchasing considerations, 350, 351
  spark plug, 358
  spool replacement, 357
  starter rope/spring breakage, 359
  starting and safe operation, 353-356
Stroboscope, 207
Stroke, 77, 92
Stud bolts, 50
Sulfating, 253
Sump, 281
Swash plate, 382
Switches, 258, 259
Switching devices, 157
Systematic troubleshooting, 206-210
  check rpm, 207, 208
  fundamental operating requirements, 206, 207
  simplest things first, 206
  test compression, 208, 209
  test differential pressure, 209, 210
Systems cooled with saltwater, 205

T

Tanks, lines, and fittings, 111, 112
Taper pins, 56, 57
Taper tap, 51
Taper tapping tool, 51
Tapered roller bearing, 302
Tapping, 45

Tapping new threads, 52
Telescoping gauges, 35-37, 275
Temperature and barometric pressure correction chart, 103
Tensile strengths, 48
Tensile stress, 94
Tetraethyl lead, 64
Thermal efficiency, 106, 107
Thermal energy, 97
Thermostat, 196-198
  operation, 197
Thickness gauges, 37-39
Thread, 43
  adhesives, 58
  designations, 50
  fit, 50
  pitch, 48
  types, 49, 50
Threaded fasteners, 43-54
  tightening and loosening, 51
Threading, 46
  tap, 51
  with die, 52
Threads, tapping new, 52
Three-point hitch, 370
Throttle, 128
Throttle system, 128
Through hole, 51
Thrust surface, 283
Timing, 266
Timing light, 266
Tinnerman speed nut, 48
Toggle switches, 259
  Tolerance and clearances, 213, 214
  chart, 214
Tolerances, 213
Tools and measuring instruments, 21-42
Toothed washers, 56
Top dead center (TDC), 79
Torque, 22, 23, 51, 102, 104, 105, 214
  and horsepower, 105
  not constant, 104, 105
  settings, tightening to specific, 51
  specifications, 23, 214
  wrench, 22
Toxic fumes, 15
Track chain replacement, snow thrower, 411-413
Track drive snow thrower, 411
Track lock lever, snow thrower, 413
Track tension adjustment, snow thrower, 413
Tractor operating safety, 365-367
  avoid tipping mower, 366
  avoiding injury adjusting blades, 366
  considerations when mowing, 366
  parking tractor safely, 366
  protect children, 365
  protective clothing, 366
  transporting tractor, 367
Trailer for hauling, pulling, 371
Transaxles, 374

Transfer-type measurement
    instruments, 34
Transistor controlled ignition (TCI)
    system, operation, 165, 166
    components, 166
Transmissions, 374-384
    hydrostatic transmissions, 377-384
    variable-speed transmission,
        374-377
Transporting PWC, 458, 459
    towing to shore, 459
Trigger module, 247
Triple beam balance, 94
Troubleshooting charts, 211-213
Tubing wrench, 22
Tungsten, 157
Two cycle, 77-90, 182, 219
    and four-cycle engines, 77-90
    engine lubrication systems, 182
    engine reeds, testing, 219
    four-stroke engine, 77-80
    small engine identification, 77-80
    two-stroke cycle engine, 80-87
Two-cylinder PWC engine, 442
Two-speed transmission system with
    variable-speed drive, 380
Two-stage snow blowers, 370
Two-stroke and four-stroke engines,
    characteristics, 89
Two-stroke cycle engine, 80-87
    exhaust, 83, 84
    fuel transfer, 83
    ignition-power, 83
    intake into crankcase, 83
    new engine break-in, 449
    principles of operation, 83
    PWC, 445, 446
    PWC, reed valve engines, 446
    PWC, rotary valve engines, 445
    reed valve engine, 85-87
    rotary disc valve engine, 84, 85
    scavenging and tuning, 84
    variations in design, 82, 83
Type of oil, 179

U

UNC, 50
Under square, 92
Undercharging, battery failure, 256
Unified National Coarse (UNC), 49
Unified National Fine (UNF), 49
Unified threads, 50
Updraft carburetors, 126, 129

V

Vacuum, 124
Vacuum governors, 135, 139, 140

Valve clearance
    adjusting, 320
    setting, 317
Valve guides, 71, 307-309
    alternate valve guide replacement,
        308, 309
    inspecting, 307, 308
    reaming, 308
Valve keepers, 306
Valve lifter, 72, 73
    or tappet, 72, 73
Valve lifter-to-stem clearance,
    316, 317
Valve overlap, 80
Valve refacing, 317
Valve seat,
    angle and width, 310-312
    counterbore insert cylinder, 315
    cutter kit, instructions, 313
    cutting tool kit, 312
    inserts, 312, 315, 316
    installing new (aluminum alloy
        block), 316
    installing new (cast iron block),
        315, 316
    removing, 315
    width, 311
Valve service, 305-320
    adjusting valve clearance, 319, 320
    inspecting valve springs, 307
    inspecting valves and seats, 306, 307
    lapping valves, 312
    rebushing worn aluminum guides,
        309, 310
    refacing valves, 317-319
    removing valve assembly, 306
    replacing worn brass/sintered iron
        guides, 310
    valve guides, 307-309
    valve lifter-to-stem clearance, 316,
        317
    valve seat angle and width, 310-312
    valve seat inserts, 312, 315, 316
Valve spring,
    assembly, 71, 72
    compressor, 306
    retainers, installing, 323
    tension tester, 39
Valve timing, 79, 80
Valves and seats, inspecting, 306, 307
Vapor locked, 116
Vapor return fuel systems, 116
Vari-volume pumps, 195
Variable-speed drive pulley, 377
Variable-speed pulley, 374
Vented, 231
Ventilation, maintain, 15
Venturi, 125, 126
    principle, 125, 126

Vernier caliper,
    digital, 35
    metric 25 division scale, 35
    metric 50 division scale, 35
    reading, 34, 35
    standard 25 division scale, 34
    standard 50 division scale,
        34, 35
Vernier micrometer, 31-33
Viscosity, 179, 180
    grade, 179
    recommendations, 180
Vise, 28
Vise-grip pliers, 23
Voltage, checking, 255
Voltage regulator service, 265-267
    adjusting current voltage,
        265, 266
    cutout relay service, 266
Volts (V), 149, 150
Volumetric efficiency, 105, 106

W

Washers, 55, 56
    common, 56
    flat, 55
    kantlink, 55
    lock, 55
    toothed, 56
    wide bearing, 56
Water cooling, 193-195
    water pumps, 193-195
Water cooling systems,
    cooler with saltwater, 205, 206
    maintaining, 205, 206
    outboard water circulation systems,
        195-198
Water jackets, 193
Water-cooled engines, storing, 206
Wear block (cam follower), 248
Wear-in, 292
Welch plugs, 221
Wheel puller, 244
Wide bearing lock washer, 56
Wing nuts, 47
Wire-type thickness gauges, 242
Wiring, 258
Woodruff key, 251
Work, 96
Worn brass/sintered iron guides,
    replacing, 310
Wrenches, 21-23
    adjustable, 22
    box-end, 21
    combination, 22
    open-end wrench, 21
    tubing, 22
Wrist pin, 70, 445